Electronics
ALL-IN-ONE
FOR DUMMIES®

A Wiley Brand

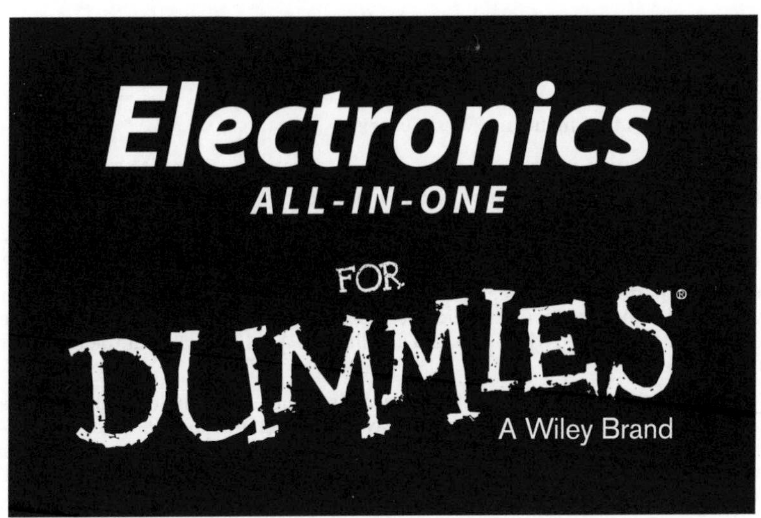

by Doug Lowe and Dickon Ross

Electronics All-in-One For Dummies®

Published by: **John Wiley & Sons, Ltd.,** The Atrium, Southern Gate, Chichester, www.wiley.com

This edition first published 2014

Registered office

John Wiley & Sons Ltd, The Atrium, Southern Gate, Chichester, West Sussex, PO19 8SQ, United Kingdom

For details of our global editorial offices, for customer services and for information about how to apply for permission to reuse the copyright material in this book please see our website at www.wiley.com.

For general information on our other products and services, please contact our Customer Care Department within the U.S. at 877-762-2974, outside the U.S. at (001) 317-572-3993, or fax 317-572-4002.

For technical support, please visit www.wiley.com/techsupport.

A catalogue record for this book is available from the British Library.

ISBN 978-1-118-58973-1 (pbk); ISBN 978-1-118-58970-0 (ebk); ISBN 978-1-118-58971-7 (ebk)

Printed in the UK

Contents at a Glance

Table of Contents

Book IV: Getting into Alternating Current.................. 383

Book V: Working with Radio and Infrared.................. 421

Introduction

Electronics changed the world during the 20th century, from radios, telephones and cinema to modern computing and the so-called Information Age. This revolution may turn out to be as important as the Industrial Revolution. The Internet was recently voted the greatest invention of all time – and yet it's not even 25 years old. Each new electronic gadget is adopted quicker than the last (you may be reading this book on an e-reader or tablet computer) and mobile phones now outnumber people on the planet.

Yet despite using electronics every day, many people have virtually no understanding of what's going on under their thumbs and behind those screens; the process may just as well be magic to them. Fortunately, you don't have to understand completely the inner workings of every component, even if you want to design an amazing new electronic gizmo.

This book is for people who've always been fascinated by electronics but didn't make a career out of it. In these pages, you find clear and concise explanations of the most important concepts that form the basis of all electronic devices, such as the nature of electricity (if you think you know what it is, you're kidding yourself); the difference between voltage, amperage and wattage; how basic components such as resistors, capacitors, diodes and transistors work; and how you can use some super-complex components such as integrated circuits to realise your own dream project.

You not only gain an appreciation of the electronic devices that are part of everyday life, but also discover how to build simple circuits that impress your friends, are great fun and may even be the prototype for the invention that makes your fortune!

About This Book

Electronics All-in-One For Dummies is a practical reference containing the most important topics you need to know when you dabble in building your own electronic circuits. It's a big book made up of eight smaller ones, which we call *minibooks.* Each of these minibooks covers the basics of one key topic for working with electronics, such as circuit-building techniques, how electronic components work or using integrated circuits.

Throughout these minibooks, we include dozens of simple, practical projects which you can build to demonstrate the operation of typical circuits. For example, in the chapter on transistors (Book II, Chapter 6), you find several simple projects that demonstrate common uses for transistors, such as driving an LED, inverting an input and creating an oscillator.

Reading about electronics circuits is one thing, but to understand how a circuit works, you need to build it and see it in operation. Most of the projects are simple enough that you can build them in 20–30 minutes, assuming you have the parts on hand.

This book doesn't pretend to be a comprehensive reference for every detail on every possible topic related to electronics. Instead, it shows you how to get up and running quickly so that you have more time to do the things you really want to do. Designed using the easy-to-follow *For Dummies* format, this book helps you get the information you need without labouring to find it.

To make its use as easy as possible, we design this book with multiple access points to help you find what you want. At the beginning of the book is a detailed table of contents that covers the entire book. Plus, each minibook begins with a minitable of contents that shows you at a mini-glance what chapters are included in that minibook. Useful running heads appear at the top of each page to point out the topic we discuss on that page, and handy thumb-tabs run down the side of the pages to help you find each minibook quickly. At the back, a comprehensive index lets you find information anywhere in the entire book.

This book isn't the kind you pick up and read from start to finish, as if it were a novel. If we see you reading it like this at the beach, we'll kick sand in your face. Beaches are for reading romance novels or murder mysteries, not electronics books. Although you can read this book straight through from start to finish, it's designed for you to pick up, open to just about any page and start reading.

You don't have to memorise anything in this book. It's a 'need-to-know' book: you pick it up when you need to know something. Want a reminder on how to calculate the correct load resistor for an LED circuit? Pick up the book. Can't remember the pinouts for a 555 timer IC? Pick up the book. After you find what you need, put the book down and get on with your life.

Foolish Assumptions

This book assumes that you're curious about electronics, but you really don't know much, if anything, about its inner workings. You chose this book, rather than a book consisting exclusively of recipes for electronic circuits, and therefore we assume that you want to discover more about how parts such as resistors, capacitors and transistors actually work.

You don't need to be well-versed in physics or mathematics to benefit from reading this book, although a little bit of school algebra is helpful (but we do our best to refresh that possibly painful memory).

We assume you may want to jump around this book a bit, diving deep into a topic or two that holds special interest for you, and possibly skimming through other topics. For this reason, we provide loads of chapter cross-references to point you to information that can fill in any gaps or refresh your memory on a topic.

Staying Safe

Most of the electronic circuits we describe in this book are perfectly safe: they run from common AAA or 9 V batteries and therefore don't work with voltages large enough to hurt you.

Occasionally, however, you come across circuits that work with higher voltages, which can be dangerous. You need to consider any project involving mains voltage (that you plug into an electrical outlet) as potentially dangerous and handle it with the utmost care. In addition, even battery-powered circuits that use large capacitors can build up charges that can deliver a potentially painful shock.

When you work with electronics, you also encounter dangers other than those posed by electricity. Soldering irons are hot and can burn you. Wire cutters are sharp and can cut you. Plus, plenty of the small parts you use can fall on the floor and find themselves in the mouths of children or pets.

Safety is such an important topic that we devote a whole chapter to it. We strongly urge you to read Book I, Chapter 4 *before* you build anything.

Icons Used in This Book

Like all *For Dummies* books, this one is chock-full of helpful icons that draw your attention to items of particular importance. You find the following icons throughout this book.

Pay special attention to this icon; it lets you know that a particularly useful fact is at hand.

We can't recall what we use this icon for, sorry. Oh, yes, important things to note and bear in mind!

Hold it – overly technical material is just around the corner. We use this icon for those paragraphs that go into greater depth, down into explaining how something works under the bonnet. Although enlightening, if these paragraphs go deeper than you want to know, just move on. You can also see this icon as a reminder not to get bogged down in details and instead focus on the larger point.

Danger, danger! This icon highlights information that may help you avert damage to your circuits or even personal disaster. Definitely pay attention to these warning icons, because they let you know about potential safety hazards.

Beyond the Book

As you walk your journey of discovery into the world of electronics, you can augment what you read here by checking out some of the access-anywhere extra goodies that we've hosted for you online.

You can find the book's e-cheat sheet online, at `www.dummies.com/cheatsheet/electronicsaiouk`. The at-a-glance, essential info that we serve up in this cheat sheet can be very handy sources of reference when printed out and pinned up near your workspace.

Additionally, you can also find a variety of great bonus content online, at `www.dummies.com/extras/electronicsaiouk`.

Where to Go from Here

This book works like a reference source. To discover the basics of electronics, peruse Book I. If you have a topic in mind that you want to find out about, look for it in the table of contents (which is detailed enough to find most topics) or turn to the index, where you can find even more detail.

The book is loaded with information and you can stay swimming in the shallow end or dive as deep as you desire. If you want to take a brief dip into a topic, you're more than welcome. If you want to know the big picture on digital electronics, for instance, read Book VI, Chapter 1. If you want to learn about logic gates, read Chapter 2 in Book VI. Or if you want to focus in on XOR gates, use the index to find the specific section in Book VI, Chapter 2.

Whatever your needs, with this book in hand you're ready to immerse yourself in the exciting hobby of electronics. Browse through the table of contents and decide where you want to start. Be bold! Be courageous! Be adventurous! Be careful! But above all, have fun!

Book I

Getting Started with Electronics

getting started with electronics

For Dummies can help you get started with lots of subjects. Visit www.dummies.com to learn more and do more with *For Dummies*.

Contents at a Glance

Chapter 1

Entering the Exciting World of Electronics

In This Chapter

- Understanding electricity
- Distinguishing between electrical and electronic devices
- Outlining the most common uses for electronics
- Looking at a typical electronic circuit board

Electronic devices are everywhere: for example (and incredibly), the number of mobile phones on the planet exceeds the number of people. Plus, no one uses film to take photos anymore because cameras have become electronic devices, and at any given moment young people in particular are engrossed in sending text messages while simultaneously listening to music on their smartphones.

Without electronics, life today would be extremely different.

If you've ever wondered what makes these electronic devices tick, this chapter's for you. Here we lay some important groundwork that helps the rest of this book to make sense. We examine the bits and pieces that make up the most common types of electronic devices, and take a look at the basic concept that underlies all electronics: electricity.

Having a bright idea

In January 1880, Thomas Edison filed a patent for a new type of device that created light by passing an electric current through a carbon-coated filament contained in a sealed glass tube. Historians say that Edison didn't really invent the light bulb, just improved on previous ideas, but that's another story.

Edison's light bulb patent was approved, but he still had a lot of work to do before manufacturing a commercially viable light bulb. The biggest problem with his design was that the bulbs dimmed the more they were used; when the carbon-coated filament inside the bulb became hot, it shed little particles of carbon that stuck to the inside of the glass. These particles resulted in a black coating on the inside of the bulb, which obstructed the light.

But one day, someone on Edison's team noticed that the black carbon came off only one end of the filament. The team thought that maybe some type of electric charge was coming out of the filament. To test this theory, they introduced a third wire into the bulb to see whether it would catch some of this electric charge.

It did. The team soon discovered that an electric current flowed from the heated filament to this third wire, and that the hotter the filament became the more electric current flowed. This discovery, today called *the Edison Effect,* marks the beginning of the technology known as *electronics.*

What Is Electricity?

We promise not to bore you with loads of tedious or complicated physics concepts, but in order to discover how electronics works at a level that lets you design and build your own electronic devices, you need to have a basic idea of what electricity is. After all, the whole purpose of electronics is to get electricity to do useful and interesting things.

Introducing electricity: Common knowledge

The concept of electricity is an odd one, familiar and mysterious. Here are a few of the familiar facts about electricity, based on practical experience:

- Electricity comes from power plants that burn coal, catch the wind or harness nuclear reactions. It travels from the power plants to people's houses in big cables hung high in the air or buried in the ground. At your home, it flows through wires in the walls (like water flows through a pipe) until it gets to electrical outlets. You plug in power cords to get the electricity into the electrical devices you depend on, such as ovens, toasters and vacuum cleaners.

- Electricity is valuable and not free. You know this because a power company asks for payment for it every month, and if you don't pay the bill it turns off your electricity.
- Electricity can be stored in batteries, which contain a limited amount of electricity that can be used up. When the batteries die, all their electricity is gone.

  Certain kinds of batteries, like the ones in mobile phones, are *rechargeable:* when drained of all their electricity, you can put more electricity back into them by plugging them into a charger, which transfers electricity from an electrical outlet into the battery. You can fill rechargeable batteries repeatedly, but eventually they lose their ability to be recharged – and then you have to replace them.
- Electricity is the stuff that makes lightning strike in a thunderstorm. Perhaps you were taught about Ben Franklin's experiment involving a kite and a key, and why you shouldn't try it at home!
- Electricity can be measured in three ways:
  - **Volts (abbreviated V):** Household electricity in the UK is 230 V. Portable batteries are 1.5 V and car batteries are 12 V.
  - **Watts (abbreviated W):** Traditional incandescent light bulbs are typically 60, 75 or 100 W. Modern compact fluorescent lights have somewhat smaller wattage ratings. Microwave ovens and hair dryers can be 1,000 W or more. The more watts, the brighter the light or the faster your pizza reheats and your hair dries. (Just to be clear, we don't recommend drying your hair in a microwave – it's not a good look!)
  - **Amps (abbreviated A):** A typical household electrical outlet is 13 A.

  Most people don't really know the difference between volts, watts and amps, but you can find out by reading Chapter 2 of this minibook.
- *Static electricity* is a special kind of electricity, which seems to sort of hang around in the air. You can transfer it to yourself by dragging your feet on a carpet, rubbing a balloon against your arms or forgetting to put an antistatic sheet in the tumble dryer.

- Electricity can be very dangerous. So dangerous in fact that it has been used to administer the death penalty in America for well over a century.

Understanding electricity basics

We devote Chapter 2 of this minibook to a deeper look at the nature of electricity, but here we introduce you to three very basic concepts of electricity: electric charge, electric current and electric circuit.

Electric charge

Electric charge refers to a fundamental property of matter still being discussed by the cleverest of physicists. Although a huge simplification, in essence two of the tiny particles that make up atoms – protons and electrons – are the bearers of electric charge. Two types of charge exist: *positive* and *negative*. Protons have positive charge and electrons have negative charge.

Electric charge is one of the basic forces of nature that hold the universe together. Positive and negative charges are irresistibly attracted to each other. Thus, the attraction of negatively-charged electrons to positively-charged protons holds atoms together.

If an atom has the same number of protons as it has electrons, the positive charge of the protons balances out the negative charge of the electrons, and the atom itself has no overall charge.

If an atom loses one of its electrons, however, the atom has an extra proton, which gives the atom a net positive charge. When an atom has a net positive charge, it 'goes looking' for an electron to restore its balanced charge.

Similarly, if an atom somehow picks up an extra electron, the atom has a net negative charge. When this happens, the atom 'goes looking' for a way to get rid of the extra electron to restore balance.

Atoms don't really 'look' for anything. They don't have eyes and they don't have minds that are troubled when they're short an electron or have a few too many. But the natural attraction of negative to positive charges causes atoms that are short an electron to be attracted to atoms that are long an electron. When they find each other, something almost magical happens. The atom with the extra electron gives its electron to the atom that's missing an electron. Thus, the charge represented by the electron moves from one atom to another, which brings us to the second important concept: current.

Electric current

Electric current refers to the flow of the electric charge carried by electrons as they move between the atoms. The concept is very familiar: when you turn on a light switch, electric current flows from the switch through the wire to the light and the room is instantly illuminated.

Electric current flows more easily in some types of materials than in others:

- **Conductors:** Materials that let current flow easily.
- **Insulators:** Materials that don't let current flow easily.

Electrical wires are made of conductors and insulators, as illustrated in Figure 1-1. Inside the wire is a conductor, such as copper or aluminium. The conductor provides a channel for the electric current to flow through. Surrounding the conductor is an outer layer of insulator, such as plastic or rubber.

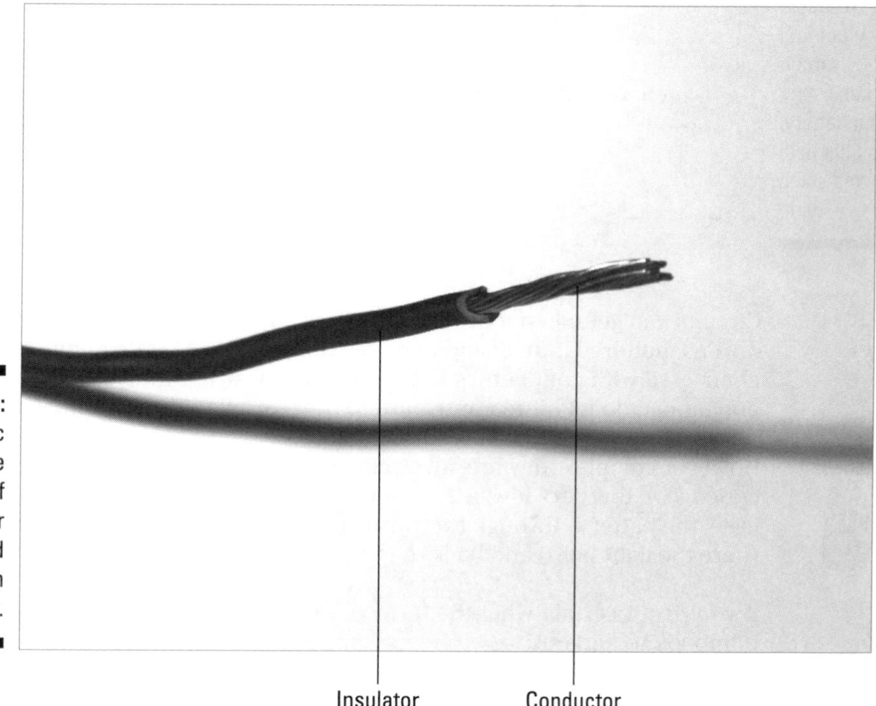

Figure 1-1: An electric wire consists of a conductor surrounded by an insulator.

The insulator serves two purposes:

- It prevents you from touching the wire when current is flowing, stopping you from receiving a nasty shock.
- It prevents the conductor inside the wire from touching the conductor inside a nearby wire. If the conductors touch, the result is a *short circuit,* which brings us to the third important concept.

Electric circuit

An *electric circuit* is a closed loop made of conductors and other electrical elements through which electric current can flow. For example, Figure 1-2 shows a simple electrical circuit that consists of three elements: a battery, a light bulb and an electrical wire that connects the two.

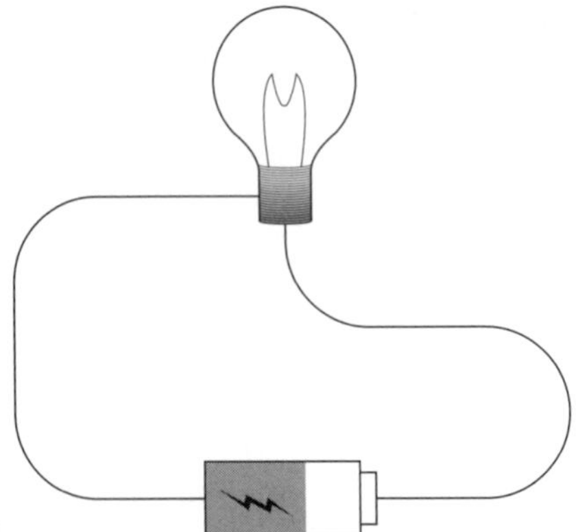

Figure 1-2: A simple electrical circuit consisting of a battery, a light bulb and some wire.

Circuits can get much more complex that this very simple one, consisting of dozens, hundreds, thousands or even millions of separate components, all connected with conductors in precisely orchestrated ways so that each component can do its bit to contribute to the overall purpose of the circuit.

However complex they are, all circuits have to obey the basic principle of a *closed loop* that provides a complete path from the source of voltage (in this case, the battery) through the various components that make up the circuit (here the light bulb) and back to the source (the battery).

If you're not certain what the term voltage means, flip to Chapter 2 of this minibook to find out.

Discovering the Difference Between Electrical and Electronic Devices

Electrical devices are common pieces of equipment, such as light bulbs, vacuum cleaners and toasters. But what exactly is the difference between *electrical* devices and *electronic* devices?

The answer lies in how devices manipulate electricity to do their work:

- **Electrical devices:** Take the energy of electric current and transform it in simple ways into some other form of energy – most likely light, heat or motion. For example, light bulbs turn electrical energy into light so that you can stay up late at night reading this book. The heating elements in a toaster turn electrical energy into heat so that you can burn your toast. And the motor in your vacuum cleaner turns electrical energy into motion that drives a pump that sucks the burnt toast crumbs out of your carpet.
- **Electronic devices:** Do much more that just convert electrical energy into heat, light or motion. They manipulate the electrical current itself to coax it into doing interesting and useful things.

One of the most common things that electronic devices do is manipulate electric current in a way that adds meaningful information to the current. For example, audio electronic devices add sound information to an electric current so that you can listen to music or talk on a mobile phone. And video devices add images to an electric current so you can watch films such as *Monty Python's Life of Brian, Airplane!* or *This is Spinal Tap* over and over again until you know every line by heart.

The distinction between electrical and electronic devices is a bit blurry. These days, simple electrical devices often include some electronic components. For example, your toaster may contain an electronic thermostat that attempts to keep the heat at just the right temperature to make perfect toast, and your washing machine has electronics so that you can choose the right programme to wash your clothes. And even the most complicated electronic devices have simple electrical components in them. For example, although your TV set's remote control is a pretty complicated little electronic device it contains batteries, which are simple electrical devices.

Electrical devices lead the way

You may be surprised to know that electrical devices had already been around for at least 100 years before Thomas Edison's innovations created electronics (check out the earlier sidebar 'Having a bright idea'). For example:

- The first electric batteries were invented by a fellow named Alessandro Volta in 1800. Volta's contribution is so important that the common *volt* is named for him.
- The electric telegraph was invented in the 1830s and popularised in America by Samuel Morse, who invented the famous Morse code used to encode the alphabet and numerals into a series of short and long clicks to transmit via telegraph. In 1866, a telegraph cable was laid across the Atlantic Ocean allowing instantaneous communication between the United States and Europe.
- Englishman Michael Faraday began experiments in 1831 on the relationship between magnetism and electricity that ultimately led to generators, motors and generally making electricity useful to society.

Using the Power of Electronics

The amazing thing about electronics is that it's being used today to do things that weren't even imaginable just a few years ago. And of course, in another few years you'll be using electronic devices that haven't even been thought up yet.

The following sections provide a very brief overview of some of the basic things you can do with electronics.

Making some noise

One of the most common applications for electronics is making noise, often in the form of music though the distinction between *noise* and *music* is often debatable. Electronic devices that make noise are often referred to as *audio devices.* These devices convert sound waves to electrical current, store, amplify and otherwise manipulate the current, and eventually convert the current back to sound waves you can hear.

Most audio devices contain these three parts:

- **Source:** The input into the system. The source can be a microphone, which converts sound waves into an electrical signal. The subtle fluctuations in the sound waves are translated into subtle fluctuations in the electrical signal. Thus, the electrical signal that comes from the source contains audio information.

  The source may also be a recorded form of the sound, such as sound recorded on a CD or in an MP3.
- **Amplifier:** Converts the small electrical signal that comes from the source into a much larger electrical signal that you can listen to, when sent to speakers or headphones.

  Some amplifiers are small, because they need to boost the signal only enough to be heard by a single listener wearing headphones. Other amplifiers are large, because they need to boost the signal enough so that thousands of people can hear, for example, the opening ceremony of the London 2012 Olympics.
- **Speakers:** Convert electrical current into sound you can hear. They can huge or small enough to fit in your ear.

Painting with light

Another common use of electronics is to produce light. The simplest electronic light circuits are LEDs, which are the electronic equivalent of a light bulb.

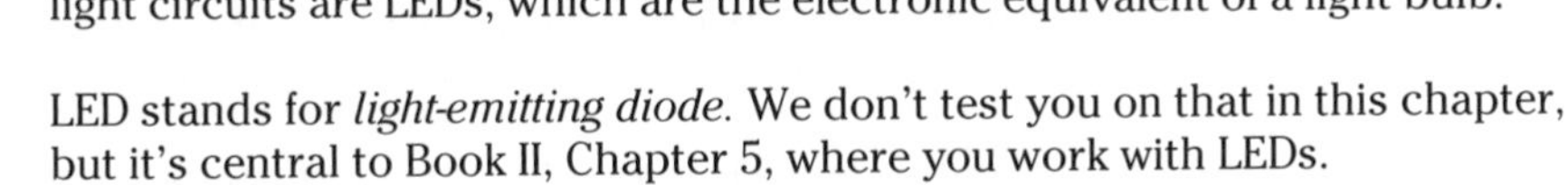

LED stands for *light-emitting diode.* We don't test you on that in this chapter, but it's central to Book II, Chapter 5, where you work with LEDs.

Video electronic devices are designed to create not just simple points of light, but complete images you can look at. The most obvious examples are television sets.

Some types of electronic devices work with light that you can't see. The most common are TV remote controls, which send infrared light to your television set whenever you push a button (assuming you can find the remote). The electronics inside the remote control manipulate the infrared light in a way that sends information from the remote control to the TV, telling it to turn up the volume, change channels or go to standby.

Transmitting to the world

Radio refers to the transmission of information without wires. Originally, radio was used as a wireless form of telegraph, broadcasting nothing more than audible clicks. Next, radio transmitted sound. In fact, to this day the term is usually associated with audio-only transmissions, such as music stations. However, the transmission of video information – in other words, television – is also a form of radio, as are wireless networking and cordless and mobile phones.

You can find out much more about radio electronics in Book V.

Computing

One of the most important applications of electronics in the last 50 years has been the development of computer technology. In just a few short decades, computers have gone from simple calculating devices to machines that have changed lives at work and home.

Computers are the most advanced form of a whole field of electronics known as *digital electronics,* which is concerned with manipulating data in the binary language of zeros and ones. You can discover digital electronics in Books VI and VII.

Looking Inside Electronic Devices

If you've ever taken apart an electronic device that no longer works, such as an old clock radio or VHS tape player, you know that inside is usually a *circuit board* (or *circuit card*): a flat, thin board with electronic gizmos mounted on it.

One or both sides of the circuit board are populated with tiny devices that look like little buildings. These components make up the electric circuit – the resistors, capacitors, diodes, transistors and integrated circuits that do the work the circuit is destined to do. (We discuss the first four of these components in Book II, Chapters 2, 3, 5 and 6, respectively, and cover integrated circuits in Book III.) In between those components the circuit board is painted with little lines of copper that look like streets. These conductors connect all the components so that they can work together.

An electronic circuit board looks like a city in miniature! For example, have a look at the circuit board pictured in Figure 1-3, which happens to be a board with components on only one side. The top of the board is shown, logically enough, at the top; it's populated with a variety of common electronic components. The underbelly of the circuit board is shown at the bottom of the figure; it has the typical shiny streaks of conductors that connect the components topside so that they can perform useful work.

Here's the essence of what's going on with these two sides of the circuit board:

- **Component side, with the little 'buildings':** Holds a collection of electronic components whose sole purpose in life is to bend, turn and twist electric current to get it to do interesting and useful things. Some of those components restrict the flow of current, like speed bumps on a road. Others make the current stronger. Some work like one-way street signs that allow current to flow in only one direction. Still others try to smooth out any ripples or variations in the current, resulting in smoother traffic flow.
- **Circuit side, with the shiny lines:** Provides the conductive pathways for the electric current to flow from one component to the other in a certain order.

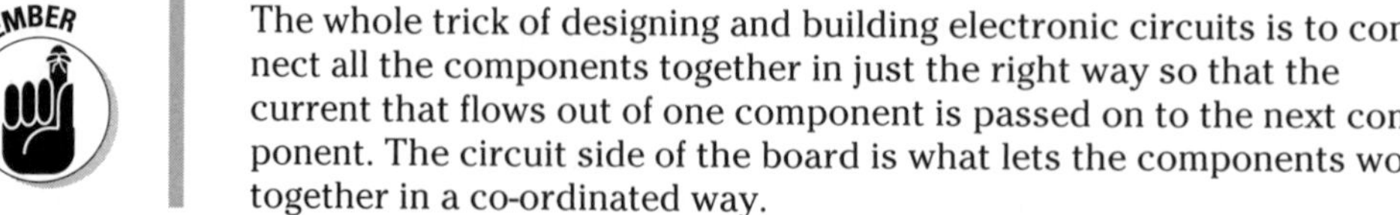

The whole trick of designing and building electronic circuits is to connect all the components together in just the right way so that the current that flows out of one component is passed on to the next component. The circuit side of the board is what lets the components work together in a co-ordinated way.

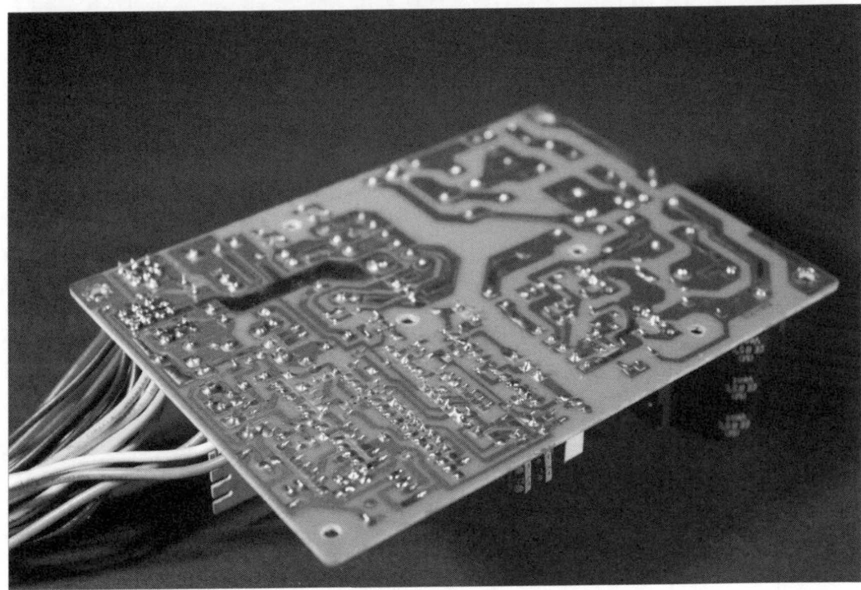

Figure 1-3: A typical electronic circuit board.

Don't under any circumstances plunge carelessly into the disassembly of old electronic circuits until you're certain you know what you're doing.

The little components on a circuit board such as the one shown in Figure 1-3 can be dangerous, even when they're unplugged. In fact, the two tall cylindrical components near the back edge of this circuit board are called *capacitors*. They can contain stored electrical energy that can deliver a powerful – even fatal – shock long after you've unplugged the power cord. Please read Chapter 4 of this minibook before you begin disassembling anything!

Chapter 2

Understanding Electricity

In This Chapter

- Unravelling the mystery of matter
- Discovering current, voltage and power

Before you can do anything interesting with electronics, you need to have a basic understanding of what electricity is and how it works. Unfortunately, understanding electricity is a tall order, and so frankly the title of this chapter is a bit ambitious. Don't let this discourage or dissuade you, though: the smartest physicists in the world are still making discoveries about electricity.

In this chapter, you take a look at the very nature of electricity: what it is and what causes it. This section may remind you of a school science lesson, as you delve into atoms, protons, neutrons and electrons.

We also introduce you to three things you have to know about electricity if you want to design and build your own circuits: current, voltage and power – the Huey, Dewey and Louie of electricity, or the Groucho, Harpo and Chico, or the – well, you get the idea.

Wondering about the Nature of Electricity

The exact nature of electricity is one of the mysteries of the universe. Although even experts don't know everything about electricity, they do know a lot about what it does and how it behaves.

Strange as it sounds, your understanding of electricity improves right away if you avoid using the term 'electricity' to describe it, because this word isn't sufficiently precise. People use this word to refer to any of several different but related things, each of which has a more accurate name, such as electric charge and electric current (check out the later sections 'Charging ahead' and 'Keeping current'), electric energy, electric field and so on. All these things are commonly called *electricity.*

But electricity is less a specific thing and more a phenomenon with many different faces. So to avoid confusion, wherever possible we use more precise terms such as 'charge' or 'current' in this book.

We prefer to think of electricity more as a wonder than a phenomenon, because the latter sounds so scientific. Electricity really does qualify as one of the great wonders of the universe. To the so-called Seven Wonders of the Ancient World we'd like to add a list called 'The Seven Wonders of the Universe', including matter, gravity, time, light, life, pizza and electricity.

Looking for electricity

One of the most amazing things about electricity is that it is, literally, everywhere. We don't mean that electricity is commonplace or plentiful, or even that the universe has an abundant supply of electricity. Instead, we mean that electricity is a fundamental part of everything.

Consider a common misconception about electric current. Many people think that wires carry electricity from place to place. When you plug in a vacuum cleaner and turn on the switch, you may think that electricity enters the vacuum cleaner's power cord at the electrical outlet, travels through the wire to the vacuum cleaner and then turns the motor to make the vacuum cleaner suck up dirt, grime and dog hair. But that's not the case.

Tracing the origins of the word 'electricity'

In the film *Jurassic Park,* scientists discover the DNA of dinosaurs locked inside bits of amber. Amber is fossilised tree resin, and it plays a key role in the history of electricity.

Since the days of the Ancient Greeks, people have known that if you rub sticks of amber with fur, the amber can be used to raise the hair on your head and that lightweight objects (such as feathers) stick to it. The Ancient Greeks had no idea why this happened, but they spotted that it did.

The Greek word for amber is *elektron* and the Latin version of the word is *electricus.* At the beginning of the 17th century, an English scientist named William Gilbert began to study electricity. He used these ancient words to describe the phenomena he was investigating, including the Latin *electricus.* The influence of Gilbert's book, which was written in Latin, led to the word 'electricity' becoming embedded in the English language.

The truth is that the electricity is already in the wire. It's always in the wire, even when the vacuum cleaner is turned off or the power cord not plugged in. That's because electricity is a fundamental part of the copper atoms that make up the wire inside the power cord. Electricity is also a fundamental part of the atoms that make up the rubber insulation that protects you from being electrocuted when you touch the power cord. And it's a fundamental part of the atoms that make up the tips of your finger that the rubber keeps from touching the wires.

In short, electricity is a fundamental constituent of the atoms that make up all matter. So, to understand what electricity is, we must look at atoms.

Peering inside atoms

All matter is made up of unbelievably tiny bits called *atoms.* They're so tiny that the full stop at the end of this sentence contains several trillion of them.

Getting your head around numbers as large as trillions is tricky. For the sake of comparison, suppose that you were able to enlarge that full stop until it was about five times the size of England. Then, each atom would be about the size of – you guessed it – the full stop at the end of this sentence.

The word 'atom' comes from an Ancient Greek fellow named Democritus. Contrary to what you may expect, the word doesn't mean 'really small': it means 'undividable'. Atoms are the smallest part of matter that can't be divided without changing it to a different kind of matter. In other words, if you divide an atom of a particular element, the resulting pieces are no longer the same thing.

For example, imagine that you have a handful of some basic element such as copper and you cut it in half. You now have two pieces of copper. Toss one of them aside, and cut the other one in half. Again, you have two pieces of copper. You keep dividing your piece of copper into ever smaller halves. But eventually, you get to the point where your piece of copper consists of just a single copper atom.

If you try to cut that single atom of copper in half, the resulting pieces aren't copper. Instead, you have a collection of the basic particles that make up atoms: the three such particles are neutrons, protons and electrons.

The neutrons and protons in each atom are clumped together in the middle of the atom, in what's called the *nucleus.* The electrons spin around the outside of the atom.

Although even today children are still taught that electrons orbit around the nucleus much like planets orbit around the sun in the solar system (building models like the one in Figure 2-1), in fact that's a really bad analogy. Instead, the electrons whiz around the nucleus in a cloud that's called, appropriately enough, the *electron cloud.* Electron clouds have weird shapes and properties, and strangely figuring out exactly where in its cloud an electron is at any given moment is next to impossible.

Figure 2-1: A common model of an atom.

Examining the elements

Here's the deal with elements: an *element* is a specific type of atom, defined by the number of protons in its nucleus. For example, hydrogen atoms have just one proton in the nucleus, an atom with two protons in the nucleus is helium and atoms with three protons are called lithium.

The number of protons in the nucleus of an atom is the *atomic number.* Thus, the atomic number of hydrogen is 1, the atomic number of helium is 2 and lithium is 3. Copper – an element that plays an important role in electronics – is atomic number 29. Thus, it has 29 protons in its nucleus.

What about neutrons, the other particle found in the nucleus of an atom? Neutrons are extremely important to chemists and physicists, but they don't play that big a role in the way electric current works, and so we can safely ignore them in this chapter (phew!). Suffice it to say that in addition to protons, the nucleus of each atom (except hydrogen) contains neutrons – and most atoms have a few more neutrons than protons.

The third particle that makes up atoms is the electron. Electrons are the most interesting particle for this book, because they're the source of electric current. They're unbelievably small: a single electron is about 200,000 times smaller than a proton. To gain some perspective on that, if a single electron were the size of the full stop at the end of this sentence, a proton would be almost the size of a football pitch.

Atoms usually have the same number of electrons as protons, and thus an atom of the element copper has 29 protons in a nucleus that's 'orbited' by 29 electrons. When an atom picks up an extra electron or finds itself short of an electron, things get interesting because of a special property of protons and electrons called *charge,* which we explain in the next section.

Charging ahead

Two of the three particles that make up atoms – electrons and protons – have a very interesting characteristic called *electric charge*. Charge can be one of two *polarities*: negative or positive. Electrons have a *negative* polarity and protons have a *positive* polarity.

Strong-arming protons

You may be wondering how the nucleus of an atom can stay together if it consists of two or more protons that have positive charges. After all, don't like charges repel? Yes they do, but the electrical repellent force is overcome by a much more powerful force called, for lack of a better term, the *strong force.* The strong force holds protons (and neutrons) together in spite of the protons' natural tendency to avoid each other.

The strong force doesn't affect electrons, and so you never see electrons clumped together the way protons do in the nucleus of an atom. The electrons in an atom stay well away from each other.

The most important thing to know about charge is that opposite charges attract and similar charges repel. Negative attracts positive and positive attracts negative, but negative repels negative and positive repels positive. As a result, electrons and protons are attracted to each other, but electrons repel other electrons and protons repel other protons.

The attraction between protons and electrons is what holds the electrons and the protons of an atom together. This attraction causes the electrons to stay in their 'orbits' around the protons in the nucleus.

Charge is a property of one of the fundamental forces of nature known as *electromagnetism.* The other three forces are gravity, the strong force (check out the nearby sidebar 'Strong-arming protons' for a little more) and the weak force. As we say in the preceding section, an atom normally has the same number of electrons as protons, because the electromagnetic force causes each proton to attract exactly one electron. When the number of protons and electrons is equal, the atom itself has no net charge. In this case, it's said to be *neutral.*

An atom can, however, pick up an extra electron. When it does, the atom has a net negative charge because of that extra electron. An atom can also lose an electron, which causes the atom to have a net positive charge, because it has more protons than electrons.

Conducting and Insulating Elements: Current, Voltage and Power

Some elements (which we introduce earlier in 'Examining the elements') don't hold on to some of their electrons as tightly as other elements. These elements (called *conductors*) frequently lose electrons or pick up extra electrons, and so they often get bumped off neutral and become negatively or positively charged (check out the preceding section for more on charges). The metals silver, copper and aluminium are the best conductors.

In contrast, other elements hold on to their electrons more tightly. In these elements (called *insulators*), prying loose an electron or forcing another electron in is harder. These elements almost always stay neutral.

In a conductor, electrons are constantly skipping around between nearby atoms. An electron jumps out of one atom – call it Atom A – into a nearby atom, which we call Atom B. This movement creates a net positive charge in Atom A and a net negative charge in Atom B. But almost immediately, an electron jumps out of another nearby atom – Atom C – into Atom A. Thus, Atom A again becomes neutral and now Atom C is negative.

This skipping around of electrons in a conductor happens constantly. Atoms are in perpetual turmoil, giving and receiving electrons and constantly cycling their net charges from positive to neutral to negative and back to positive.

Ordinarily, this movement of electrons is completely random. One electron may jump left, but another one jumps right. One goes up, another goes down. One goes east, the other goes west. The net effect is that although all the electrons are moving, collectively they aren't going anywhere. They're like the Keystone Kops, running around aimlessly in every direction, bumping into each other, falling down, picking themselves back up and then running around some more. When this randomness stops and the Keystone Kops get organised, the result is electric current (the subject of the next section).

Keeping current

Electric *current* is what happens when the random exchange of electrons that occurs constantly in a conductor becomes organised and begins to move in the same direction.

When current flows through a conductor such as a copper wire, all those electrons that were previously moving about randomly get together and start moving in the same direction. A very interesting effect then happens: the electrons transfer their electromagnetic force through the wire almost instantaneously. The electrons themselves all move relatively slowly – around a few millimetres per second. But as each electron leaves an atom and joins another atom, that second atom immediately loses an electron to a third atom, which immediately loses an electron to the fourth atom and so on trillions upon trillions of times.

The result is that even though the individual electrons move slowly, the current itself moves at nearly the speed of light. Thus, when you flip a light switch, the light turns on immediately, no matter how much distance separates the light switch from the light bulb.

One way to illustrate this principle is to line up 15 balls on a pool table in a perfectly straight line, as shown in Figure 2-2. If you hit the cue ball on one end of the line, the ball on the opposite end of the line almost immediately moves. The other balls move a little, but not much (assuming you line them up straight and strike the cue ball straight).

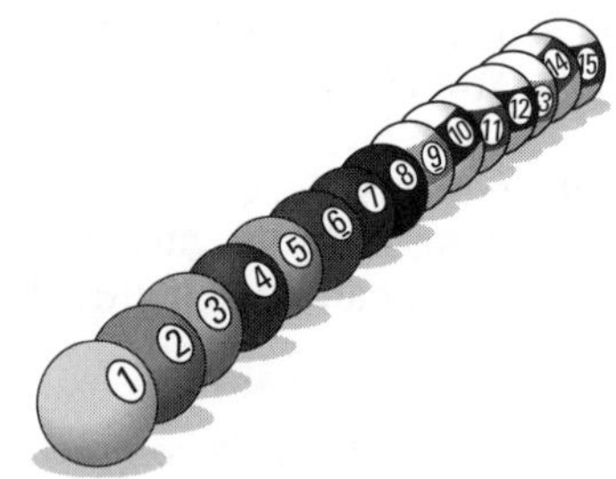

Figure 2-2: Electrons transfer current through a wire much like a row of pool balls transfer motion.

This effect is similar to what happens with electric current. Although each electron moves slowly, the ripple effect as each atom loses and gains an electron is lightning fast (literally!).

Here are a few additional points to help you understand the nature of current:

- Most electric incandescent light bulbs have about a quarter of an amp of current flowing through them when they're turned on. A hair dryer uses about 5 A.
- Current in electronic circuits is usually much smaller than current in electrical devices such as light bulbs and hair dryers (if you're unclear of the difference between electronic and electrical devices, flip to Chapter 1 of this minibook). The current in an electronic circuit is often measured in thousandths of amps (*milliamps,* abbreviated mA).
- Current is often represented by the letter *I* (for *intensity*) in electrical equations.
- The fact that moving water is also called current is no coincidence. Many early scientists who explored the nature of electricity believed that electricity was a type of fluid, and that it flowed in wires in much the same way that water flows in a river.

Pushing electrons around: Voltage

In its natural state, the electrons in a conductor such as copper freely move from atom to atom, but in a completely random way. To get them to move together in one direction, all you have to do is give them a push. The technical term for this push is *electromotive force* (abbreviated EMF, or sometimes simply E). You know it more commonly as *voltage.*

Amping things up

The strength of an electric current is measured with a unit called the *ampere,* sometimes used in the short form amp or abbreviated A. The ampere is nothing more than a measurement of how many charge carriers (in most cases, electrons) flow past a certain point in one second. One ampere is equal to 6,240,000,000,000,000,000 electrons per second. That's 6.24 trillion electrons per second in the continental European numbering system (6.24 quintillion electrons per second in the American system). Either way, it's a lot.

A voltage is nothing more than a difference in charge between two places. For example, suppose that you have a small clump of metal whose atoms have an abundance of negatively charged atoms and another clump of metal whose atoms have an abundance of positively charged atoms. In other words, the first clump has too many electrons and the second clump has too few. A voltage exists between the two clumps. If you connect the two clumps with a conductor, such as a copper wire, you create a *circuit* through which electric current can flow.

This current continues to flow until all the extra negative charges on the negative side of the circuit have moved to the positive side. When that has happened, both sides of the circuit become electrically neutral and the current stops flowing.

Although current stops flowing when the two sides of the circuit have been neutralised, the electrons in the circuit don't stop moving. Instead, they simply revert to their natural random movement. Electrons are always moving in a conductor. When they get a push from a voltage, they move in the same direction. With no voltage to push them along, they move about randomly.

Whenever a difference in charge exists between two locations, a current may flow between the two locations if they're connected by a conductor. Because of this possibility, the term *potential* is often used to describe voltage. Without voltage, you can't have current. Thus, voltage creates the potential for a current to flow.

If we compare current to the flow of water through a hose, you can then see voltage as the water pressure at the tap. Water pressure causes the water to flow in the hose.

Here are some facts and figures about voltage:

- Voltage is measured using a unit called, naturally, the *volt* (usually abbreviated V). The voltage available in a standard electrical outlet in the UK is about 230 V. The voltage available in a battery is about 1.5 V and a car battery provides about 12 V.
- To find out how much voltage exists between two points, you use a *voltmeter,* a device with two wire test leads that you touch to different points in a circuit to measure the voltage between those points. Figure 2-3 shows a typical voltmeter. (In fact, this meter is a *multimeter,* which simply means that it can measure things other than voltage as well. In the figure, the multimeter is functioning as a voltmeter. For more information about using a voltmeter, refer to Chapter 8 of this minibook.)
- Voltages can be considered positive or negative, but only when compared with some reference point. For example, in a battery the voltage at the positive terminal is +1.5 V relative to the negative terminal. The voltage at the negative terminal is –1.5 V relative to the positive terminal.

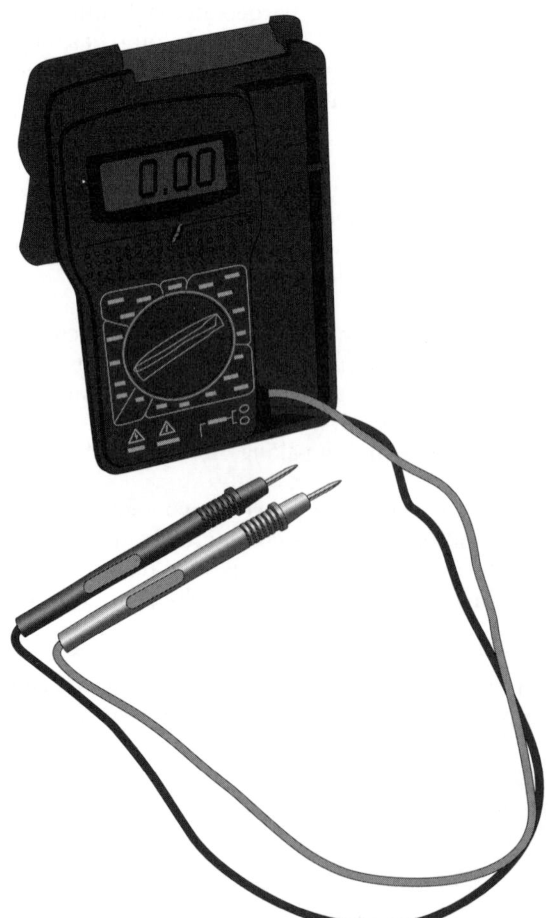

Figure 2-3: Measuring voltage with the voltmeter function of a multimeter.

Are you positive about that?

For the first 150 years or so of serious research into the nature of electricity, scientists had the concept of electric current backwards: they thought that electric current was the flow of positive charges and that electric current flowed from the positive side of a circuit to the negative side.

Not until around 1900 did scientists begin to unravel the structure of atoms. They figured out that electrons have a negative charge and that current is the flow of these negatively-charged electrons. In other words, current flows in the opposite direction to what they'd previously thought.

Old ideas die hard, however, and to this day most people think of electric current as flowing from positive to negative (sometimes called *conventional current*). Modern electronic circuits are almost always described in terms of conventional current, encouraging the assumption that current flows from positive to negative, even though the reality is that the electrons in the circuit are flowing in the opposite direction.

Comparing direct and alternating current

An electric current that flows continuously in a single direction is called a *direct current* (DC). The electrons in a wire carrying direct current move slowly, but eventually they travel from one end of the wire to the other because they keep plodding along in the same direction.

The voltage in a DC circuit needs to be constant, or at least relatively constant, to keep the current flowing in a single direction. Thus, the voltage provided by a torch battery remains steady at about 1.5 V. The positive end of the battery is always positive relative to the negative end, and the negative end of the battery is always negative relative to the positive end. This constancy is what pushes the electrons in a single direction.

The other common type of current is *alternating current* (AC). In an AC circuit, voltage periodically reverses itself. When the voltage reverses, so does the direction of the current flow. In the most common form of AC, used in most power-distribution systems throughout the world, the voltage reverses itself 50 or 60 times per second, depending on the country. In the UK, the voltage is reversed 50 times per second.

Alternating current is used in nearly all the world's power-distribution systems for the simple reason that AC is much more efficient when it's transmitted through wires over long distances. All electric currents lose power when they flow for long distances, but AC circuits lose much less power than DC circuits.

The electrons in an AC circuit don't really move along with the current flow. Instead, they sort of sit and wiggle back and forth. They move one direction for 1⁄50 of a second, and then turn around and go the other direction for 1⁄50 of a second. The net effect is that they don't really go anywhere.

A popular toy called *Newton's Cradle* can help you understand how AC works. The toy consists of a series of metal balls hung by string from a frame, such that the balls are just touching each other in a straight line, as shown in Figure 2-4. If you pull the ball on one end of the line away from the other balls and then release it, that ball swings back to the line of balls, hits the one on the end and instantly propels the ball on the other end of the line away from the group. This ball swings up for a bit, and then turns around and swings back down to strike the group from the other end, which then pushes the first ball away from the group. This alternating motion, back and forth, continues for an amazingly long time if the toy is carefully constructed.

Figure 2-4: This Newton's Cradle works like alternating current.

Alternating current works in much the same way. The electrons initially move in one direction, but then reverse themselves and move in the other direction. The back and forth movement of the electrons in the circuit continues as long as the voltage continues to reverse itself.

To see a Newton's Cradle in action, go to YouTube and search for 'Newton's Cradle'.

The reversal of voltage in a typical AC circuit isn't instantaneous. Instead, the voltage swings smoothly from one polarity to the other. Thus, the voltage in an AC circuit is always changing. It starts out at zero, increases in the positive direction for a bit until it reaches its maximum positive voltage and then decreases until it gets back to zero. At that point, it increases in the negative direction until it reaches its maximum negative voltage, at which time it decreases again until it gets back to zero. Then the whole cycle repeats itself. (Flick to Book I Chapter 9 to see what this voltage swing looks like on a graph.)

The fact that the amount of voltage in an AC circuit is always changing is incredibly useful. (To discover how, flip to Book IV, Chapter 1, where we take a deeper look at AC.)

Working out with power

The third of the three key concepts about electricity, in addition to current and voltage (see the earlier sections 'Keeping current' and 'Pushing electrons around: Voltage' respectively), is power (abbreviated *P* in equations).

Simply put, *power* is the work done by an electric circuit. Electric current becomes useful when the energy it carries is converted into some other form of energy, such as heat, light, sound or radio waves. For example, in an incandescent light bulb, voltage pushes current through a *filament,* which converts the energy carried by the current into heat and light. When this happens, the circuit is said to *dissipate* power.

Calculating the power dissipated by a circuit is often a very important part of circuit design, because electrical components such as resistors, transistors, capacitors and integrated circuits have maximum power ratings. For example, the most common type of resistor can dissipate at most ¼ watt. If you use a ¼-watt resistor in a circuit that dissipates more than ¼ watt of power, you run the risk of burning up the resistor.

As you can see, power is measured in units called *watts* (W). The definition of 1 W is simply the amount of work done by a circuit in which 1 A of current is driven by 1 V.

This relationship lends itself to a simple equation. Although we use as few equations in this book as possible, we have to include some basic ones. Fortunately, this one is pretty simple:

$$P = V \times I$$

In other words, power *(P)* equals voltage *(V)* times current *(I)*.

This formula is sometimes called Joule's Law, after the person who discovered it.

Confusingly, current is represented by the letter *I* (for intensity), not the letter C. But at least power is represented by *P*. Table 2-1 may prove helpful in keeping these abbreviations sorted out.

Table 2-1 The Three Central Concepts of Electricity

Concept	*Abbreviation in Equations*	*Unit*
Current	*I*	amp (A)
Voltage	*V* (or sometimes *E* or *EMF*)	volt (V)
Power	*P*	watt (W)

To use the equation correctly, make sure that you measure power, voltage and current using their standard units: watts, volts and amperes. For example, suppose that you have a light bulb connected to a 10-V power supply and one-tenth of an ampere is flowing through the light bulb. To calculate the wattage of the light bulb, you use the $P = V \times I$ formula as follows:

$$P = 10\text{ V} \times 0.1\text{ A} = 1\text{ W}$$

Thus, the light bulb is doing 1 W of work.

Often, you know the voltage and wattage of the circuit and you want to use those values to determine the amount of current flowing through the circuit. You can do so by turning the equation around:

$$I = \frac{P}{V}$$

For example, if you want to determine how much current flows through a lamp with a 60-W light bulb when it's plugged into a 230-V electrical outlet, use the formula like this:

$$\text{I} = \frac{60\ \text{W}}{230\ \text{V}} = 0.26\ \text{A}$$

Thus, the current through the circuit is 0.26 A.

In the earlier section 'Pushing electrons around: Voltage', we say that you need to know what power is in order to define 1 V. Now you can see that 1 V is the amount of electromotive force necessary to do 1 W of work at 1 A of current.

Chapter 3

Creating Your Own Mad-Scientist Lab

In This Chapter

- Finding a suitable place to build your laboratory
- Investing in good-quality tools
- Picking out components to get you started

We love the Frankenstein films, especially the scenes in Dr Frankenstein's laboratory that's filled with the most amazing and exotic electrical gadgets. The mad doctor's assistant, Igor, throws a giant knife switch at just the right moment and sparks fly, the music rises to a crescendo, the creature jerks to life and the crazy doctor yells 'it's *alive*!'.

Whichever of the Frankenstein movies is your favourite – the original 1931 *Frankenstein* starring Boris Karloff, Mel Brooks's 1974 *Young Frankenstein* with Gene Wilder or Tim Burton's recent *Frankenweenie* – they all have great laboratory scenes.

Fortunately, you don't need a deserted castle or an elaborate mad-scientist laboratory to construct your own basic electronic circuits. You do, however, need to build yourself a modest workplace, equipped with a basic set of tools as well as some fundamental electronic components to work with.

In this chapter, we introduce you to the stuff you need to get before you can begin building electronic circuits. You don't have to buy everything straightaway, of course. You can start with just a simple collection of tools and a small space to work in. As you become more advanced in your electronic skills, you can acquire additional tools and equipment as your needs develop.

Setting up Your Lab

To begin building your own electronic circuits, you need to create a good place to work. You can establish a fancy workbench in your garage or in a spare room (see Figure 3-1), but if you don't have that much space, you can set up an ad-hoc mad-scientist lab just about anywhere. All you need is a place for a small workbench and a chair.

Figure 3-1: Some electronic work areas do end up looking quite mad!

Here are the essential ingredients of any effective work area for electronic tinkering:

- **Adequate space:** You need to have adequate space for your work. When you're just getting started, your work area can be small – maybe just a square metre or two in the corner of the garage – but as your skills progress, you need more space.

  The location you choose for your work area must be secure, especially if you have young children around. Your work area is going to be filled with dangers – things that can cause shocks, burns and cuts, as well as things that under no circumstances should be eaten. Little hands are incredibly curious, and children are prone to put anything they don't recognise in their mouths. So be sure to keep everything safely out of reach, ideally behind a locked door. Check out Chapter 4 of this mini-book for loads more on safety.

- **Good lighting:** The ideal lighting is directly overhead instead of from the side or behind you. If possible, purchase an inexpensive fluorescent shop light and hang it directly over your work area. If your chosen spot doesn't allow you to hang lights overhead, the next best bet is a desk lamp that swings overhead, to bring light directly over your work.
- **A solid workbench:** Initially, you can get by with something as simple as a small folding table. Eventually, though, you're going to want something more permanent and substantial. You can make yourself an excellent workbench from an inexpensive but sturdy office desk.

If your only option for your workbench is your kitchen table, go to your local big box hardware store and buy up to a square metre of 1.5-cm-thick plywood. This area can serve as a good solid work surface until you acquire a real workbench.

- **Comfortable seating:** If your workbench is a folding table or desk, the best seating is a good office chair. Many workbenches stand a little taller than desk height, however, to allow you to work comfortably while standing. If your workbench is tall, you need a seat of the correct height. You can buy a bench stool from a home-improvement store or check out second-hand shops for a cheap bar stool.
- **Plenty of electricity:** Obviously you need a source of electricity nearby as you build electronic projects. A standard 13-amp electrical outlet supplies enough current capacity, but it probably doesn't provide enough electrical outlets for your needs. The easiest way to meet that need is to purchase several multi-outlet power strips and place them in convenient locations behind or on either side of your work area.
- **Plenty of storage:** You need a place to store your tools, supplies and components. The ideal storage for hand tools is a small sheet of pegboard mounted on the wall right behind your workbench. Then, you can use hooks to hang your tools within easy reach. For larger tools, such as a drill or saw, built-in cabinets are best.

For small parts, multicompartment storage boxes such as the ones shown in Figure 3-2 are best. We suggest that you get one or two to store all the little components, such as resistors, diodes, capacitors, transistors and so on. If you buy two boxes, get one with a few larger compartments and another with a greater number of smaller compartments.

Keep a few small, shallow storage bins handy. They're especially useful for storing parts for the project you're working on, and a shallow bin is better than having them scattered loose all over your work area.

Figure 3-2: Multi-compartment storage boxes are ideal for storing small components.

Equipping Your Lab

Like any hobby, electronics has its own special tools and supplies. Fortunately, you don't need to run out and buy everything at once. But the more involved you get with the hobby, the more you want to invest in a wide variety of quality tools and supplies. The following sections outline some of the essential stuff you need at your disposal.

Acquiring basic hand tools

For starters, you need a basic set of hand tools, similar to the assortment shown in Figure 3-3. Specifically, you require the following items:

- **Pliers:** Occasionally you need to use standard flat-nosed pliers, but for most electronic work you depend on *needle-nose pliers* instead, which are especially adept at working with wires – bending and twisting them, pushing them through holes and so on. Most needle-nose pliers also have a cutting edge that lets you use them as wire cutters.

  Get a small set of needle-nose pliers with thin jaws for working with small parts and a larger set for bigger jobs.

- **Screwdrivers:** Most electronic work is relatively small, and so you don't need huge, heavy-duty screwdrivers. But do get yourself a good assortment of small- and medium-sized screwdrivers, flat-blade and Phillips head.

  A set of *jeweller's screwdrivers* is sometimes very useful. The swivelling knob on the top of each one makes holding the screwdriver in a precise position easy while turning the blade.

Figure 3-3: Basic hand tools you want to have.

- **Wire cutters:** Although you can use needle-nose pliers to cut wire, you also want a few different types of wire cutters at your disposal. Get something heavy-duty for cutting thick wire and something smaller for cutting small wire or component leads.
- **Wire strippers:** Figure 3-4 shows two pieces of wire that we *stripped* (removed the insulation from). We stripped the one on top with a set of wire cutters and the one on the bottom with a set of *wire strippers.* The crimping you can see in the one at the top, where the insulation ends, was caused by using just a bit too much pressure on the wire cutters. That crimp has created a weak spot in the wire that may eventually break.

  Therefore, to avoid damaging your wires when you strip them, we suggest you purchase an inexpensive (under £10) wire-stripping tool. You'll be glad you did later.

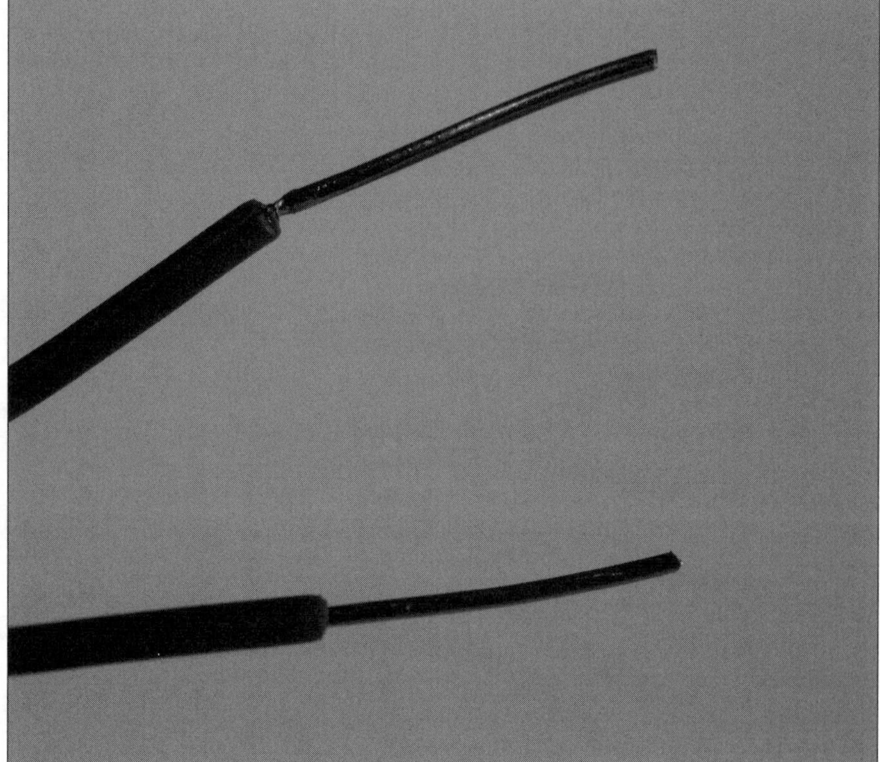

Figure 3-4: The wire on the top was stripped with wire cutters; the one on the bottom was stripped with a wire stripper.

Getting what you pay for

With tools, the old mantra 'you get what you pay for' is generally true. Good tools that are manufactured from the best materials and with the best quality fetch a premium price. Cheap tools are, well, cheap. The price range can be substantial. You can easily spend £20 or £25 on decent wire cutters, or you can buy cheap ones for £3 or £4.

Two main drawbacks apply to cheap tools:

- They don't last. The business end of a cheap tool wears out very quickly. For example, each time you cut a wire with a pair of cheap wire cutters, you ding the cutting blade a bit. Pretty soon, the cutters can barely cut through the wire.
- When cheap tools wear out, they tend to damage the materials you use them on. For example, tightening a screw with a badly worn screwdriver can strip the screw. Likewise, attempting to loosen a tight nut with a worn-out wrench can strip the nut.

We do, however, endorse spending money on cheap tools to help you get started in this fascinating hobby as inexpensively as possible. You can start with cheap tools and replace them one by one with more expensive tools as your experience, confidence, love of the hobby and, crucially, your budget increase.

Seeing clearly with magnifying glasses

One of the most helpful items you can have in your tool arsenal is a good magnifying glass. After all, most electronic stuff is small and resistors, diodes and transistors are downright tiny.

We suggest that you have at least three types of magnifying glasses on hand:

- **A handheld magnifying glass:** To inspect solder joints, read the labels on small components and so on.
- **A magnifying glass mounted on a base:** So that you can hold your work behind the glass. The best mounted glasses combine a light with the magnifying glass so that the object you're magnifying is bright.
- **Magnifying goggles:** For completely hands-free magnifying for delicate work. The ideal goggles have lights mounted on them.

Getting a firm grip on third hands and hobby vices

A *third hand* is a common tool with hobbyists. It's a small stand that has a couple of clips that you use to hold your work, freeing up your hands to do delicate work. Most third-hand tools also include a magnifying glass. Figure 3-5 shows an inexpensive third-hand tool holding a circuit card.

The most common use for a third hand in electronics is soldering (a skill we describe in this minibook's Chapter 7). You use the clips to hold the parts you want to solder, positioned behind the magnifying glass so you can get a good look.

Although the magnifying glass on the third hand is helpful, it does tend to get in the way of the work: manoeuvring your soldering iron and solder behind the magnifying glass can be awkward. For this reason, we often remove the magnifying glass from the third hand and use magnifying goggles instead (check out the preceding section).

The third hand is often helpful for assembling small projects, but it lacks the sturdiness required for larger projects. Eventually you're going to want to invest in a small hobby vice such as the one shown in Figure 3-6 (made by PanaVise; `www.panavise.com`).

Figure 3-5: A third hand can hold your stuff so both your hands are free to do the work.

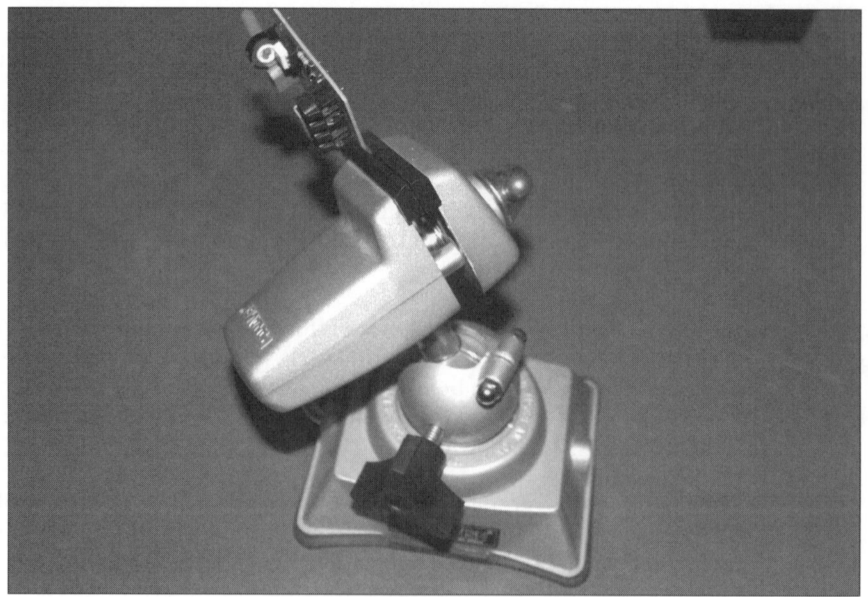

Figure 3-6: A hobby vice.

Here are a few things to look for in a hobby vice:

- **Mount:** Get a vice that has a base with the proper type of workbench mount. Three common types of mounts are available:
  - *Bolt mount:* The base has holes through which you can pass bolts or screws to attach the vice to your workbench. This mount is the most stable, but it requires that you put holes in your workbench.
  - *Clamp mount:* The base has a clamp that you can tighten to fix the base to the top and bottom of your workbench. Clamp mounts are pretty stable but can be placed only near the edge of your workbench.
  - *Vacuum mount:* The base has a rubber seal and lever you pull to create a vacuum between the seal and the workbench top. Vacuum mounts are the most portable but work well only when the top of your workbench is smooth.
- **Movement:** Select a vice with plenty of movement so that you can swivel your work into a variety of different working positions. Make sure that when you lock the swivel mount into position, it stays put. You don't want your work sliding around while you're trying to solder on it.
- **Protection:** Make sure that the vice jaws have a rubber coating to protect your work.

Making connections with a soldering iron

Soldering is one of the basic techniques you use to assemble electronic circuits. The purpose of soldering is to make a permanent connection between two conductors – usually between two wires or between a wire and a conducting surface on a printed circuit board.

The basic technique of soldering is to connect physically the two pieces to be soldered, and then heat them with a soldering iron until they're hot enough to melt *solder* (a special metal that has a low melting point). You then apply the solder to the heated parts so that it melts and flows over the parts.

After the solder flows over the two conductors, you remove the soldering iron. As the solder cools, it hardens and bonds the two conductors together.

You can discover all about soldering in Chapter 7 of this minibook. For now, however, you need just three things for successful soldering:

- **Soldering iron:** A little hand-held tool that heats up enough to melt solder. An inexpensive soldering iron is just fine to get started with. As you become more involved with electronics, you can invest in a better soldering iron that has more precise temperature control and is internally grounded.
- **Solder:** The soft metal that melts to form a bond between the conductors.
- **Soldering iron stand:** To set your soldering iron on when you aren't soldering. Some soldering irons come with stands, but the cheapest ones don't. Figure 3-7 shows a soldering iron that has a stand.

Figure 3-7: A soldering iron with a stand.

Measuring with a multimeter

In Chapter 2 of this minibook, we discuss measuring voltage with a voltmeter. You can also use meters to measure many other quantities that are important in electronics. Besides voltage, the two most common measurements you need to make are for current and resistance.

Instead of using three different meters to take these measurements, hobbyists commonly use a single instrument called a *multimeter* (like the one shown in Figure 3-8). We describe using multimeters in detail in this minibook's Chapter 8.

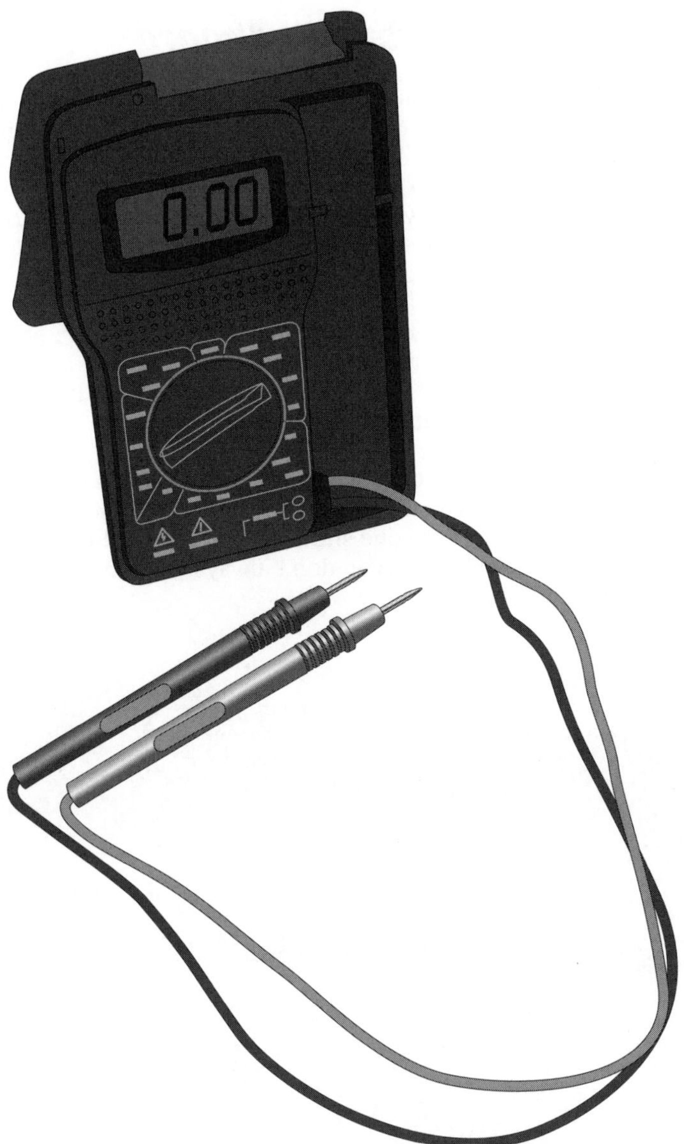

Figure 3-8: An inexpensive multimeter.

Using a solderless breadboard

A *solderless breadboard* – usually just called a *breadboard* – is a must for experimenting with circuit layouts. A breadboard is a board that has holes in which you can insert wires or electronic components, such as resistors, capacitors, transistors and so on, to create a complete electronic circuit without any soldering. When you're finished with the circuit, you can take it apart and then reuse the breadboard and the wires and components to create a completely different circuit.

Breadboards are so useful because the holes in the board are solderless connectors that are internally connected to one another in a specific, well-understood pattern. When you get the hang of working with a breadboard, you soon have no trouble understanding how it works.

Throughout this book, we show you how to create dozens of different circuits on a breadboard. As a result, you need to invest in at least one. We suggest you get one similar to the one shown in Figure 3-9, plus one or two other, smaller breadboards. That way, you don't always have to take one circuit apart to build another.

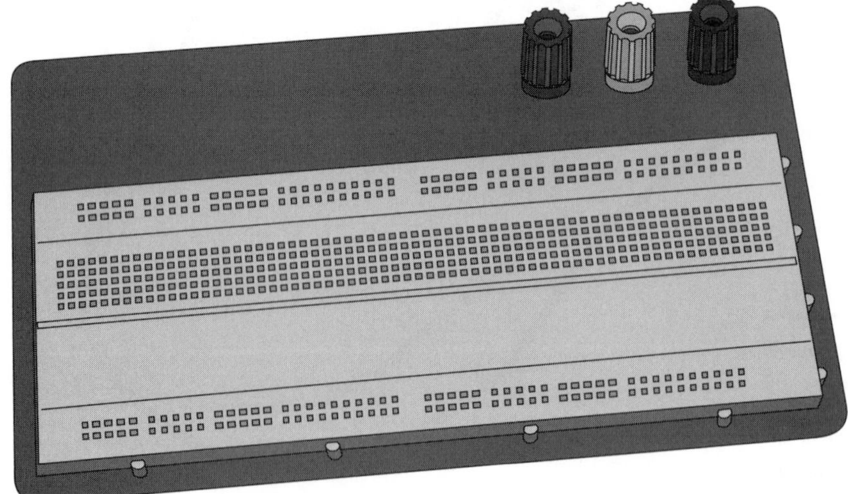

Figure 3-9: A solderless breadboard.

You can read more about working with solderless breadboards in Chapter 6 of this minibook.

Conducting electricity with wires

One of the most important items to have on hand in your lab is *wire,* which is simply a length of a conductor, usually made out of copper but sometimes made of aluminium or some other metal. The conductor is usually covered with an outer layer of insulation. In most wire, the insulation is made of polyethylene, which is the same stuff used to make plastic bags.

Wire comes in these two basic types:

- **Solid wire:** Made from a single piece of metal.
- **Stranded wire:** Made of a bunch of smaller wires twisted together.

Figure 3-10 shows both types of wire with the insulation stripped back so you can see the difference.

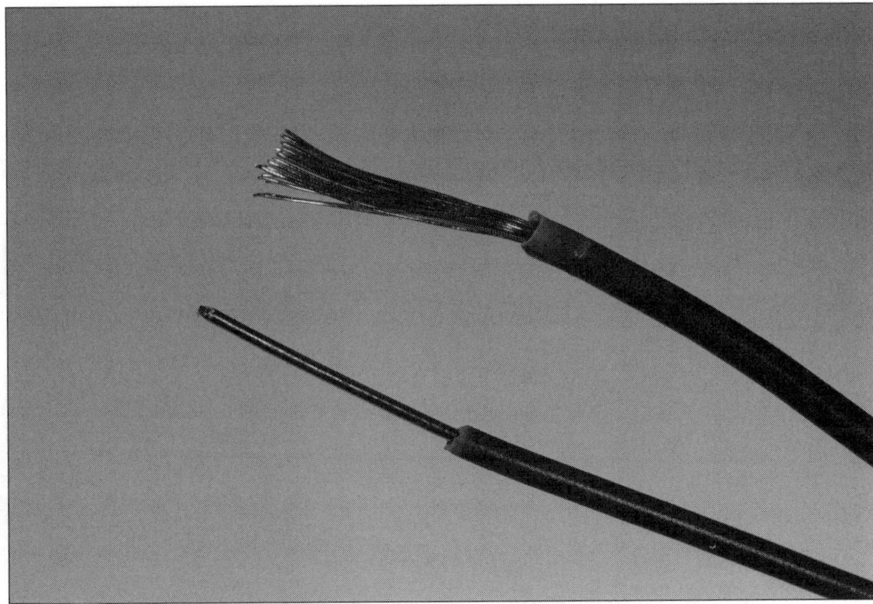

Figure 3-10: Solid and stranded wire.

For most purposes in this book, you want to work with solid wire, because inserting it into breadboard holes and other types of terminal connections is easier. Solid wire is also easier to solder. When you try to solder stranded wire, inevitably one of the tiny strands gets separated from the rest of the strands, which can create the potential for a short circuit.

Stranded wire, however, is more flexible than solid wire. If you bend a solid wire enough times, you eventually break it. For this reason, use stranded for wires that are frequently moved.

Wire comes in a variety of sizes, which are specified by the wire's *gauge,* sometimes given in millimetres for diameter or cross-sectional surface area. You may also see gauge expressed as a two-digit number in which, strangely, the larger the gauge number, the smaller the wire. For most electronics projects, you need 20- or 22-AWG wire. (AWG stands for American Wire Gauge, but is used and understood in the UK.) Plus, circuits running off household electrical power use large wires (usually 14 or 16 AWG).

You may have noticed that the insulation around a wire comes in different colours. The colour doesn't have any effect on how the wire performs, but often hobbyists use different colours to indicate the purpose of the wire. For example, in direct-current circuits red wire is commonly used for positive voltage connections and black wire for negative connections.

To get started, we suggest you purchase a variety of wires – at least four rolls: 20-AWG solid, 20-AWG stranded, 22-AWG solid and 22-AWG stranded. If you can find an assortment of colours, some much the better.

In addition to wires on rolls, you may also want to pick up *jumper wires,* which are precut, stripped and bent for use with solderless breadboards (see Figure 3-11).

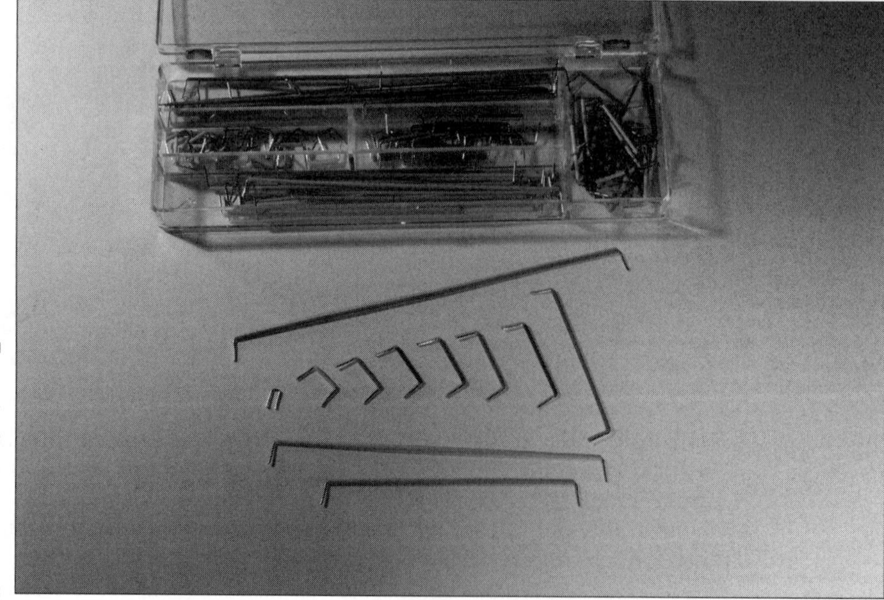

Figure 3-11: Jumper wires for working with a solderless breadboard.

Supplying power with batteries

Don't forget the batteries! Most of the circuits we cover in this book use AA or 9-volt batteries, and so you want to stock up.

You can use rechargeable batteries. They cost more initially, but you don't have to replace them when they lose their charge. Of course, if you use rechargeables, you also need a battery charger.

To connect the batteries to the circuits, get hold of several AA battery holders: one that holds two batteries and another that holds four. You should also buy a couple of 9-volt battery clips. Figure 3-12 shows these holders and clips.

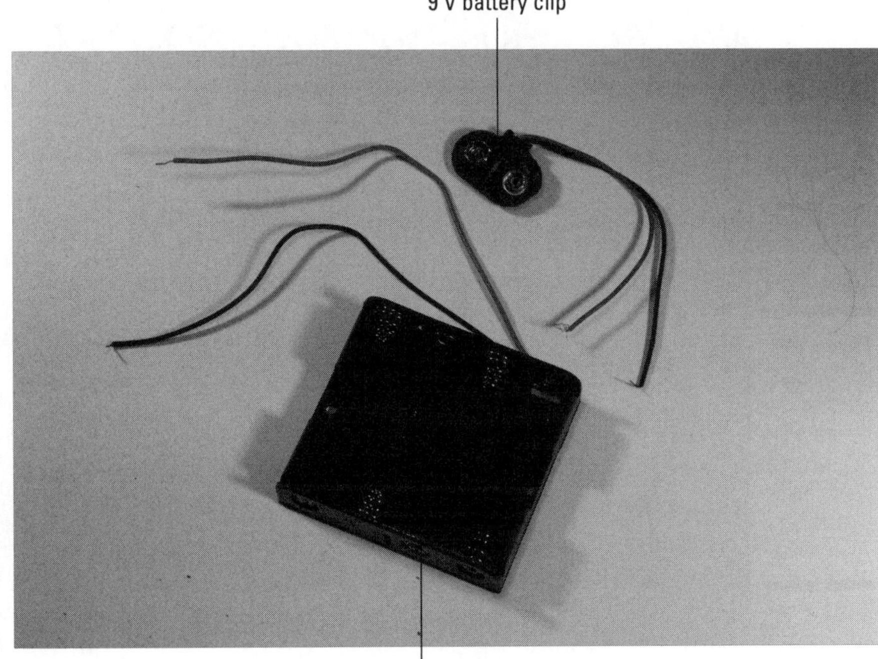

Figure 3-12: Battery holders help to deliver power to your circuits.

Stocking up on other useful items

Here are a few other items you may need to use from time to time:

- **Cable ties:** Also called *zip ties,* these little plastic ties are handy for temporarily (or permanently) holding wires and other things together.
- **Compressed air:** A small can of compressed air comes in handy for blowing dust off an old circuit board or component.
- **Electrical tape:** Get a roll or two of plain black electrician's tape for wrapping around temporary connections to hold them together and for keeping them from shorting out.
- **Jumper clips:** These short (typically 30 or 45 cm) wires have crocodile clips attached on both ends, as shown in Figure 3-13. You use them to make quick connections between components for testing purposes.

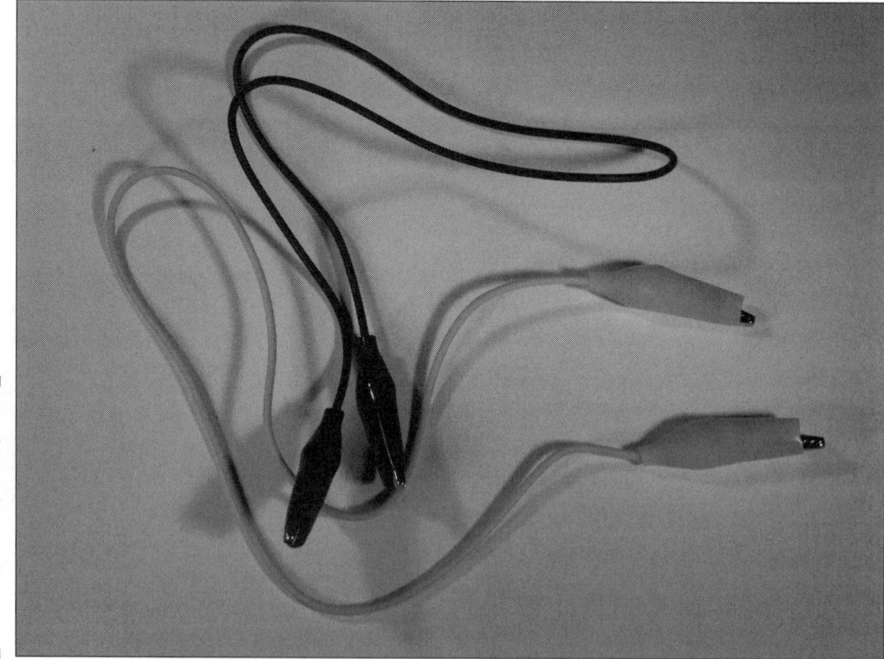

Figure 3-13: Jumper clips are great for making quick connections.

Gathering Together the Basic Electronic Components

Besides the tools and supplies we describe in the preceding sections, you also need to have a collection of inexpensive electronic components to get you started with your circuits. You don't have to buy everything immediately, but you do want to gather at least the basic parts before you go much farther in this book.

You can buy many of these components in person at a Maplin store. Alternatively, you can buy the parts online from RS Components (`http://uk.rs-online.com`), Farnell (`uk.farnell.com`), Digikey (`www.digikey.co.uk`), Mouser Electronics (`uk.mouser.com`), `www.radioshack.com` or other electronic parts distributor.

Curbing current with resistors

A *resistor* is a component that resists the flow of current. It's one of the most basic components used in electronic circuits; in fact, every single circuit in this book contains at least one resistor. Figure 3-14 shows three resistors, next to a coin so that you get an idea of how small they are. Flip to Book II, Chapter 2, for all about resistors.

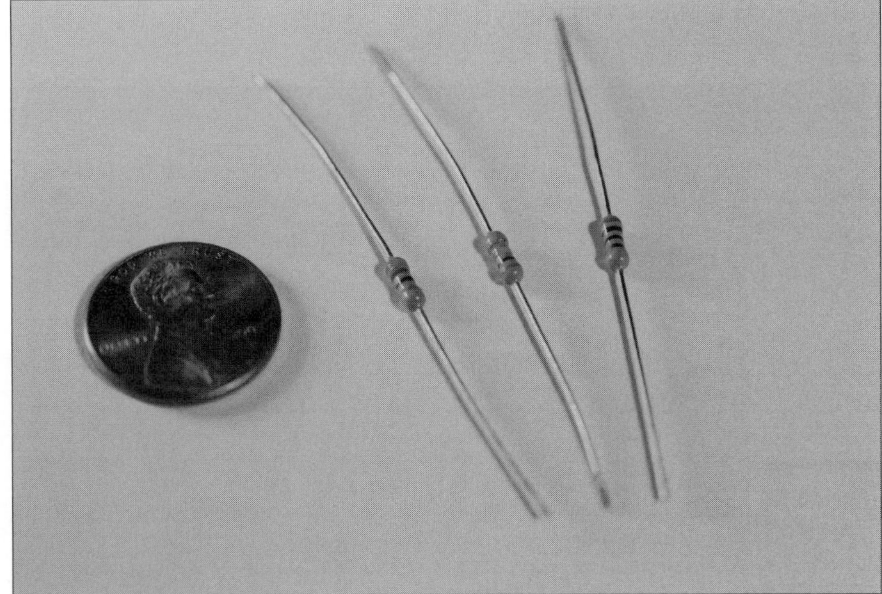

Figure 3-14: Resistors are one of the most commonly used circuit components.

Resistors come in a variety of *resistance values* (how much they resist current, measured in units called ohms and designated by the symbol Ω) and *power ratings* (how much power they can handle without burning up, measured in watts (W)).

All the circuits in this book can use resistors rated for 1.5 W. You need a wide variety of resistance values and we recommend buying at least ten each of the following 12 resistances:

470 Ω	4.7 kΩ	47 kΩ	470 kΩ
1 kΩ	10 kΩ	100 kΩ	1 MΩ
2.2 kΩ	22 kΩ	33 kΩ	220 kΩ

You can save money by purchasing a package that contains a large assortment of resistors. For example, Maplin sells several packs of hundreds of resistors for under £8.

Keeping charged with capacitors

Capacitors are probably the second most commonly used component in electronic circuits (after resistors). A *capacitor* is a device that can temporarily store an electric charge. We cover capacitors in Book II, Chapter 3 and Figure 3-15 shows a selection.

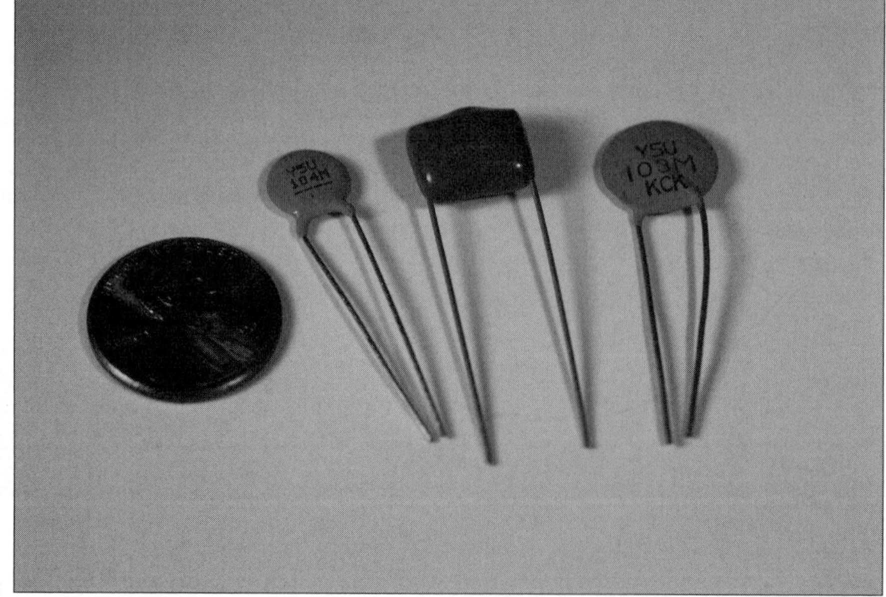

Figure 3-15: Capacitors come in many shapes and sizes.

You can buy capacitors in several different varieties, the two most common being *ceramic disc* and *electrolytic*. The amount of capacitance of a given capacitor is usually measured in *microfarads* (abbreviated μF).

As a starting assortment of capacitors, we suggest you get at least five each of the following capacitors:

- **Ceramic disc:** 0.01 μF and 0.1 μF.
- **Electrolytic:** 1 μF, 10 μF, 100 μF, 220 μF and 470 μF.

The ceramic disc capacitors you buy should have a voltage rating of 50 V. The larger-value electrolytic parts will have lower voltage ratings.

Depending on diodes to block or allow current

A *diode* is a device that lets current flow in only one direction and has two terminals: the *anode* and the *cathode*. Current flows through the diode only when positive voltage is applied to the anode and negative voltage to the cathode. If these voltages are reversed, current doesn't flow.

You discover diodes in Book II, Chapter 5. We suggest that you get at least five of the basic diodes known as the 1N4001.

Producing light with LEDs

A *light-emitting diode* (LED) is a special type of diode that emits light when current passes through it. You can read more on LEDs in Book II, Chapter 5.

Although many different types of LEDs are available, we suggest you start by purchasing five red diodes. Figure 3-16 shows typical LEDs.

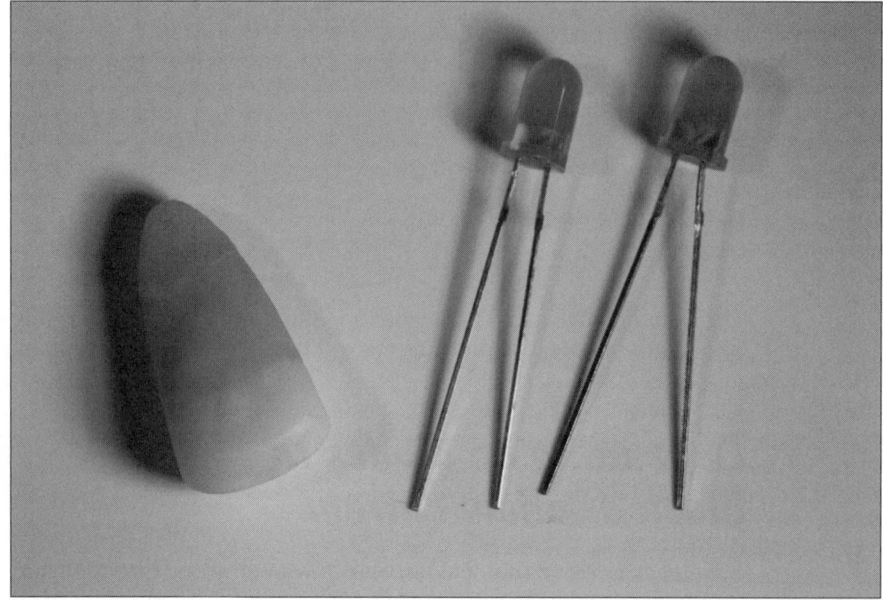

Figure 3-16: Light-emitting diodes.

Controlling current with transistors

A *transistor* is a three-terminal device in which a voltage applied to one of the terminals (called the *base*) can control current that flows across the other two terminals (called the *collector* and the *emitter*). The transistor is one of the most important devices in electronics, and we cover it in detail in Book II, Chapter 6. Get yourself a few simple 2N3904 NPN transistors, as shown in Figure 3-17, to have on hand.

Don't worry; Book II, Chapter 6, explains what the designation 2N3904 NPN means.

Figure 3-17: A look at a 2N3904 NPN transistor.

Chipping in with integrated circuits

An *integrated circuit* (also called a *silicon chip*) is a special component that contains an entire electronic circuit, complete with transistors, diodes and other elements, all photographically etched onto a tiny piece of silicon. Integrated circuits are the building blocks of modern electronic devices such as computers and mobile phones.

You discover how to work with some basic integrated circuits in Book III. To get started, you need to pick up a few each of two different types of integrated circuits: a 555 timer and an LM741 op-amp (see Figure 3-18).

Figure 3-18: Two popular integrated circuits: a 555 timer and an LM741 op-amp.

Chapter 4

Staying Safe

In This Chapter

- Understanding the risks of electric shock
- Protecting yourself from the perils of stray electricity
- Safeguarding your gear from static electricity

The possibility of electric shock is always present whenever you work with electricity, but don't forget the other potential dangers as well. If you're not careful, you can start a fire or otherwise injure yourself or other people.

In this chapter we explain how to keep yourself safe while experimenting with electronics. We discuss the different dangers that electricity poses to you and to your equipment. Please read and heed every piece of advice we give.

Facing the Shocking Truth about Electrical Dangers

The simple fact is that an electric shock, if strong enough, can kill you. So whenever you work with electricity, ensure that you take every precaution you can to avoid being the recipient of a shock strong enough to do you damage.

Thousands of people die every year from accidental electrocutions. Many of those accidents are industrial or weather-related, in which people come into contact with downed power lines. But many are completely avoidable accidents that happen in the home. In the sections that follow, we give you specific guidelines for avoiding accidental electrocution.

Heeding the warning: Household electrical current can kill you!

One of our favourite bits from the TV series *Father Ted* comes when he and Father Dougal wait for news on the condition of their parochial house guest Father Stone, the most boring priest in the world, who's struck by lightning while playing crazy golf in a thunder storm:

> Father Dougal: Who'd have thought being hit by lightning would land you in hospital?
>
> Father Ted: What? What are you talking about? Of course it can land you in hospital.
>
> Father Dougal: Well it's not usually serious is it Ted. I mean, I was hit by lightning a few times and I never had to go to hospital.
>
> Father Ted: Yes Dougal, but you're different from most people. All that happened to you was that balloons kept sticking to you.

You aren't Father Dougal, you aren't strangely immune and you aren't stupid. Electricity can kill you, whether it's a lightning strike or a shock from the household mains, and we hope we don't really need to tell you that:

The electricity in your home wiring system is more than strong enough to kill you.

You're exposed to household electrical current primarily in two places: in electrical outlets and in the lamp sockets within light fixtures. As a result, you need to be extra careful whenever you plug or unplug something into or from an electrical outlet and whenever you change a light bulb. Specifically, follow these precautions:

- **Never change a light bulb when the light is turned on**. If the light is controlled by a switch, turn the switch off. If the light isn't controlled by a switch, unplug the light from the wall outlet.
- **Discard any extension cords that become frayed or damaged in any way.** When the insulation begins to rub off an extension cord, the shock hazard is all too real.
- **Never perform electrical wiring work while the circuit is energised.** If you insist on changing your own light switches or electrical outlets, *always* turn off the power to the circuit by turning off the circuit breaker that controls the circuit before you begin. To be even surer, switch off the whole of the supply.

  People die every year because they think they can be careful enough to safely work with live power.

- **Never work on an alternating current (AC)-powered appliance when it has power applied.** Simply turning the appliance off isn't enough to be safe. If the appliance has a power cord, unplug it before you work on it. If it doesn't have a power cord, turn off the power to the appliance by throwing the circuit breaker on your home's electrical panel.
- **Take extra special precautions when working near AC circuits powered by household mains electricity.** We talk about using AC circuits in Book IV and say more about AC safety there.

Understanding that even relatively small voltages can hurt you

Most of the projects in this book work with AA batteries, usually four of them tied together to produce a total of 6 V. That's not enough voltage to do serious harm. Even if you do get a shock with 6 V, you probably barely feel it.

You can, however, injure yourself with voltages even as low as 6 V. If you create accidentally a short circuit between the two poles of a battery, a lot of current flows very quickly. The likely result is that the wire connecting the two ends of the battery gets extremely hot and the battery itself heats up. This heat can be enough to inflict a nasty burn.

Is it true that current, not voltage, kills?

An old adage says that 'it's the current that kills, not the voltage'. Although this statement may be technically true, it's also dangerously misleading. In fact, it stems from a fundamental misunderstanding of what current and voltage are and can cause you to take dangerous risks if you don't understand the relationship between current and voltage.

The danger from electric shock occurs as current passes through vital parts of your body – specifically your heart. Only a few milliamperes (mA) of current are necessary to stop your heart. At somewhere around 10 mA, muscles seize up and letting go if you're holding a live wire is almost impossible. At around 15 mA, the muscles in your chest can seize up, making breathing impossible. And at around 60 mA, your heart can stop. Only a few moments of exposure is needed for these effects to occur.

So yes, current passing through your body is what can kill you. But current is inseparable from voltage. Current can't happen without voltage, and all other things being equal, the greater the voltage, the greater the current. As a result, receiving a lethal shock from 3 volts (V) is very difficult even if you're dripping wet and standing on bare concrete. But under those conditions, 30 V may be enough to create a painful and damaging shock.

Saying 'the current kills, not the voltage' is like saying 'the lack of oxygen, not water, causes drowning', 'it's not the gun that kills, it's the bullet' or 'it's not the fall that kills you, but the sudden stop'. In each case, one leads to the other.

If the racing current goes unchecked, a possibility exists that the battery may explode. Trust us; you don't want to be nearby if that happens and you really don't want to make a trip to A&E to have fragments of an exploded battery removed from your eyes.

As a result of this danger, take the following precautions when working with the battery-powered circuits described in this book:

- **Don't connect power to the circuit until the circuit is completely finished and you've reviewed your work to ensure that everything is connected properly.**
- **Don't leave your circuits unattended when they're connected to power, and always remove the batteries before you walk away from your workbench.**
- **Touch the batteries periodically with your finger to ensure that they aren't hot, and if they're getting warm, remove the batteries and recheck your circuit to make sure that you haven't made a wiring mistake.**
- **Remove the batteries and recheck your circuit if you smell anything burning.**
- **Never use Lithium cells – those little flat batteries that you sometimes find in portable gadgets.** Lithium cells can explode in circuits without special precautions. Standard NiCd batteries are safer but be careful that other voltage sources in a circuit don't charge the batteries accidentally.
- **Wear protective eyewear at all times to protect yourself against exploding batteries, and remember that under the right circumstances other components can explode as well!**

Staying safe by staying dry

Water and electricity is an extremely bad combination, because water is an excellent conductor of electricity and it flows everywhere (read the nearby sidebar 'Wading into the subject of water and electricity' for more).

Always avoid water when working with electrical current. Here are a few safety tips to bear in mind:

- **Make sure that the floor is dry and never work on electronic or electrical devices in an area where the floor is wet.**
- **Beware of high humidity, especially if it condenses into moisture on your projects.**
- **Dry your hands well before working with electrical current, because even a small amount of sweat on your hands can lower your body's natural resistance and accentuate the danger of electrical shocks from lower voltages.**

Wading into the subject of water and electricity

No doubt you've seen murders committed on TV crime dramas by throwing a plugged-in electrical appliance such as a hair dryer into a bath filled with water and the victim. We wonder how often that really happens and how likely it is to be fatal. For example, how quickly would the circuit breaker kick in and cut power to the hair dryer? Would the special GFCI-protection devices required in all bathrooms work as designed and cut power to the hair dryer in time?

Not that we want to conduct an experiment to find out – and neither should you, under any circumstances. Strictly speaking, pure uncontaminated water is an insulator. But pure water is very rare. Most water is filled with contaminates, and those contaminates turn it into an excellent, and dangerous, conductor.

Realising that voltage can hide in unexpected places

One of the biggest shock risks in electronics comes from voltages that you don't expect to be present. Keeping your eye on the voltages you know about is easy enough, such as in your power supply or batteries, but some electronic circuits are designed to amplify voltages. So even though your circuit runs on 6 V batteries, much larger voltages may be present at specific points within your circuit.

In addition, some electrical devices can store electric charge long after you disconnect the power from your circuit. The most notorious device with this characteristic is the *capacitor,* which alternately builds up and then releases electrical charges.

Remain wary of any circuit that contains capacitors – especially if the capacitors are large. Common ceramic-disk capacitors, which are typically smaller than a tiddlywink, don't store much charge. But if your circuit has capacitors the size of batteries, be very careful when working around them. Such capacitors can hold large charges long after the power is cut off.

Here are some safety points concerning capacitors:

- **You commonly find large capacitors in the power-supply circuit.** Any electronic device that plugs into a household electrical outlet has a power-supply circuit that may contain a large capacitor. Be very careful around these capacitors. In fact, if the power-supply circuit is inside its own enclosed box, *don't open the box.* Instead, replace the entire power supply if you suspect that it's not working.

- **You often find high-voltage capacitors in flash cameras.** Even though the battery may be just 1.5 V, the capacitor that drives the flash unit may well be holding a charge at 300 V or more.
- **Always discharge the capacitor before working on a circuit that contains one.** You can discharge small capacitors by shorting out their leads with the blade of a screwdriver. Make sure that you touch only the insulated handle of the screwdriver while you short out the leads, and don't touch any other part of the circuit with your free hand.

- **Discharge larger capacitors by connecting their leads to a lamp or a large resistor.** The easiest way is to wire up a lamp holder to a pair of crocodile clips, screw a lamp into the lamp holder and then carefully connect the clips to the capacitor leads. If the capacitor is holding a charge, the lamp glows for a moment as the capacitor discharges through the lamp.

If you don't feel completely confident in what you're doing where large capacitors are concerned, walk away from the project.

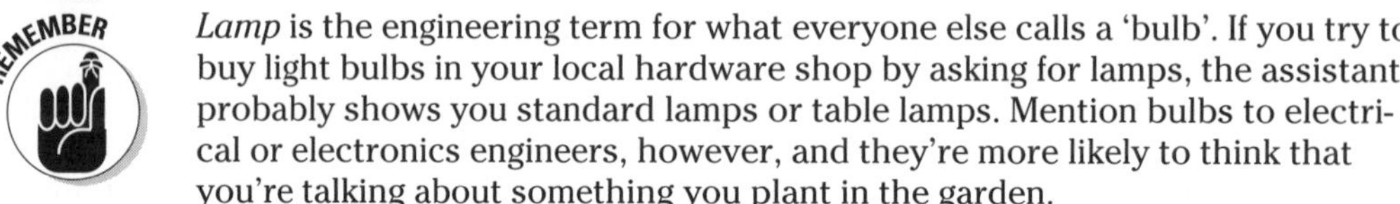

Lamp is the engineering term for what everyone else calls a 'bulb'. If you try to buy light bulbs in your local hardware shop by asking for lamps, the assistant probably shows you standard lamps or table lamps. Mention bulbs to electrical or electronics engineers, however, and they're more likely to think that you're talking about something you plant in the garden.

Considering Other Ways to Stay Safe

Electric shock (which we discuss in the earlier 'Facing the Shocking Truth about Electrical Dangers' section) isn't the only danger you encounter when you work with electronics. This section summarises a few of the other risks you can be exposed to and describes the precautions to take to minimise those risks:

- **Soldering poses an obvious fire hazard.** If your soldering iron is hot enough to melt solder (we describe the process in this minibook's Chapter 7), it's also hot enough to ignite combustible materials such as paper, wire insulation and so on. Therefore:
    - *Always be aware of when your soldering iron is on.* Don't plug it in until you need it, and unplug it when you're finished soldering.

    - *Never set down a hot soldering iron directly on your workbench.* Instead, get a soldering iron stand to hold the soldering iron safely while it's hot. Figure 4-1 shows a soldering iron resting in a simple stand. As you can see, this stand keeps the business end of the soldering iron safely elevated away from the work surface.

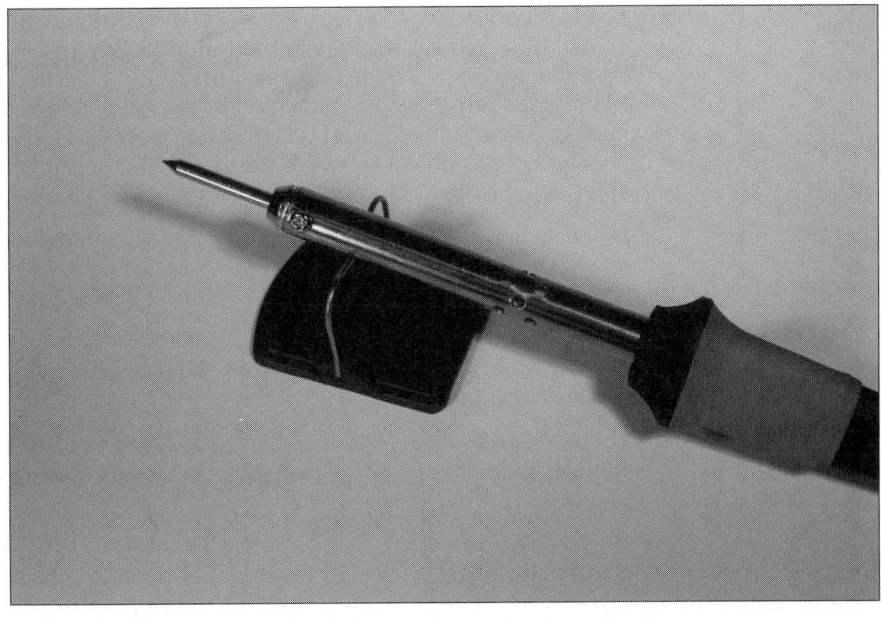

Figure 4-1: A soldering iron resting on a stand.

- *Give your soldered joints a few minutes to cool down before you handle them.*
- *Watch out for the soldering iron's electrical cord.* Obviously, you want to avoid burning the cord with the soldering iron. As ridiculous as it sounds, putting a hot iron down on its own cord, or in such a way that it can slip onto its own cord, is an easy mistake to make.

  Make sure that the soldering iron's power cord is placed safely away from your stuff so that you can't bump it as you work, knocking it out of its stand and perhaps causing a nasty burn.

- *Be sure to wear eye protection when you solder.* As solder melts, it occasionally boils and splatters little globules of hot solder through the air. You really don't want molten metal anywhere near your eyes.

- **Electronics – and especially soldering – can create a chemical hazard.** When you solder, fumes are released into the air. Therefore:
  - *Always work in a well-ventilated place.*
  - *Wash your hands after working with solder or any other electronic components before you touch your face, mouth, nose or eyes.* Small amounts of substances are bound to get on your hands. Wash them frequently to keep whatever gunk they pick up from getting into your body.

  - *Keep your soldering tools away from children.* Young children (and animals) love to stick things in their mouths. If you leave solder or little electronic parts, such as resistors or diodes, sitting loose on top of your workbench, youngsters or pets may decide to make a meal of them. So keep such things safely stored in boxes or cabinets and, if possible, keep your entire work area safely off-limits and behind closed doors.
  - *Don't get into the habit of sticking parts into your mouth to hold them while you're working.* We've seen people hold a dozen resistors in their mouth while soldering each one into a printed circuit board. That's definitely a bad idea.

- **Working with sharp tools such as knives, wire cutters and power drills creates a risk of injury:**

  - *Think before you cut.* Make sure that you know exactly where you want to make the cut and exactly where all your fingers are before you start cutting.
  - *Let the tool do the work.* Don't apply excessive force to coerce a tool into making a bigger, deeper or wider cut than it's designed to do.
  - *Keep your tools sharp.* Working with dull tools causes you to use extra force, which often results in the tool slipping and lodging in your finger.
  - *Remove jewellery such as rings, wristwatches and long dangling necklaces before you start – especially if you're working with power tools.*
  - *Wear safety goggles whenever you're cutting, sawing or drilling.* Little pieces of the work or blade can easily break off and hit you in the face. Add bits of insulation, copper wire and broken drill bits to the growing list of things you don't want in your eyes.

Keeping Safety Equipment on Hand

In spite of every precaution you take, accidents are bound to happen as you work with electronics. Other than preventing an accident from happening in the first place, the best strategy for dealing with one is to act like a scout and 'be prepared'.

Keep the following items nearby whenever you're working with electronics:

- **Fire extinguisher:** So you can quickly put out any fire that may start before it gets out of hand.

- **First-aid kit:** For treating small cuts and abrasions as well as small burns. The kit should include bandages, antibacterial creams or sprays, and burn ointments.
- **Friend:** A friend could get you help in case you get shocked.
- **Phone:** So that you can call for assistance in case something goes seriously wrong.

Protecting Your Equipment from Static Discharges

Static electricity – more properly called *electrostatic charge* – results when electric charge (that is, voltage) builds up in the absence of a circuit that allows current to flow. Your own body is frequently the carrier of static charge, which can be created by a variety of causes: the most common is friction from simple activities such as walking across a carpet. Your clothes can also pick up static charge, and usually do when they're tossed them around in a tumble dryer.

Static charge accumulated in your body usually discharges itself over time. However, if you touch a conductor – such as a brass doorknob – while you're charged up, the charge dissipates itself quickly in an annoying shock.

If the conductor happens to be a sensitive electronic component, such as a transistor or an integrated circuit, the discharge can be more than annoying; it can fry the innards of the component, rendering it useless for your projects. For this reason, you need to protect your stuff from static discharge when working on your electronic projects.

The easiest way to do so is to make sure that you're properly discharged before you start your work. If you have a metal workbench or a large metal tool such as a drill press or grinder near your workbench, simply reach out and touch it after you're settled in your seat and before you begin work.

A more reliable way to protect your gear from static discharge is to wear a special *antistatic wristband* on one wrist, as shown in Figure 4-2. Wear the wristband tightly so that it's in good solid contact with your skin all the way around your wrist. Then, plug the crocodile clip into a metal surface such as your workbench frame or a nearby drill press.

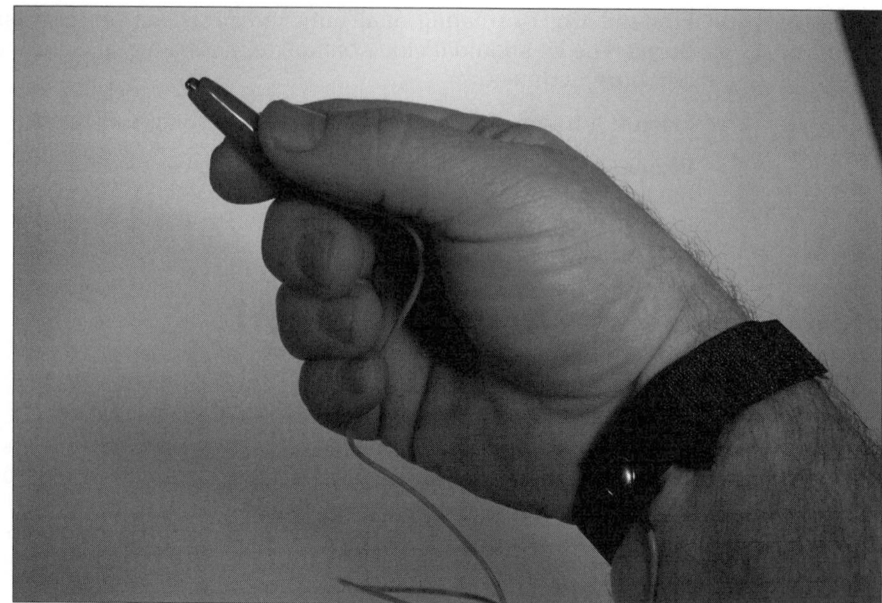

Figure 4-2: An antistatic wristband.

For best results, the crocodile clip on your antistatic wristband needs to be connected to a proper *earth ground,* by clamping a long length of wire (long enough to reach from the pipe to your workbench) to a metal water pipe. Carefully route the wire from the pipe to your workbench, strip off a couple of centimetres or so of insulation, and staple or clamp the wire to the workbench, leaving the stripped end free so that you can attach the crocodile clip from your antistatic wristband to it. (Note that this technique works only if the building uses metal pipes throughout. If it uses plastic pipes, the water pipe doesn't provide a proper ground.)

Some people suggest connecting the wristband to an earth ground using the ground receptacle of a properly grounded electrical outlet. We're definitely not fans of this method, because the key to its operation lies in the term 'properly grounded electrical outlet'. One stupid wiring mistake, or one wire that works loose, and suddenly that ground wire may not be a ground wire anymore – it may be energised. Call us paranoid if you like, but we can't possibly recommend strapping a conductor around your wrist and then plugging it into an electrical outlet.

Chapter 5

Reading Schematic Diagrams

In This Chapter

- Depicting circuits in simplified diagrams
- Introducing common component symbols
- Representing voltage supply and common ground circuits
- Seeing how components are typically labelled

You come across many different types of maps: roadmaps to take on holiday, political maps of the countries of the world and geological maps of underlying rocks, to mention just a few. When you think of a map, you may well visualise a bird's eye view, showing the precise distances and physical relationships between places, features and landmarks.

Yet one of the world's most famous, useful and influential maps doesn't do that at all. Harry Beck's colour-coded diagrammatic map of the London Underground doesn't depict exact locations or distances, but focuses instead on the important aspects of the order and connections (check out the nearby sidebar 'Finding your way underground').

Electronics uses its own form of map: the *schematic diagram*. Like Beck's map these diagrams show how all the different parts that make up an electronic circuit are connected, instead of attempting to show an accurate bird's eye view.

Schematic diagrams use special symbols to represent the different parts of a circuit, such as batteries, resistors and diodes, and they have conventions that are almost always used. For example, positive voltages are usually shown at the top of a schematic diagram, just as north is usually placed at the top of a traditional map.

In this chapter, you find out how to decipher schematic diagrams, including their symbols, labels and conventions.

Finding your way underground

Engineering draftsman Harry Beck drew up the London Underground in his spare time in 1931. He realised that when travelling in dark tunnels underneath the city streets, passengers need to know the order of stations and see clearly where to change lines or exit stations, instead of their precise location. London Underground was a little sceptical about Beck's revolutionary map at first, but the public loved it and his brilliantly simple idea is still in use today. Today it's an iconic diagram.

The fact that Harry Beck was an engineer by training is no coincidence.

Introducing Simple Schematic Diagrams

In a computer programming book, the first complete computer program usually described is called Hello World, which simply displays the text "Hello World!" on a screen and then quits. This computer program is pretty much the simplest one you can write. It doesn't do anything useful but is a great starting point for learning how to write computer programs.

Figure 5-1 shows a schematic diagram that's the electronic equivalent of the Hello World program. This diagram is about the simplest schematic diagram possible that does something: it lights a lamp, thus announcing to the world that a circuit is indeed working.

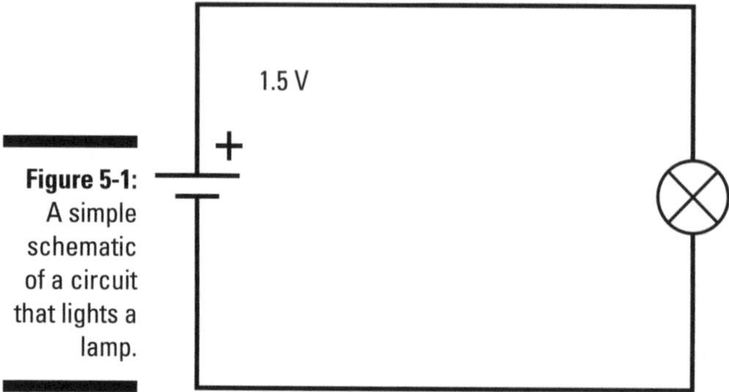

Figure 5-1: A simple schematic of a circuit that lights a lamp.

The diagram in Figure 5-1 contains two symbols representing the two components in the circuit: a 1.5 V battery and an incandescent lamp. The lines that connect the two components represent conductors, which can be actual wires or traces of copper in a printed circuit board.

In the circuit in this schematic, the positive side of the battery is connected to one lead from the lamp, and the other lead from the lamp is connected to the negative side of the battery. When these connections are made, current flows from the battery to the lamp, through the lamp's filament to produce light, and then back to the battery.

Schematic diagrams always depict *conventional current flow,* which as we state in Chapter 2 of this minibook means that current flows from positive to negative. Thus, the current flows from the positive terminal of the battery through the lamp and then back to the negative terminal of the battery.

In reality, conventional current flow is the opposite of the actual flow of electrons through the circuit. The negative side of the battery has an excess of negatively charged particles (extra electrons) whereas the positive side has an excess of positively charged particles (missing electrons). Thus, the electric charge flows through the conductor from the negative side of the battery, through the lamp and back to the positive side. (For more about the difference between real current flow and conventional current flow, refer to Chapter 2 of this minibook.)

As it passes through the lamp, the resistance of the lamp's filament causes the current to heat the filament, which in turn causes the filament to emit visible light.

Laying out a circuit

One of the most important things to realise about a schematic diagram is that the arrangement of components in the diagram doesn't necessarily correspond to the physical arrangement of parts in the circuit when you build it in the real world.

For example, the circuit in Figure 5-1 shows the battery on the left side of the circuit and the lamp on the right. It also shows the battery oriented so that the positive terminal is at the top and the negative terminal is at the bottom. That doesn't mean, however, that you have to build the circuit that way. If you want, you can put the lamp on the left and the battery on the right or place the battery at the top and the lamp on the bottom.

In that sense, as we say in the earlier sidebar 'Finding your way underground', schematic diagrams are more like the map of the London Underground than a road map or an ordnance survey map for hiking.

The physical arrangement of the circuit doesn't matter as long as the component connections remain the same as shown in the schematic. Thus, in the example in the earlier Figure 5-1, no matter how you arrange the components physically, you must connect the positive terminal of the battery to one lead of the lamp and the negative terminal to the other lead.

In fact, Figure 5-1's circuit contains only two components and two conductors and so messing up the connections would be pretty difficult. In a more complicated circuit, however, with dozens of components and dozens of connections, laying out the circuit and making sure that all the connections exactly match the connections indicated in the schematic can be a challenge. You have to check each connection carefully to ensure that it's correct.

Connecting or not connecting

One of the goals when laying out a schematic circuit diagram is to keep the diagram as simple as possible. But the lines in all but the simplest schematic diagrams do at some places need to cross over each other. When they do, you have to be able to tell whether the lines that cross represent actual connections (also called *junctions*) between the conductors or whether the lines cross over each other but don't physically connect.

Unfortunately, no clear and universally used standard exists to indicate whether crossed lines represent a junction. Figure 5-2 illustrates some of the ways for showing crossed wires with or without junctions.

The three examples on the left side of Figure 5-2 show how junctions are indicated. The example at the top left shows the most common way to indicate a junction: by placing a conspicuous dot at the point where the wires cross. Any time you see a dot where two lines intersect, you know that the two lines form a junction.

In the two junction styles shown in the middle-left and bottom-left examples in Figure 5-2, the vertical lines are angled to avoid coming together at the same spot on the horizontal line. With or without the dot, junctions are clearly indicated in both of these examples.

The three examples on the right side of Figure 5-2 show how lines that cross but don't connect to form junctions are most commonly shown. In the top two examples, one line 'hops' over the other and one of the lines is broken at the spot where it crosses the other.

The example in the bottom-right corner of Figure 5-2 is a bit ambiguous. Here, the lines cross each other. The diagram, however, contains no hop or break to indicate that no junction is present and no dot to indicate that a junction should be present. So, is a junction here or not? The answer is, in most cases, no. You can usually assume that a junction is *not* present when lines cross but no dot is shown. But you need to examine the rest of the diagram to make sure. If you find other places in the diagram where a hop or a break is used to indicate non-junctions, the crossed lines without the hop or break may indeed depict a junction.

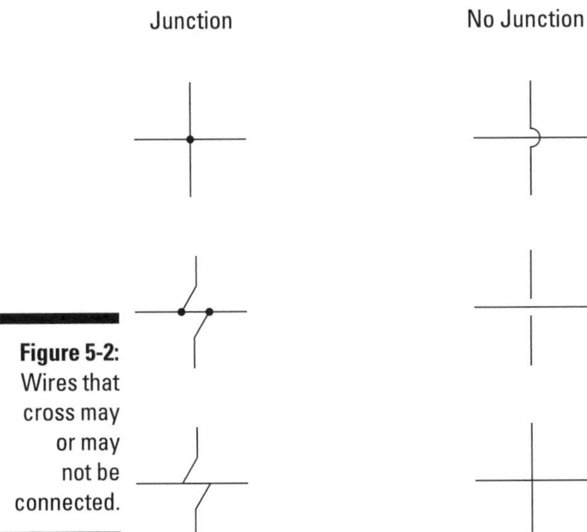

Figure 5-2: Wires that cross may or may not be connected.

To make reading schematics as clear as we can for you, the schematic diagrams in this book always use a dot to indicate a junction and a hop to indicate a non-junction. You never see lines simply crossing without a hop or a dot. But be aware that you do see them used – and quite correctly – in many other published circuit diagrams.

Interpreting Symbols in Schematic Diagrams

The circuit shown in the earlier Figure 5-1 has just two components: a battery and a lamp. Most electronic circuits, of course, have additional components. Hundreds of different types of electronic components exist and each has its own unique schematic diagram symbol, as we describe in this section.

Looking at commonly used symbols

Fortunately, you need to know only a few basic component symbols to get started in electronics and we summarise them in Table 5-1. (Note that when used in an actual circuit diagram, the symbols are often rotated.)

Table 5-1 Common Symbols for Schematic Diagrams

Symbol	*Description*
	Battery
	Capacitor
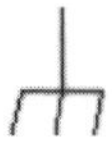	Chassis ground or frame
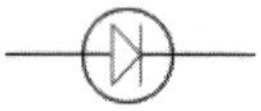	Diode
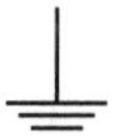	Earth ground
	Inductor (coil)
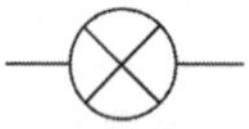	Lamp
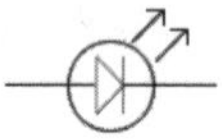	Light-emitting diode (LED)
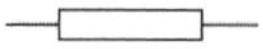	Resistor
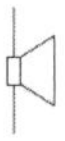	Speaker
	Switch
	Transformer
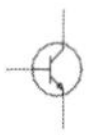	Transistor (NPN)

Symbol	*Description*
	Transistor (PNP)
	Variable resistor (potentiometer)

We discuss resistors, capacitors, diodes and transistors in Book II, Chapters 2, 3, 5 and 6, respectively.

Figure 5-3 shows a schematic diagram that includes several of these components. Don't worry – you don't need to understand this diagram right now. We just want you to get an idea of what real-world schematic diagrams look like and how to read them.

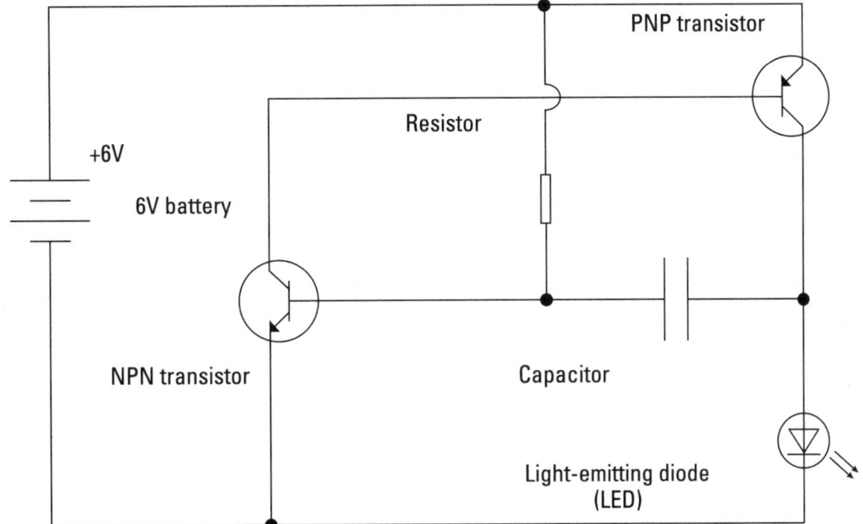

Figure 5-3: A typical schematic diagram.

As you can see, the circuit in Figure 5-3 contains six components. Working from left to right, they are:

- 6 V battery
- NPN transistor
- Resistor
- Capacitor

- PNP transistor (at the top right)
- Light-emitting diode (at the bottom right)

Throughout the course of this book, we use these and other symbols in the schematic diagrams that describe the circuits. Whenever we use a symbol for the first time, we explain what it is and how it works.

Representing integrated circuits in a schematic diagram

One important symbol that isn't shown in Table 5-1 is the one for integrated circuits (ICs, which you can read more about in Book III). ICs are small assemblies that usually have multiple leads, called *pins,* which connect to various parts of the circuit contained within the assembly. Some ICs have as few as six or eight pins; others have dozens or even hundreds. These pins are numbered, beginning with pin 1.

Each pin in an IC has a distinct purpose, and so connecting to the correct pins in your circuit is vital to the circuit's proper operation. If you connect to the wrong pins, your circuit doesn't work and you may damage the IC.

The most common way to depict an IC in a schematic diagram is as a simple rectangle with leads coming out of it to represent the various pins.

The arrangement of the pins in the schematic diagram doesn't necessarily correspond to the physical arrangement of pins on the IC itself. Instead, the pins are positioned to provide the simplest circuit paths in the diagram. The pins in the diagram are numbered to indicate the correct pin to use.

For example, Figure 5-4 shows a schematic diagram that uses a popular IC (called a 555 timer IC) to make an LED flash. The 555 has eight pins, and you can see that the schematic calls for connections on all eight. However, the pins in the diagram are arranged in a manner that simplifies the connections you need to make to the pins. In an actual 555 IC, the pins are arranged in numerical order on either side of the IC, with pins 1 through 4 on one side and pins 5 through 8 on the other side.

You can discover the details of the operation of this circuit in Book III, Chapter 2. We include it here only so that you can see how ICs are depicted in a schematic diagram.

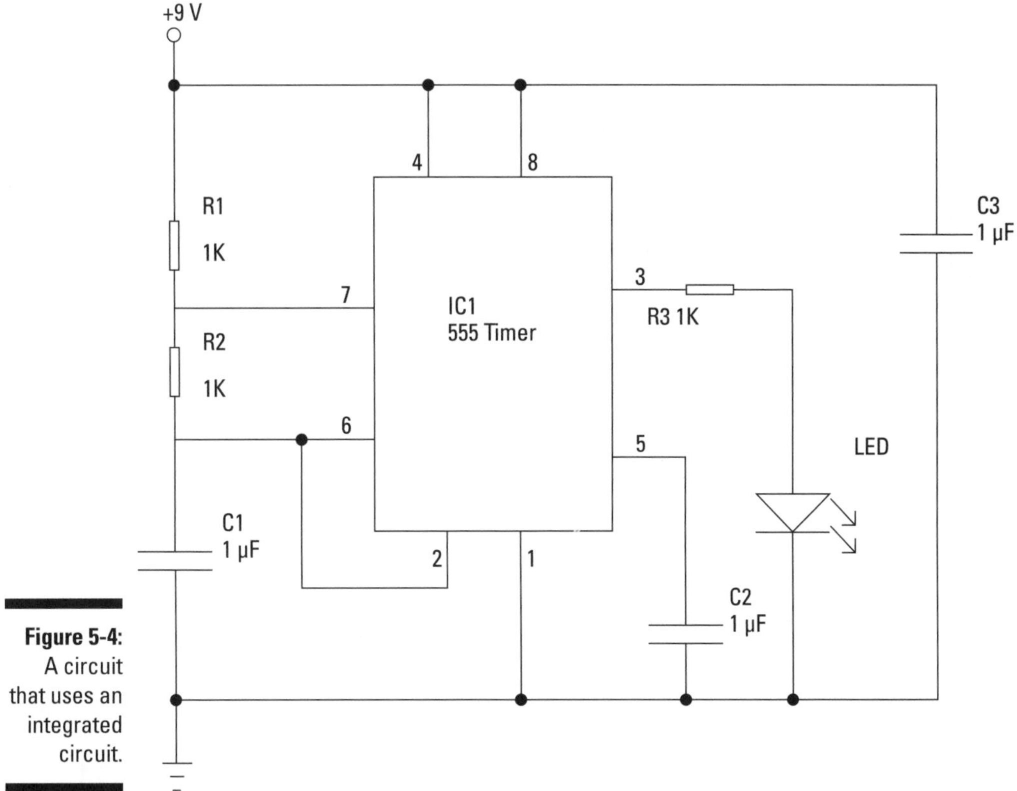

Figure 5-4: A circuit that uses an integrated circuit.

Simplifying Ground and Power Connections

In many electronic circuits, the distribution of voltage connections is one of the most complicated aspects of the circuit. For example, about half of the connections in the schematic diagram shown in the earlier Figure 5-3 are used to connect the resistor, transistors and LED to the positive or negative terminal of the battery.

More complicated circuits can have hundreds of power connections. If all the lines representing those connections had to be drawn to the positive or negative side of the battery symbol, schematic diagrams would quickly be overwhelmed by the power connections.

Most circuits have a common path by which current returns to its source. In the case of Figure 5-3, the conductor at the very bottom of the diagram collects current from the LED and the resistor and returns it to the battery. This conductor is necessary to complete the circuit so that current can flow in a complete loop from the battery through the various components and then back to the battery.

This common return path is often called the *ground,* and can be replaced by the earth ground symbol shown earlier in Table 5-1. You may also see the chassis ground symbol used to represent common ground.

Figure 5-5 shows a schematic diagram that uses ground symbols instead of a line to show the path by which current returns to the battery. (The circuit in Figure 5-5 is identical to the one shown in the earlier Figure 5-3.)

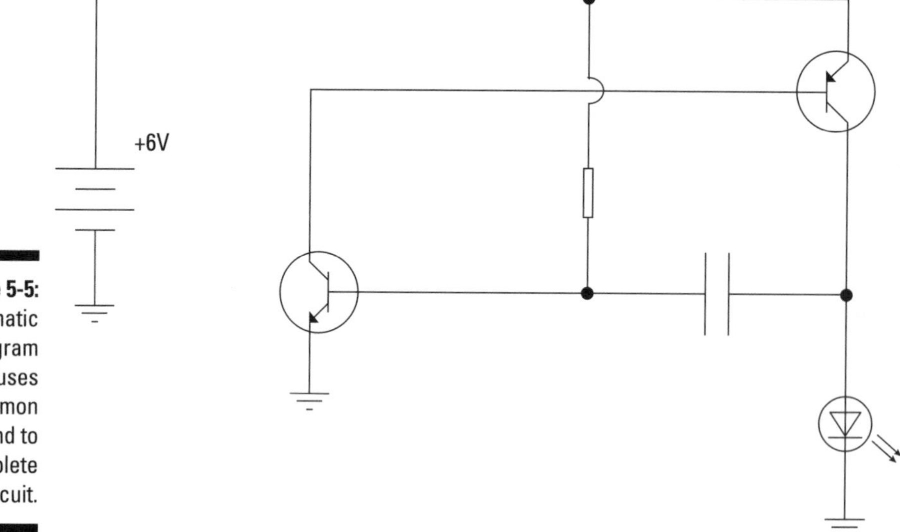

Figure 5-5: A schematic diagram that uses a common ground to complete the circuit.

In addition to a common ground path, most circuits also have a common voltage path. In the case of the circuit shown in Figures 5-3 and 5-5, the common voltage path goes from the battery to the resistor and on to the second transistor. This conductor can be replaced by symbols representing voltage sources that appear wherever voltage is required in a circuit.

The symbol used for a voltage source is an open circle or an arrow. The quantity of voltage is always indicated next to the circle or arrow. When a voltage source symbol is used in a schematic diagram, the symbol for the battery (or other power source if the circuit isn't powered by a battery) is omitted. Instead, the presence of voltage source symbols implies that voltage is provided by some means, by a battery or some other device such as a solar cell or by a power supply plugged into an electrical outlet.

Figure 5-6 illustrates a schematic diagram for the same circuit shown in Figures 5-3 and 5-5, but with voltage source symbols instead of a battery symbol. As you can see, +6 V is required at two places in the circuit: at the resistor and at the second transistor. This circuit is functionally identical to the circuits in Figures 5-3 and 5-4.

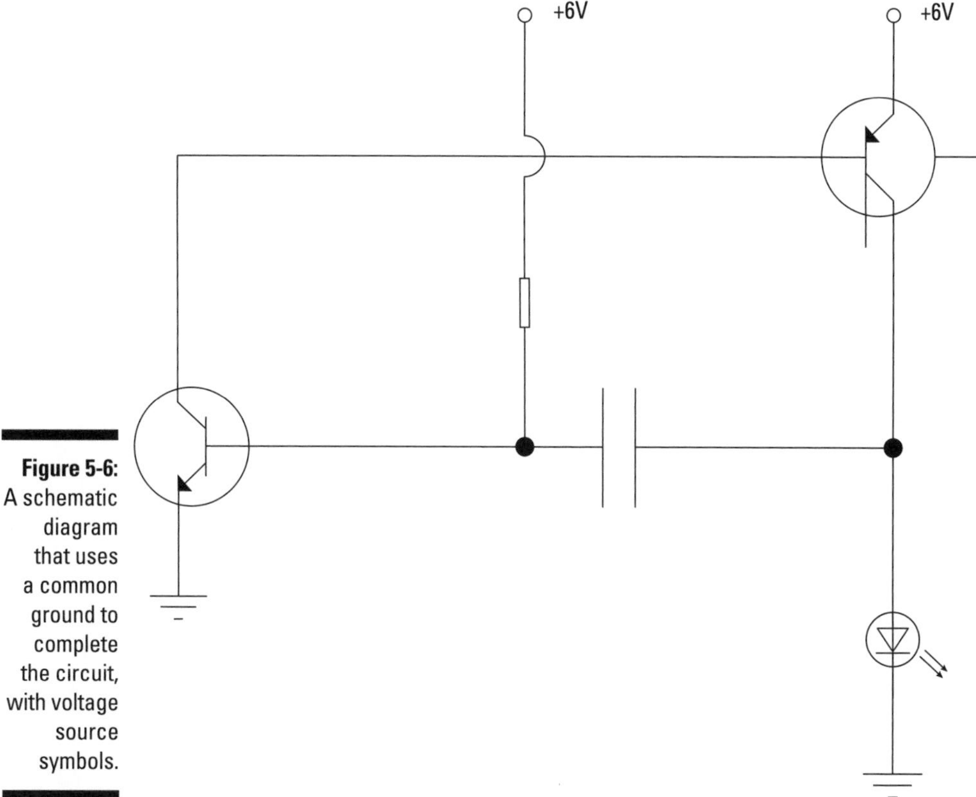

Figure 5-6: A schematic diagram that uses a common ground to complete the circuit, with voltage source symbols.

Although the circuit shown in Figure 5-6 has a positive voltage source and the ground is negative, this isn't always the case. You can also use the voltage source symbol to refer to negative voltage. In that case, the ground actually carries positive voltage back to the source.

In some cases, a circuit may require positive and negative voltages at different places within the circuit. As we explain in Chapter 2 of this minibook, voltages are always measured with respect to two points in a circuit. Thus, voltages are always relative. For example, the positive pole of an AAA battery is +1.5 V relative to the negative pole. At the same time, the negative pole of the battery is –1.5 V relative to the positive pole.

Now suppose that you connect two AAA batteries end to end. Then, the voltage at the positive terminal of the first battery is +3 V relative to the voltage at the negative terminal of the second battery. But the voltage at the positive pole of the first battery is +1.5 V relative to the point between the batteries, and the voltage at the negative pole of the second battery is –1.5 V relative to the point between the batteries.

Figure 5-7 shows how this arrangement may be drawn in a schematic diagram, with a pair of resistors connected across each battery to the middle point. The diagram on the left shows the batteries and connections to them. The diagram on the right shows the same circuit using ground and voltage source symbols instead.

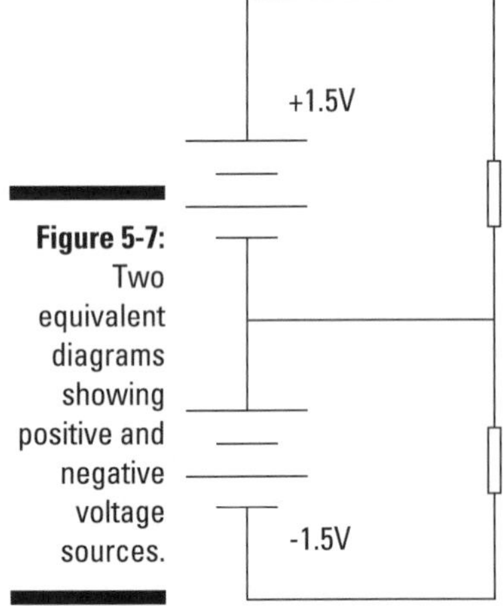

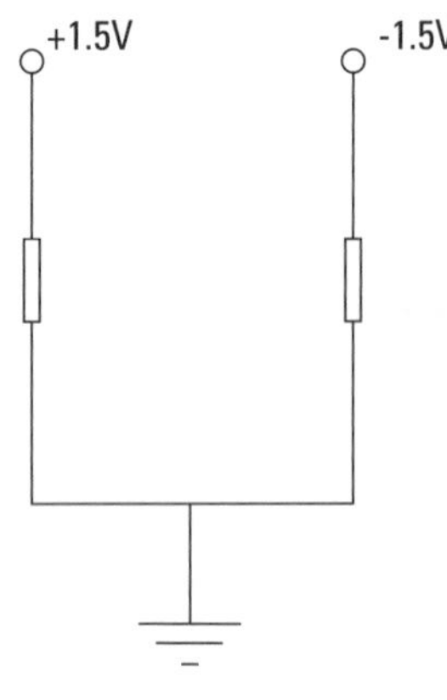

Figure 5-7: Two equivalent diagrams showing positive and negative voltage sources.

Labelling Components in a Schematic Diagram

A symbol alone isn't usually enough information to identify an electronic component completely in a schematic diagram. Further information is often included with text that's placed adjacent to the symbol, as shown in Figure 5-8. This additional information may include the following:

- **Reference identifier:** Each component is usually labelled with a letter that designates the type of component followed by a number that helps identify each component of the same type. For example, if a circuit has four resistors, the resistors are identified as R1, R2, R3 and R4. We show the most commonly used letters in Table 5-2.

Table 5-2 Commonly Used Reference Identifiers

Letter	*Meaning*
R	Resistor
C	Capacitor
L	Inductor
D	Diode
LED (or D)	Light-emitting diode
Q	Transistor
SW	Switch
IC (or U)	Integrated circuit

- **Value or part number:** For components such as resistors and capacitors, the value is given in ohms (for resistors) and microfarads (for capacitors). Thus, a 470 Ω resistor has the number 470 next to it and a 100 μF capacitor has the number 100 next to it. (We discuss the resistor and capacitor measurements and abbreviations in Chapter 3 of this minibook.)

  The letters K and M are used to denote thousands and millions. For example, a 10,000 Ω resistor is identified as 10K in a schematic.

  Components such as diodes, transistors and integrated circuits don't have values; instead, they have manufacturer part numbers. Thus, you may find a part number such as 1N4001 (for a diode), 2N2222 (for a transistor) or 555 (for an integrated circuit) next to one of these components.

  TIP

  In some cases, the value or part number is omitted from the schematic diagram itself and instead included in a separate *parts list* that identifies the value or part number of each referenced part that appears in the schematic. In such cases, to find the value or part number of a particular component, you look up the component by its reference identifier in the parts list.

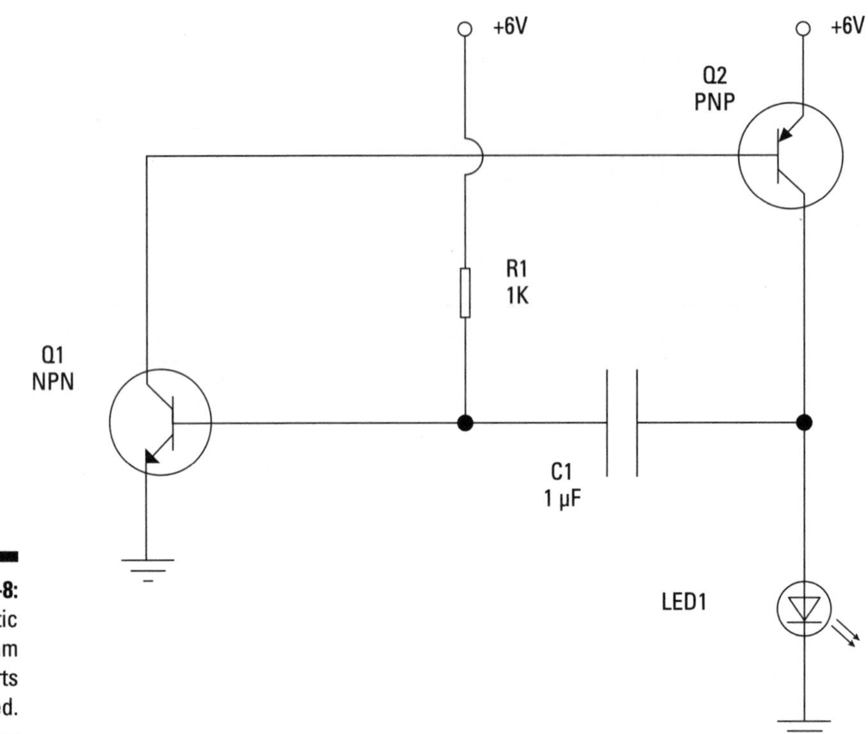

Figure 5-8: A schematic diagram with parts labelled.

Chapter 6

Building Your Own Electronic Projects

In This Chapter

- Fleshing out your idea
- Creating a workable circuit design
- Building a prototype on a solderless breadboard
- Creating a permanent circuit on a printed circuit board

For this chapter, we assume that you know something about circuits and components, have your workbench set up and have read the vital Chapter 4 on safety – you have, haven't you? Seriously, if not, do it now: electricity can kill.

All the theory and preparation are fine, of course, but the heart and fun of electronics is getting your hands dirty building real working circuits and putting them to use. Throughout this book, we back up theoretical explanations with simple construction projects you can build to demonstrate the theory in practice. In this chapter, you discover the basic construction techniques needed to build these projects.

Specifically, you find out how to design your project and create a prototype of a circuit using a handy device called a solderless breadboard. We also describe creating a more permanent version of the circuit, in which you solder all the components and the circuit's interconnections on a circuit board. Plus you discover how to mount your circuit board in a project box or other enclosure.

Building an Electronic Project in Five Steps

Electronic projects such as the ones in this book typically follow the following predictable sequence of general steps from start to finish:

1. **Decide what you want to build.**

   Before you can design or build an electronic project, you need a solid idea in mind for what you expect the project to do, what you want it to look like and how human beings are going to interact with it.

2. **Design the circuit.**

   When you've settled on what you want to build, you have to design an electronic circuit that gets the job done. The end result of this step is a schematic diagram.

3. **Build a prototype.**

   Before you invest in the time and materials needed to construct a permanent circuit, building a prototype first is a good idea. This prototype lets you test the circuit quickly to make sure it works. Usually, you build the prototype on a solderless breadboard (as we describe in the later aptly named section 'Prototyping Your Circuit on a Solderless Breadboard').

4. **Build a permanent circuit.**

   When your prototype is working, you can build a permanent version of the circuit, usually by soldering components onto a printed circuit board (PCB).

5. **Finish the project.**

   To complete the project, you mount the circuit board along with any other necessary components such as batteries, switches or light-emitting diodes (LEDs) in a suitable enclosure.

This chapter describes each of these steps in greater detail.

Envisioning Your Project

Before you get lost in the details of designing and building your project, you need to step back and look at the big picture to make sure that you have a solid idea for your project. Think about why you want to build it, what it's going to do, who will use it and why.

Perhaps the project is something you don't really need, but may be fun and instructive to make, like a flashing Christmas tree decoration or a giant Halloween jack-in-the-box to scare trick-or-treaters. Or it may be an extraordinary new invention of your own that's about to make you a billionaire. We can't show you a picture of the latter (because you haven't built it yet!), but we do show you the giant Halloween jack-in-the-box in Figure 6-1.

Figure 6-1: A very scary electronics project.

When you have a general idea for a project, you can start to flesh out the details. Consider questions such as the following:

- What will its *user interface* be? In other words, how will a person work with the device to get it to do what it's supposed to do?
- Will it be a stand-alone device or interact with other devices?
- Will it be powered by batteries or plug into a wall outlet to get its power? Or will it be solar powered?
- How big will it be? Does it need to be small enough to hold in your hand or fit in your pocket? Or will it sit on a shelf?

We take a fairly simple project to start with: an electronic decision-maker. You'll appreciate this project if you've ever resorted to tossing a coin to make a difficult decision, because it's an electronic version of a coin toss. Instead of flipping a coin into the air to decide, say, which team kicks off, you can build an electronic device that does the equivalent of a coin toss for you.

The specifications for the coin-toss project are as follows:

- The device has two LED indicators to indicate heads and tails.
- The device has two small metal contacts, which users can touch with their fingers. When a person touches both of the posts, the LEDs start flashing, alternating back and forth, much like a coin flips end over end when tossed into the air. When the user stops touching the two metal contacts, one of the two lights stays lit, indicating whether the result of the coin toss is heads or tails. Which light stays lit is essentially random.
- The device has an on/off pushbutton to conserve battery life. The user depresses the pushbutton to make the device work; when the button is released, the device is turned off.
- The device is battery powered and contained in an enclosure small enough to hold in your hand.

As you flesh out the details for your project, you may want to start drawing diagrams to show how it's going to look. Figure 6-2 shows a hand-drawn sketch for the electronic coin tosser.

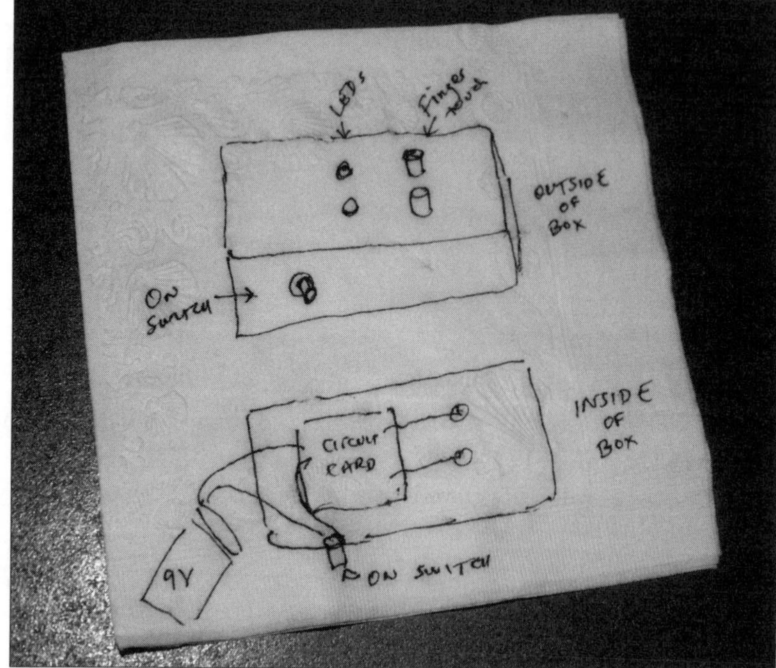

Figure 6-2: A hand-drawn sketch for an electronic coin tosser.

Designing Your Circuit

After you have an idea for a project (as we describe in the preceding section), you need to design a circuit that meets the project's needs. At first, you may have difficulty designing your own circuits, but don't worry: simply turn to books like this one or to the Internet to find other people's circuit designs. With a bit of Google searching, you can probably find a schematic diagram that's very close to what your project requires.

In many cases, you won't be able to find exactly the circuit you're looking for. Instead, you find a circuit that's close and need to make minor modifications to it fit your project's needs. At first, making modifications to a circuit may seem beyond your abilities, but as you gain experience you soon find yourself tweaking circuits all the time to fit specific applications.

One helpful strategy for designing circuits is to break complex requirements down into simpler parts. For example, consider the pop-up jack-in-the-box Halloween prop we mention in the preceding section. The complete circuit for this project requires several different elements, including:

- A circuit to detect when someone has entered the room to trigger the prop's action.
- A circuit to open and close the jack-in-the-box.
- A circuit to time how long the jack-in-the-box is to stay open.
- A circuit that plays a screaming sound.
- A circuit that provides a 30-second delay before the prop is activated again.

The coin-toss project is much simpler than the jack-in-the-box project. In fact, a quick Google search turns up several possible circuits that do almost exactly what the coin-toss project requires. For example, Figure 6-3 shows the schematic diagram for a typical coin-toss circuit you can find on the Internet. This circuit diagram uses a 555 timer integrated circuit (IC), four resistors, two LEDs, one capacitor, a switch and a 9 V power supply (most likely a 9 V battery).

You may have noticed that the schematic diagram shown in Figure 6-3 differs from this project's needs in just two ways: it doesn't have an on/off switch and it uses a pushbutton instead of the user's fingers to start and stop the LEDs from flashing.

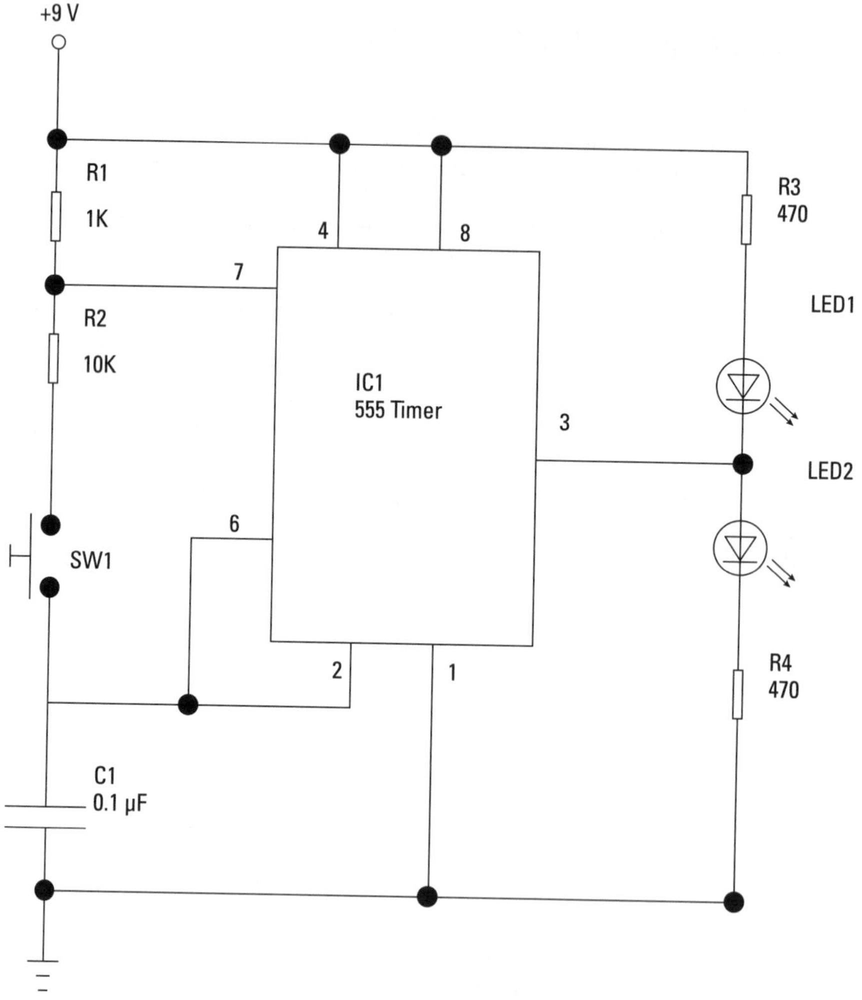

Figure 6-3: A schematic diagram for a simple coin-toss circuit.

Figure 6-4 shows the schematic after we make these modifications. As you can see, we add a pushbutton switch to be pressed to provide the +9 V voltage needed to run the circuit and we replace the pushbutton in the original schematic with two open terminals. When the user touches these two terminals, the resistance of the fingers completes the circuit.

Please don't worry at all if you don't understand how the circuit in Figure 6-4 works. We don't expect you to at this point! Understanding how a circuit works and building that same circuit are two entirely different things; you can (and probably will) build plenty of circuits whose operation you don't understand. The only thing to focus on at this point is how the schematic diagram indicates the various connections between the parts in the circuit. You discover the details of how this circuit works in Book III, Chapter 2.

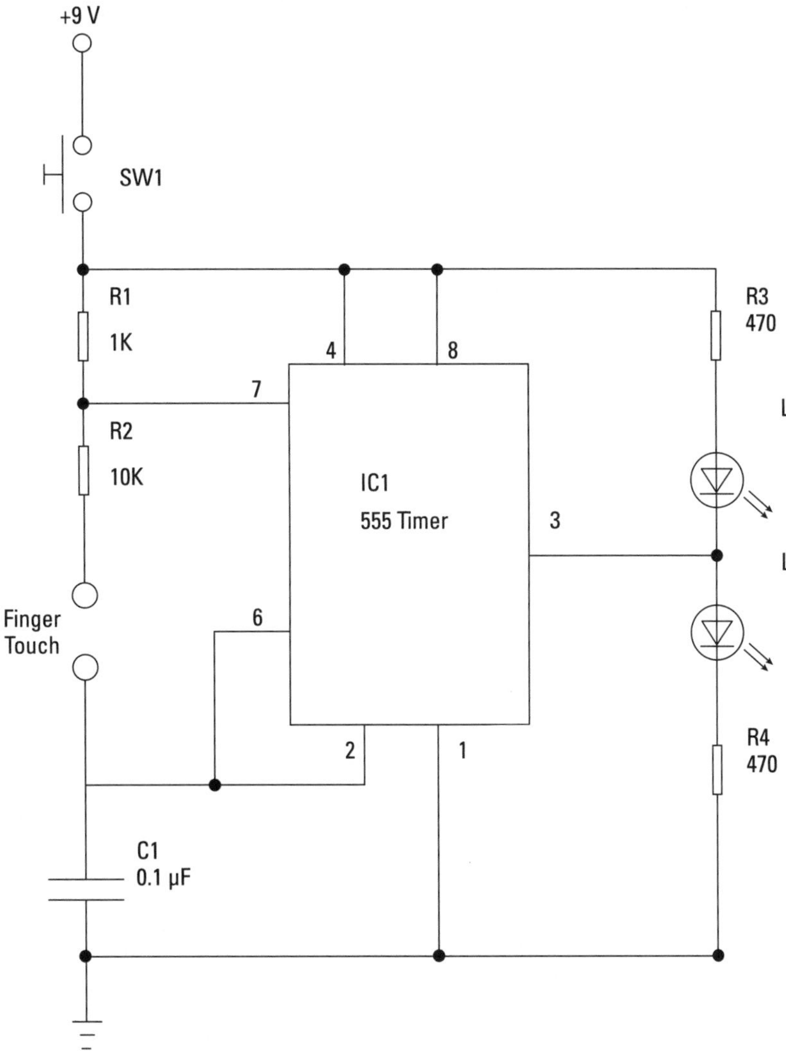

Figure 6-4: The schematic diagram for the coin-toss circuit after we modify it a bit for this project.

When designing a circuit, consider creating a final version of the schematic diagram that indicates what components are to be mounted on your final circuit board and what components won't be on the circuit board. This diagram comes in handy later, when you're ready to create the circuit board that becomes the permanent home of your circuit.

For example, Figure 6-5 shows a version of the coin-toss circuit that uses a dashed line to delineate the items that won't be mounted on the circuit board: the battery power supply (that is, the +9 V voltage source and the ground), the pushbutton power switch, the two metal finger contacts and the two LEDs. Instead, they're going to be mounted separately within the project box. Thus, the circuit board needs to hold only six components: the 555 timer IC, the four resistors and the capacitor.

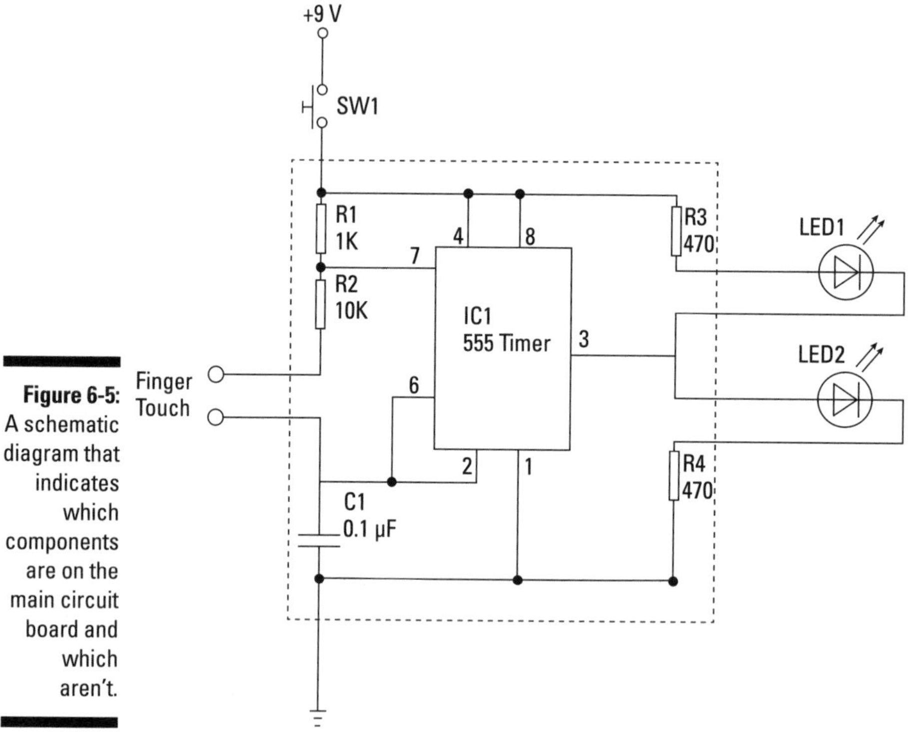

Figure 6-5: A schematic diagram that indicates which components are on the main circuit board and which aren't.

After you complete your circuit design, compile a list of all the parts you need to build the circuit. Then, you can rummage through your parts bin to figure out what parts you already have at your disposal and what parts you need to purchase. Here's a list of the components you need to build the coin-toss circuit:

Part ID	***Description***
R1	1 KΩ, 0.25 W resistor
R2	10 KΩ, 0.25 W resistor
R3	470 Ω, 0.25 W resistor

R4	470 Ω, 0.25 W resistor
C1	0.1 μF capacitor
LED1	5 mm red LED
LED2	5 mm green LED
IC1	555 timer IC
SW1	Momentary-contact, normally open pushbutton

Prototyping Your Circuit on a Solderless Breadboard

Before you commit your circuit to a permanent circuit board, you want to make sure that it works. The easiest way to do so is to build the circuit on a *solderless breadboard,* which lets you assemble the components of your circuit quickly without soldering anything. Instead, you just push the bare wire leads of the various components you need into the holes on the breadboard and then use *jumper wires* (short pieces of connecting wire) to connect the components together.

The beauty of working with a solderless breadboard is that if the circuit doesn't work the way you expect it to, you can make changes to it simply by pulling components or jumper wires out and inserting new ones in their place. If you discover that your schematic diagram is missing an important connection, you can add another jumper wire to create the missing connection. Or if you want to see how the circuit may work with a different resistor or capacitor, you can pull out the original resistor or capacitor and insert a different one in its place. Figure 6-6 shows a standard solderless breadboard.

Figure 6-6: A typical solderless breadboard.

Understanding how solderless breadboards work

Although many different manufacturers make solderless breadboards, they all work pretty much the same way. The board consists of several hundred little holes called *contact holes* that are spaced 0.1" apart (2.54 mm). This is the standard spacing for the pins that come out of the bottom or sides of most integrated circuits and allows you to insert all the pins of even a large integrated circuit directly into a solderless breadboard.

Beneath the plastic surface of the solderless breadboard, the contact holes are connected to one another inside the breadboard. These connections are made according to a specific pattern that's designed to make constructing even complicated circuits easy. Figure 6-7 shows how this pattern works.

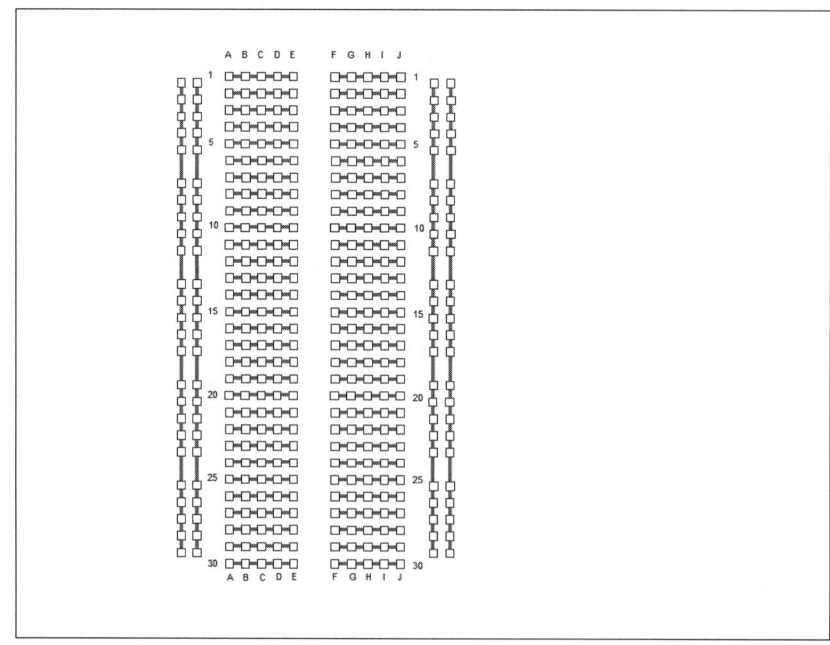

Figure 6-7: The contact holes in typical solderless breadboards are internally connected following this pattern.

The holes in the middle portion of a solderless breadboard are connected in groups of five that are called *terminal strips.* These terminal strips are arranged in two groups, with a long open slot between the two groups, like a little ditch. You connect components such as resistors, capacitors, diodes and integrated circuits in these holes.

The rows of holes aren't connected across the ditch. Thus, each row comprises two electrically separate terminal strips: one connecting the holes labelled A to E, the other connecting the holes labelled F to J.

The breadboard is designed so that integrated circuits can be placed over the top of the ditch, with the pins on either side of the integrated circuit pushed into the holes on either side of the ditch.

The holes on the outside edges of the breadboard are called *bus strips* and you find two bus strips on either side of the breadboard. For most circuits, you use the bus strips on one side of the breadboard for the voltage source and the bus strips on the other side of the board for the ground circuit.

Most solderless breadboards use numbers and letters to designate the individual connection holes in the terminal strips. In Figure 6-7, the rows are labelled with numbers from 1 to 30 and the columns are identified with letters A to J. Thus, the connection hole in the top-left corner of the terminal strip area is A1 and the hole in the bottom-right corner is J30. The holes in the bus strips aren't typically numbered.

Solderless breadboards come in several different sizes. Small breadboards usually have about 30 rows of terminal strips and about 400 holes altogether. But you can get larger breadboards, with 60 or more rows with 800 or more holes.

Solderless breadboards are designed for low-voltage DC circuits only. *Never* use one with 230 VAC household mains electricity, because doing so is very dangerous indeed. It can fry the board in a best case scenario or you in the worst case.

Laying out your circuit

The most difficult challenge of creating a circuit on a solderless breadboard is translating a schematic diagram into a layout that you can assemble on the breadboard. Only in rare cases does a circuit assembled on a breadboard look like the circuit's schematic diagram. Most often, the components are arranged differently and jumper wires are required to connect the components together.

The key when assembling a circuit on a solderless breadboard is to ensure that every connection represented in the schematic diagram is faithfully recreated on the breadboard. For example, the schematic diagram in Figure 6-4 indicates that pin 1 of the 555 timer IC must be connected to ground. Thus, when you build the circuit on a breadboard, you must ensure that this connection is properly made.

One of the first tasks you face when building a circuit on a breadboard is connecting the pins on an integrated circuit. In a schematic diagram, the pin connections on an integrated circuit are rarely drawn in numerical order. For example, in the schematic diagram shown in Figure 6-4, the pin connections on the 555 timer IC are listed in this order, going counterclockwise from the top left: 7, 6, 2, 1, 3, 8 and 4 (pin 5 isn't used).

But the pins on an actual 555 timer IC chip are arranged in numerical order starting at the top-left corner of the chip, as shown in Figure 6-8. Notice also that pins appear on the left and right side of the chip but none on the top or bottom. (The dot imprinted on the top of the chip is used to identify pin 1.)

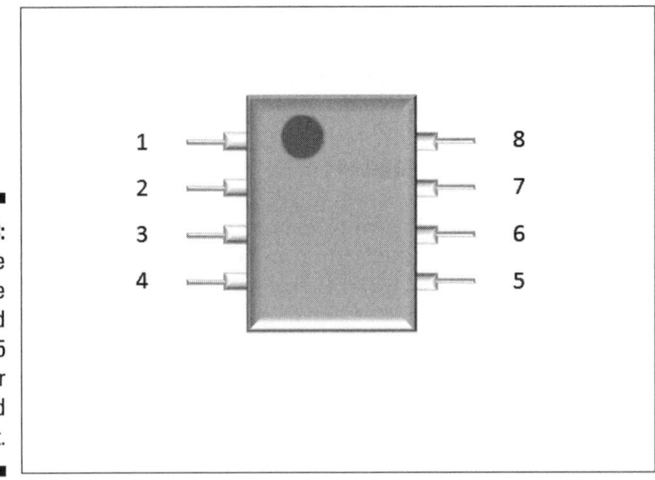

Figure 6-8: How the pins are numbered on a 555 timer integrated circuit.

You have to use your wits to recreate a circuit represented by a schematic diagram on a solderless breadboard. Here are some pointers to get you started:

1. **Start by designating the top row of bus strips as the positive power supply and the bottom row as the ground.** Connect your battery connector to holes in one end of these bus strips, but don't yet connect the battery; applying power to your circuit before you finish assembling it is never a good idea.
2. **Insert any ICs required for the circuit.** Place them so that they straddle the ditch in the middle of the terminal rows and, if your circuit has more than one IC, orient them all the same. You only confuse yourself if pin 1

is on the bottom-left corner of some of your ICs and on the top-right corner of others.

Each pin of each IC is connected to a terminal strip that has four additional connection holes, and so you can connect as many as four additional components or jumper wires to each pin. If your circuit requires more than four component connections to a single pin, use a jumper wire to extend the pin's terminal strip to an unused row anywhere on the breadboard.

3. **Use jumper wires to connect the voltage source and ground pins for each IC.**

   Attach them to the nearest available connection hole in the voltage and ground buses.

4. **Work your way around the rest of the pins for each IC, connecting each component as needed.** If one end of a component connects to an IC pin and the other end connects to the voltage source or ground, plug one end of the component into an available connection hole on the terminal strip for the IC pin and plug the other end into the nearest available connection hole on the voltage supply or ground buses.

5. **Trim the leads of the various components if you want so that the parts fit closer to the breadboard.** Doing so results in a neater breadboard circuit and, with fewer bare leads sticking up high above the breadboard, you reduce the likelihood of leads accidentally coming in contact with each other and creating a short circuit. In a complex circuit, keeping the component leads away from each other without cutting them down to size can be difficult.

Assembling the coin-toss circuit on a solderless breadboard

We now describe the complete procedure for assembling the coin-toss circuit from the earlier sections on a small solderless breadboard. With all your materials gathered together, you should be able to complete this project in about an hour.

You can purchase all the parts required to build this prototype circuit from your local electronics shop or you can order them online from any electronic parts supplier. Components are available in different tolerances. This list doesn't specify these, but for this project a tolerance of 5 per cent is perfectly fine – you don't need to spend more of your hard-earned cash on the more expensive and precise 0.1 per cent parts.

For your convenience, here's a complete list of the parts you need to build this prototype circuit:

Quantity	*Description*
1	Small solderless breadboard
1	Solderless breadboard jumper wire kit
1	555 timer IC
1	1 kΩ, 0.25 W resistor
1	10 kΩ, 0.25 W resistor
2	470 Ω, 0.25 W resistor
1	0.1 μF polyester film capacitor
1	Red LED 5 mm
1	Green LED 5 mm
1	9 V battery snap connector
1	9 V battery

You don't need many tools for this project. In fact, you can probably assemble it without any tools at all, but you may want to keep your wire cutters, wire strippers and tweezers handy.

The steps that follow identify specific holes in the terminal strip area of the breadboard using numbers and letters. If you're using a breadboard that's different to the one we're using, you may encounter a different numbering system. If so, refer to Figure 6-7 and translate the numbers given in the steps for your breadboard.

Inserting the integrated circuit

When you have everything you need ready, follow the steps below to assemble the circuit. The first four steps involve inserting the IC and connecting it to power:

1. **Insert the 555 timer IC.**

   Take a close look at the 555 timer IC. On the top, notice a small dot in one corner; this dot marks the location of pin 1. Carefully insert the leads of the 555 timer IC into the breadboard near the middle of the board, inserting pin 1 into hole E14 and pin 8 in hole F14. The IC straddles the ditch that runs down the centre of the board.

2. **Connect pin 1 of the 555 timer IC to the ground bus.**

   Insert one end of a small jumper wire into hole A14 and the other end into the nearest available hole in the bottommost bus strip.

3. **Connect pin 8 of the 555 timer IC to the +9 V bus.**

   Insert one end of a small jumper wire into hole J14 and the other end into the nearest available hole in the topmost bus strip.

4. **Connect pins 2 and 6 of the 555 timer together.**

   Insert one end of a small jumper wire into hole C15 and the other end in hole H16. The jumper wire reaches over the top of the 555 timer chip.

Figure 6-9 shows what the breadboard looks like after these four steps.

Figure 6-9: The breadboard after the IC has been inserted and connected to the power buses.

Wiring the LEDs

The next five steps connect the LEDs and resistors R3 and R4. The LEDs use the terminal strips in rows 19 and 21:

1. **Connect pin 3 of the IC to row 19.**

   Insert one end of a short jumper wire into hole C16 and the other end into hole C19.

2. **Connect the two segments of row 19.**

   Insert one end of a short jumper wire into hole E19 and the other end into hole F19. This jumper wire bridges the gap between the two terminal strips in row 19, effectively making them a single terminal strip.

3. **Insert the red LED.**

   If you look carefully at the red LED, you can see that one lead is a bit shorter than the other. This short lead is called the *cathode.* The longer lead is called the *anode.* Insert the cathode (shorter lead) into hole D21 and then insert the anode (longer lead) into hole D19.

4. **Insert the green LED.**

   The green LED also has a short cathode lead and a longer anode lead. Insert the anode (long) lead in hole G21 and the cathode (short) lead in hole G19.

   Note that the leads of the two LEDs are installed reversed from one another: the red LED's anode and the green LED's cathode are inserted into row 19 and the red LED's cathode and the green LED's anode are inserted into row 21. A very good technical reason does exist for this method, but for now take it on trust that you must install the two LED's reversed like this for the circuit to work. (You discover more about LEDs, cathodes and anodes in Book II, Chapter 5.)

5. **Insert resistors R3 and R4.**

   Both of these resistors are 470 Ω. You can identify them by looking at the three colour stripes painted on the resistors: they're yellow, purple and brown. Insert one end of the first resistor in hole B21 and the other end in the nearest available hole in the bottommost bus strip (the ground bus). Then, insert one end of the other resistor in hole I21 and the other end in the nearest available hole in the topmost bus strip (the +9 V bus).

Figure 6-10 shows what the breadboard looks like after these steps.

Figure 6-10: The breadboard after the LEDs have been connected.

Connecting the contact jumpers

The next five steps connect the finger-touch circuit that lets the user activate the coin toss by touching the two metal contacts. For the purposes of this prototype, you connect one end of a pair of jumper wires to the circuit and leave the other ends protruding from the end of the breadboard. Touching the bare ends of these wires with your fingers simulates touching the metal contacts that you use in the final version of the circuit. You insert the two jumper wires into holes in row 9:

1. **Insert resistor R1 from pin 7 of the IC to the +9 V bus.**

   Resistor R1 is the 1 kΩ resistor, which you connect between pin 7 of the IC and the +9 V bus. This resistor has stripes in the following sequence: brown, black and red. Insert one end of this resistor into hole J15 and the other end into the nearest available hole in the topmost bus strip.

2. **Insert capacitor C1 from pin 2 of the IC to the ground bus.**

   Insert one lead of the capacitor (it doesn't matter which) into hole B15, and then insert the other into the nearest available slot in the bottom-most bus strip.

3. **Insert resistor R2 from pin 7 of the IC to one of the metal contacts.**

   This resistor is the 10 kΩ. Connect it between pin 7 of the IC and one of the metal contacts that the user touches to activate the coin-toss action. This resistor has the following sequence of colour stripes: brown, black and orange. Insert one end of it into hole H15 and the other end into hole H9.

4. **Connect a jumper wire from pin 2 of the IC to the other metal contact.**

   Insert one end of a short jumper wire into hole B15 and the other end into hole B9.

5. **Insert the two jumper wires that simulate the metal contacts.**

   Pick out a couple of jumper wires long enough to reach from row 9 and dangle 2–3 cm over the edge of the breadboard. Insert one end of these wires into holes E9 and F9 and leave the other ends free. Separate the ends of the two jumper wires to ensure that they're not touching; make them about 12 mm apart.

Figure 6-11 shows what the breadboard looks like after these steps.

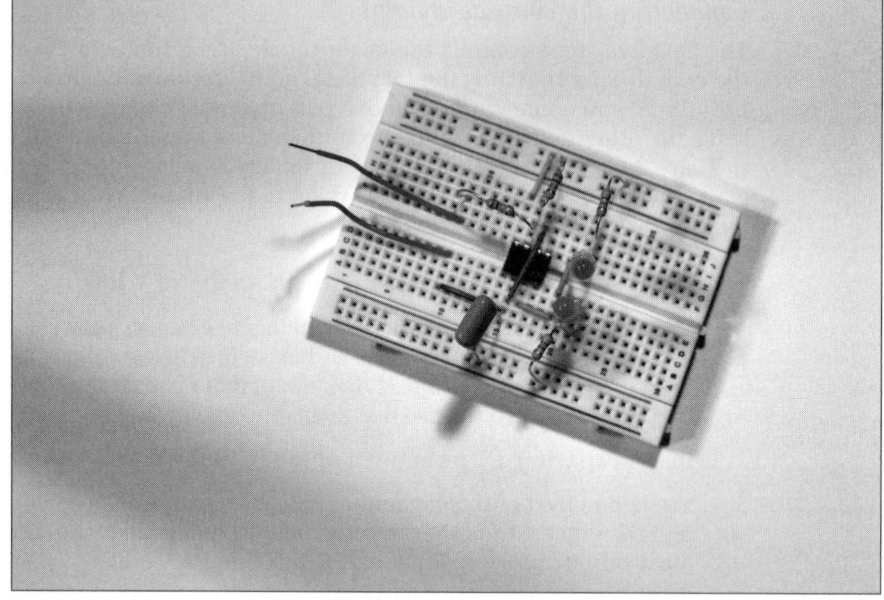

Figure 6-11: The breadboard after the finger contact jumpers are connected.

Powering up

The remaining two steps complete the circuit by connecting the power supply:

1. **Connect the battery snap connector.**

   The leads on the battery snap connector use stranded rather than solid wire, so you need to prepare them a bit before you insert them into the breadboard:

   a. Use your wire strippers to strip off about 1 cm of insulation from the end of both leads.

   b. Use your fingers to twist the leads as tightly as you can, so that no individual strands are protruding from the very tip of the wire.

   c. Insert the red lead into the last hole of the topmost row and insert the black lead into the last hole of the bottommost row.

2. **Connect the 9 V battery to the snap connector.**

   The red LED should immediately light up. (If not, see the troubleshooting tips in the next section.)

You can now test the circuit by touching both of the two free jumper wires. Pinch them between your thumb and index finger, but don't let the wires actually touch each other. The resistance in your skin conducts enough current to complete the circuit, and the LEDs start alternately flashing: red, green, red, green and so on. They continue to flash until you let go of the jumper wires. Then, one or the other stays lit. When you touch the wires again, the flashing resumes.

Figure 6-12 shows the completed circuit in operation.

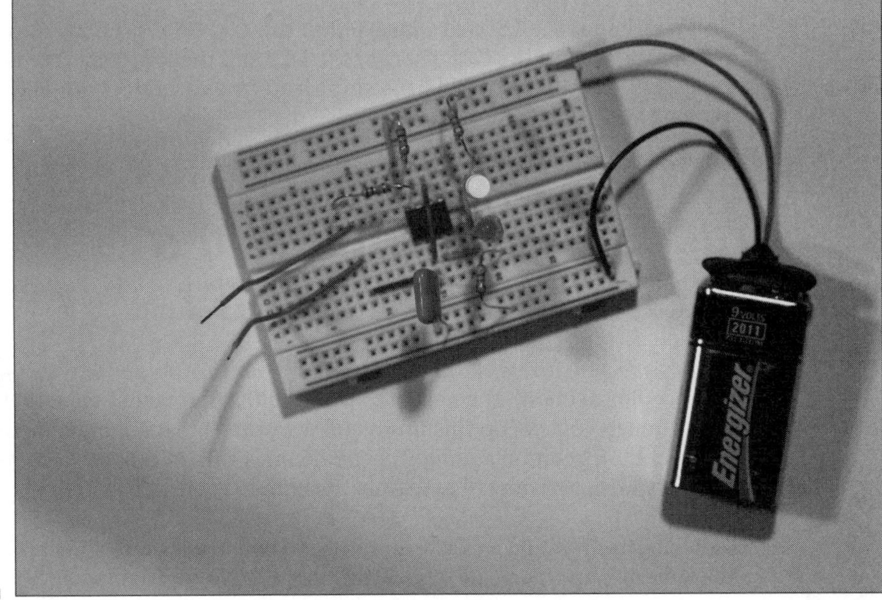

Figure 6-12: The prototype of the coin tosser in operation.

Notice that if you squeeze the wires tightly, the rate at which the LEDs flash increases. If you squeeze tight enough, the LEDs flash so fast that they appear to be on constantly. The LEDs are still flashing alternately, but they're flashing faster than the ability of your eyes to discern the difference.

Troubleshooting if your circuit fails to work

If your circuit doesn't work, here are some troubleshooting tips to help you find out why and correct the problem:

- **Examine all the component leads to make sure that none are touching each other.** If any of the leads are touching, gently adjust them so that they're not touching.
- **Make sure that the circuit is getting power.** Use your multimeter to test the battery voltage (see Chapter 8 of this minibook for information on

how to do that), and verify that the leads from the battery snap connector are inserted properly into the solderless breadboard.

- **Double-check your wiring carefully.** Make sure that you've inserted every jumper and every component in the correct spot.
- **Verify the orientation of the 555 timer IC.** You need to ensure that pin 1 is in hole E14 and pin 8 is in hole F14.
- **Ensure that the LEDs are inserted in the correct direction.** For the red LED, the short lead (the cathode) goes in D21 and the long lead (the anode) goes in D19. For the green LED, the short lead (the cathode) goes in G19 and the long lead (the anode) goes in G21.

Constructing Your Circuit on a Printed Circuit Board (PCB)

When you're satisfied with the operation of your prototype (as we describe in the preceding section), the next step is to build a permanent version of the circuit. Although you can do this in several ways, the most common method is to construct the circuit on a *printed circuit board* (PCB). In this section, you discover how PCBs work and how to assemble the coin-toss circuit on one.

Assembling a circuit on a PCB requires that you know how to solder. To find out about this all-important skill, check out Chapter 7 of this minibook.

Understanding how PCBs work

A PCB is made from a layer of insulating material such as plastic or some similar material. Copper circuit paths are bonded to one side of the board. The circuit paths consist of *traces,* which are like the wires that connect components, and *pads,* which are small circles of copper to which you can solder the component leads. Figure 6-13 shows a typical PCB.

You can purchase two basic styles of PCBs:

- **Through-hole:** A PCB in which the copper circuits are on one side of the board and the components are installed on the opposite side. In a through-hole PCB, small holes are drilled through the board at the centre of the copper pads. You mount components to the blank side of the board by passing their leads through the holes and soldering the leads to the copper pads on the other side of the board. When the solder joint is completed, you trim away any excess wire lead.

- **Surface-mount:** A PCB in which the components are installed on the same side of the board as the copper circuits. No holes are drilled.

Figure 6-13: A printed circuit board.

Surface-mount PCBs are easier for large-scale automated circuit assembly. However, they're much more difficult to work with as a hobbyist because the components tend to be smaller and the leads are closer together. All the PCBs we use in this book are of the through-hole variety.

Using a preprinted circuit board

The easiest way to work with PCBs is to purchase a preprinted board from an electronics shop or order one from an electronics parts supplier online. Maplin carries several different preprinted circuit boards in its shops and at `http://www.maplin.co.uk/components/pcb-development` on its website.

Preprinted circuit boards come in a wide variety of shapes and sizes. The most useful, from a hobbyist's point of view, are the ones that mimic the terminal-strip and bus-strip layout of a solderless breadboard (check out the earlier section 'Understanding how solderless breadboards work'). For example, Figure 6-14 shows a PCB that has 550 holes laid out in a standard breadboard arrangement. A preprinted circuit board with a breadboard layout allows you to transfer your breadboard prototype circuit to the PCB without having to come up with an entirely new layout.

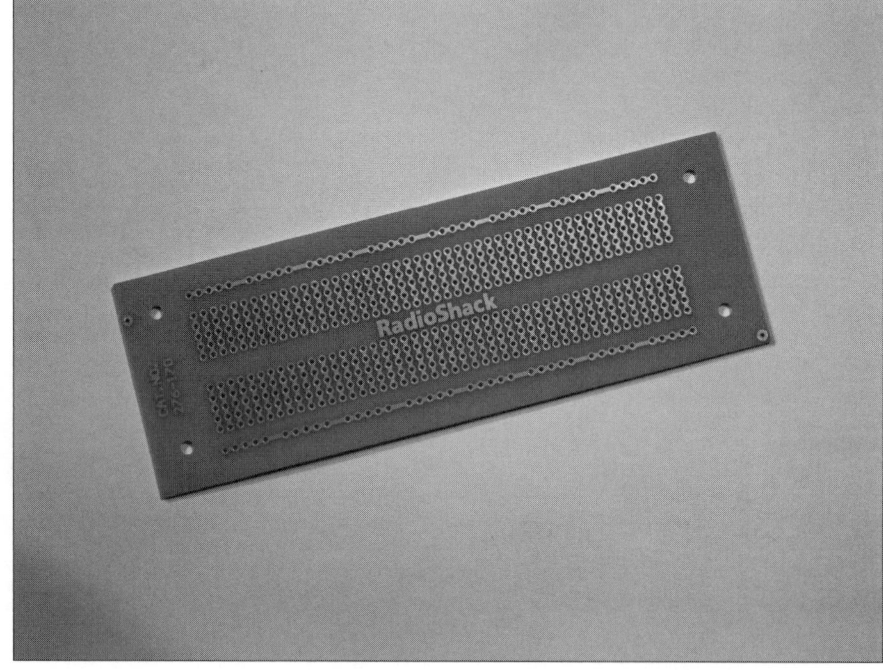

Figure 6-14: A preprinted circuit board that uses a standard breadboard layout.

Some preprinted circuit boards have layouts that are similar to standard breadboard layouts, but not identical. Check carefully before you build; you may have to make minor adjustments to your circuit layout to accommodate the PCB you're using.

If necessary, you can cut a larger PCB to a smaller size. One way to do so is to score it on both sides with a heavy-duty utility knife, and then snap it at the score.

Building the coin-toss circuit on a PCB

This section presents a complete procedure for building the coin-toss circuit (that we introduce in the earlier section 'Envisioning Your Project') on a small preprinted circuit board. With your materials together, you can complete this project in about an hour.

Creating a custom PCB

As you become more advanced in your electronics skills, you may find that you want to create your own custom PCBs, instead of forcing your circuit designs to fit the limited variety of available preprinted circuit boards. Although it isn't a trivial or inexpensive process, you can make your own PCBs customised for your circuit.

This overview is intended only to give you an idea of what's involved – consult the PCB manufacturers for more detailed instructions – but here are the basic steps for creating your own PCBs:

1. **Purchase a blank PCB. The entire surface of one side of this board is completely coated with copper.**
2. **Create a mask on the copper surface that indicates the circuit layout.**

   You can create a mask in several ways. For simple circuits, simply hand-draw the circuit onto the copper using an abrasive pen designed for the purpose. For more complicated circuits, purchase stickers that are shaped like pads and common traces and place the stickers directly on the copper. Or you can design the circuit on your computer using any graphics-drawing software, print the design onto special paper and transfer the design to the copper using a hot iron.
3. **Etch the board by dipping it into a chemical that eats away all the copper that isn't covered by your mask.**

   This process is pretty nasty and involves hazardous chemicals. It needs to be done in the right containers, outside in a well-ventilated area, while wearing gloves, a face mask and goggles. When the etching is finished, all the copper that wasn't covered by the mask is gone.
4. **Wash the board to get off all that nasty copper etching solution.**
5. **Scrub away the mask to reveal the beautiful copper circuit pattern left behind.**
6. **Drill holes in the centre of each pad, and then assemble your circuit.**

To find out more about making your own PCBs, search the Internet for 'make printed circuit board'. Look out for the tutorials and videos that go into the detail, and the companies that make small runs of boards to your circuit designs.

You can buy most of the parts required to build this circuit from your local electronics shop or order them online from any electronics parts supplier. You can order the specific Radioshack preprinted circuit board that we use from `www.t2retail.co.uk`. Here's a complete list of the parts you need to build this circuit (the list is similar to the earlier one in 'Assembling the coin-toss circuit on a solderless breadboard' and so by all means reuse any parts here):

Quantity	***Description***
1	Radioshack dual general-purpose, IC PC board
1	555 timer IC
1	1 kΩ, 0.25 W resistor (5 per package)

1	10 kΩ, 0.25 W resistor, 5% tolerance (5 per package)
2	470 Ω, 0.25 W resistor, 5% tolerance (5 per package)
1	0.1 μF polyester film capacitor
1	Red LED 5 mm
1	Green LED 5 mm
1	Normally open, momentary-contact pushbutton
1	9 V battery snap connector
1	9 V battery

You also need about 30 cm each of 22-AWG (American Wire Gauge) solid insulated wire and 22-AWG stranded insulated wire. The colour doesn't matter.

Figure 6-15 shows the layout of the preprinted circuit board. Before you start building the circuit, study this layout to familiarise yourself with it. As you can see, this board doesn't contain bus strips like those found on a breadboard. However, the overall layout is similar to the layout of the terminal strips on a breadboard. The centre portion of the board contains a total of 20 terminal strips, 10 on each side of the ditch. Each strip has three holes but is also connected to a second strip of two holes along the edge of the board. Thus, each strip effectively has five holes.

The strips aren't numbered on the board, but we number them in Figure 6-15. We use numbers 1 to 10 for the strips on the left side of the board and numbers 11 to 20 for the strips on the right. In the instructions that follow, we use these numbers to indicate which holes you attach components or jumper leads to. To keep things simple, we just specify the terminal strip number and leave it up to you to decide which of the five holes in the strip to use.

The numbers used in the PCB layout are relative to the bottom of the board: that is, the surface of the board with the copper traces and pads. Thus, the numbers 1 to 10 are on the left and the numbers 11 to 20 are on the right. When you flip the board over to insert components from the top of the board, you have to reverse the numbers mentally: 1 to 10 is on the right and 11 to 20 on the left.

When installing components onto the PCB, use a crocodile clip as a temporary clamp to hold the components flush against the board. Doing so enables you to turn the board upside down so you can solder the leads to the pads. If you don't clamp the component to the board, the component falls out when you turn the board upside down.

Don't be tempted to hold the component in place with one finger while you solder the leads: resistors get really hot when you solder them.

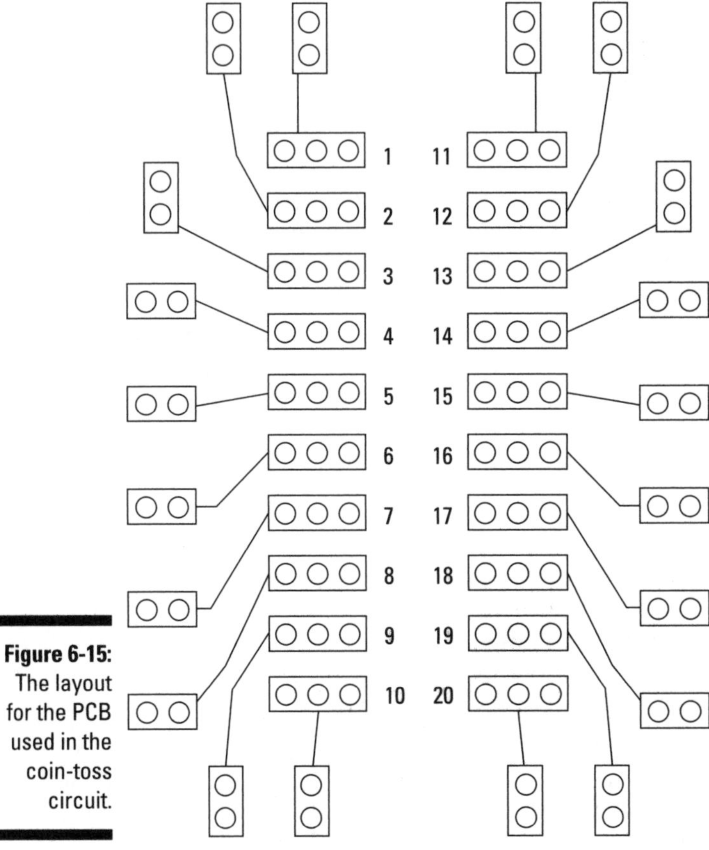

Figure 6-15: The layout for the PCB used in the coin-toss circuit.

Here are the steps for building the coin-toss circuit on a preprinted circuit board:

1. **Break the PCB in half.**

   The preprinted circuit board comes with two identical sections. You need just one of those sections for this project, and so break the board in half and save the other half for another project. (To break the board, just grab an end in each hand and snap it in two.)

2. **Insert the 555 timer IC.**

   Remember that the dot or notch on the 555 IC marks pin 1. Install the chip so that pin 1 is in strip 4 and pin 8 is in strip 14. Then solder the chip carefully into place. (See Chapter 7 of this minibook for tips on soldering.)

3. **Install the jumper wires.**

This circuit needs a total of nine jumper wires. Cut the jumper wires from the 22-AWG solid wire and carefully strip the insulation from each end. Use needle-nosed pliers to bend the bare end of each jumper wire down, insert both ends into the appropriate holes, solder the leads to the pads and then use your wire cutters to snip the excess of the end of each lead.

The following minitable provides the PCB strip locations for each jumper wire. Use your own judgement to determine in which hole in the indicated strip to place the jumper wire. Whenever possible, use the shortest possible path for each jumper wire.

Jumper #	*From Strip*	*To Strip*
1	9	10
2	19	20
3	5	16
4	4	10
5	7	19
6	2	6
7	1	5
8	2	12
9	14	19

4. **Install the resistors.**

   You need to install four resistors. Use the following minitable to install each resistor into its correct location. Bend the leads down and insert each resistor into the correct holes, solder the resistor in place and then snip off the excess wire from the ends of the leads.

Resistor #	*Value*	*Colours*	*From Strip*	*To Strip*
R1	1 kΩ	Brown, black, red	9	10
R2	10 kΩ	Brown, black, orange	11	15
R3	470 Ω	Yellow, purple, brown	13	19
R4	470 Ω	Yellow, purple, brown	3	10

5. **Install the capacitor.**

   Install the capacitor into strips 5 and 10. Push the capacitor all the way in until it's flush with the board. Then solder the leads to the pad and trim off the excess wire.

6. **Connect the two segments of row 19.**

   Insert one end of a short jumper wire into hole E19 and the other end into hole F19. This jumper wire bridges the gap between the two terminal strips in row 19, effectively making them a single terminal strip.

7. **Install the LEDs.**

   Remember that LEDs are directional and you have to install them in the correct direction or they don't work. One lead is shorter than the other to help you tell which lead is which. This short lead is the cathode and the longer lead is the anode.

   Here's where to install the LEDs:

LED Colour	***Cathode (Short Lead)***	***Anode (Long Lead)***
Red	12	13
Green	3	2

   When you install the LEDs, *don't* push the LED in until it's flush with the circuit board. Instead, push just a little bit of the leads into the holes so that the LED stands up about 25 mm from the top of the board.

8. **Install the jumper wires for the metal contacts.**

   Cut two, 5-centimetre lengths of stranded wire and strip about 1 centimetre of insulation from each end. Solder one end of each wire into holes in strips 1 and 11 and leave the other ends free. When the circuit is installed in its final enclosure (see the later section 'Mounting Your Circuit'), you connect the ends of these wires to the metal posts that the user touches to activate the coin-toss circuit.

   Feeding the stranded wire through the holes in the circuit board can be tricky. Carefully twist the loose strands until no stragglers protrude from the end of the wire, and then carefully push the wire through the hole. If any of the strands get caught and refuse to go through the hole, pull the wire out and try again. Alternatively, you can 'tin' the twisted wire by adding a thin coat of solder to the end, which maintains the flexibility of stranded wire whilst making it easier to feed through small holes.

9. **Connect the pushbutton.**

   Cut a 5 cm length of stranded wire and strip about 1 cm of insulation from each end. Solder one end to either terminal on the pushbutton (it doesn't matter which). Push the other end through a hole in strip 10 and solder it in place.

10. **Connect the battery snap connector.**

    Strip off about 1 cm of insulation from the end of both leads. Then solder the black lead to the free terminal on the pushbutton (the terminal you didn't use in Step 9) and solder the red lead to a hole in strip 20 on the PCB.

11. **Connect the 9 V battery to the snap connector.**

    The red LED immediately lights up, indicating that the circuit is ready to do its decision-making work. If the LED doesn't light up, something isn't right; use the troubleshooting guide for the prototype to help you work out what has gone wrong.

12. **Turn off your soldering iron.**

    Congrats . . . you're done!

Test the circuit by pinching both of the free jumper wires between your fingers. The LEDs alternately flash until you let go, at which time one or the other remains lit.

Figure 6-16 shows the completed circuit in operation.

Figure 6-16: The completed coin-tosser PCB.

Mounting Your Circuit

When your circuit board is finished, the final step to completing your project is to mount it in a nice enclosure such as a plastic, metal or wooden box. You can purchase boxes specifically designed for electronics projects from most parts suppliers. Maplin stocks various sizes of plastic, a few metal and several waterproof enclosures in its shops and you can find loads more online. See Figure 6-17 for some options.

Figure 6-17: Project boxes come in a variety of shapes and sizes.

Finding a suitable enclosure for your circuit

If you want something a bit different to a standard electronic project box, here are a few alternative ways to find the perfect enclosure for your project:

- **Search discount department stores for small storage boxes.** You may find one that's just the right size and shape for much less money than an official project box of the same size.
- **Check out the electrical department of any hardware store, where you can find inexpensive plastic and metal boxes designed for household wiring.** You can adapt many of these boxes for your electronic projects.
- **Take a quick look at boxes that used to contain any old electronic gizmos before you throw them away.** If you think a box may be useful for a project someday, take it apart and discard all the innards, keeping only the empty carcass.

Be careful whenever you disassemble any electronic device. Make sure that you first completely remove the power source and watch out for large capacitors that may be holding on to their charge.

- **Keep an eye out at car-boot sales.** You never know when you may find something suitable to adapt.
- **Make one out of Lego.** The bricks are too small and hard to drill for inlets and outlets, but you can build in the entry holes where you need them.

Working with a project box

Most project boxes are made of plastic or metal and have a detachable lid that's held on with four screws, one at each corner of the lid. To gain access to the insides of the box, you simply remove the screws to release the lid.

The inside of the box may contain ridges or mounting studs designed to make mounting components inside the box easier. If it's just smooth, however, you have to devise your own method of attaching the various bits and pieces that need to go inside. Here are some tips:

- **Gather together a good assortment of small drill bits to drill holes through the box to mount your components.** You need to drill holes to mount the circuit board as well as such things as battery holders, switches, LEDs, speakers and whatever else your project requires.
- **Make a good sketch of your project box and how its parts are to be arranged before you start drilling holes.** When you're sure that you have everything laid out the way you want, use a marker to indicate the exact position of the holes you need to drill.
- **Mount the circuit board using standoffs to provide some empty space between the board and the case.** A *standoff* is a screw that allows you to mount the board so that it's raised above the bottom of the project box.

  If you have an ample supply of nuts and bolts, you can fashion your own standoffs simply by cutting a short length of plastic tubing and feeding a long bolt through it.
- **Consider mounting the circuit board on the back of the lid instead of inside the body of the box.** This approach sometimes frees up more room within the box for larger items such as batteries or a speaker.
- **Attach switches to the box by drilling a hole large enough to allow the neck of the switch to pass through.** The switch comes with a nut that you can tighten over the neck of the screw to secure the screw to the box.

- **Use a small amount of epoxy or other glue to set components without mounting nuts in place.** The glue helps to secure them to the box.
- **Employ stranded wire for connections within the project box.** Stranded wire holds up better to the handling it occasionally gets when you open the box, for example, to change the batteries.

Chapter 7

Uncovering the Secrets of Successful Soldering

In This Chapter

- Assembling your soldering toolkit
- Soldering successfully
- Distinguishing a good solder joint from a not-so-good one
- Repairing a bad solder joint

Soldering is one of the basic skills of building electronic projects. Although you can use solderless breadboards to build test versions of your circuits (check out Chapter 6 of this minibook), sooner or later you want to build permanent versions of your circuits. To do that, you need to know how to solder.

In this chapter, you discover the basics of soldering: how soldering works, what tools and equipment you need and how to create the perfect solder joint. You also find out how to correct your mistakes when (not if) you make them.

Soldering is a skill that takes a bit of practice to master. Although not nearly as hard as playing the French horn, soldering does have a learning curve. When you're first getting started, you may feel like you're all thumbs, and your solder joints may look less than perfect. But stick with it – with a little practice, you get the hang of soldering in no time at all.

Understanding How Solder Works

Before we get into the nuts and bolts of making a solder joint, spend a few minutes thinking about what soldering is and what it isn't.

Soldering refers to the process of joining two or more metal objects by heating them, and then applying solder to the joint. *Solder* is a soft metal traditionally made from a combination of tin and lead (it still is in America). European law now permits only lead-free solder, comprising tin and various other metals. When the solder melts, at several hundred degrees Celsius, it flows over the metals to be joined. When the solder cools, it locks the metals together in a connection.

Several hundred degrees Celsius is pretty hot – certainly hot enough to burn your skin instantly on contact – and so soldering is an inherently dangerous task. It isn't extremely dangerous, because the amount of solder you use and the tools you use are fairly small, and so any burns you do receive are likely to be small. But they can be painful, and so you need to take great care whenever you're soldering. And take special care of your eyes.

Soldering is especially useful for electronics because it not only creates a strong physical connection between metals, but also creates an excellent conductive path for electric current to flow from one conductor to another: solder itself is an excellent conductor. For example, you can create a reasonably good connection between two wires simply by stripping the insulation off the ends of each wire and twisting them together. However, current can flow through only the areas that are physically touching. Even when twisted tightly together, most of the surface area of the two wires aren't touching. But when you solder them, the solder flows through and around the twists, filling any gaps while connecting the entire surface area of both wires.

Soldering isn't the same as braising or welding:

- **Braising** is similar to soldering, but instead of solder, metals with a higher melting point (usually 450°C or more) are used. Braising forms a stronger bond than soldering.
- **Welding** is an entirely different process in which the metals to be joined are melted. In liquid form, the metals intermix and cool to form an extremely strong bond.

Procuring What You Need to Solder

To solder, you need to get together the equipment we describe in the following sections.

Buying a soldering iron

A *soldering iron* is the basic tool for soldering. Figure 7-1 shows a typical soldering iron.

Figure 7-1: A soldering iron.

Here are some things to look for when purchasing a soldering iron:

- **The wattage rating needs to be between 20 and 50 watts.** Note that the wattage doesn't control how hot the soldering iron gets. Instead, it controls how fast it heats up and how fast it regains its normal operating temperature after completing each solder joint.

  Each time you solder, the tip of the soldering iron cools a bit as it transfers its heat to the wires you're joining and to the solder itself. A higher-wattage soldering iron can maintain a stable temperature longer while you're soldering a connection and can reheat itself faster in between.

- **The tip needs to be replaceable.** When you purchase your soldering iron, buy a few extra tips so that you have replacements handy when you need them.

- **Although you can buy a soldering iron by itself for under £10, we suggest you spend more on a *soldering station* that includes an integrated stand for around £20.** A secure place to rest your soldering iron when not in use is essential. Without a good stand, we guarantee that your workbench is soon covered with unsightly burn marks. (The soldering iron in the earlier Figure 7-1 includes a soldering station.)
- **More expensive soldering irons have built-in temperature control.** Although not necessary, temperature control is a nice feature if you plan to do a lot of soldering.

Stocking up on solder

Although solder is wound on spools and looks like wire, in fact it's a thin hollow tube with a thin core of rosin in the centre. This rosin, called the *flux,* plays a crucial role in the soldering process. It has a slightly lower melting point than the metal alloy, and so it melts just a few moments before the metal mixture melts. The flux prepares the metals by cleaning and lubricating the surfaces to be joined.

Getting together the other goodies you need

Besides a soldering iron and solder, you also require a few additional things on hand for successful soldering:

- **A vice:** You need at least three hands to solder: one to hold the items you're soldering, one to hold the soldering iron and one to hold the solder. Unless you're from the planet tri-mittia (where the inhabitants are helpful but pick-pocketing is a terrible problem!) and have three hands, you need to use a third-hand tool, a vice or some other resourceful device to hold the items you're soldering, to allow you to wield the soldering iron and the solder. (Refer to Chapter 3 of this minibook for photographs of a third-hand tool and a hobby vice.)
- **A sponge:** To clean the tip of the soldering iron.
- **Crocodile clips:** These serve two purposes when soldering:
  - As a clamp to hold a component in place while you solder it.
  - As a heat sink to avoid damaging a sensitive component when soldering the component's leads. (A *heat sink* is simply a piece of

metal attached to a heat-sensitive component that helps dissipate heat radiated by the component.)

- **Eye protection:** Always wear eye protection when soldering. Sometimes hot solder pops and flies through the air. Your eye and melted solder aren't a good mix.
- **Magnifying glass:** Soldering is much easier if you do it through a magnifying glass so that you can get a better look at your work. You can use a table-top magnifying glass, a magnifying glass attached to a third-hand tool or special magnifying goggles.
- **Desoldering braid and desoldering pump:** These tools are used to undo soldered joints when necessary to correct mistakes. For more information, see the section 'Undoing Your Work: Desoldering' later in this chapter.

Preparing to Solder

Before you start soldering, follow these steps to prepare your soldering iron:

1. **Turn on your soldering iron.**

   It takes about a minute to heat up.

2. **Clean the tip when the iron is hot.**

   To clean your soldering iron, wipe its tip on a damp sponge. As you work, wipe the iron on the sponge frequently to keep the tip clean.

3. **Tin the soldering iron.**

   Tinning refers to the process of applying a light coat of solder to the tip of the soldering iron. Tinning the tip of your soldering iron helps the solder flow more freely after it heats up. To tin your soldering iron, melt a small amount of solder on the end of the tip and then wipe the tip dry with your sponge.

Clean the tip of your soldering iron frequently as you work – ideally, immediately after every joint you solder. In addition, tin the soldering iron tip occasionally. Usually, tinning the iron once at the start of each project is sufficient. But if the solder disappears from the tip of the iron, you should tin it again.

The Ten Soldering Commandments

In truth, the Ten Soldering Commandments are more like guidelines than commandments. But if you heed them, things go well with your solder joints, your children's solder joints and your children's children's solder joints. So let it be written; so let it be done:

I. Thou shalt wear eye protection whenever thou solderest, lest thy get molten solder in thine eye.

II. Thou shalt not touch the heated end of thy soldering iron, lest thy burn thyself.

III. Thou shalt not fashion molten solder into false globs.

IV. Thou shalt wash thy hands after thou solderest, to remove vile contamination from upon thy hands before thou eatest.

V. Thou shalt provide bright illumination upon thine objects which thou solderest, that thou might see clearly the way unto which the solder may be applied.

VI. Thou shalt not spill thy excess solder upon thy neighbour's pad lest thy create unintended pathways through which current may flow.

VII. Thou shalt not leave thine hot soldering iron unattended.

VIII. Thou shalt not covet thy neighbour's professional-grade temperature-controlled soldering station.

IX. Thou shalt not apply solder directly upon thine soldering iron, but shalt instead apply solder to the objects which thou solderest, that their heat may melteth thy solder.

X. Thou shalt always place thine hot soldering iron in a suitable holder.

Soldering a Solid Solder Joint

The most common form of soldering when creating electronic projects is soldering component leads to copper pads on the back of a printed circuit board (PCB), to which we introduce you in Chapter 6 of this minibook. If you can carry out that task, you'll have no trouble with other types of soldering, such as soldering two wires together or soldering a wire to a switch terminal.

The following steps outline the procedure for soldering a component lead to a PCB:

1. **Pass the component leads through the correct holes.**

   Check the circuit diagram carefully to be sure that you have installed the component in the correct location. If the component is polarised (such as a diode, electrolytic capacitor or integrated circuit) verify that the component is oriented correctly. You don't want to solder it in backward.

2. **Secure the component to the PCB.**

   If the component is near the edge of the board, the easiest way to secure it is with a crocodile clip. You can also secure the component with a bit of tape.

3. **Clamp the PCB in place with your vice.**

   Turn the board so that the copper-plated side is up. If you're using a magnifying glass, position the board under the glass.

4. **Make sure that you have adequate light.**

   If you have a desktop lamp, adjust it now so that it shines directly on the connection to be soldered.

5. **Touch the tip of the soldering iron to the pad and the lead at the same time.**

   You have to touch the tip of the soldering iron to the copper pad and the wire lead. The idea is to heat them both so that solder flows and adheres to both.

   The easiest way to achieve the correct contact is to use the tip of the soldering iron to press the lead against the edge of the hole, as shown in Figure 7-2.

Figure 7-2: Positioning the soldering iron.

6. **Let the lead and the pad heat up for a moment.**

   They take only a few seconds to heat up sufficiently.

7. **Apply the solder to the lead on the opposite side of the tip of the soldering iron, just above the copper pad.**

   The solder begins to melt almost immediately.

   Figure 7-3 shows the correct way to apply the solder.

Figure 7-3: Applying the solder.

Don't touch the solder directly to the soldering iron. If you do, the solder melts immediately, and you can end up with an unstable connection, often called a *cold joint,* where the solder doesn't properly fuse itself to the copper pad or the wire lead.

8. **Feed just enough solder to cover the pad when the solder begins to melt.**

   As the solder melts, it flows down the lead and then spreads out onto the pad. You want just enough solder to cover the pad completely, but not enough to create a big glob on top of the pad.

Be stingy when applying solder: too much solder is a more common problem than too little. Adding a little more solder later if you don't get quite enough coverage is a lot easier than removing too much solder.

9. **Remove the solder and soldering iron and let the solder cool.**

   Be patient – the solder takes a few seconds to cool. Don't move anything while the joint is cooling. If you inadvertently move the lead, you create an unstable cold joint that you have to resolder.

10. **Trim the excess lead by snipping it with wire cutters right above the top of the solder joint.**

    Use a small pair of wire cutters so that you can trim it close to the joint.

Checking Your Work

After you complete a solder joint, inspect it to make sure that the joint is good. Look at it under a magnifying glass, and gently wiggle the component to see if the joint is stable. A good solder joint should be shiny and fill but not overflow the pad, as shown in Figure 7-4.

Figure 7-4: A good solder joint.

Nearly all bad solder joints are caused by one of three things: not allowing the wire and pad to heat sufficiently, applying too much solder or melting the solder with the soldering iron instead of with the wire lead. Here are some indications of a bad solder joint:

- **The pad and lead aren't completely covered with solder, enabling you to see through one side of the hole the lead passes through.** Either you didn't apply quite enough solder, or the pad wasn't quite hot enough to accept the solder.
- **The lead is loose in the hole or the solder isn't firmly attached to the pad.** One possible reason is that you moved the lead before the solder completely cooled.
- **Solder overflows the pad and touches an adjacent pad.** Sometimes this happens if you apply too much solder. It can also occur if the pad doesn't get hot enough to accept the solder, which can cause the solder to flow off the pad and onto an adjacent pad. If solder spills over from one pad to an adjacent pad, your circuit may not work correctly.

Undoing Your Work: Desoldering

Desoldering refers to the process of undoing a soldered joint. You may have to desolder if you discover that a solder joint is less than satisfactory, if a component fails or if you connect your circuit incorrectly.

To desolder a solder connection, you need a desoldering pump and a desoldering braid (see Figure 7-5).

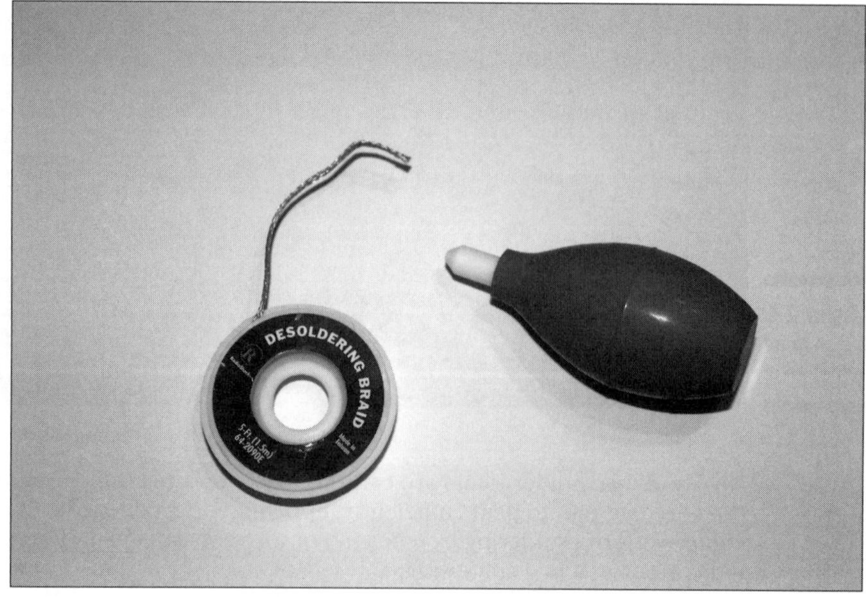

Figure 7-5: A desoldering pump and a desoldering braid.

Here are the steps for undoing a solder joint:

1. **Apply the hot soldering iron to the joint you need to remove.**

   Wait a second for the solder to melt.

2. **Squeeze the desoldering pump to expel the air it contains, and then touch the tip of the desoldering pump to the molten solder joint and release the pump.**

   As the pump expands, it sucks the solder off the joint and into the pump.

3. **Apply heat again if the desoldering pump doesn't completely free the lead, and touch the remaining molten solder with the desoldering braid.**

   The desoldering braid is specially designed to draw up solder, much like a candle wick draws up wax.

4. **Use needle-nose pliers or tweezers to remove the lead.**

   Don't try to remove the lead with your fingers after you desolder the connection. The lead remains hot for a while after you desolder it.

Chapter 8

Measuring Circuits with a Multimeter

In This Chapter

- Familiarising yourself with a multimeter
- Measuring current, voltage and resistance
- Looking at your first electronics equation: Ohm's Law

Just think how great it would be if every circuit you built worked right the first time you built it. You'd quickly develop a reputation as an electronics genius and in no time at all be the president of Intel.

But in the real world, a circuit doesn't always work correctly the first time. Then you have two options: you can scratch your head and stare at it for hours on end, or you can pull out your test equipment and analyse the circuit to find out what's wrong.

In this chapter, you discover how to use one of the electronics engineer's favourite tools – the *multimeter,* which contains several different meters that can measure current, voltage and resistance. Learn how to use it well because it's going to be your trusty companion throughout your adventures in electronics.

Looking at Multimeters

You know those late-night TV ads that try to sell you amazing kitchen gadgets that are like a combination blender, juicer, food processor, mixer and bottle opener all in one? A multimeter is kind of like one of those daft gadgets, except that a multimeter really *can* do all the things it claims without dozens of confusing attachments. And, unlike the daft kitchen gadgets, the things a multimeter can do turn out to be genuinely useful.

Along with a good soldering iron, a good multimeter is the most important item in your toolbox.

Figure 8-1 shows a simple, inexpensive multimeter. You can buy this sort of model from your local electronics shop or via the Internet for under £10 and it's sufficient when you're just getting started. Eventually, you probably want to invest a little more money in a better-quality multimeter.

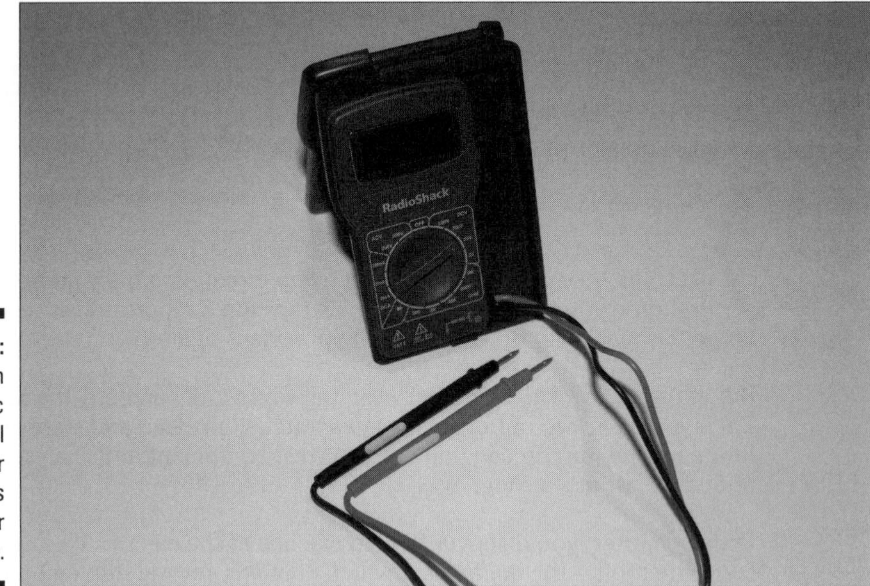

Figure 8-1: You can buy a basic digital multimeter like this one for under £10.

The multimeter in Figure 8-1 is a *digital* one, which displays its values using a digital display that shows the numbers for the measurements being taken. The alternative to a digital multimeter is an *analogue* one, which shows its readings by moving a needle across a printed scale. To determine the value of a measurement, you simply read the scale behind the needle.

Figure 8-2 shows a typical analogue multimeter. This one happens to be one of our favourites. Even though it's old enough to be called vintage, it's still an excellent and accurate meter. One of the benefits of spending a little more to buy quality equipment is that you get many years – sometimes decades – of reliable service.

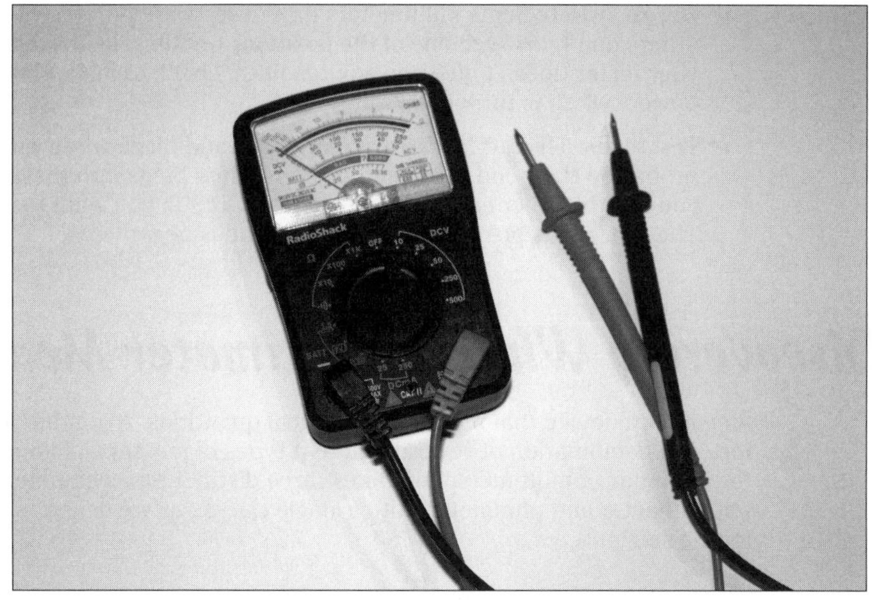

Figure 8-2: An analogue multimeter.

The following list describes the various parts that make up a typical multimeter:

- **Display or meter:** Indicates the value of the measurement being taken. In a digital multimeter, the display is a number that indicates the amperage (current), voltage or resistance being measured. In an analogue meter, a needle moves across a printed scale to indicate the current, voltage or resistance. To get the value, you look straight down at the needle and read the scale printed behind it.
- **Selector:** Most multimeters – digital or analogue – have a dial that you can turn to tell the meter what you want to measure. The various settings on this dial indicate not only the type of measurement you want to make (voltage, current or resistance), but also the range of the expected measurements. The range is indicated by the maximum amount of voltage, current or resistance that the meter can measure.

Higher ranges let you measure higher values, but with less precision. For example, the analogue multimeter in Figure 8-2 has the following ranges for reading direct current (DC) voltage: 2.5 V, 10 V, 50 V, 250 V and 500 V. If you use the 2.5 V range, you can easily tell differences of a tenth of a volt, such as the difference between 1.6 and 1.7 V. But when the range is set to the 500 V range, you're lucky to pick out differences of 10 V.

- **On/off switch:** Some multimeters have a separate on/off switch but others don't. Instead, one of the positions on the selector dial is 'Off'. If your meter doesn't give you any readings, check to make sure that the power switch is turned on.
- **Test leads:** The *test leads* are a pair of red and black wires with metal probes on their ends. One end of these wires plugs into the meter and you use the other end to connect to the circuits you want to measure. The red lead is positive and the black lead is negative.

Discovering What a Multimeter Measures

A *meter* is a device that measures electrical quantities. A multimeter, therefore, is a combination of several different types of meters all in one box. At the minimum, a multimeter combines three distinct types of meters (ammeter, voltmeter and ohmmeter) into a single device, as we describe in the following sections.

Keeping current with the ammeter

As you find out in Chapter 2 of this minibook, *current* is the flow of electric charge through a conductor. Current is measured in units called *amperes* and a meter that measures amperage is called an *ammeter.*

Very few electronic circuits have currents so strong that they can be measured in actual amperes. Therefore, ammeters usually measure current in *milliamperes* (also called a *milliamp* and usually abbreviated mA). One milliamp is 1,000th of an ampere; in other words, an amp contains 1,000 milliamps.

A happy accident

The world's first ammeter was invented by a Dutch physicist named Hans Christian Oersted in 1821, when he accidentally left a compass next to a wire that had an electric current flowing through it. Hans noticed that when the current flowed, the needle moved away from its normal northerly orientation and pointed towards the wire. This is because current moving through a wire creates a magnetic field around the wire, and the magnetic field was strong enough to attract the magnetised end of the compass needle.

After fooling around with it for a bit, Hans discovered that the more current he ran through the wire, the further the needle strayed from north. Soon he figured out that he could use his discovery to measure the amount of current flowing through a circuit. Analogue ammeters work by this same principle even today.

Calculating voltage with the voltmeter

In Chapter 2 of this minibook, you find out all about a second fundamental quantity of electricity, *voltage,* a term that refers to the difference in electric charge between two points. If those two points are connected to a conductor, a current will flow through the conductor. Thus, voltage is the instigator of current.

The unit of voltage is, naturally, the *volt,* and a device that measures voltage is called a *voltmeter.*

All other things being equal, a change in the amount of voltage between two points results in a corresponding change in current. Thus, if you can keep things equal, you can measure voltage by measuring current, and as we discuss in the preceding section you can measure current with an ammeter.

The basic difference between an ammeter and a voltmeter is that in an ammeter, you let current run directly through the meter so that you can measure the amount of current. In a voltmeter, the current is first run through a very large resistor and then through the ammeter, and the device makes the necessary calculations as follows.

In Book II, Chapter 2, you find out that a direct relationship exists between voltage, resistance and current in an electrical circuit. In particular, if you know any two of these quantities, the third one is easy to calculate. In a voltmeter, a large fixed resistance is used, and the ammeter measures the current. Because you know the amount of the fixed resistance and the amount of current, you can easily calculate the amount of voltage across the circuit.

Don't worry; you don't have to do any maths to calculate this voltage. The voltmeter does that for you. In an analogue voltmeter, the calculation is built in to the scale that's printed on the meter, so all you have to do is look at the position of the needle on the scale to read the voltage. In a digital voltmeter, the voltage is automatically calculated and displayed digitally.

For a brief explanation of the relationship between current, voltage and resistance, see the later sidebar 'A quick look at Ohm's law'.

Sensing resistance with the ohmmeter

A *resistor* is a material that resists the flow of current. How much the current is restricted is a function of the amount of resistance in the resistor, which is measured in units called *ohms.* The symbol for ohms is the Greek letter omega, Ω. A device that measures resistance is called an *ohmmeter.*

Like voltage, resistance can also be measured with an ammeter. As we state in the preceding section, a direct relationship exists between voltage, resistance and current in any circuit, and if you know any two of these quantities you can easily calculate the third. So to measure voltage, a voltmeter provides a fixed resistance, uses an ammeter to measure the current and then uses the resistance and current to calculate the voltage.

To measure the resistance of a circuit, an ohmmeter provides a fixed amount of voltage across the circuit, uses an ammeter to measure the current that flows through the circuit and then uses the amount of voltage the meter provides and the amount of current the meter reads to calculate the resistance.

As with voltage, you don't have to do this calculation (phew!); it's automatically made by digital multimeters and is built in to the meter scale for analogue multimeters. Thus, all you have to do is read the display or the needle on the meter to determine the resistance.

A quick look at Ohm's law

In Book II, Chapter 2, you take a close look at resistors and at Ohm's law, which is one of the most important mathematical relationships in electronics. Here we give you a brief overview without going into all the details.

Ohm's law describes a fundamental relationship between current, voltage and resistance in an electrical circuit. Remember that voltage is the difference in electrical charge between two points, and that if a conductor connects those two points current will flow through the conductor.

Other than exotic superconductors (which exist only in laboratory experiments), no conductor is perfect. All conductors have a certain amount of resistance that inhibits the flow of current. The greater this resistance, the less the current will flow. The less this resistance is, the more the current will flow.

Ohm's law is a mathematical formula that describes the relationship between current, voltage and resistance. Here's the formula:

$$I = \frac{V}{R}$$

In other words, the amount of current running through a circuit is equal to the amount of voltage across the circuit divided by the amount of resistance in the circuit. The amount of current in amperes is represented by the letter I (don't ask why; it just is). V represents voltage in volts and R represents resistance in ohms.

With basic maths you can use this equation to calculate the voltage if you know the current and the resistance. Then, the formula becomes:

$$V = IR$$

In other words, voltage is equal to current times resistance.

Similarly, you can calculate resistance if you know the current and the voltage. Then, the formula becomes:

$$R = \frac{V}{I}$$

In other words, resistance equals the voltage divided by the current.

Meeting some other measurements

All multimeters can measure current, voltage and resistance, but some can perform other types of measurements as well. For example, certain meters can measure the capacitance of capacitors and some can test diodes or transistors. These features are handy, but not essential.

Reading schematic symbols for meter functions

Ammeter, voltmeter and ohmmeters are often included in schematic diagrams. When they are, the following symbols are used:

Symbol	*Meaning*
—(A)—	Ammeter
—(V)—	Voltmeter
—(Ω)—	Ohmmeter

Using Your Multimeter

In this section, we show you how to use your multimeter to measure current, voltage and resistance in a simple circuit. The circuit being measured consists of just three components: a 9 V battery, a light-emitting diode (LED) and a resistor. Figure 8-3 shows the schematic for this circuit.

Building a circuit to measure

If you want to follow along with the measuring procedures detailed in the following sections, you can build this circuit on a solderless breadboard. You need the following parts:

- Small solderless breadboard
- 470 Ω, 1/4W resistor, 5 per cent tolerance
- Red LED, 5 mm

- 9 V battery snap connector
- 9 V battery
- Short length of jumper wire (25 mm or less)

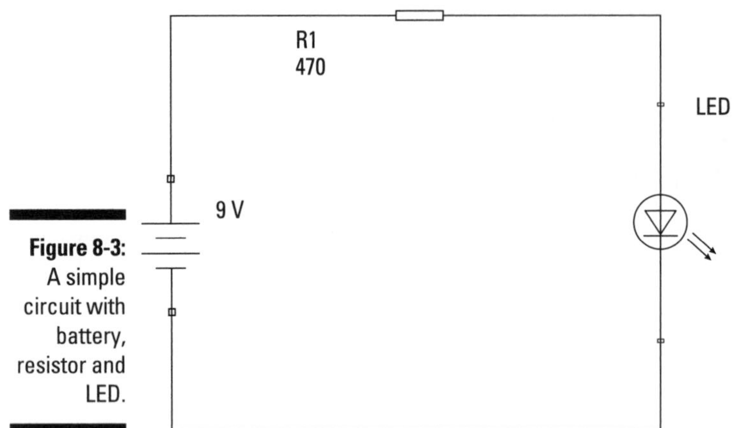

Figure 8-3: A simple circuit with battery, resistor and LED.

Figure 8-4 shows the circuit installed on the breadboard. Here are the steps for building this circuit:

1. **Connect the battery snap connector.**

   Insert the red lead in the top bus strip and the black lead in the bottom bus strip. Any hole is fine, but connecting the battery at the very end of the breadboard makes sense.

2. **Connect the resistor.**

   Insert one end of the resistor into any hole in the bottom bus strip. Then, pick a row in the nearby terminal strip and insert the other end into a hole in that terminal strip.

3. **Connect the LED.**

   Notice that the leads of the LED aren't the same length; one lead is shorter than the other. Insert the short lead into a hole in the top bus strip, and then insert the longer lead into a hole in a nearby terminal strip.

   Insert the LED into the same row as the resistor. In Figure 8-4, the LED and the resistor are in row 26.

4. **Use the short jumper wire to connect the terminal strips into which you inserted the LED and the resistor.**

The jumper wire straddles the gap that runs down the middle of the breadboard.

5. **Connect the battery to the snap connector.**

   The LED lights up. If it doesn't, double-check your connections to make sure that the circuit is assembled correctly. If it still doesn't light up, try reversing the leads of the LED (you may have inserted it backwards). If that doesn't work, try a different battery.

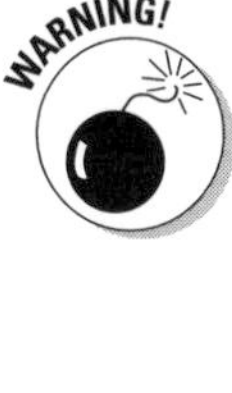

Do *not* connect the LED directly to the battery without a resistor. If you do, the LED flashes brightly – it may even go bang – and then it's dead forever.

Figure 8-4: The LED circuit assembled on a breadboard.

Measuring current

Electric current is measured in amperes, but in most electronics work you measure current in milliamps (or mA). To measure current, you need to connect the two leads of the ammeter in the circuit so that the current flows through the ammeter. In other words, the ammeter must become a part of the circuit itself.

The only way to measure the current flowing through the LED circuit that's shown in Figure 8-3 is to insert your ammeter into the circuit. Figure 8-5 shows one way to do so. Here, the ammeter is inserted into the circuit between the LED and the resistor.

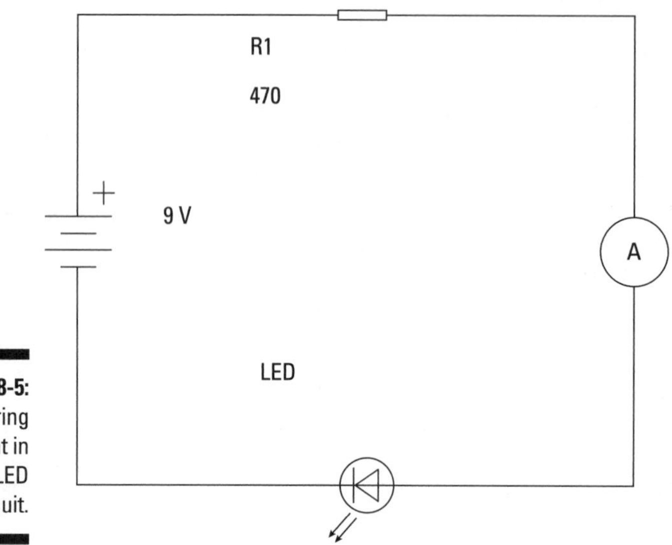

Figure 8-5: Measuring current in the LED circuit.

Note that where in this circuit you insert the ammeter doesn't matter. You get the same current reading whether you insert the ammeter between the LED and the resistor, between the resistor and the battery, or between the LED and the battery.

To measure the current in the LED circuit, follow these steps:

1. **Set your multimeter's range selector to a DC milliamp range of at least 20 mA.**

   This circuit uses DC and so you need to ensure that the multimeter is set to a DC current range.

2. **Remove the jumper wire that connects the two terminal strips.**

   The LED goes dark, because removing the jumper wire breaks the circuit.

3. **Touch the black lead from the multimeter to the LED lead that connects to the terminal strip (not the bus strip).**

4. **Touch the red lead from the multimeter to the resistor lead that connects to the terminal strip (not the bus strip).**

The LED lights up again, because the ammeter is now a part of the circuit and current can flow.

5. **Read the number on the multimeter display.**

   It should read between 12 and 13 mA. (The precise reading depends on the exact resistance value of the resistor. Resistor values aren't exact, and so even though you're using a 470 Ω resistor in this circuit, the actual resistance of the resistor may be anywhere from around 446 to 494 Ω. For more about this effect, check out Book II, Chapter 2.)

6. **Congratulate yourself!**

   You've made your first official current measurement.

7. **Replace the jumper wire you removed in Step 2 (after a suitable celebration).**

   If you forget to replace the jumper wire, the procedure we describe in the next section for measuring voltage doesn't work.

Don't connect the ammeter as follows in this circuit:

- Don't connect the ammeter directly across the two battery terminals. Doing so effectively shorts out the battery, which gets really hot, really fast.
- Don't connect one lead of the ammeter to the positive battery terminal and the other directly to the LED lead. That bypasses the resistor, which probably blows out the LED.
- Don't try to measure too big a current. Doing so will usually blow the fuse in a digital multimeter or trip a re-settable switch in an analogue meter.

If you want to experiment a little more, try measuring the current at other places in the circuit. For example, remove the battery snap connector from the battery and then reconnect it so that just the negative battery terminal is connected. Then, touch the red meter lead to the positive battery terminal and the black lead to the lead of the resistor that's connected to the bus strip (not the lead that's connected to the terminal strip). This measures the current by inserting the ammeter between the resistor and the battery. You get the same value as when you measure between the LED and the resistor.

You can use a similar method to measure the current between the LED and the negative battery terminal. Again, the result should be the same.

Measuring voltage

Measuring voltage is a little easier than measuring current because you don't have to insert the meter into the circuit. Instead, all you have to do is touch the leads of the multimeter to any two points in the circuit. When you do, the multimeter displays the voltage that exists between those two points.

Figure 8-6 shows how you can insert a voltmeter into the LED circuit to measure voltage. In this case, the voltage is measured across the battery. It probably reads in the vicinity of 8.3 V (9 V batteries rarely provide a full 9 V).

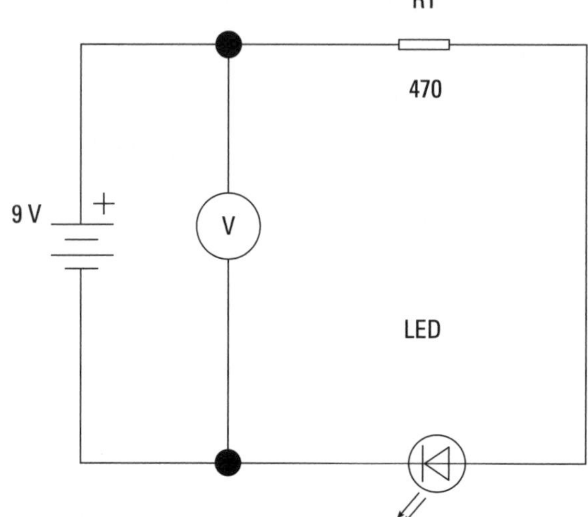

Figure 8-6: Measuring voltage in the LED circuit.

To measure voltages in the LED circuit, first put the circuit back together (if you took it apart to measure currents in the preceding section). Then, spin the multimeter dial to a range whose maximum is at least 10 V. Now touch the leads to different spots in the circuit. To measure the voltage across the entire circuit as shown in Figure 8-6, touch the black lead to the LED lead that's inserted into the negative bus strip and touch the red lead to the resistor lead that's inserted into the positive bus strip.

Here's an interesting exercise. Write down the following three voltage measurements:

- **Across the battery:** Connect the red meter lead to the resistor lead that's inserted into the positive bus strip and the black meter lead to the LED lead that's inserted into the negative bus strip.
- **Across the resistor:** Connect the red meter lead to the resistor lead that's inserted into the positive bus and the black meter lead to the other resistor lead.

- **Across the LED:** Connect the black meter lead to the LED lead that's inserted into the negative bus and the red meter lead to the other LED lead.

What do you notice about these three measurements? (The question's a little bit of a puzzle, and so we don't give the answer here; you can find it in Book II.)

Measuring resistance

Measuring resistances is similar to measuring voltages, with one key difference.

Before starting, disconnect all voltage sources from the circuit whose resistance you want to measure. That's because the multimeter injects a known voltage into the circuit so that it can measure the current and then calculate the resistance. If any outside voltage sources are in the circuit, the voltage isn't fixed and so the calculated resistance is wrong.

Here are the steps for measuring resistance in the LED circuit:

1. **Remove the battery.**

   Just unplug it from the battery snap connector and set the battery aside.

2. **Turn the meter selector dial to one of the resistance settings.**

   If you have an idea of what the resistance is, pick the smallest range that's greater than the value you're expecting. Otherwise, pick the largest range available on your meter.

3. **Calibrate the meter if you're using an analogue one.**

   You have to calibrate analogue meters before they can give an accurate resistance measurement. To calibrate an analogue meter, touch the two meter leads together. Then, adjust the meter's calibration knob until the meter indicates 0 resistance.

4. **Touch the meter leads to the two points in the circuit for which you want to measure resistance.**

   For example, to measure the resistance of the resistor, touch the meter leads to the two leads of the resistor. The result should be in the vicinity of 470 Ω.

You can discover loads more about measuring resistances in Book II, Chapter 2, but here are a few additional thoughts to tide you over:

- When you measure the resistance of an individual resistor or of circuits consisting of nothing other than resistors, the direction the current flows through the resistor doesn't matter. Thus, you can reverse the multimeter leads and still get the same result.

- Some components such as diodes pass current better in one direction than in the other. In that case, the direction of the current does matter. You can read about this effect in Book II, Chapter 5.
- Resistors aren't perfect. Thus, a 470 Ω resistor rarely provides exactly 470 Ω of resistance. The usual tolerance for resistors is 5%, which means that a 470 Ω resistor has somewhere between 446.5 and 493.5 Ω of resistance. For most circuits, this level of imprecision doesn't matter. But in circuits where it does, you can use the ohmmeter function of your multimeter to determine the exact value of a particular resistor. Then, you can adjust the rest of your circuit accordingly. (We write more on this subject in Book III, Chapter 2.)

Chapter 9

Catching Waves with an Oscilloscope

In This Chapter

- Learning what an oscilloscope is
- Getting started with an oscilloscope
- Calibrating your scope
- Looking at waveforms

Electronics can whip up some gnarly waves, dude. You can surf sine waves that roll along nice and easy, like corduroy on the horizon. But sawtooth waves are hairy: they ride up slow and then drop you way fast. And of course, square waves . . . they're just, you know, so *square.*

Your basic multimeter, which you can read about in this minibook's Chapter 8, is essential. You can't do electronics without one. In this chapter, we tell you about another incredibly useful tool to have on your workbench called an *oscilloscope.* Although the voltmeter in your multimeter can give you a simple number that represents voltage, an oscilloscope can draw a picture of voltage. And, as they say, a picture's worth a thousand words – er, numbers.

A picture is much more valuable than a number as regards voltage, because in all but the simplest circuits voltage is always in motion; it's always changing, and an oscilloscope is the perfect tool for observing voltage in motion.

Unfortunately, oscilloscopes are expensive. A decent one costs at least a few hundred pounds, and really good ones cost thousands. So although oscilloscopes are useful, most hobbyists get by without one.

The purpose of this chapter is twofold: first, to show you how to use an oscilloscope if you manage to get your hands on one; second, and perhaps more importantly, to convince you to start saving your pocket change so that someday you're able to afford one. After you get an oscilloscope for your workbench, you wonder how you ever managed without it.

Understanding Oscilloscopes

Figure 9-1 shows a typical oscilloscope. This one is an older analogue model, but although oscilloscope technology has changed over the years even older ones are useful for basic circuit testing. Invest in an oscilloscope and you have a tool that lasts for many years.

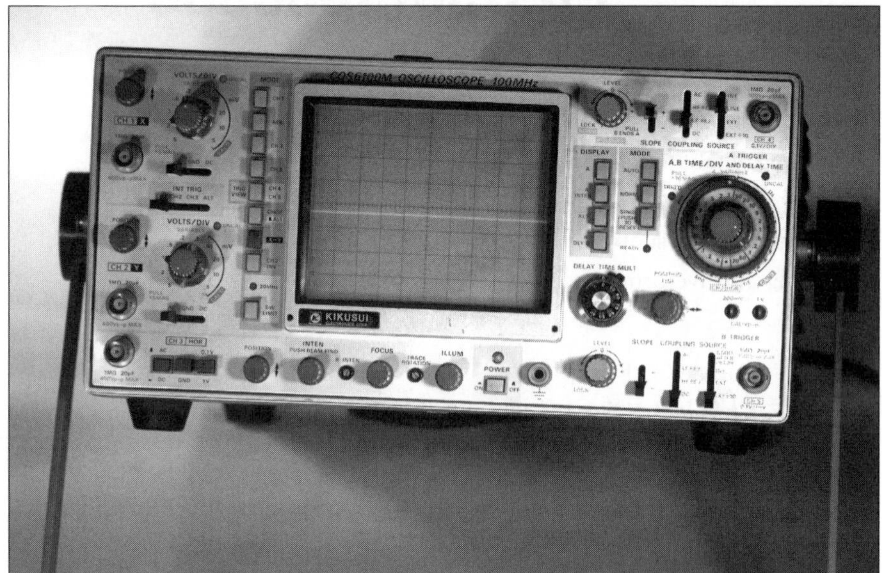

Figure 9-1: A typical oscilloscope.

The most obvious feature of any oscilloscope is its screen. On older oscilloscopes, the screen is a cathode-ray tube (CRT) similar to an older television or computer monitor. On newer digital models, the screen is an LCD display like a flat-screen computer monitor.

Whether CRT or LCD, the purpose of the screen is the same: to display a simple graph of an electrical signal. This graph, called a *trace,* shows how voltage changes over time. The horizontal axis of this graph, reading from left to right, represents time. The vertical axis, going up and down, represents voltage.

Figure 9-2 contains a typical oscilloscope display showing a very common type of trace known as a *sine wave.* we talk about the sine wave in the next section, but here are a few things to notice about the display:

- Gridlines are printed on the display. On most oscilloscopes, these lines are 1 cm apart, with ten horizontal and eight vertical divisions.

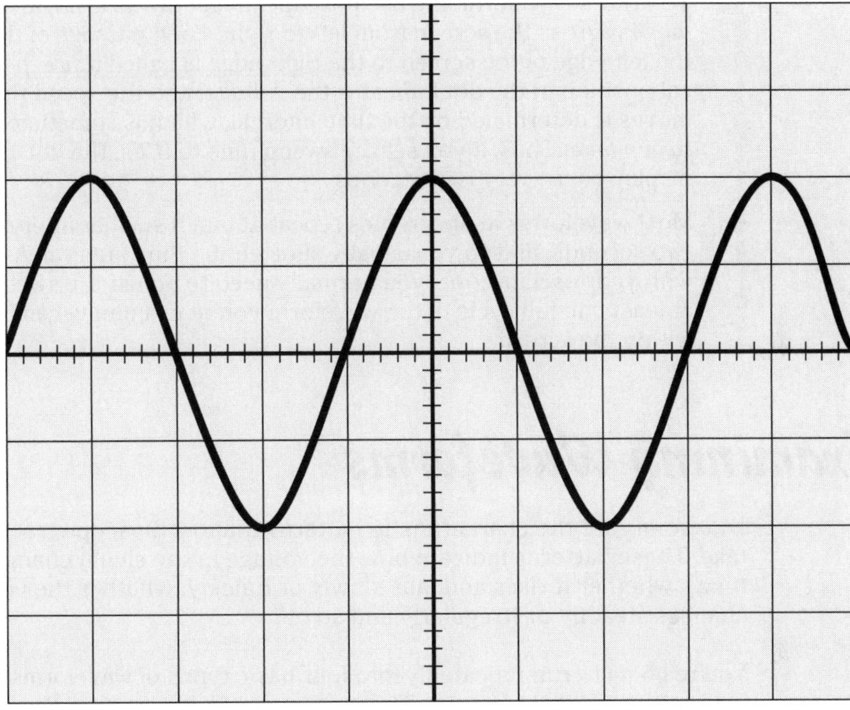

Figure 9-2: An oscilloscope trace showing a sine wave.

- The vertical and horizontal lines in the middle are thicker than the other lines and include hash marks, usually 2 mm apart. These hash marks help you pinpoint the exact position of the trace between the major intervals.
- Various knobs and dials on the oscilloscope let you set the *scale* at which the graph of the waveform is plotted.
- The vertical divisions represent voltage. On most oscilloscopes, you can set the voltage scale to as little as 5 mV (millivolts) and as much as 10 V (volts) or more. The oscilloscope usually represents 0 V by the horizontal line in the middle – so lines in the top half of the display represent positive voltage and lines in the bottom half are negative voltage. Thus, if the voltage scale is set to 1 V, the display can show voltages between +4 V and –4 V. If you set the scale to 2 V, the display can show voltages between +8 V and –8 V. Oscilloscopes often have a control knob to move the zero line up or down.

- The horizontal divisions represent time. The maximum time per division is typically 0.2 s (seconds) and the minimum time is typically 0.55 μs – that's half a microsecond. A single second has a million microseconds.

  To draw a waveform, the oscilloscope in fact draws a single dot that moves across the screen from left to right. Each passage of the dot from the left edge of the screen to the right edge is called a *sweep.* The vertical position of the dot indicates the voltage, and the speed that the dot moves is determined by the time interval, which is sometimes called the *sweep time.* Thus, if you set the sweep time to 0.2 s, the dot sweeps the display once every two seconds.

  Most waveforms in electronics repeat at much smaller intervals than two seconds, and so you usually shorten the time interval. As you work with your oscilloscope, you normally need to adjust the sweep time until at least one full cycle of the waveform you're examining can be shown within the screen.

Examining Waveforms

Waveforms are the characteristic patterns that oscilloscope traces usually take. These patterns indicate how the voltage in the signal changes over time – whether it rises and falls slowly or quickly, whether the voltage changes steadily or irregularly and so on.

You're going to run repeatedly into four basic types of waveforms as you work with electronic circuits. These four waveforms are (see Figure 9-3):

- **Sine wave:** The voltage increases and decreases in a steady curve. If you remember your trigonometry class from school, you may recall a trig function called *sine,* which has to do with angles measured in right-angle triangles.

  We're sure that you don't want to go back to school trigonometry (we certainly don't) and so that's all we say about the mathematics behind sine waves, except that they're found everywhere in nature. For example, sine waves can be found in sound waves, light waves, ocean waves – even a slinky.

  Most importantly from the standpoint of electronics, the alternating current (AC) voltage that's provided in the public power grid is in the form of a sine wave. In an AC sine wave, voltage increases steadily until a peak voltage is reached. Then, the voltage begins to decrease until it reaches zero. At that point, the voltage becomes negative, which causes the current flow to reverse direction. When it's negative, the voltage continues to change until it reaches its peak negative voltage, and then it begins to increase until it reaches zero again. The voltage then becomes positive, the current reverses and the sine wave cycle repeats.

The number of times a sine wave (or any other wave for that matter) repeats itself is called its *frequency.* Frequency is measured in units called *hertz,* abbreviated *Hz.* The AC available from a standard electrical outlet changes 50 times a second. Thus, the frequency of household AC is 50 Hz. Most waveforms found in electronic circuits have a much higher frequency than household AC, typically in the range of several thousand hertz (*kilohertz* or *kHz*) or millions of hertz (*megahertz* or *MHz*).

Sine wave

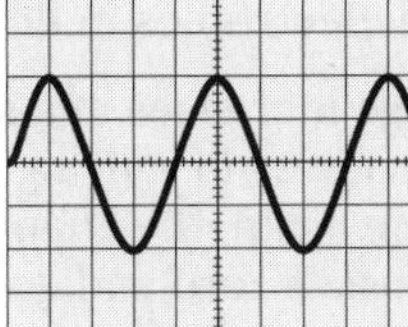

Square wave

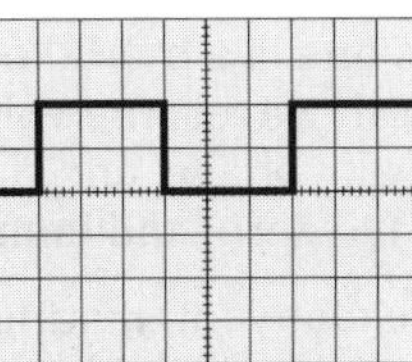

Triangle wave

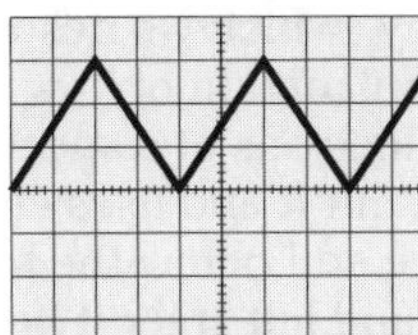

Sawtooth wave

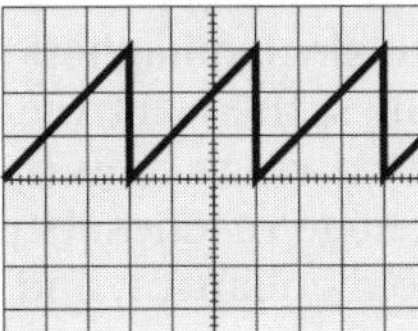

Figure 9-3: Four common waveforms.

- **Square wave:** Represents a signal in which a voltage simply turns on, stays on for a while, turns off, stays off for a while and then repeats. The graph of such a wave shows sharp, right-angled turns, which is why it's called a square wave.

  In practice, most circuits that attempt to create square waves don't do their job perfectly. As a result, the voltage rarely comes on absolutely instantly and rarely shuts off absolutely instantly. Thus, the vertical parts of the square wave in Figure 9-3 aren't vertical in the real world. In addition, sometimes the initial voltage overshoots the target voltage by a little bit, and so the initial vertical uptake goes a little too high for a very brief moment, and then settles down to the right voltage.

You can find square waves in many electronic circuits. For example, the 555 timer IC in the coin-toss project that we present in Chapter 7 of this minibook produces square waves that flash the light-emitting diodes on and off. Also, digital logic circuits (for example, computer circuits) rely almost exclusively on square waves to represent the ones and zeros of digital electronics. (We tell you more about the 555 timer IC in Book III, Chapter 2, and digital electronics in Book VI.)

- **Triangle wave:** Voltage increases in a straight line until it reaches a peak value, and then it decreases in a straight line. If the voltage reaches zero and then begins to rise again, the triangle wave is a form of direct current. If the voltage crosses zero and goes negative before it begins to rise again, the triangle wave is a form of alternating current.
- **Sawtooth wave:** This one is a hybrid of a triangle wave and a square wave. In most sawtooth waves, the voltage increases in a straight line until it reaches its peak voltage, and then drops instantly (or as close to instantly as possible) to zero, and immediately repeats.

Sawtooth waves have many interesting applications. One of the most appropriate for the purposes of this chapter is within an oscilloscope that has a CRT display (which we describe in the preceding section). Here's a simplified explanation of how the CRT in an oscilloscope works: it shoots a beam of electrons at a specially coated glass surface that glows when electrons hit it and uses electromagnets to steer the beam. Electromagnets above and below the beam steer it vertically; electromagnets to the right and left steer it horizontally.

To create the sweep of the electron beam from left to right, a sawtooth wave is applied to the electromagnets on the left and right of the beam. As the voltage increases, the electromagnet produces an increasingly stronger magnetic field, which pulls the beam toward the right side of display. When the voltage reaches its peak and drops instantly back to zero, the magnetic field collapses, and the electron beam snaps back to the left side of the display.

Changing the sweep rate of the oscilloscope is simply a matter of changing the frequency of the sawtooth wave applied to the horizontal electromagnets in the oscilloscope's CRT.

Calibrating an Oscilloscope

Quick question: what were the first words spoken by the astronauts after the moon landing?

You may be surprised to know that the words 'That's one small step for [a?] man' came several hours later. They weren't even: 'Houston, Tranquility Base here; the Eagle has landed'.

The very first words formed a much more prosaic conversation: 'Shutdown / Okay. Engine Stop. ACA out of detent / Out of detent. Auto. . .' and so on. Before Neil Armstrong was able to make his historic announcement that the Eagle had landed on the moon, the astronauts had to verify the settings of some key controls within the lunar module to ensure that everything was working well.

In the same way, before you make your historic first waveform measurement you need to verify the settings of some key controls on your oscilloscope to ensure that everything's working effectively. The exact steps you need to follow to set up your oscilloscope vary depending on the type and model of your scope, and so be sure to read your scope's instruction manual. The general steps are as follows:

1. **Examine all the controls on your scope and set them to normal positions.**

   For most scopes, all rotating dials need to be centred, all pushbuttons need to be out and all slide switches and paddle switches need to be up.

2. **Turn your oscilloscope on.**

   If it's the old-fashioned CRT kind, give it a minute or two to warm up.

3. **Set the VOLTS/DIV control to 1.**

   Doing so sets the scope to display 1 V per vertical division. Depending on the signal you're displaying, you may need to increase or decrease this setting, but 1 V is a good starting point.

4. **Set the TIME/DIV control to 1 millisecond (ms).**

   This control determines the time interval represented by each horizontal division on the display. Try turning this dial to its slowest setting. (On our scope, the slowest setting is half a second, and so for the dot to travel across the screen takes a full five seconds.) Then, turn the dial one notch at a time and watch the dot speed up until it becomes a solid line.

5. **Set the Trigger switch to Auto.**

   The Auto position enables the oscilloscope to stabilise the trace on a common trigger point in the waveform. If the trigger mode isn't set to Auto, the waveform may drift across the screen, making it difficult to watch.

6. **Connect a probe to the input connector.**

   If your scope has more than one input connector, connect the probe to the one labelled A.

   Oscilloscope probes include a probe point, which you connect to the input signal and a separate ground lead. The ground lead usually has a crocodile clip. When testing a circuit, you can connect this clip to any common ground point within the circuit. In some probes, the ground lead is detachable so that you can remove it when not needed.

7. **Touch the end of the probe to the scope's calibration terminal.**

   This terminal provides a sample square wave that you can use to calibrate the scope's display. Some scopes have two calibration terminals, labelled 0.2 V and 2 V. If your scope has two terminals, touch the probe to the 2 V terminal.

   For calibrating, using a crocodile clip test probe is best. If your test probe has a pointy tip instead of a crocodile clip, you can usually push the tip through the little hole in the end of the calibration terminal to hold the probe in place.

   You don't need to connect the ground lead of your test probe for calibration.

8. **Adjust the TIME/DIV and VOLTS/DIV controls, if necessary, until the square wave fits nicely within the display.**

   For an example, check out Figure 9-4.

9. **Adjust the Y-POS control to centre the trace vertically, if necessary.**

10. **Adjust the X-POS control to centre the trace horizontally, if necessary.**

11. **Adjust the Intensity and Focus settings to get a clear trace, if necessary.**

12. **Congratulate yourself!**

    You're now ready to begin viewing the waveforms of actual electronic signals.

The controls of every oscilloscope make and model are unique. Read carefully the owner's manual that comes with your oscilloscope to see whether you need to follow any other setup or calibration procedures before feeding real signals into your scope.

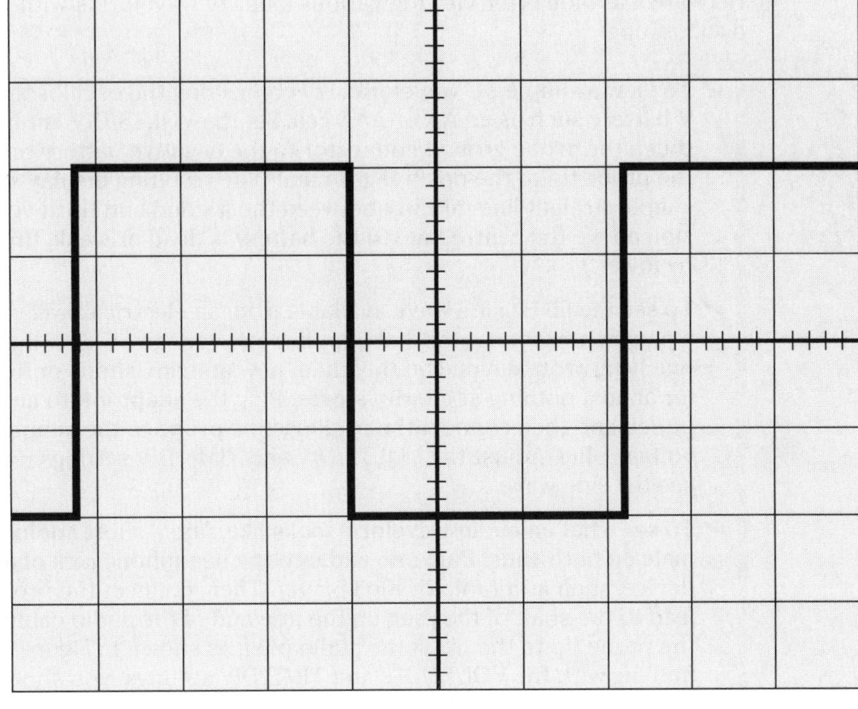

Figure 9-4: An oscilloscope trace showing a square wave.

Displaying Signals

The basic procedure for testing a circuit with an oscilloscope is to attach the ground connector of the scope's test lead to a ground point in the circuit, and then touch the tip of the probe to the point in the circuit that you want to test.

For example, if you want to verify that the output from a pin of an integrated circuit is emitting a square wave, touch the oscilloscope probe to the pin and look at the display on the scope. Note that you may need to adjust the VOLTS/DIV and TIME/DIV settings on the scope (as we describe in the preceding section) to see the waveform clearly. But after you get those settings adjusted correctly, you should be able to visualise the square wave. If the square wave doesn't appear, you probably have a problem with the circuit.

Never connect the oscilloscope probe directly to an electrical outlet. You're likely to kill your scope or yourself.

Here are a few ideas for viewing various kinds of waveforms with your oscilloscope:

- To view a simple DC waveform, try connecting the oscilloscope to a 1.5 V battery such as an AA or AAA cell. Set the VOLTS/DIV knob to 2 V and touch the probe ground connector to the negative battery terminal and the probe tip to the positive terminal. The resulting display should be a simple straight line midway between the second and third vertical division above the centre line. (If the battery is dead or weak, this line may be lower.)
- To see the 50 Hz sine wave available from an electrical wall outlet, find a plug-in power adapter that supplies low-voltage AC. If you don't have one lying around, you can buy them new at many shops or find them for almost nothing at charity stores. Plug the adaptor into an electrical outlet, and then connect the oscilloscope probe to the adaptor's low-voltage plug. Adjust the VOLTS/DIV and TIME/DIV settings until you can see the sine wave.
- To see what an audio waveform looks like, find a short audio cable that's male on both ends. Plug one end into the headphone jack of any audio device, such as a radio or mp3 player. Then, connect the probe's ground lead to the shaft of the plug on the free end of the audio cable and touch the probe tip to the tip of the audio plug, as shown in Figure 9-5. After fiddling with the VOLTS/DIV and TIME/DIV settings, you should see a display of the jumbled waveform that's typical of audio signals.

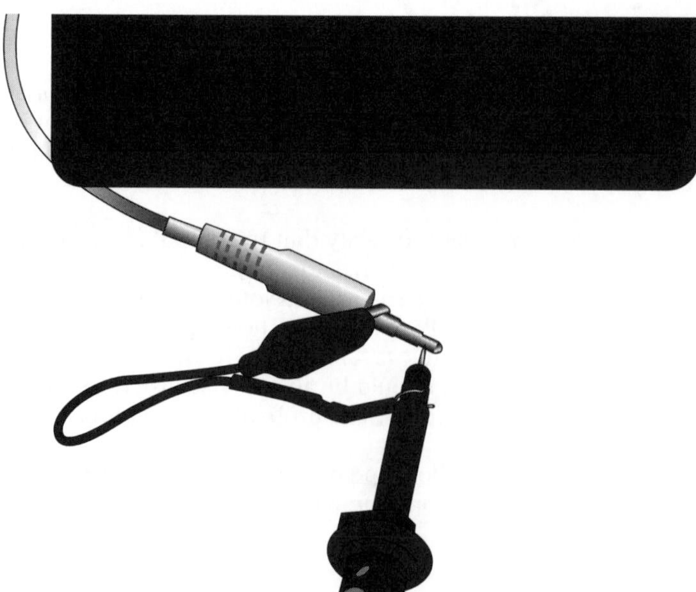

Figure 9-5: Connecting an oscilloscope probe to an audio plug.

As you build circuits while working your way through this book, keep your oscilloscope handy. Don't hesitate at any time to pick up the oscilloscope probe and check out the signals that are being generated at various points within your circuit. Connect the probe's ground clip to any ground point in the circuit, and then touch the probe tip to every loose wire and exposed pin to see what's going on inside the circuit.

Book II

Working with Basic Electronic Components

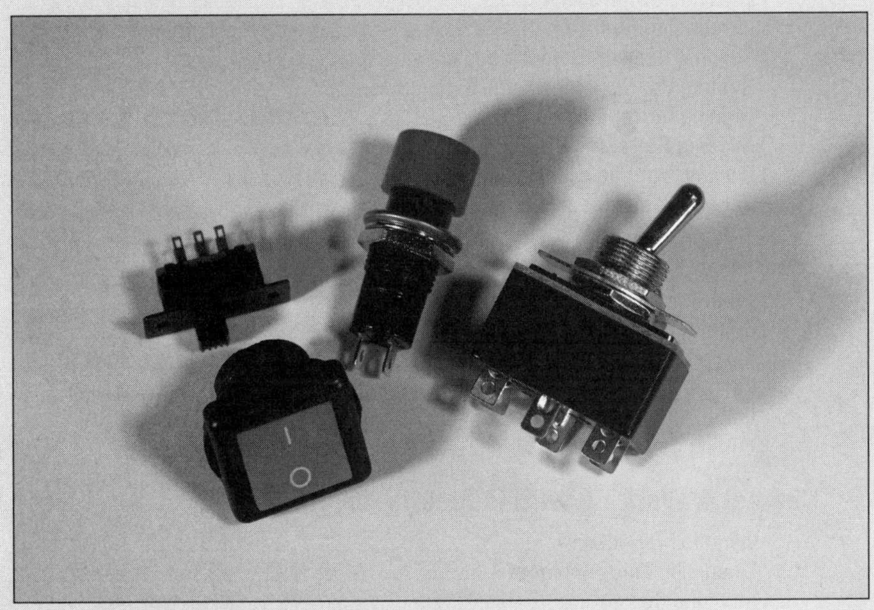

Discover how to protect your delicate electronics components from static discharge-induced damage, at www.dummies.com/extras/electronicsaiouk.

Contents at a Glance

Chapter 1

Working with Basic Circuits

In This Chapter

- Examining the basic elements of circuits
- Powering your circuits with different batteries
- Categorising switches for controlling your circuits
- Using series and parallel circuits and switches

This chapter explores the basic concepts of an electronic circuit. The circuits we cover are very simple, consisting of nothing other than batteries that supply voltage, wires that carry current and lamps that consume power (because they light up in the process!). We also throw in some switches that let you turn the circuit on and off.

Although simple, these circuits are a great introduction to the more complicated ones that we describe throughout this minibook, where we add additional elements such as resistors, capacitors, coils and diodes in Chapters 2, 3, 4 and 5, respectively, and transistors (in Chapter 6) and integrated circuits in Book III. Reading this minibook helps you to become familiar with all the basic components of electronic circuits.

What Is a Circuit?

A *circuit* is a complete course of conductors through which current can travel. Circuits provide a path for current to flow. To be a circuit, this path must start and end at the same point. In other words, a circuit has to form a loop.

For example, Figure 1-1 shows a simple circuit that includes two components: a battery and a *lamp* – which is the electrical engineer's term for what is more commonly known as a light bulb. The circuit allows current to flow from the battery to the lamp, through the lamp and then back to the battery. Thus, the circuit forms a complete loop.

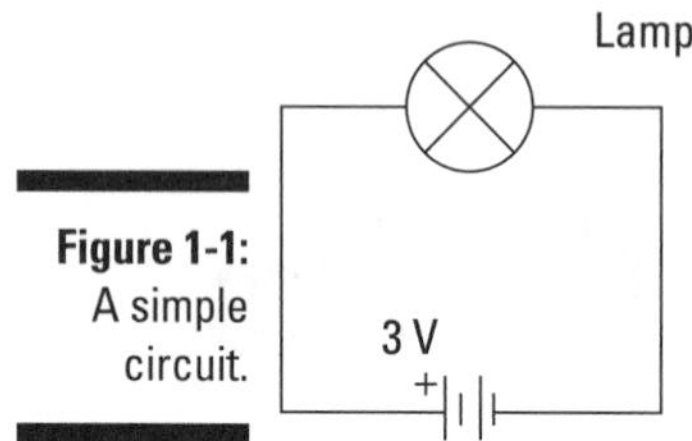

Figure 1-1: A simple circuit.

Of course, circuits can be more complex, but all circuits can be distilled down to three basic elements:

- **Voltage source:** A voltage source causes current to flow. In Figure 1-1, the voltage source is the battery.
- **Load:** The load consumes power; it represents the actual work done by the circuit. Without the load, you don't have use for a circuit.

  In Figure 1-1, the load is the lamp. In complex circuits, the load is a combination of components, such as resistors, capacitors, transistors and so on.
- **Conductive path:** The conductive path provides a route through which current flows. This route begins at the voltage source, travels through the load and then returns to the voltage source. This path must form a loop from the negative side of the voltage source to the positive side of the voltage source.

  In Figure 1-1, the two lines that travel between the battery and the lamp represent the conductive path. This path can be intricate in a complex circuit, but it must still form a loop from the negative side of the voltage source to the positive side.

Here are a few additional interesting points to keep in mind as you ponder the nature of basic circuits:

- **Closed circuit:** This term indicates that a circuit is complete and forms a loop that allows current to flow.
- **Open circuit:** This term indicates that some part of the circuit is disconnected or disrupted so that a loop isn't formed and current can't flow.

  Open circuit is an oxymoron. After all, the components must form a complete path to be considered a circuit: if the path is open, it isn't a circuit. Therefore, the term is usually used to describe a circuit that has become broken, perhaps on purpose (by the use of a switch, which we discuss later in this chapter in 'Working with Switches') or by some error, such as a loose connection or a damaged component.

- **Short circuit:** This term refers to a circuit that doesn't have a load. For example, Figure 1-2 shows a short circuit; the lamp is connected to the circuit but a direct connection is also present between the battery's negative terminal and its positive terminal.

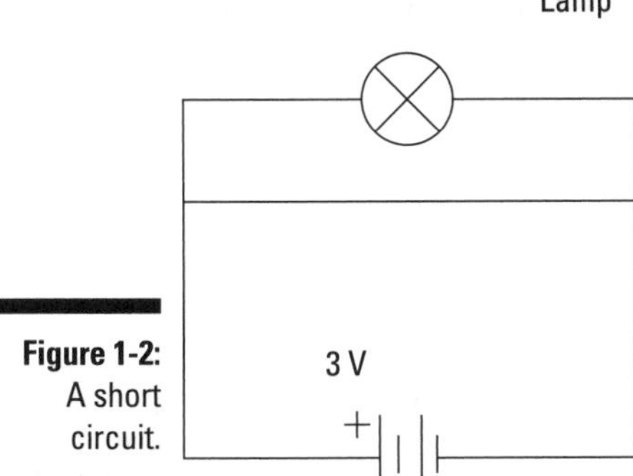

Figure 1-2: A short circuit.

Current in a short circuit can flow at dangerously high levels. Short circuits can damage electronic components, cause a battery to explode or maybe start a fire.

The short circuit shown in Figure 1-2 illustrates an important point about electrical circuits: they commonly have multiple pathways for current to flow. In Figure 1-2, the current can flow through the lamp as well as through the path that connects the two battery terminals directly.

Current flows everywhere it can. If your circuit has two pathways through which current can flow, the current doesn't choose one over the other; it chooses both. However, not all paths are equal, and so current doesn't flow equally through all paths. For example, in the circuit in Figure 1-2, current flows much more easily through the short circuit than through the lamp. Thus, the lamp doesn't glow because nearly all the current bypasses the lamp in favour of the easier route through the short circuit. Even so, a small amount of current does flow through the lamp.

To determine how much current flows through a given path, you use the mathematical formula that you can read about in Chapter 2 of this minibook. Nevertheless, when one of the available paths is a short circuit, you don't need the formula because nearly all the current will flow through the short circuit.

Imagine electric current flowing through a circuit from the positive side of the voltage source to the negative side, because this is usually how you visualise a circuit when you study it. For example, in Figure 1-1, think of the current as flowing in a clockwise direction, starting at the positive terminal of the battery, flowing through the path at the left side of the diagram to the lamp at the top of the diagram, through the lamp and then through the path at the right side of the diagram, finally returning to the negative terminal of the battery.

As you can see in Book I, Chapter 2, this way of thinking about current flow is called *conventional current.* In reality, the electric charge – and the electrons – in the circuit flow from the negative side of the voltage source through the circuit to the positive side of the voltage source.

Using Batteries

The easiest way to provide a voltage source for a circuit is to include a battery. You can use plenty of other ways to provide voltage, including AC adaptors (which you plug into the wall) and solar cells (which convert sunlight to voltage), but batteries remain the most practical source of juice for most of the circuits you build in this book.

A *battery* is a device that converts chemical energy into electrical energy in the form of voltage, which in turn can cause current to flow. A battery works by immersing two plates made of different metals into a special chemical solution called an *electrolyte.* The metals react with the electrolyte to produce a flow of charges that accumulate on the negative plate, called the *anode.* The positive plate, called the *cathode,* is sucked dry of charges. This creates a voltage between the two plates. These plates are connected to external terminals to which you can connect a circuit to cause current to flow.

Figure 1-3 shows a simplified diagram of how a battery works. A bowl filled with the right kind of chemical that includes an anode and cathode made of the right metal gives you a working battery.

Technically, Figure 1-3 shows a *cell,* not a battery. A *battery* is a combination of two or more cells.

Introducing the battery

Batteries come in many different shapes and sizes, but for the purposes of this book, you need concern yourself only with a few standard types of batteries that you can buy at any supermarket or hardware shop. Figure 1-4 shows the most common sizes available.

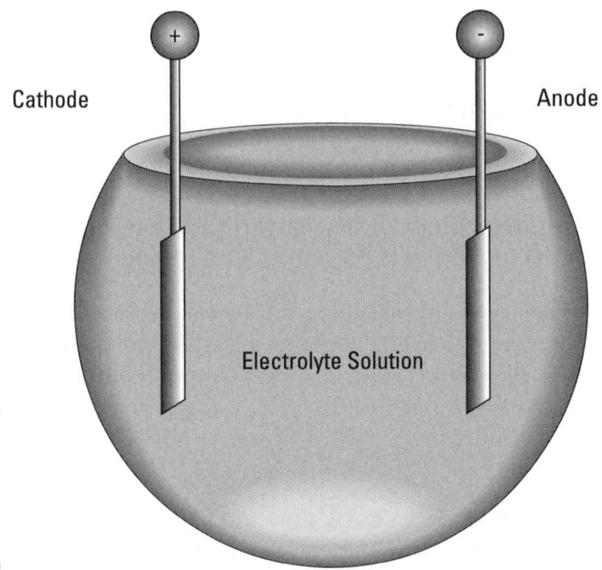

Figure 1-3: What goes on inside a battery.

Figure 1-4: Common batteries.

Cylindrical batteries come in four standard sizes: AAA, AA, C and D. Regardless of the size, these batteries provide 1.5 V each; the only difference between the smaller and larger sizes is that the larger batteries can provide more current. Technically, AAA, AA, C and D cells are just that – single cells, not batteries. But if you ask for a cell in your local hardware shop, people are likely to think you should be in one!

The cathode, or positive terminal, in a cylindrical battery is the end with the metal bump. The flat metal end is the anode, or negative terminal.

The rectangular battery in Figure 1-4 is a 9 V battery. It truly is a battery because that little rectangular box contains six small cells, each about half the size of an AAA cell. The 1.5 volts produced by each of these small cells combine to create a total of 9 volts.

Here are a few more battery facts:

- **Available sizes:** Besides AAA, AA, C, D and 9 V batteries, many other battery sizes are available. Most of these batteries are designed for special applications, such as digital cameras, hearing aids, laptop computers and so on.

- **Chemicals:** All batteries contain chemicals that are toxic to you and to the environment. Treat them with care and don't leave them unused for too long so that they end up leaking chemicals. When they're dead put them in a recycling bin.
- **Measuring voltage:** You can (and should) use your multimeter (a piece of equipment we describe in Book I, Chapter 8) to measure the voltage produced by your batteries. Set the multimeter to an appropriate DC voltage range (such as 20 V). Then, touch the red test lead to the positive terminal of the battery and the black test lead to the negative terminal. The multimeter tells you the voltage difference between the negative and positive terminals. For cylindrical batteries (AAA, AA, C or D) the reading should be about 1.5 V. For 9 V batteries, it should be about 9 V. It will be lower if they are exhausted through use or age.
- **Rechargeable batteries:** These cost more than non-rechargeable batteries but last longer, because you can recharge them when they go dead. However, they start life with a slightly lower voltage and this gets worse with the number of recharge cycles.

- The technical term for a cell that can't be recharged is a *primary cell.* A rechargeable cell is properly called a *secondary cell.*

The easiest way to use batteries in an electronic circuit is to use a battery holder, which is a little plastic widget designed to hold one or more batteries.

Sizing up the missing batteries

Have you ever wondered why you can buy AAA, AA, C and D cells, but not A or B? In fact, A and B cell batteries do exist, but they never really caught on in consumer devices, and so they aren't readily available in shops.

Building a lamp circuit

Project 1-1 shows you how to build a simple circuit that uses a battery to light a lamp. Although this circuit is quite simple, it helps illustrate the basic principles we're explaining. Figure 1-5 shows the circuit assembled.

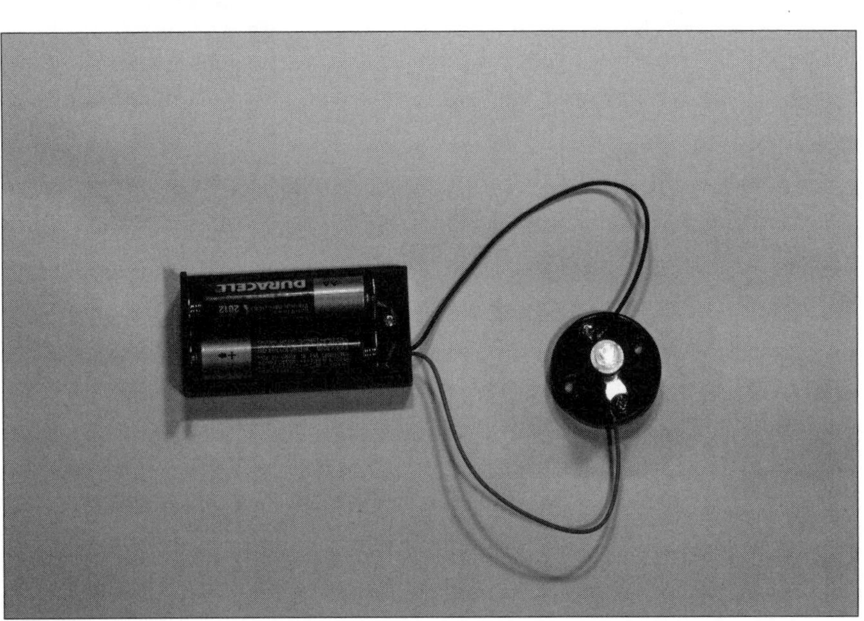

Figure 1-5: A simple lamp circuit.

Project 1-1: A Simple Lamp Circuit

In this project, you build a simple circuit that connects a lamp to a battery. You also use a multimeter to measure the voltage and current in the circuit. To assemble and test the circuit, you need a small Phillips-head screwdriver and a multimeter.

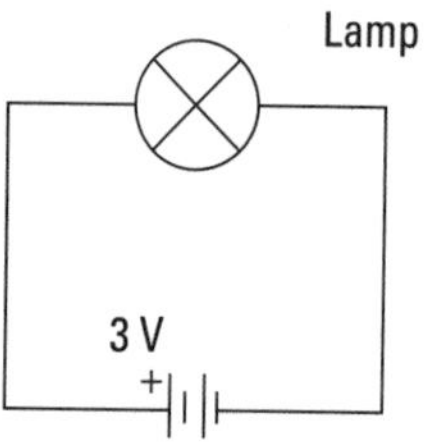

Parts List
2 AA batteries
1 Battery holder
1 Lamp holder
1 3V lamp

Steps

1. **Attach the red lead from the battery holder to one of the screw terminals on the lamp holder.**

   Loosen the screw terminal a bit to create a gap under the screw. Next, bend the stripped end of the red lead into a hook shape (you can do this easily by wrapping the wire around the tip of the screwdriver). Insert the stripped end of the lead beneath the terminal screw and then tighten the screw to secure the wire.
2. **Attach the black lead to the other terminal on the lamp holder.**
3. **Insert the lamp into the lamp holder.**
4. **Insert the batteries into the holder.**

   The lamp should light.

1. Set the multimeter range to the lowest DC Voltage setting that measures at least 3 volts.

   Our meter has a 5VDC range, so we use that setting.
2. Touch the red multimeter lead to the lamp holder terminal that the red battery holder lead is attached to and then touch the black lead to the other terminal.

   The meter should read about 3 volts.
3. Disconnect the red battery lead from the lamp holder.

   This breaks the circuit, and so the lamp goes out.
4. Turn the multimeter dial to your highest DC mA setting.

   On ours, the highest setting is 1,000 DC mA.
5. Touch the stripped end of the red battery lead to the tip of the red multimeter lead, then touch the tip of the black multimeter lead to the unconnected terminal on the lamp holder.

   The meter should read approximately 250 mA. If the largest DC mA range on your multimeter is less than 250 mA, you may not get an accurate reading. However, you should get an indication that the current exceeds the maximum for the range. If the meter's maximum current range is much less than 250 mA and the meter has fuse protection, making the measurement could blow the meter's fuse.

   Additionally, when you test the current, the lamp comes back on because the meter completes the circuit.

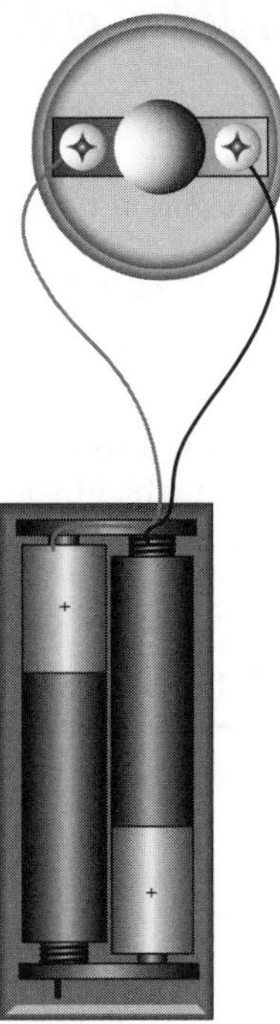

Working with Switches

Switches are an important part of most electronic circuits. In the simplest case, most circuits contain an on/off switch to turn the circuit (unsurprisingly!) on and off. Many circuits also contain additional switches that control how the circuit works or to activate different features of the circuit.

Switches are mechanical devices with two or more *leads* (or *terminals*) that are internally connected to metal contacts, which the person operating the switch can open or close. When the switch is in the On position, the contacts are brought together to complete the circuit so that current can flow.

When the contacts are together, the switch is closed. When the contacts are apart, the switch is open and current can't flow.

The following sections describe two ways to categorise switches: by the method used to operate the switch and by the connections made by the switch.

Moving switches in different ways

You can categorise switches by the movement a person uses to open or close the contacts. Figure 1-6 shows many different switch designs. Here are the most common types:

- **Knife switch:** The kind of switch Igor throws in a Frankenstein film to reanimate the creature. In a knife switch, the contacts are exposed for everyone to see.
- **Pushbutton switch:** Features a knob that you push to open or close the contacts. In some pushbutton switches, you push the switch once to open the contacts and then push again to close the contacts. In other words, each time you push the switch, the contacts alternate between opened and closed.

  Other pushbutton switches are *momentary contact switches,* where contacts change from their default state only when the button is pressed and held down. The two types of momentary contact switches are:

  - *Normally open (NO):* The default state of the contacts is open. When you push the button, the contacts are closed. When you release the button, the contacts open again. Thus, current flows only when you press and hold the button.

- *Normally closed (NC):* The default state of the contacts is closed. Thus, current flows until you press the button. When you press the button, the contacts are opened and current doesn't flow. When you release the button, the contacts close again and current resumes.

- **Rocker switch:** Has a seesaw action. You press one side of the switch down to close the contacts and press the other side down to open the contacts.
- **Rotary switch:** Contains a knob that you turn to open and close the contacts. The switch in the base of many tabletop lamps is an example of a rotary switch.
- **Slide switch:** Features a knob that you can slide back and forth to open or close the contacts.
- **Toggle switch:** Contains a lever that you flip up or down to open or close the contacts. Common household light switches are examples of toggle switches.

Figure 1-6: You can find a switch for every need.

Making connections with poles and throws

You can classify switches by the connections they make. If you were under the impression that switches simply turn circuits on and off, guess again. Two important factors determine what types of connections a switch makes:

- **Poles:** A switch *pole* refers to the number of separate circuits that the switch controls:
  - A *single-pole* switch controls just one circuit.
  - A *double-pole* switch controls two separate circuits.

  A double-pole switch is like two separate single-pole switches that are mechanically operated by the same lever, knob or button.
- **Throw:** The number of *throws* indicates how many different output connections each switch pole can connect its input to. Figure 1-7 shows the two most common types:
  - A *single-throw* switch is a simple on/off switch that connects or disconnects two terminals. When the switch is closed, the two terminals are connected and current flows between them. When the switch is opened, the terminals are not connected and so current doesn't flow.
  - A *double-throw* switch connects an input terminal to one of two output terminals. Thus, a double-pole switch has three terminals. One of the terminals is called the *common terminal.* The other two terminals are often referred to as *A* and *B.* When the switch is in one position, the common terminal is connected to the A terminal, and so current flows from the common terminal to the A terminal but no current flows to the B terminal. When the switch is moved to its other position, the terminal connections are reversed: current flows from the common terminal to the B terminal, but no current flows though the A terminal.

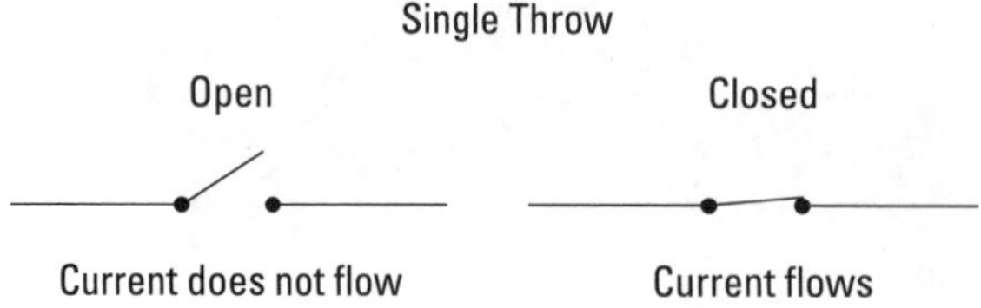

Double Throw
Common
A
B
Current flows through terminal A
Common
A
B
Current flows through terminal B

Figure 1-7: Single and double-throw switches.

Switches vary in the number of poles and the number of throws. In theory, any number of poles and any number of throws is possible. Most switches, however, have one or two poles and one or two throws. The result is four common combinations, as described in the following list. We show the symbols used in schematic diagrams for each of these switches in the margin:

- **SPST (single pole, single throw):** A basic on/off switch that turns a single circuit on or off. An SPST switch has two terminals: one for the input and one for the output.
- **SPDT (single pole, double throw):** An SPDT switch routes one input circuit to one of two output circuits. This type of switch is sometimes called an A/B switch, because it lets you choose between two circuits, called A and B. An SPDT switch has three terminals: one for the input and two for the A and B outputs.
- **DPST (double pole, single throw):** A DPST switch turns two circuits on or off. A DPST switch has four terminals: two inputs and two outputs.
- **DPDT (double pole, double throw):** A DPDT switch routes two separate circuits, connecting each of two inputs to one of two outputs. A DPDT switch has six terminals: two for the inputs, two for the A outputs and two for the B outputs.

Check out these other facts about poles and throws:

- Switches with more than two poles or more than two throws aren't commonplace, but they do exist. Rotary switches lend themselves especially well to having many throws. For example, the rotary switch in a multimeter typically has 16 or more throws, one for each range of measurement the meter can make.
- A common variation of a double throw switch is to have a middle position that doesn't connect to either output. Often called *centre open,* this type of switch has three positions, but only two throws. For example, an SPDT centre open switch can switch one input between either of two outputs, but in its centre position, neither output is connected.

- When you're stocking up on switches to have on hand, buy DPDT switches instead of single pole or single throw switches, because a DPDT can be used when a circuit calls for a simpler SPST, SPDT or DPST switch. You can use a DPDT switch when a simpler type is specified, because no law says that you have to wire all the contacts on the switch. For example, to use a DPDT switch as an SPST switch, you just use one of the poles and one of the throws and leave the other connections unused.

Building a switched lamp circuit

Project 1-2 presents a simple construction project that lets you explore the use of a simple on/off switch to control a lamp. Figure 1-8 shows the assembled project.

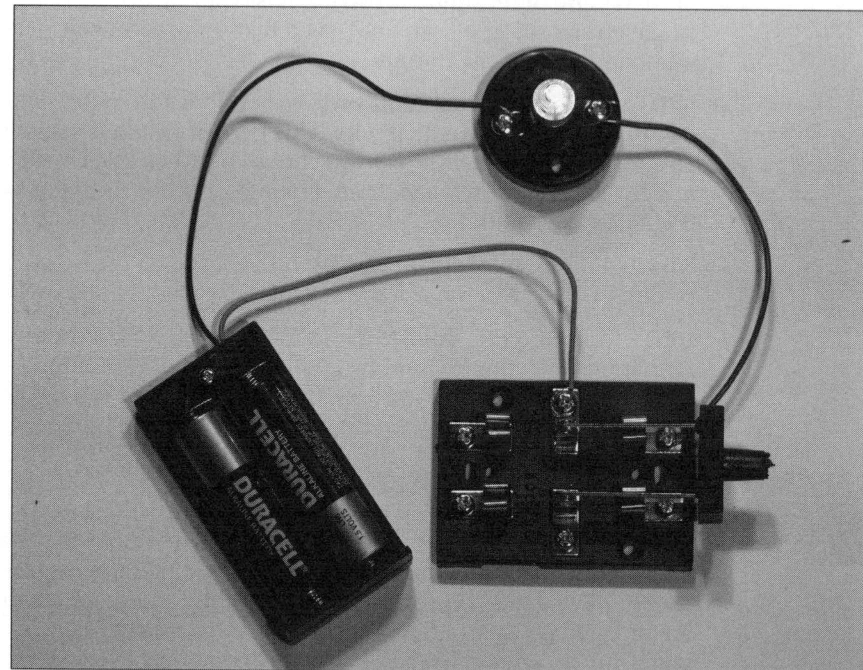

Figure 1-8: The switched lamp project.

This project and the remaining projects in this chapter use a RadioShack DPDT knife switch, pictured in Figure 1-9, which you can order online from `www.t2retail.co.uk`. You can use a DPDT toggle or rocker switch instead, which you can buy at your high-street electronics shop. In fact you're unlikely to use a knife switch in an actual electronic circuit, but one like this is useful when practising, because it's entirely exposed and you can see how it works. Plus, they come on their own base and have screw terminals, which makes connecting them in temporary circuits simple because you don't have to do any soldering.

As you can see in Figure 1-9, the knife switch is a double pole, double throw (DPDT) switch, which means it operates like two SPDT switches that are mechanically linked. We number the six terminals on the switch 1X, 1A, 1B, 2X, 2A and 2B:

- **1 and 2:** Designate which of the two circuits is being switched.
- **X terminals:** The input terminals in the centre of the switch.
- **A and B terminals:** For the two possible outputs.

Thus, when the switch is flipped one way, 1X is connected to 1A and 2X is connected to 2A. When the switch is flipped the other way, 1X is connected to 1B and 2X is connected to 2B.

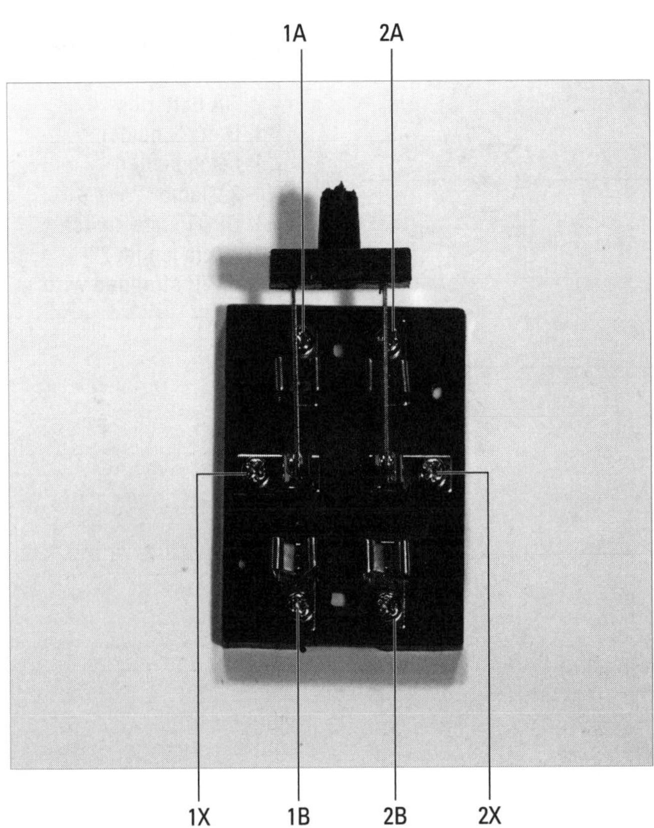

Figure 1-9: The DPDT knife switch.

Project 1-2: A Lamp Controlled by a Switch

In this project, you build a simple circuit that connects a lamp to a battery and uses a switch to turn the lamp on and off. To assemble and test the circuit, you need a small Phillips-head screwdriver, wire cutters and a wire stripper.

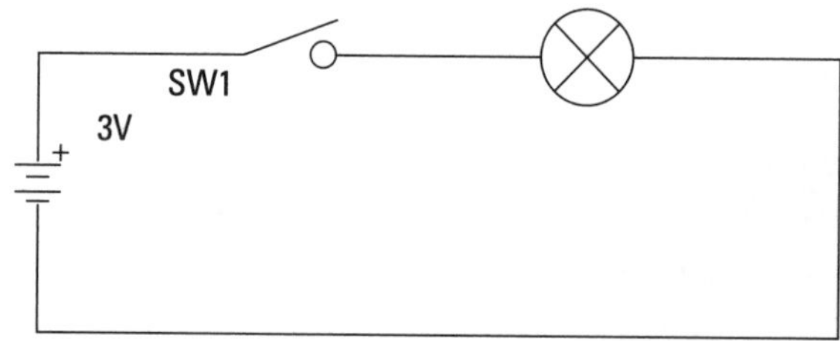

Parts List	
2	AA batteries
1	Battery holder
1	Lamp holder
1	3 V lamp
1	DPDT knife switch
1	15 cm length 22-AWG stranded wire

Steps

1. Cut a 15 cm length of wire and strip 2–3 cm of insulation from each end.
2. Cut a 15 cm length of wire and strip 2–3 cm of insulation from each end.
3. Attach the red lead from the battery holder to terminal 1X on the knife switch.
4. Attach the black lead to one of the terminals on the lamp holder.
5. Use the 15 cm wire to connect terminal 1A of the knife switch to the other terminal of the lamp holder.
6. Insert the batteries into the holder.
7. Close the knife switch on the `A´ side.

   The lamp should light up.

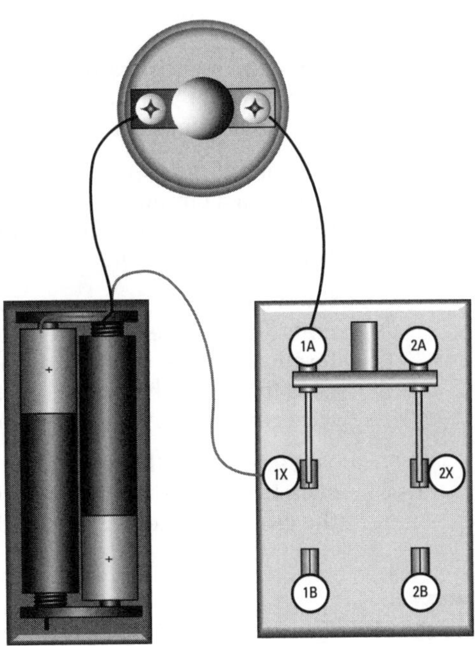

If you want to experiment with different variations of this project, try the following:

- **Move the switch from the positive side of the circuit to the negative side.** In other words, connect the red lead from the battery holder to the lamp and connect the black lead to the 1X terminal on the switch.

  The circuit functions the same, which shows that the location of a switch in a circuit often doesn't matter. If the circuit is broken anywhere, current can't flow. Thus, whether the switch is before or after the lamp doesn't matter.

- **Cut a second 15 cm piece of wire and strip the insulation from both ends.** Then, wire the circuit so that the red battery lead goes to switch terminal 1X, the black lead goes to switch terminal 2X, one of the wires goes from switch terminal 1A to one of the lamp terminals, and the other wire goes from switch terminal 2A to one of the lamp terminals.

  You've created the circuit shown in Figure 1-10. In this circuit, the knife switch is used as a DPST (double pole, single throw) switch to interrupt the circuit on the negative and the positive side of the lamp.

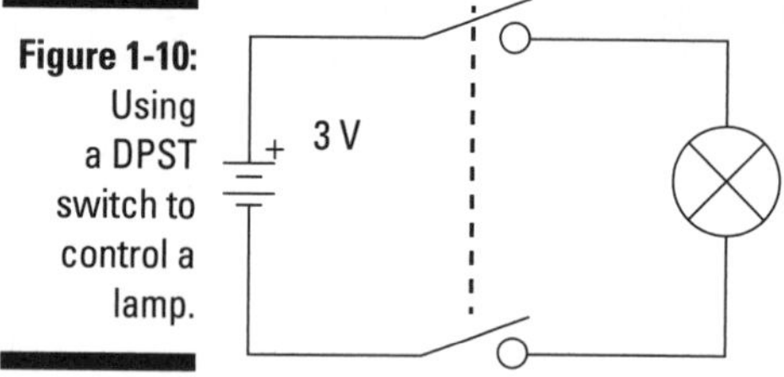

Figure 1-10: Using a DPST switch to control a lamp.

Understanding Series and Parallel Circuits

Whenever you have circuits that consist of more than one component, those components have to be linked together. The two ways to connect components in a circuit are in series and in parallel. Figure 1-11 illustrates how you can use series and parallel circuits to connect two lamps in a single circuit.

In a *series* connection, components are connected end to end, so that current flows first through one and then through the other. As you can see in the top circuit in Figure 1-11, the current goes through one lamp and then the other. The lamps are strung together end to end.

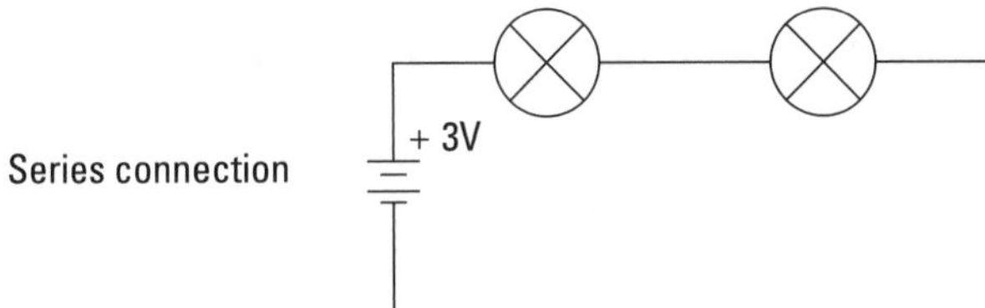

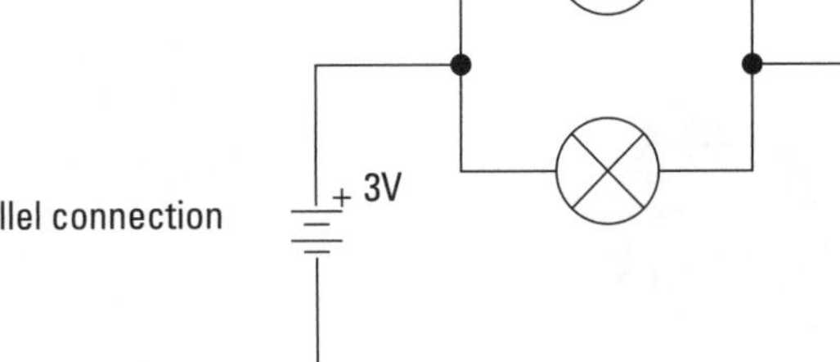

Figure 1-11: Lamps connected in series and in parallel.

A drawback of series connections is that if one component fails in a way that results in an open circuit, the entire circuit is broken and none of the components work. For example, if one of the lamps in the series circuit in Figure 1-11 burns out, neither lamp works because current must flow through both lamps for the circuit to be complete.

In the *parallel* connection shown in Figure 1-11, each lamp has its own direct connection to the battery. This arrangement avoids the if-one-fails-they-all-fail nature of series connections. In a parallel connection, the components don't depend on each other for their connection to the battery. Thus, if one lamp burns out, the other continues to burn.

An interesting thing happens with voltage when components are connected in series: the voltages present at each component are divided up. For example, in a circuit with a 3 V battery and two identical lamps connected in series, each lamp sees only 1.5 V. If you connect three identical lamps in series, each lamp sees only 1 V.

You can measure the voltage seen by any component in a circuit by setting your multimeter to an appropriate voltage range and then touching the leads to both sides of the component. The voltage you measure is called the component's *voltage drop*.

Building a series lamp circuit

In Project 1-3, you build a simple circuit that connects two lamps in series. You then use your multimeter to measure the voltages at various points in the circuit. The completed project is shown in Figure 1-12.

Figure 1-12: Lamps connected in series.

Project 1-3: A Series Lamp Circuit

In this project, you connect two lamps together in a series circuit. The lamps are powered by a pair of AA batteries. To build this project, you need a small Phillips-head screwdriver, wire cutters, wire strippers and a multimeter.

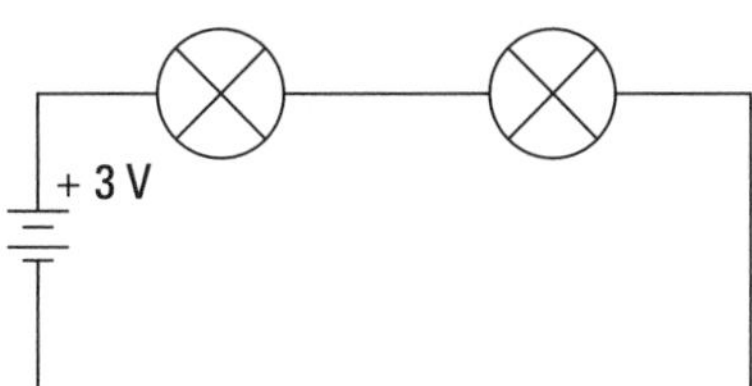

Parts List
2 AA batteries
1 Battery holder
2 Lamp holders
2 3 V lamps
1 15 cm length 22-AWG stranded wire

Steps

1. **Cut a 15 cm length of wire and strip 2–3 cm of insulation from each end.**
2. **Attach the red lead from the battery holder to one of the terminals on one of the lamp holders.**
3. **Attach the black lead to one of the terminals on the other lamp holder.**
4. **Use the 15 cm wire to connect the unused terminal of the first lamp holder to the unused terminal of the second lamp holder.**
5. **Insert the batteries into the holder.**

   Both lamps light.

   Notice that the lamps are dim. That's because in a series circuit made with two identical lamps, each of the two lamps sees only half the total voltage.
6. **Remove one of the lamps from its holder.**

   The other lamp goes out. This is because in a series circuit, a failure in any one component breaks the circuit so none of the other components work.
7. **Replace the lamp you removed in Step 6.**
8. **Set your multimeter to a DC voltage range that can read at least 3 volts.**
9. **Touch the leads to the two terminals on the first lamp holder.**

   The multimeter should read approximately 1.5 V. (If you're using an analogue meter and the needle moves backwards, just reverse the leads.)
10. **Touch the leads to the two terminals on the other lamp holder.**

    Again, the multimeter should read approximately 1.5 V.
11. **Touch the red lead of the meter to the terminal that the red lead from the battery is connected to and touch the black meter lead to the terminal that the black battery lead is connected to.**

    This measures the voltage across both lamps combined. The meter indicates 3V.

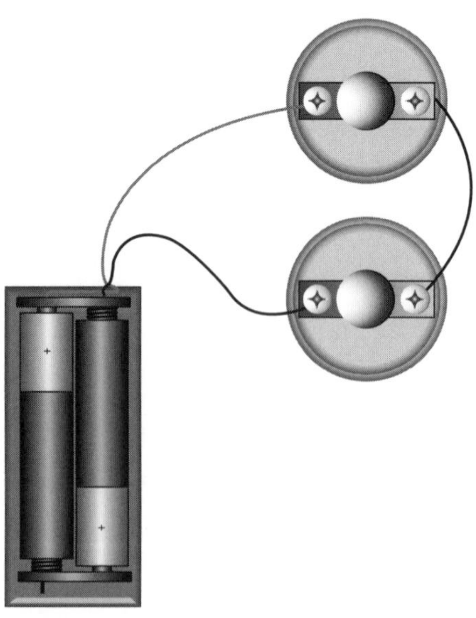

Building a parallel lamp circuit

In Project 1-4, you build a circuit that connects two lamps in parallel and use your multimeter to measure voltages within various points in the circuit. The completed project is shown in Figure 1-13.

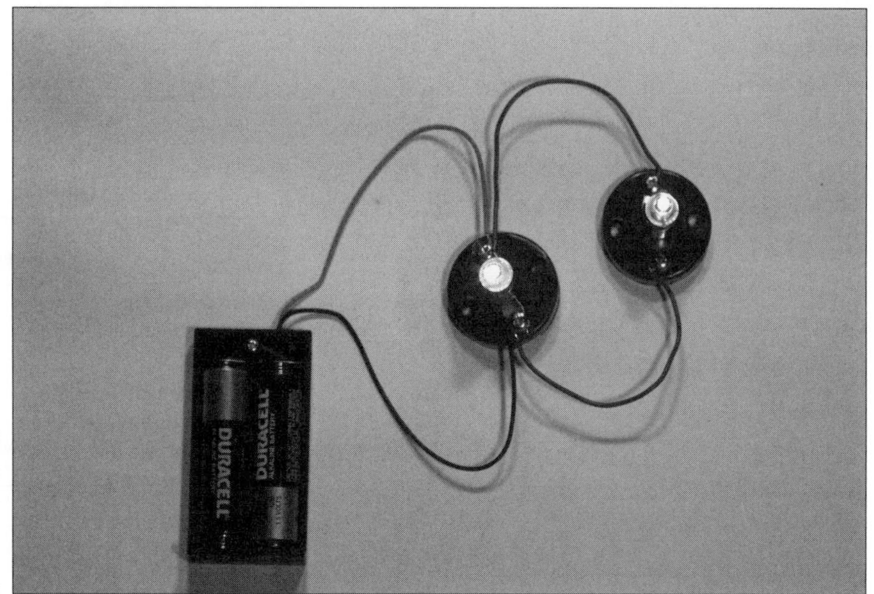

Figure 1-13: Lamps connected in parallel.

Project 1-4: A Parallel Lamp Circuit

In this project, you connect two lamps together in a series circuit. The lamps are powered by a pair of AA batteries. To build this project, you need a small Phillips-head screwdriver, wire cutters, wire strippers and a multimeter.

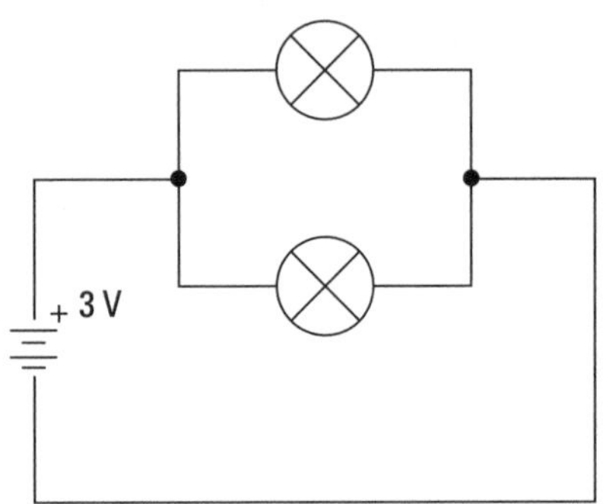

Parts List
2 AA batteries
1 Battery holder
2 Lamp holders
2 3 V flashlight lamps
2 15 cm length 22-AWG stranded wires

Steps

1. **Cut two 15 cm lengths of wire and strip 2–3 cm of insulation from each end.**
2. **Attach the red lead from the battery holder to one of the terminals on the first lamp holder.**
3. **Attach the black lead to the other terminals on the first lamp holder.**
4. **Use the two wires to connect each of the terminals on the first lamp holder to the terminals on the second lamp holder.**

   This wiring connects the two lamp holders in parallel.
5. **Insert the batteries.**

   The lamps light, brighter than when you connect them in series in Project 1-3.
6. **Remove one of the lamps.**

   Notice that the other lamp remains lit.
7. **Replace the lamp you removed in Step 6.**
8. **Set your multimeter to a DC voltage range that can read at least 3 volts.**
9. **Touch the leads of your multimeter to the two terminals on the first lamp holder.**

   Make sure that you touch the red meter lead to the terminal that the red battery lead is connect to and the black meter lead to the terminal that the black battery lead is attached to.

   Note that the voltage reads a full 3 V.
10. **Touch the meter leads to the terminals on the second lamp stand.**

    Note that the voltage again reads 3 V. When components are connected in parallel, the voltage isn't divided among them. Instead, each component sees the same voltage. That's why the lamps light at full intensity in the parallel circuit.

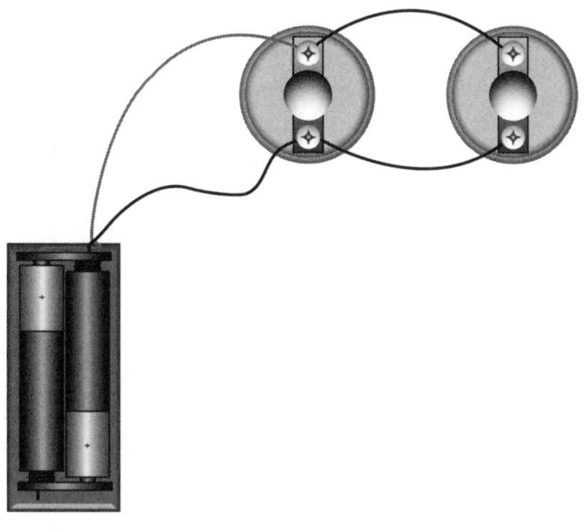

Using Switches in Series and Parallel

Just as with lamps (see the preceding section), you can also connect switches in series or parallel. For example, Figure 1-14 shows two circuits that each use a pair of single pole, single throw (SPST) switches to turn a lamp on or off (check out the earlier section 'Making connections with poles and throws' for the different switch types). In the top circuit, the switches are wired in series. In the lower one, the switches are wired in parallel.

The interesting thing to note about wiring switches in series is that both switches must be closed in order to complete the circuit. A great example of switches wired in series is in the typical nuclear-war film, where two people have to flip a switch in order to launch the missiles. Switches wired in series means that Denzel Washington *and* Gene Hackman must agree to launch the missiles (a sensible precaution when one is as mad as a hatter!).

When switches are wired in parallel, closing either switch completes the circuit. Thus, parallel switches are often used when you want the convenience of controlling a circuit from two different locations. If the nuclear-missile switches were wired in parallel, either Denzel Washington *or* Gene Hackman on his own can fire the missiles.

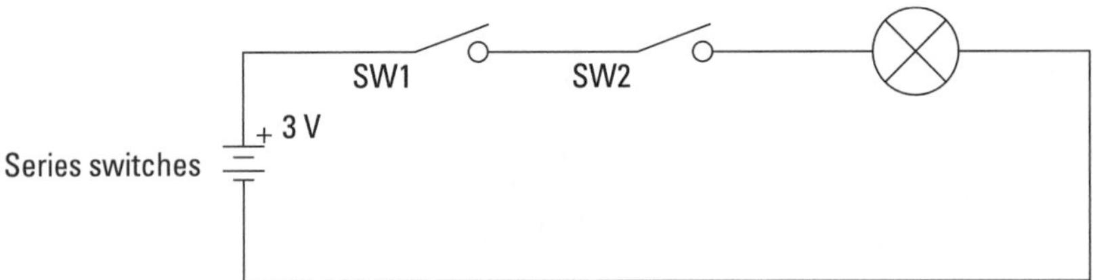

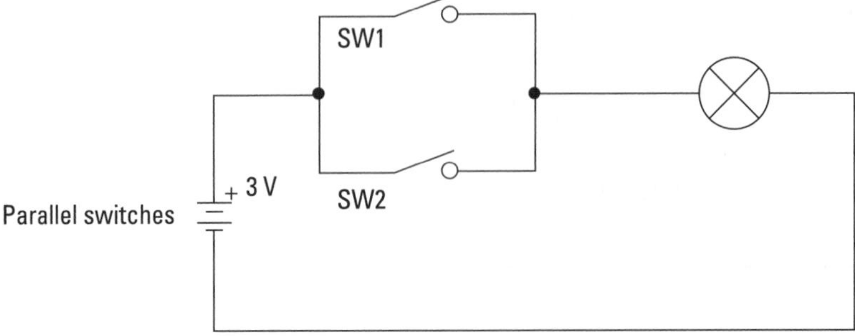

Figure 1-14: Schematic diagrams for series and parallel switch circuits.

Building a series switch circuit

Project 1-5 presents a simple project that uses two switches to open or close a circuit that lights a lamp. The switches are wired in series, and so both switches have to be closed to light the lamp. Figure 1-15 shows the completed project.

Figure 1-15: The assembled series switch circuit.

Project 1-5: A Series Switch Circuit

In this project, you build a simple circuit that uses two knife switches to control a single lamp. To complete this project, you need a small Phillips-head screwdriver, wire cutters and wire strippers.

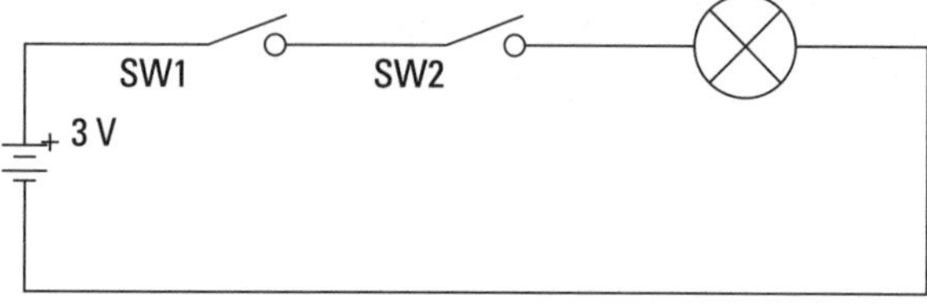

Parts List

- 2 AA batteries
- 1 Battery holder
- 1 Lamp holder
- 1 3 V flashlight lamp
- 2 DPDT knife switches
- 2 15 cm length 22-AWG stranded wire

Steps

1. **Cut two 15 cm lengths of wire and strip 2–3 cm of insulation from each end.**
2. **Open both switches.**

   Move the handles to the upright position so the contacts aren't connected.
3. **Attach the red lead from the battery holder to terminal 1X of one of the switches.**
4. **Attach the black lead to one of the terminals on the lamp holder.**
5. **Connect one of the 15 cm wires from terminal 1A of the first switch to terminal 1X of the second switch.**
6. **Connect the other 2–3 cm wire from terminal 1A of the second switch to the unused terminal of the lamp holder.**
7. **Insert the batteries into the holder.**
8. **Close the first switch.**

   Notice the lamp doesn't light.
9. **Close the second switch.**

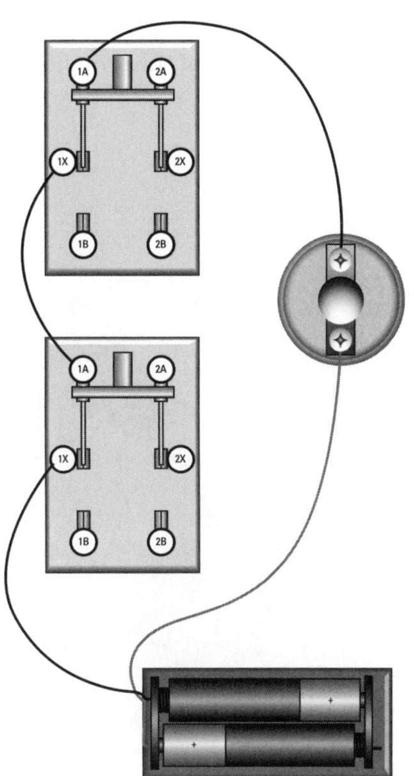

Building a parallel switch circuit

In Project 1-6, you build a simple circuit that uses two switches wired in parallel to control a lamp. Because the switches are wired in parallel, the lamp lights if either of the switches is closed. Figure 1-16 shows the completed project.

Figure 1-16: The assembled parallel switch circuit.

Project 1-6: A Parallel Switch Circuit

This project is a circuit that uses two switches wired in parallel to control a lamp. To complete this project, you need a small Phillips-head screwdriver, wire cutters and wire strippers.

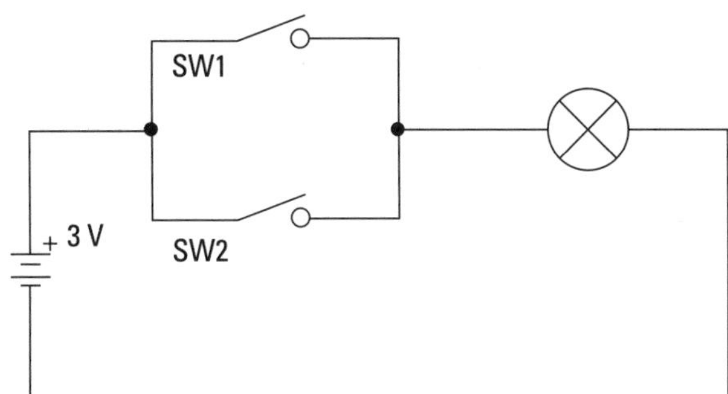

Parts List

- 2 AA batteries
- 1 Battery holder
- 1 Lamp holder
- 1 3 V lamp
- 2 DPDT knife switches
- 3 15 cm length 22-AWG stranded wire

Steps

1. **Cut three 15 cm lengths of wire and strip 2–3 cm of insulation from each end.**
2. **Open both switches.**

   Move the handles to the upright position so the contacts aren't connected.
3. **Attach the red lead from the battery holder to terminal 1X of the first switch.**
4. **Attach the black lead to the first terminal on the lamp holder.**
5. **Use the first 15 cm wire to connect terminal 1X of the first switch to terminal 1X of the second switch.**

   Be sure to leave the red battery wire in place at terminal 1X on the first switch. Terminal 1X on the first switch should have two wires: the red battery terminal and the wire going to terminal 1X of the second switch.
6. **Use the second 15 cm wire to connect terminal 1A of the first switch to the second terminal of the lamp holder.**
7. **Use the third 15 cm wire to connect terminal 1A of the second switch to the second terminal of the lamp holder.**

   In other words, the unused terminal of the lamp holder should be connected to terminal 1A on both knife switches.
8. **Insert the batteries.**
9. **Close one of the switches.**

   The lamp lights.
10. **Open the switch you closed in Step 9 and then close the other switch.**

    Again, the lamp lights. When switches are connected in parallel, current flows through the circuit when either of the switches is closed.
11. **Close both switches.**

    The lamp remains lit because current continues to flow when both switches are closed.
12. **Open both switches.**

    The lamp goes out. With switches wired in parallel, at least one of the switches must be closed for the current to flow.

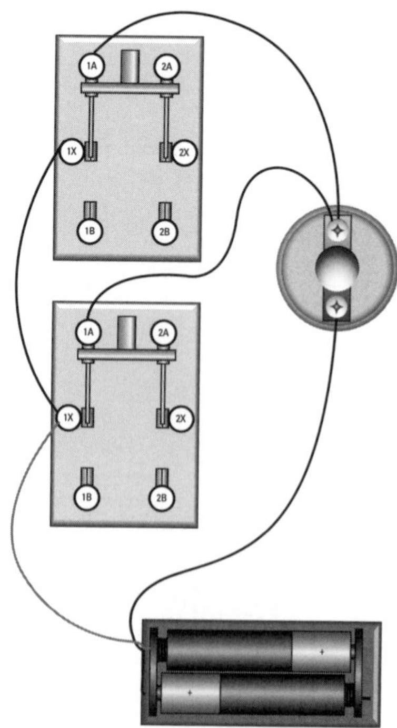

Switching between two lamps

In Project 1-7, you build a simple circuit that uses a single pole, double throw (SPDT) to switch a circuit between one of two lamps. In other words, one of two lamps lights depending on the position of the switch. This type of switching is a common requirement in electronic circuits. Figure 1-17 shows the completed circuit.

Figure 1-17: The switch controls two lamps.

Project 1-7: Controlling Two Lamps with One Switch

In this project, you build a circuit that uses a single switch to control two lamps. When the switch is in the first position, the first lamp lights and the second lamp is dark. When the switch is in the second position, the second lamp lights and the first lamp is dark. You need a small Phillips screwdriver, wire cutters and wire strippers to build this project.

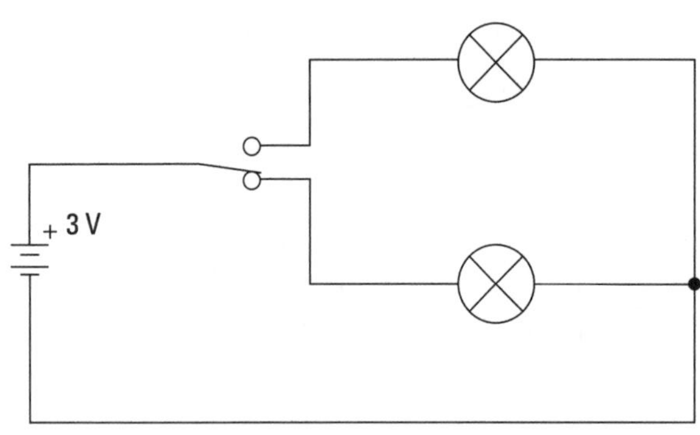

Parts List
2 AA batteries
1 Battery holder
1 Lamp holder
1 3 V flashlight lamp
2 DPDT knife switches
3 15 cm length 22-AWG stranded wire

Steps

1. **Cut three 15 cm lengths of wire and strip 2–3 cm of insulation from each end.**
2. **Open the switch.**

   Lift the handle into the upright position so the contacts aren't connected.
3. **Attach the red lead from the battery holder to terminal 1X of the switch.**
4. **Attach the black lead to one of the terminals on the first lamp holder.**
5. **Connect one of the 15 cm wires from terminal 1A of the switch to one terminal of the second lamp holder.**
6. **Connect another 15 cm wire from terminal 1B of the switch to the unused terminal of the first lamp holder.**
7. **Use the third 15 cm wire to connect the unused terminal of the second lamp holder to the terminal of the first lamp holder that the black battery lead is connected to.**
8. **Insert the lamps into the lamp holders.**
9. **Insert the batteries into the holder.**
10. **Flip the switch to the A position.**

    The second lamp lights.
11. **Change the switch to the B position.**

    The first lamp lights.

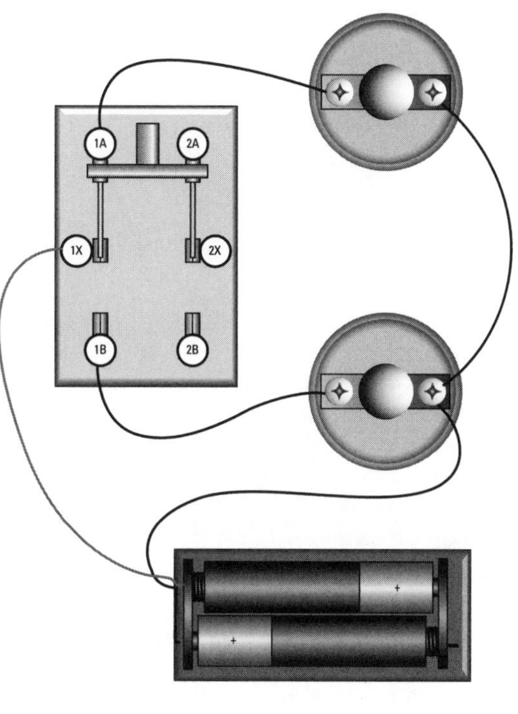

An interesting variant of the circuit in Project 1-7 uses both poles of the DPDT knife switch (from Project 1-2 in the earlier section 'Building a switched lamp circuit') to switch the circuit on the negative and positive sides of the lamp. For this circuit, you need four 15 cm wires. Connect the six terminals of the DPDT switch as follows:

Terminal	*Connect To*
1X	Red battery lead
1A	Terminal 1 of second lamp
1B	Terminal 1 of first lamp
2X	Black battery lead
2A	Terminal 2 of second lamp
2B	Terminal 2 of first lamp

Figure 1-18 shows this circuit assembled.

Figure 1-18: Another way to control two lamps.

Building a three-way lamp switch

Many homes and offices have hallways with a light switch at both ends. You can turn the light on or off by flipping either switch. This kind of switching arrangement is called a *three-way switch*.

If you think about it, how these three-way switches work is a bit puzzling. If the light is on, flipping either switch turns it off. If the light is off, flipping either switch turns it on. Imagine that you flip one switch to its On position and the light goes on, and then you go to the other switch, flip it to turn the light off and come back to the first switch. It's still in its On position, but the light is off. To turn the light back on, you can flip the first switch again.

In other words, sometimes the light is on when the switch is up, sometimes it's on when the switch is down. How can this be?

The answer is that both switches are single pole, double throw (SPDT) switches (flip to the earlier section 'Making connections with poles and throws' for a definition), and they're wired in series so that both switches must be up or both must be down to complete the circuit. If one switch is up and the other is down, the circuit is open.

Sometimes electricians install one of the switches upside down or wire the three-way switch backwards just to confuse you. Then, the switches work backwards: if both switches are up or if both switches are down, the circuit is open and the lamp lights only when one switch is up and the other is down. But that's not the normal way to wire a three-way switch.

In Project 1-8, you build a simple circuit that uses two SPDT switches to show how a three-way light switch works. Figure 1-19 shows the completed project.

Figure 1-19: The completed three-way light switch circuit.

Project 1-8: A Three-Way Light Switch

In this project, you create a three-way switch circuit in which a single lamp is controlled by either one of two switches. You need a small Phillips-head screwdriver, wire cutters and wire strippers to complete this project.

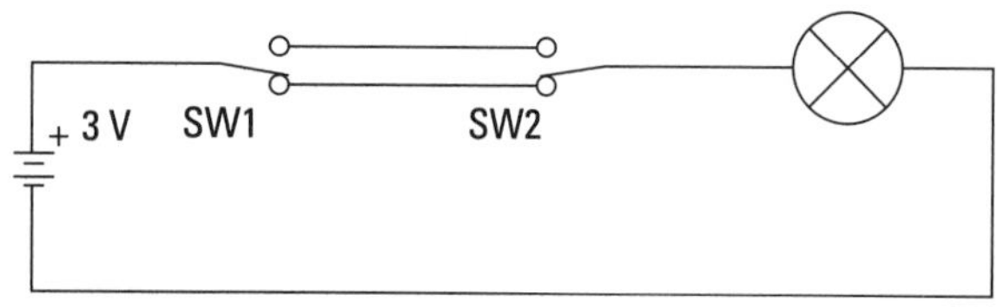

Parts List

- 2 AA batteries
- 1 Battery holder
- 1 Lamp holder
- 1 3 V lamp
- 2 DPDT knife switches
- 3 15 cm length 22-AWG stranded wires

Steps

1. **Cut three 15 cm lengths of wire and strip 2–3 cm of insulation from each end.**
2. **Open both switches.**

   Move the handles to their upright positions so the contacts aren't connected.
3. **Attach the red lead from the battery holder to terminal 1X of the first switch.**
4. **Attach the black lead to one of the terminals on the lamp holder.**
5. **Connect one of the 15 cm wires from terminal 1A of the first switch to terminal 1A of the second switch.**
6. **Connect another 15 cm wire from terminal 1B of the first switch to terminal 1B of the second switch.**
7. **Connect the last 15 cm wire from terminal 1X of the second switch to the unused terminal of the lamp socket.**
8. **Insert the lamps into the lamp holder.**
9. **Insert the batteries into the holder.**
10. **Flip the switches to see the operation of the three-way switch.**

    The lamp lights only when both switches are in the A position or when both switches are in the B position.

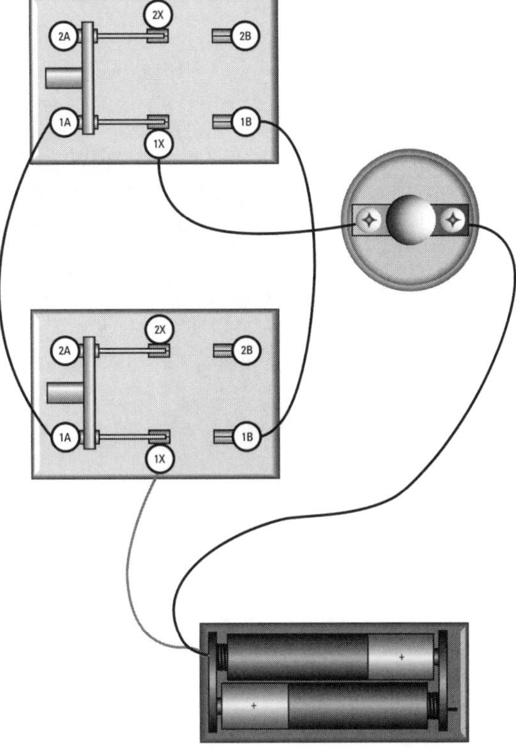

Reversing Polarity

Project 1-9 shows you a common trick: using a double pole, double throw (DPDT) switch (which we explain in the earlier section 'Making connections with poles and throws') to reverse the polarity of a circuit.

One common use for this technique is powering a DC motor with the circuit. When you reverse the polarity of a DC motor, the motor spins in the opposite direction. Therefore, you can use a DPDT switch to control the direction in which a DC motor turns. Figure 1-20 shows the assembled project.

Figure 1-20: The assembled polarity-reversing circuit.

Project 1-9: A Polarity-Reversing Circuit

In this project, you build a circuit that uses a DPDT switch to reverse a circuit's polarity. In other words, flipping the switch from one position to the other reverses the direction in which current flows through the circuit. To build this circuit, you need a small Phillips-head screwdriver, wire cutters and wire strippers.

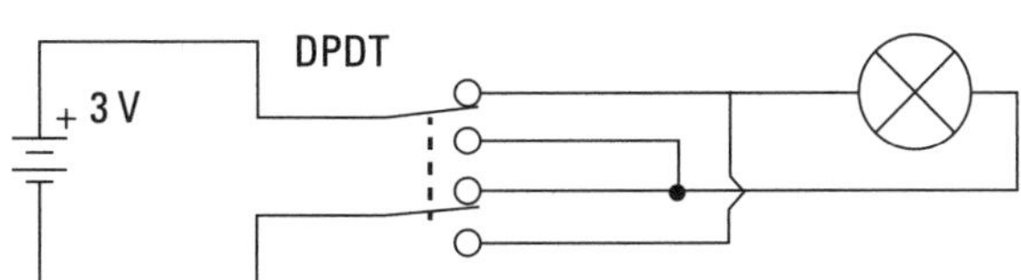

Parts List
2 AA batteries
1 Battery holder
1 Lamp holder
1 3 V lamp
1 DPDT knife switch
4 15 cm length 22-AWG stranded wires

Steps

1. **Cut four 15 cm lengths of wire and strip 2–3 cm of insulation from each end.**
2. **Open the switch.**

   Move the handles to their upright positions so the contacts aren't connected.
3. **Attach the red lead from the battery holder to terminal 1X of the switch.**
4. **Attach the black lead from the battery to terminal 2X of the switch.**
5. **Connect the first 15 cm wire from terminal 1B of the switch to terminal 2A of the switch.**
6. **Connect the second 15 cm wire from terminal 2B of the switch to terminal 1A of the switch.**
7. **Connect the third 15 cm wire from terminal 1A of the switch to one terminal of the lamp socket.**
8. **Connect the fourth 15 cm wire from terminal 2A of the switch to the other terminal of the lamp socket.**
9. **Insert the lamps into the lamp holder.**
10. **Insert the batteries into the battery holder.**
11. **Flip the switch to the A position and measure the voltage at the lamp.**

    Touch the red meter probe to the lamp terminal that's connected to switch terminal 1A and touch the black meter probe to the other lamp terminal. You should read approximately +3 V.
12. **Flip the switch to the B position and measure the voltage again.**

    You should read approximately –3 V. (If you're using an analogue meter, you have to reverse the probes to read the negative voltage.)

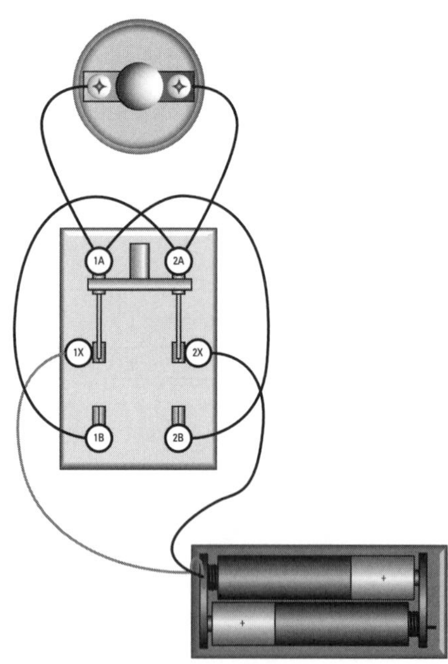

Chapter 2

Working with Resistors

In This Chapter

- Understanding and measuring resistance
- Calculating resistance with Ohm's law
- Determining resistor values and tolerance
- Working with resistors in series, parallel and combination
- Creating a voltage-divider circuit
- Looking at how a potentiometer varies resistance

Sci-fi villains the Borg from *Star Trek: The Next Generation* had a saying: 'Resistance is futile'. Captain Picard and the rest of the crew of the *Enterprise* eventually proved the Borg wrong of course.

Resistance isn't useful only in battling hostile aliens – it's also vital in electronics, which is all about manipulating the flow of current. One of the most basic ways to do so is to reduce it through resistance. Without resistance, current flows unregulated with no way to coax it into doing useful work.

In this chapter, you find out what resistance is and how to work with *resistors,* which are little devices that let you introduce resistance intentionally into your circuits. Along the way, you discover a fundamental relationship in the nature of electricity between voltage, current and resistance. This relationship is expressed in a simple mathematical formula called *Ohm's law.* (Don't worry, the maths isn't complicated; if you know how to multiply and divide, you can understand Ohm's law.) We also show you the most common ways in which resistors are used in circuits.

What Is Resistance?

As we discuss in Book I, a *conductor* is a material that allows current to flow and an *insulator* is a material that doesn't. Good conductors allow current to flow with abandon, without impediments. Examples of good conductors include the metals copper and aluminium. Carbon is also an excellent conductor. Good insulators, on the other hand, erect solid walls that completely block current. Examples of good insulators include glass, Teflon and plastic.

The key factor that determines whether a material is a conductor or an insulator is how readily its atoms give up electrons to move charge along. Most atoms are very possessive of their electrons, and are therefore good insulators. But some atoms don't have a strong hold on their outermost electrons. Those atoms are good conductors.

Creating resistance

If a conductor and an insulator are mixed together, the result is a compound that conducts current, but not very well. Such a compound has *resistance* – that is, it resists the flow of current. The degree to which the compound resists current depends on the exact mix of elements that make up the compound.

For example, a conducting material such as carbon may be mixed with an insulating material such as ceramic. If the mix is mostly carbon, the overall resistance of the mixture is low. If the mix is mostly ceramic, the overall resistance is high.

The truth is that *all* materials have some resistance. Even the best conductors have a small but measurable amount of resistance. The only exceptions are certain materials called *superconductors* that, when chilled to unbelievably low temperatures, conduct with 100% efficiency. Unfortunately, you can't buy superconductors at your local electronic shop, and even if you could, you'd never get a freezer powerful enough to chill the stuff down to absolute zero.

Measuring resistance

Resistance is measured in units called *ohms,* represented by the Greek letter omega (Ω). The standard definition of *1 ohm* is simple: the amount of resistance required to allow 1 ampere of current to flow when 1 V of voltage is applied to the circuit. In other words, if you connect a 1 ohm resistor across the terminals of a 1 V battery, 1 amp of current flows through the resistor (unless it is rated at less than 1 W, in which case it will go pop!).

A single ohm (1 Ω) is in fact a very small amount of resistance. Electronic circuits usually call for resistances in the hundreds, thousands or even millions of ohms.

Owing up to the discovery of ohms

The unit *ohm* is named after the famous German physicist Georg Ohm, who first explained the relationship between voltage, current and resistance in 1827.

The discovery itself, however, was made by a British scientist named Henry Cavendish more than 45 years earlier, but Cavendish never published his work. If he had, resistance would be measured in 'cavens', not ohms!

In Book I, Chapter 8, you discover that you can measure resistance of a circuit using an *ohmmeter,* which is a standard feature found in most multimeters. The procedure is simple: you disconnect all voltage sources from the circuit and then touch the ohmmeter's two probes to the ends of the circuit and read the resistance (in ohms) on the meter.

Here are a few other points to consider about resistance and ohms:

- **Abbreviations:** The abbreviations k (for *kilo*) and M (for *mega*) are used for thousands and millions of ohms. Thus, a 1,000-ohm resistance is written as 1 kΩ, and a 1,000,000-ohm resistance is written as 1 MΩ.

- **Zero resistance:** For the purposes of most electronic circuits, you can assume that the resistance value of ordinary wire is zero ohms (0 Ω). In reality, however, only superconductors have a resistance of 0 Ω. Even copper wire has some resistance. Therefore, the resistance of wire is usually measured in terms of ohms per kilometre. Though, of course, electronic circuits usually deal with wires that are at most a few centimetres or metres long, not kilometres.

  Short circuits essentially have zero resistance.

- **Infinite resistance:** Just as you can think of ordinary wire and short circuits as having zero resistance, insulators and open circuits can be considered to have infinite resistance, though in reality no such thing exists as completely infinite resistance.

  If you connect two wires to the terminals of a battery and hold the wires apart, a voltage exists between the ends of those two wires, and a very small current travels between them – even through the air because air doesn't have infinite resistance. This current is extraordinarily small – too small to even measure – but it's present nonetheless. Electric currents are literally everywhere.

Looking at Ohm's Law

The term *Ohm's law* refers to one of the fundamental relationships found in electric circuits: for a given resistance, current is directly proportional to voltage. In other words, if you increase the voltage through a circuit whose resistance is fixed, the current goes up. If you decrease the voltage, the current goes down.

Ohm's law expresses this relationship as a simple mathematical formula:

$$V = I \times R$$

In this formula, *V* stands for voltage (in volts), *I* stands for current (in amperes) and *R* stands for resistance (in ohms).

Ohm's law is incredibly useful because it lets you calculate an unknown voltage, current or resistance. In short, if you know two of these three quantities you can calculate the third.

Here's an example of how to calculate voltage in a circuit with a lamp powered by the two AA cells. Suppose you already know that the resistance of the lamp is 12 Ω, and the current flowing through the lamp is 250 mA (which is the same as 0.25 A). Then, you can calculate the voltage as follows:

$$V = I \times R$$

Go back (if you dare) to your school algebra class and remember that you can rearrange the terms in a simple formula such as Ohm's law to create other equivalent formulas. In particular:

- If you don't know the voltage, you calculate it by multiplying the current by the resistance:

  $$V = I \times R$$

- If you don't know the current, you can calculate it by dividing the voltage by the resistance:

  $$I = \frac{V}{R}$$

- If you don't know the resistance, you can calculate it by dividing the voltage by the current:

  $$R = \frac{V}{I}$$

To convince yourself that these formulas work, look again at the circuit with a lamp that has 12 Ω of resistance connected to two AA batteries for a total voltage of 3 V. Then you can calculate the current flowing through the lamp as follows:

$$I = \frac{V}{R} = \frac{3V}{12\Omega} = 0.25A$$

Why *I* stands for current

If you wonder why the symbols for voltage and resistance are *V* and *R,* which make perfect sense, but the symbol for current is *I,* which makes no sense, you can blame history. The unit of measure for current – the ampere – is named after André-Marie Ampère, a French physicist who was one of the pioneers of early electrical science. The French word he used to describe the strength of an electric current was *intensité* – in English, *intensity.* So *I* is for intensity.

By the way, the term *volt* is named after the Italian scientist Alessandro Volta, who invented the first electric battery in 1800. (His full name was Count Alessandro Giuseppe Antonio Anastasio Volta, but that isn't on the test!)

If you know the battery voltage (3 V) and the current (250 mA, which is 0.25 A), you can calculate the resistance of the lamp like this:

$$R = \frac{V}{I} = \frac{3\text{V}}{0.25\text{A}} = 12\Omega$$

The most important thing to remember about Ohm's law is that you must always do the calculations in terms of volts, amperes and ohms. For example, if you measure the current in milliamps (which you usually do in electronic circuits), you have to convert the milliamps to amperes by dividing by 1,000. For example, 250 mA is 0.25 A.

As we say in the preceding section, the definition of 1 ohm is the amount of resistance that allows 1 ampere of current to flow when 1 V of potential is applied to it. This definition is based on Ohm's law. If *V* is 1 and *I* is 1, *R* must also be 1:

$$R = \frac{V}{I} = \frac{1\text{V}}{1\text{A}} = 1\Omega$$

Introducing Resistors

A *resistor* is a small component that's designed to provide a specific amount of resistance in a circuit. Because resistance is an essential element of nearly every electronic circuit, you're going to use resistors in just about every circuit you build.

Although resistors come in a variety of sizes and shapes, the most common type for hobby electronics is the *carbon film resistor,* shown in Figure 2-1. These resistors consist of a layer of carbon laid down on an insulating material and contained in a small cylinder, with wire leads attached to both ends.

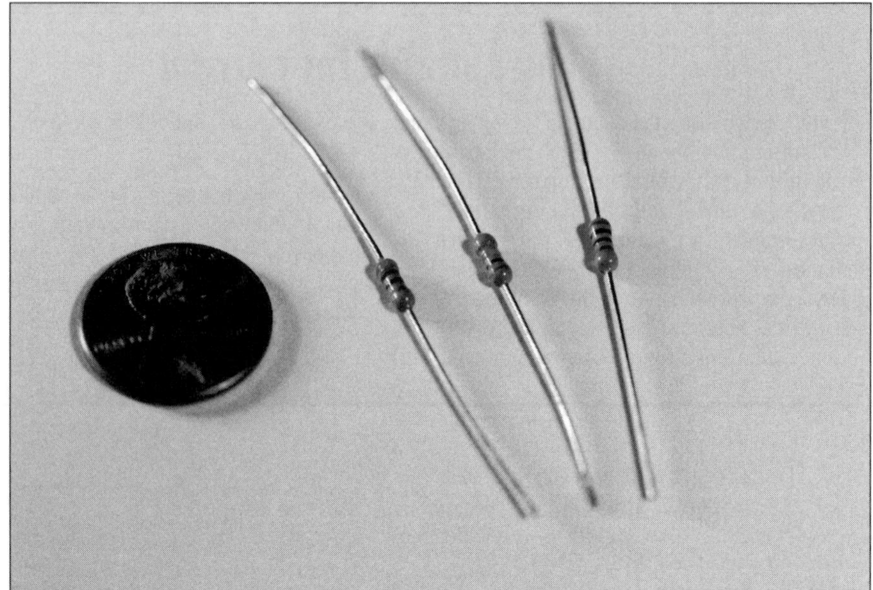

Figure 2-1: Carbon film resistors.

Resistors are blind to the polarity in a circuit. Thus, you don't have to worry about installing them backwards. Current can pass equally through a resistor in either direction.

A resistor appears in schematic diagrams as an open rectangle, like the one shown here in the margin. The resistance value is typically written next to the resistor symbol. In addition, an identifier such as R1 or R2 is also sometimes written next to the symbol.

In some schematics, particularly those drawn in America, the jagged line symbol as shown in the margin is used instead of the rectangle.

Resistors are used for many reasons in electronic circuits. The three most popular uses are:

- **Limiting current:** You can use resistors to introduce resistance into a circuit to limit the amount of current that flows through the circuit. In accordance with Ohm's law, if the voltage in a circuit remains the same, the current decreases if you increase the resistance.

Many electronic components have an appetite for current that has to be regulated by resistors. One of the best known is the light-emitting diode (LED), which is a special type of diode that emits visible light when current runs through it. Unfortunately, LEDs don't know when to step away from the table when it comes to consuming current, because they have very little internal resistance. Plus LEDs don't have much tolerance for current, so too much current burns them out. (Flip to the later section 'Limiting Current with a Resistor' for more on limiting currents and LEDs.)

As a result, always place a resistor in series with an LED to keep the LED from burning itself up. (You can find out a whole lot more about LEDs in Chapter 5 of this minibook.)

You can use Ohm's law to your advantage when using current-limiting resistors. For example, if you know what the supply voltage is and you know how much current you need, you can use Ohm's law to determine the right resistor to use for the circuit as we explain in the preceding section.

- **Dividing voltage:** You can use resistors to reduce voltage to a level that's appropriate for specific parts of your circuit. For example, suppose your circuit is powered by a 3 V battery but a part of your circuit needs 1.5 V. You can use two resistors of equal value to split this voltage in half, yielding 1.5 V. For more information, see the section 'Dividing Voltage' later in this chapter.
- **Resistor/capacitor networks:** You can use resistors in combination with capacitors for a variety of interesting purposes. Read about this use of resistors in Chapter 3 of this minibook.

Reading Resistor Colour Codes

You can determine the resistance provided by a resistor by examining the *colour codes* that are painted on the resistor. These little stripes of bright colours indicate two important factoids about the resistor: its resistance in ohms and its *tolerance,* which indicates how close to the indicated resistance value the resistor actually is.

Most resistors have four stripes of colour. The first three stripes indicate the resistance value and the fourth one indicates the tolerance. Some resistors have five stripes of colour, with four representing the resistance value and the last one the tolerance.

If you're uncertain from which side of the resistor to read the colours, start with the side closest to the colour stripe. The first stripe is usually painted very close to the edge of the resistor; the last stripe isn't as close to the edge.

Noticing standard resistor values

In theory, 100 different combinations of colours exist for the first two bands, covering the range of values from 00 through 99. In practice, however, you commonly encounter only a few colour combinations. These combinations represent standardised values that allow manufacturers to produce resistors that are useful in a wide variety of applications.

For example, take the value 47, represented by the colour code yellow-violet. 47 happens to be one of the preferred resistor numbers, and so you can easily obtain resistors of 4.7 Ω, 47 Ω, 470 Ω, 4.7 KΩ, 47 KΩ, 470 KΩ and 4.7 MΩ.

In contrast, 45 isn't one of the preferred values and so you don't find 45 Ω or 450 Ω resistors.

Although several different systems are available for standardising preferred resistor values, the most common system uses 12 different standard values:

First Two Colours	*Standard Value*
Brown-black	10
Brown-red	12
Brown-green	15
Brown-grey	18
Red-red	22
Red-violet	27
Orange-orange	33
Orange-white	39
Yellow-violet	47
Green-blue	56
Blue-grey	68
Grey-red	82

These values are standardised and designed to provide a wide range of resistance values. Although not exact, each value is approximately 1.2 times larger than the previous value.

Working out a resistor's value

To read a resistor's colour code, check out Table 2-1. Here's the procedure for determining the value of a resistor with four stripes:

1. **Turn the resistor so you can read the stripes properly.**

   Read the stripes from left to right. The first stripe is the one that's closest to one end of the resistor. If this stripe is on the right side of the resistor, turn the resistor around so the first stripe is on the left.

2. **Look up the colour of the first stripe to determine the value of the first digit.**

   For example, if the first stripe is yellow, the first digit is 4.

3. **Look up the colour of the second stripe to determine the value of the second digit.**

   For example, if the first stripe is violet, the second digit is 7.

4. **Look up the colour of the third stripe to determine the multiplier.**

   For example, if the third stripe is brown, the multiplier is 10.

5. **Multiply the two-digit value by the multiplier to determine the resistor's value.**

   For example, 47 times 10 is 470. Thus, a yellow-violet-brown resistor is 470 Ω.

If a resistor has five stripes, the first three stripes are the value digits and the fourth stripe is the multiplier. The fifth stripe is the tolerance, as described in the next section.

Table 2-1 Resistor Colour Codes (Resistance Values)

Colour	*Digit*	*Multiplier*
Black	0	1
Brown	1	10
Red	2	100
Orange	3	1 k
Yellow	4	10 k
Green	5	100 k
Blue	6	1 M
Violet	7	10 M
Grey	8	100 M
White	9	1,000 M
Gold		0.1
Silver		0.01

Here are a few examples to help you understand how to read resistor codes:

Colour Stripes	*Digit Values*	*Multiplier (in Ohms)*	*Resistor Value*
Brown-black-brown	10	10	100 Ω
Brown-black-red	10	100	1 kΩ
Red-red-orange	22	1 k	22 kΩ
Red-red-yellow	22	10 k	220 kΩ
Yellow-violet-black	47	0.1	47 Ω

Understanding resistor tolerance

The value indicated by the stripes painted on a resistor provides an estimate of the actual resistance. The exact resistance varies by a percentage that depends on the *tolerance* factor of the resistor.

For example, a 22 kΩ resistor with a 5% tolerance actually has a value somewhere between 5% above and 5% below 22 kΩ, which works out to somewhere between 20.9 and 23.1 kΩ. A 470 Ω resistor with a 10% tolerance has an actual value somewhere between 423 and 517Ω.

Why the approximations? Simply because manufacturing resistors to very close tolerances costs more money, and for most electronic circuits a 5% or 10% margin of error is perfectly acceptable. For example, if you're building a circuit to limit the current flowing through a component to 200 mA, the actual current being a little above or below 200 mA probably doesn't matter much. Thus, a 5%- or 10%-tolerance resistor is acceptable.

If your application demands higher precision, you can spend a bit more money to buy higher-tolerance resistors. But 5%- or 10%- tolerance resistors are fine for most work, including all the circuits we present in this book (unless otherwise indicated).

The tolerance of a resistor is indicated in the resistor's last colour stripe, as shown in Table 2-2.

Table 2-2 Resistor Colour Codes (Tolerance Values)

Colour	*Tolerance (%)*
Brown	1
Red	2
Yellow	5

Colour	Tolerance (%)
Gold	5
Silver	10
None	20

Heating Up: Resistor Power Ratings

Resistors are like brakes for electric current. They work by applying the electrical equivalent of friction to flowing current. This friction inhibits the flow of current by absorbing some of the current's energy and dissipating it in the form of heat. Whenever you use a resistor in a circuit, you need to make sure that the resistor is capable of handling the heat.

The *power rating* of a resistor indicates how much power a resistor can handle before it becomes too hot and burns up. As you can discover in Book I, Chapter 2, power is measured in units called *watts*. The more watts a resistor can handle, the larger and more expensive the resistor is.

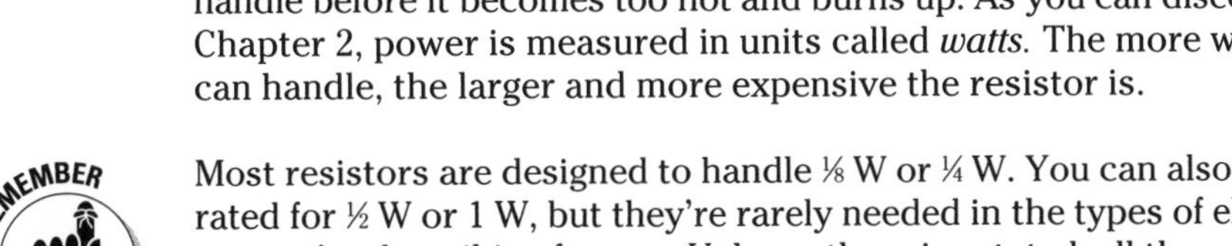

Most resistors are designed to handle ⅛ W or ¼ W. You can also find resistors rated for ½ W or 1 W, but they're rarely needed in the types of electronic projects we're describing for you. Unless otherwise stated, all the resistors used in this book are rated at ¼ W.

Unfortunately, you can't tell a resistor's power rating just by looking at it. Unlike resistance and tolerance, wattages don't have colour codes (see the preceding section). The size of the resistor, however, is a good indicator of its power rating and the ratings are written on the packaging when you buy new resistors. After you work with them for a while, you can quickly recognise the size difference between resistors of different power ratings.

If you want to be safe, you can calculate the power demands required of a particular resistor in your circuits. To start, use Ohm's law to calculate the voltage across the resistor and the current that's going to pass through the resistor. For example, if a 100 Ω resistor will have 3 V across it, you can calculate that 30 mA of current will flow through the resistor by dividing the voltage by the resistance (3 V ÷ 1,000 Ω = 0.03 A, which is 30 mA).

When you know the voltage and the current, you can calculate the power that's going to be dissipated by the resistor by using the power formula we describe in Book I, Chapter 2:

$$P = I \times V$$

Thus, the power dissipated by the resistor is just 0.09 W, well under the maximum that a ¼ W (0.25 W) resistor can handle. (A ⅛ W resistor should be able to handle this amount of power too, but with power ratings erring on the large side is best.)

Limiting Current with a Resistor

One of the most common uses for resistors is to limit the current flowing through a component. Some components, such as light-emitting diodes, are very sensitive to current. A few milliamps of current is enough to make an LED glow; a few hundred milliamps is enough to destroy the LED.

Project 2-1 shows you how to build a simple circuit that demonstrates how a resistor can be used to limit current to an LED. The finished circuit, which you assemble on a small solderless breadboard, is shown in Figure 2-2.

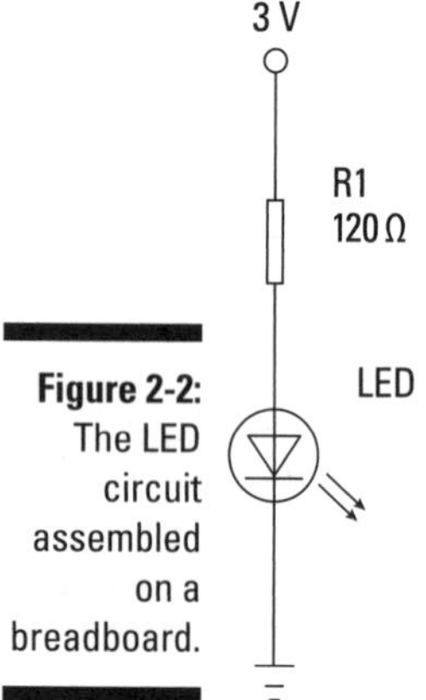

Figure 2-2: The LED circuit assembled on a breadboard.

Before we get into the construction of the circuit, here's a simple question: why use a 120 Ω resistor? Why not a larger or a smaller value? In other words, how do you determine what size resistor to use in a circuit like this?

The answer is simple: Ohm's law, which can easily tell you what size resistor to use. But you first have to know the voltage and current (see the earlier section 'Looking at Ohm's Law'). In this case, the voltage is easy to figure out: you know that two AA batteries provide 3 V. To figure out the current, you just need to decide how much current is acceptable for your circuit. The technical specifications of the LED tell you how much current the LED can handle. In the case of a standard 5 mm red LED (the kind you can buy for about 60 pence), the maximum allowable current is 28 mA. To be safe and make sure that you don't destroy the LED with too much current, round the maximum current down to 25 mA.

To calculate the desired resistance, you divide the voltage (3 V) by the current (0.025 A). The result is 120 Ω.

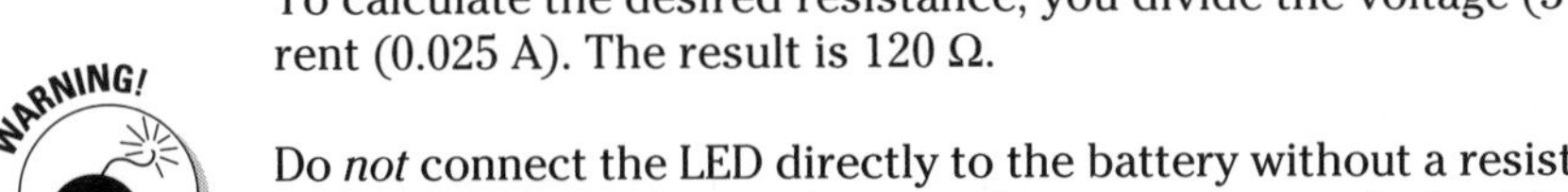

Do *not* connect the LED directly to the battery without a resistor. If you do, the LED flashes brightly – and may go bang – and then it's dead forever.

Project 2-1: Using a Current-Limiting Resistor

In this project, you build a simple circuit that uses a resistor to limit the current to an LED. Without this current-limiting resistor, too much current would flow through the LED and the LED would be destroyed.

The only tools you need for this project are a small Phillips-head screwdriver to open the battery holder, and wire cutters and strippers to cut and strip the jumper wire.

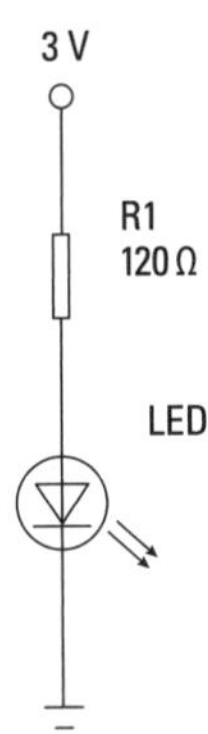

Parts List	
2	AA batteries
1	Battery holder
1	Red LED
1	120 Ω resistor

Steps

1. **Connect the battery holder.**

   Position the breadboard as shown in the diagram, so that hole A30 is near the top left. Then, insert the black lead in the top bus strip and the red lead in the bottom bus strip. Any location on the correct bus is fine, but we recommend you insert the leads into the holes at the very end of the breadboard.

2. **Connect the resistor.**

   Insert one end of the resistor into the breadboard in hole G25, and insert the other lead into any nearby hole in the positive voltage bus strip (the strip that's connected to the red battery wire).

3. **Connect the LED.**

   Notice that the leads of the LED aren't the same length; one lead is shorter than the other. Insert the longer lead (called the *anode*) into hole A25 and the shorter lead (called the *cathode*) into any nearby hole in the negative voltage bus strip (the one that the black battery lead is inserted into).

   Note that the circuit doesn't work if you insert the LED backward. The short lead must be in the negative voltage bus strip.

4. **Use the short jumper wire to connect the terminal strips into which you inserted the LED and the resistor.**

   Insert the jumper in holes E25 and F25, so that the jumper straddles the gap that runs down the middle of the breadboard. This connects the resistor to the LED.

5. **Insert the batteries.**

   The LED lights up. If it doesn't, double-check your connections to make sure that the circuit is assembled correctly. If it still doesn't light up, try a different battery.

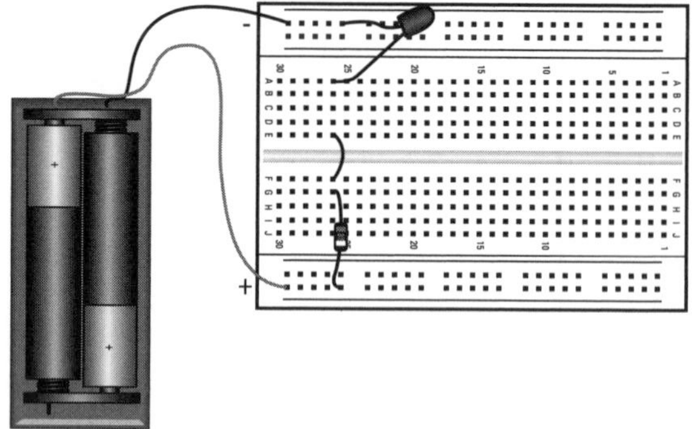

Building Resistance in Combination

Suppose that you design the perfect circuit, and it calls for a 1,100 Ω resistor in a critical spot. Then you discover that you can't find a 1,100 Ω resistor anywhere. You can buy 1 kΩ resistors and 100 Ω resistors, but no one seems to have any 1,100 Ω resistors.

Do you have to settle for a 1 kΩ resistor and hope that it's close enough? Certainly not!

All you have to do is use two or more resistors in combination to create the necessary resistance. Such a combination of resistors is sometimes called a *resistor network.* You can freely substitute a resistor network for a single resistor whenever you want.

You can combine resistors in two basic ways: in series (strung end to end) and in parallel (side by side). This section explains how you calculate the total resistance of a network of resistors in series and in parallel, and shows you how to mix the two arrangements.

Combining resistors in series

Calculating the total resistance for two or more resistors strung end to end – that is, in series – is straightforward: you simply add the resistance values to get the total resistance.

For example, if you need 1,100 ohms of resistance and can't find a 1,100 Ω resistor, you can combine a 1,000 Ω resistor and a 100 Ω resistor in series. Adding these two resistances together gives you a total resistance of 1,100 Ω.

You can place more than two resistors in series if you want. You just keep adding up all the resistances to get the total resistance value. For example, if you need 1,800 Ω of resistance, you can use a 1 kΩ resistor and eight 100 Ω resistors in series.

Figure 2-3 shows how serial resistors work. Here, the two circuits have identical resistances. The circuit on the left accomplishes the job with one resistor; the circuit on the right does it with three. Therefore, the circuits are equivalent.

Any time you see two or more resistors in series in a circuit, you can substitute a single resistor whose value is the sum of the individual resistors.

Similarly, any time you see a single resistor in a circuit, you can substitute two or more resistors in series as long as their values add up to the desired value.

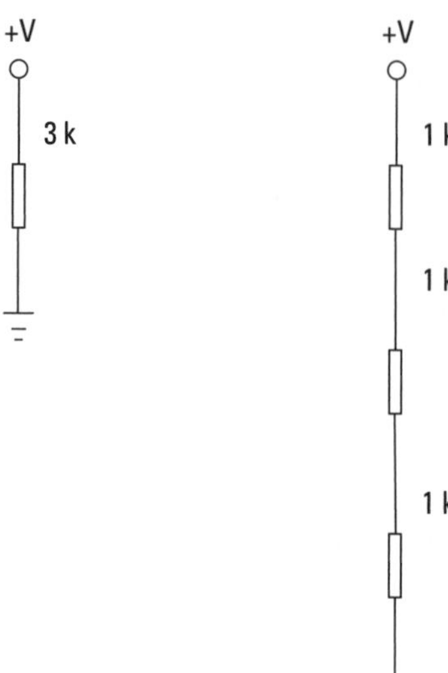

Figure 2-3: Combining resistors in series.

The total resistance of resistors in series is always greater than the resistance of any of the individual resistors, because each resistor adds its own resistance to the total.

Combining resistors in parallel

You can combine resistors in parallel to create equivalent resistances, but calculating the total resistance for such resistors is a bit more complicated than calculating the resistance for resistors in series, and so you may need to dust off your thinking cap. Although Ohm's law is simple enough, the calculations required to work out parallel resistors can get a little complex. The maths isn't horribly complicated, but it isn't trivial either.

As the circuit in Figure 2-4 illustrates, when you combine two resistors in parallel, current can flow through both resistors at the same time. Although each resistor does its job to hold back the current, the total resistance of two resistors in parallel is always less than the resistance of either of the individual resistors because the current has two pathways through which to go.

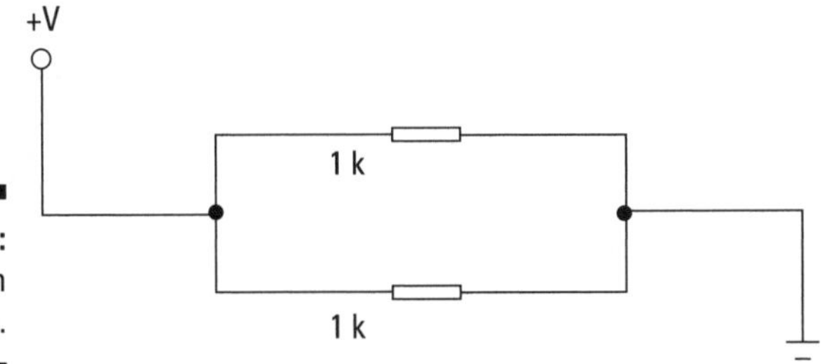

Figure 2-4: Resistors in parallel.

So how do you calculate the total resistance for resistors in parallel? Very carefully. Here are the rules:

- **Resistors of equal value in parallel:** In this, the simplest case, you can calculate the total resistance by dividing the value of one of the individual resistors by the number of resistors in parallel. For example, the total resistance of two 1 kΩ resistors in parallel is 500 Ω and the total resistance of four 1 kΩ resistors is 250 Ω.

  Unfortunately, this is the only case that's simple. The maths when resistors in parallel have unequal values is more complicated.

- **Two resistors of different values in parallel:** With only two involved, the calculation isn't too bad:

  $$R_{total} = \frac{R1 \times R2}{R1 + R2}$$

  In this formula, $R1$ and $R2$ are the values of the two resistors.

  Here's an example, based on a 2 kΩ and a 3 kΩ resistor in parallel:

  $$R_{total} = \frac{R1 \times R2}{R1 + R2} = \frac{2{,}000\Omega \times 3{,}000\Omega}{2{,}000\Omega + 3{,}000\Omega} = 1{,}200\Omega$$

- **Three or more resistors of different values in parallel:** Here the calculation begins to look like rocket science:

  $$R_{total} = \frac{1}{\frac{1}{R1} + \frac{1}{R2} + \frac{1}{R2} \cdots}$$

  The dots at the end of the expression indicate that you keep adding up the reciprocals of the resistances for as many resistors as you have.

  In case you're crazy enough to want to do this kind of maths, here's an example for three resistors whose values are 2 kΩ, 4 kΩ and 8 kΩ:

  $$R_{total} = \frac{1}{\frac{1}{2{,}000\ \Omega} + \frac{1}{4{,}000\ \Omega} + \frac{1}{8{,}000\ \Omega}} = \frac{1}{0.000875\ \Omega} = 1{,}142.857\ \Omega$$

  As you can see, the final result is 1,142.857 Ω. That's more precision than you can possibly want, and so you can probably safely round it off to 1,143 Ω, or maybe even 1,150 Ω.

Conducting your way through parallel resistors

The parallel resistance formula makes more sense if you think about it in terms of the opposite of resistance, which is called *conductance.* Resistance is the ability of a conductor to block current; conductance is the ability of a conductor to pass current. Conductance has an inverse relationship with resistance: when you increase resistance, you decrease conductance, and vice versa.

Because the pioneers of electrical theory had a nerdy sense of humour, they named the unit of measure for conductance the *mho,* which is *ohm* spelled backwards. The mho is the reciprocal (the inverse) of the ohm. To calculate the conductance of any circuit or component (including a single resistor), you just divide the resistance of the circuit or component (in ohms) into 1. Thus, a 100 Ω resistor has 1⁄100 mho of conductance.

When circuits are connected in parallel, current has multiple pathways it can travel through. Fortunately, the total conductance of a parallel network of resistors is simple to calculate: you just add up the conductances of each individual resistor. For example, if you have three resistors in parallel whose conductances are 0.1 mho, 0.02 mho and 0.005 mho (the conductances of 10 Ω, 50 Ω and 200 Ω resistors, respectively), the total conductance of this circuit is 0.125 mho (0.1 + 0.02 + 0.005 = 0.125).

One of the basic rules of doing maths with reciprocals is that if one number is the reciprocal of a second number, the second number is also the reciprocal of the first number. Thus, because mhos are the reciprocal of ohms, ohms are the reciprocal of mhos. To convert conductance to resistance, you just divide the conductance into 1. Thus, the resistance equivalent to 0.125 mho is 8 Ω (1 ÷ 0.125 = 8).

Remembering how the parallel resistance formula works is easier when you realise that what you're really doing is converting each individual resistance to conductance, adding them up and then converting the result back to resistance. In other words, you convert the ohms to mhos, add them up and then convert them back to ohms. That's how – and why – the resistance formula works in practice.

Mixing series and parallel resistors

Resistors can be combined to form complex networks in which some of the resistors are in series and others are in parallel. For example, Figure 2-5 shows a network of three 1 kΩ resistors and one 2 kΩ resistor. These resistors are arranged in a mixture of serial and parallel connections.

To calculate the total resistance of this type of network, you divide and conquer. Look for simple series or parallel resistors, calculate their total resistance and then substitute a single resistor with an equivalent value. For example, you can replace the two 1 kΩ resistors that are in series with a single 2 kΩ resistor. Now, you have two 2 kΩ resistors in parallel.

Remembering that the total resistance of two resistors with the same value is half the resistance value, you can replace these two 2 kΩ resistors with a single 1 kΩ resistor. You're now left with two 1 kΩ resistors in series. Thus, the total resistance of this circuit is 2 kΩ.

Deceptively simple, isn't it?

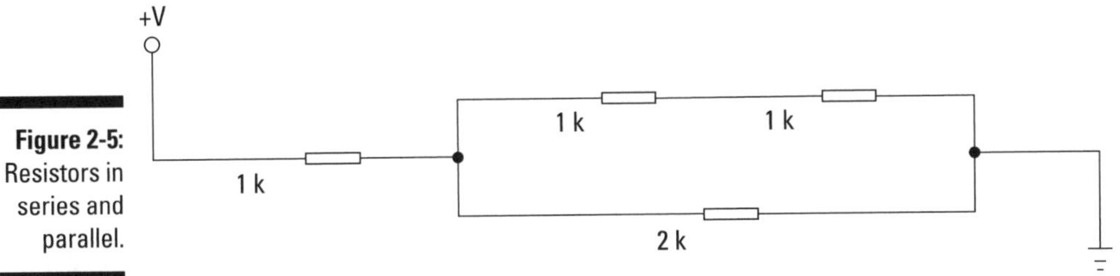

Figure 2-5: Resistors in series and parallel.

Assembling resistors in series and parallel

Project 2-2 lets you do a little hands-on work with some simple series and parallel resistor connections so that you can see firsthand how the calculations we describe in the previous three sections work in the real world.

You may well find that the individual variations of resistors (due to their manufacturing tolerances) mean that the calculated resistances don't always match the resistance of the circuits. But in most cases, the variations aren't significant enough to affect the operation of your circuits.

In this project, you assemble five resistors into three different configurations: the first has all five resistors in series; the second has them all in parallel; and the third creates a network of two sets of parallel resistors that are connected in series. Figure 2-6 shows how these three configurations appear when assembled.

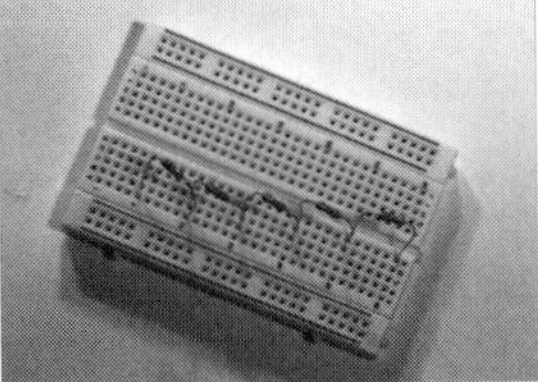
Resistors in series

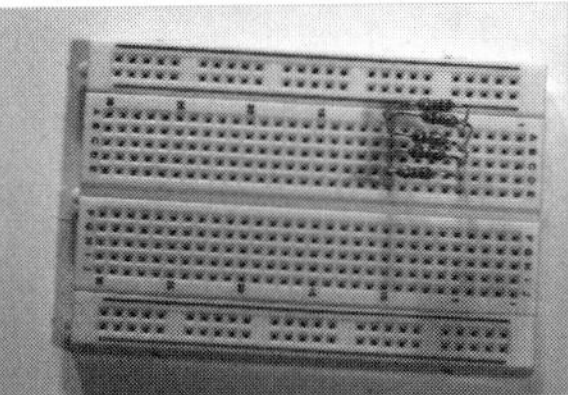
...in parallel

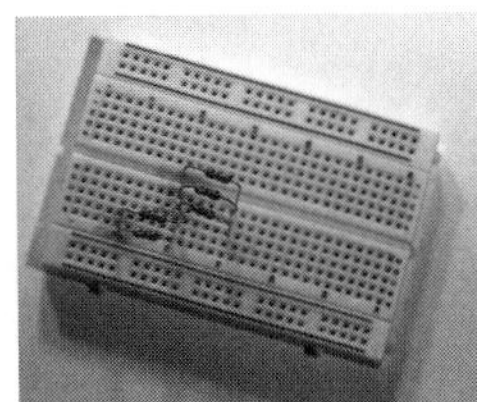
...and in series and parallel

Figure 2-6: The assembled resistors for Project 2-2.

Project 2-2: Resistors in Series and Parallel

In this project, you experiment with resistors in series and in parallel. You need a multimeter with an ohmmeter function to measure the resistances.

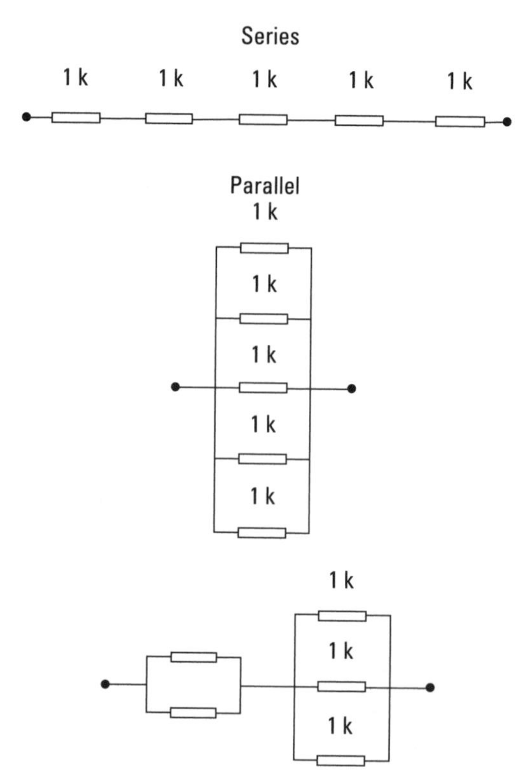

Parts List

- 1 Solderless breadboard
- 5 1 kΩ resistors (brown-black-red)
- 1 Multimeter with an ohmmeter function

Steps

1. **Set your multimeter to its ohmmeter function with a range large enough to measure at least 5 kΩ of resistance.**

   On our ohmmeter, the closest range is 20 kΩ.

2. **Insert the five 1 kΩ resistors in series.**

   Use the following holes on the breadboard:

Resistor	First Lead	Second Lead
1	A5	A10
2	B10	B15
3	C15	C20
4	D20	D25
5	E25	E30

   Refer to the Series Layout diagram for the layout of these resistors.

3. **Using your multimeter, measure the resistance of each resistor individually, and then measure the total resistance of two, three, four and five resistors in series.**

   To do these measurements, place the meter probes on the leads at the breadboard holes indicated in the later Series Resistance Measurements table. Write your actual measurements in the column on the far right

4. **Rearrange the resistors into a parallel circuit.**

   Remove the resistors and reinsert them into the following holes:

Resistor	First Lead	Second Lead
1	A5	A10
2	B5	B10
3	C5	C10
4	D5	D10
5	E5	E10

   Refer to the Parallel Layout diagram for the layout of these resistors.

5. **Use your ohmmeter to measure the resistance of the parallel resistor circuit.**

   Because there are five 1 kΩ resistors, your measurement should be approximately 200 Ω.

6. **Rearrange the resistors into a series/parallel network.**

   Remove resistors 3, 4 and 5 and insert them as follows:

Resistor	First Lead	Second Lead
1	A5	A10
2	B5	B10
3	C10	C15
4	D10	D15
5	E10	E15

   Refer to the Series/Parallel Layout diagram for the layout of these resistors.

   This configuration has the first two resistors in one parallel circuit and the other three in a second parallel circuit. The two parallel circuits are connected together to form a series circuit.

7. **Use your ohmmeter to measure the resistance of two parallel circuits and the entire circuit.**

   Record your measurements in the Series/Parallel Resistance Measurements table.

8. **You're done!**

   Unless you're a glutton for punishment. In that case, feel free to experiment with other combinations of series and parallel resistor circuits. Grab some resistors with other values and throw them into the mix. Each time, do the maths to predict what the resulting resistance should be. With some practice, you'll get good at calculating resistor networks.

Red Lead	Black Lead	Number of Resisters	Expected Measurement	Your Measurement
Series Resistance Measurements				
A5	A10	1	1 kΩ	
B10	B15	1	1 kΩ	
C15	C20	1	1 kΩ	
D20	D25	1	1 kΩ	
E25	E30	1	1 kΩ	
A5	B15	2	2 kΩ	
A5	C20	3	3 kΩ	
A5	D25	4	4 kΩ	
A5	E30	5	5 kΩ	
Series/Parallel Resistance Measurements				
A5	A10	2	500 kΩ	
C10	C15	3	333 kΩ	
A5	C15	5	833 kΩ	

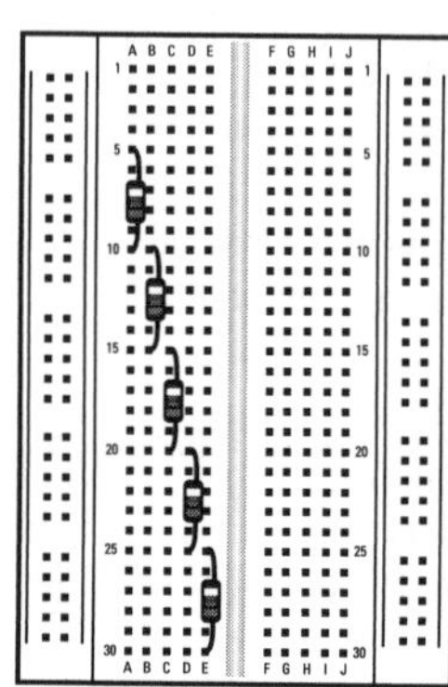

Series Layout

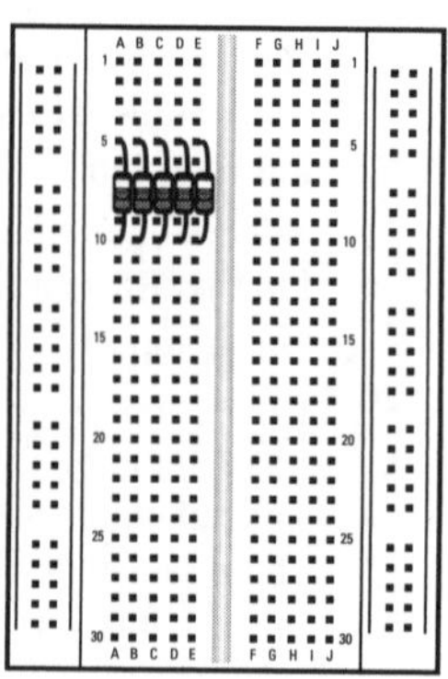

Parallel Layout

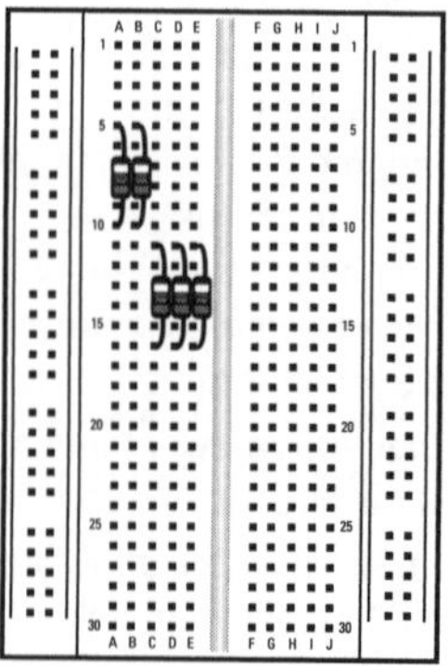

Series/Parallel Layout

Dividing Voltage with Resistors

One interesting and useful property of resistors is that if you connect two resistors together in series, you can tap into the voltage at the point between the two resistors to get a voltage that's a fraction of the total voltage across both resistors. This type of circuit is called a *voltage divider,* and is a common way to reduce voltage in a circuit. Figure 2-7 shows a typical voltage-divider circuit.

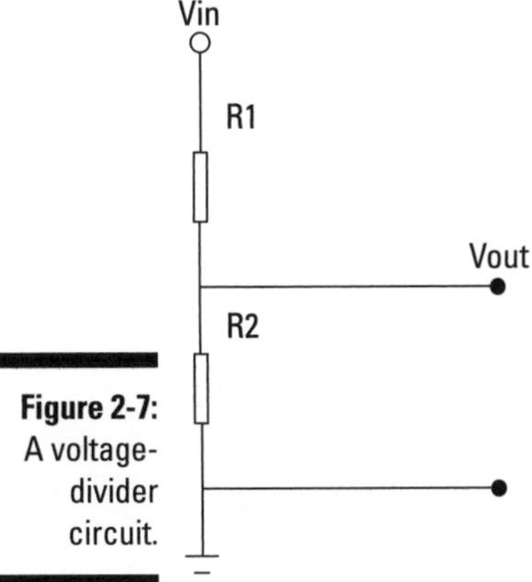

Figure 2-7: A voltage-divider circuit.

When the two resistors in the voltage divider are of the same value, the voltage is cut in half. For example, imagine that your circuit is powered by a 9 V battery, but your circuit only needs 4.5 V. You can use a pair of resistors of equal value across the battery leads to provide the necessary 4.5 V.

When the resistors are of different values, you need to do a little maths to calculate the voltage at the centre of the divider. The formula is as follows:

$$V_{out} = \frac{V_{in} \times R2}{R1 + R2}$$

For example, suppose that you're using a 9 V battery, but your circuit requires 6 V. In this case, you can create a voltage divider using a 1 kΩ resistor for *R*1 and a 2 kΩ resistor for *R*2. Here's the maths:

$$V_{out} = \frac{9\text{V} \times 2{,}000\ \Omega}{1{,}000\ \Omega + 2{,}000\ \Omega} = \frac{18{,}000\text{V}}{3{,}000} = 6\text{V}$$

As you can see, these resistor values cut the voltage down to 6 V.

In Project 2-3, you build a simple-voltage divider circuit on a solderless breadboard to provide either 3 V or 6 V from a 9 V battery. The assembled circuit is shown in Figure 2-8.

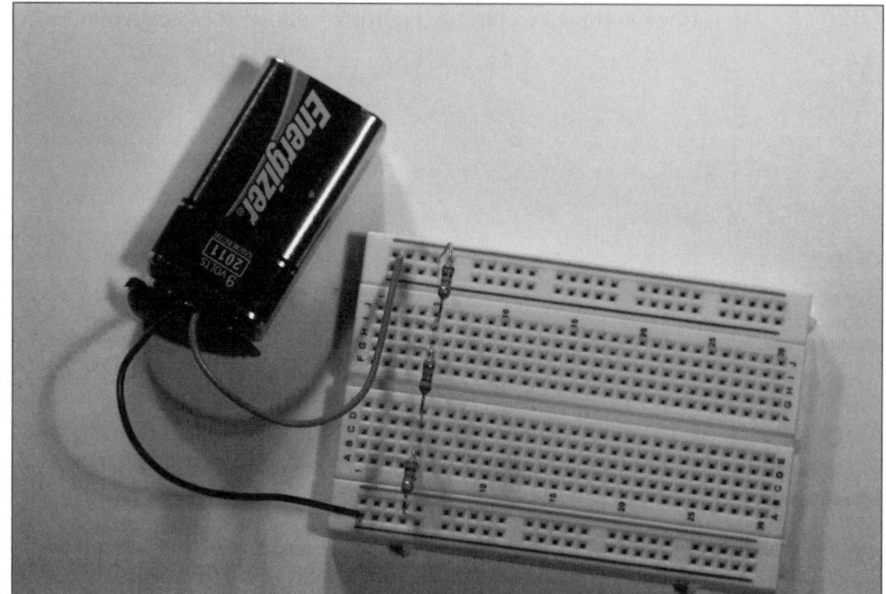

Figure 2-8: The assembled voltage-divider circuit.

Project 2-3: A Voltage Divider Circuit

In this project, you build a simple voltage divider circuit using three 1 KΩ resistors and you use the ohmmeter function of your multimeter to measure the effect of the divider.

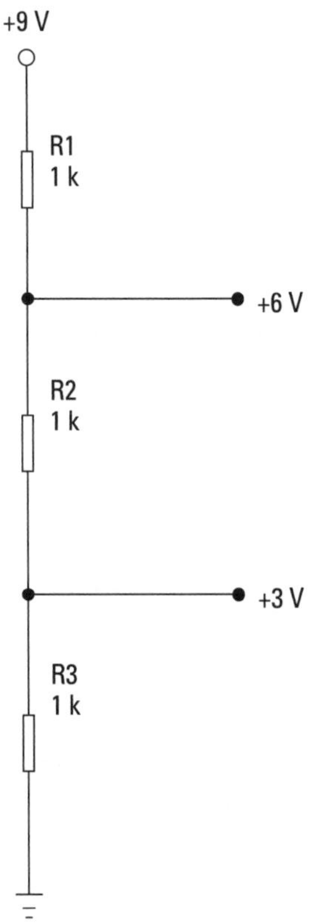

Parts List	
1	Solderless breadboard
1	9 V battery
1	9 V battery snap holder
3	1 kΩ resistor (brown-black-red)
1	Multimeter with a voltmeter function

Steps

1. **Connect the battery holder.**

   Insert the black lead into the first hole in the ground bus (on the side nearest column A). Then, insert the red lead in the first hole in the positive voltage bus on the side nearest column J).

2. **Insert the resistors.**

   Place the resistor leads into the holes indicated in the following table:

Resistor	First Lead	Second Lead
1	Ground bus near row 5	B5
2	D5	G5
3	I5	Positive bus near row 5

3. **Connect the battery.**

4. **Use the voltmeter to observe how the voltage is divided.**

   Set the voltmeter to a range that measures at least 10 V DC. Then take the following measurements by touching the meter leads to the resistor leads that are plugged into the holes indicated in the table.

5. **You're done!**

   Be sure to unplug the battery from the circuit. If you leave it plugged in, current continues to flow through the series resistor circuit, and the battery soon goes dead.

Series Resistance Measurements

Black Lead	Red Lead	Expected Measurement	Your Measurement
Ground bus	Positive bus	9 V	
Ground bus	H5	6 V	
Ground bus	A5	3 V	
A5	H5	3 V	

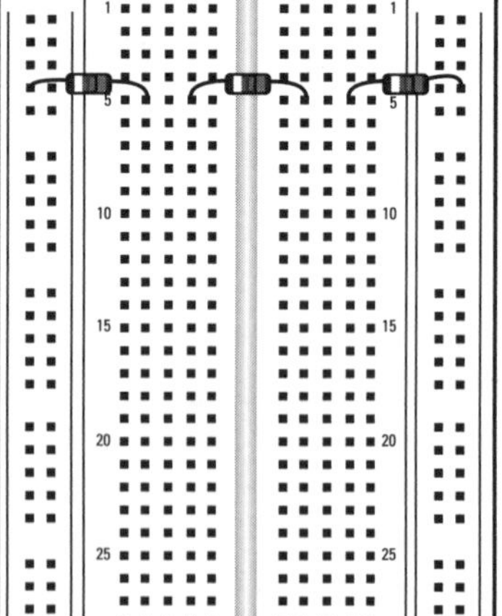

Varying Resistance with a Potentiometer

Many circuits call for a resistance that the user can vary. For example, most audio amplifiers include a volume control that lets you turn the volume up or down, and similarly you can create a simple light dimmer by varying the resistance in series with a lamp.

A variable resistor is called a *potentiometer* (or just *pot* for short). A potentiometer is simply a resistor with three terminals. Two of the terminals are permanently fixed on each end of the resistor, but the middle terminal is connected to a wiper that slides in contact with the entire surface of the resistor. Thus, the amount of resistance between this centre terminal and either of the two side terminals varies as the wiper moves.

Figure 2-9 shows how a typical potentiometer looks from the outside. The resistive track and slider (properly called the *wiper*) are enclosed within the metal can and the three terminals are beneath it. The rod that protrudes from the top of the metal can is connected to the wiper so that when the user turns the rod, the wiper moves across the resistor to vary the resistance.

Figure 2-9: A potentiometer.

Figure 2-10 shows how a potentiometer works on the inside. Here, you can see that the resistor is made of a semicircular piece of resistive material such as carbon. The two outer terminals are connected to either end of the resistor. The wiper, to which the third terminal is connected, is mounted so that it can rotate across the resistor. When the wiper moves, the resistance between the centre terminal and the other two terminals changes.

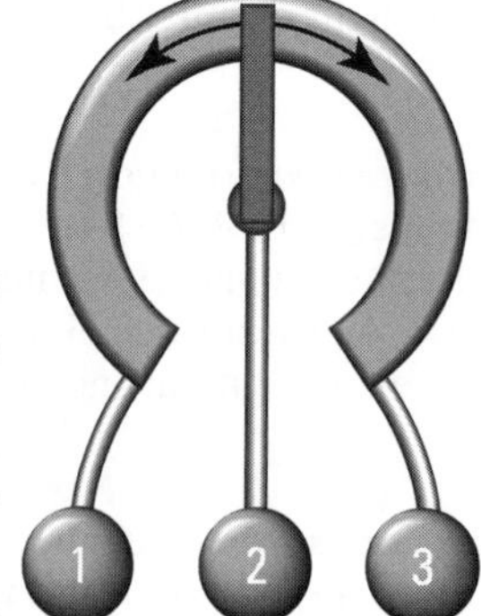

Figure 2-10: How a potentiometer works.

The symbol used for a potentiometer in schematic diagrams is shown in the margin. As you can see, the centre tap of the resistor is indicated by an arrow that's meant to reflect that the value of the resistance at this terminal varies when the wiper moves.

Potentiometers are rated by their total resistance. The resistance between the centre terminal and the two other terminals always adds up to the total resistance rating of the potentiometer. For example, the two resistances split by a 100 kΩ potentiometer always add up to 100 Ω. When the dial is exactly in the centre, both resistances are 50 Ω. As you move the wiper one way or the other, one resistance increases while the other decreases; but in all cases, the total of the two resistances always adds up to 100 Ω.

Here are a few other thoughts to keep in mind about potentiometers, plus a few different types:

- **Variety:** Potentiometers come in a wide variety of shapes and sizes. With a little hunting around in shops or on the Internet, you can find the perfect potentiometer for every need.
- **Trim pots:** These potentiometers are very small and can be adjusted only by the use of a tiny screwdriver. They're designed to make occasional fine-tuning adjustments to your circuits.
- **Switches:** Some potentiometers have switches incorporated into them, so that when you turn the knob all the way to one side or pull it out, the switch operates to open or close the circuit.

- **Wipers:** When the wiper reaches one end of the resistor or the other, the resistance between the centre terminal and the terminal on that end is essentially zero. Keep this point in mind when you're designing circuits. Putting a small resistor in series with a potentiometer is a common way to avoid circuit paths with no resistance.
- **Linear tapers:** In these potentiometers, the resistance varies evenly as you turn the dial. For example, if the total resistance is 10 kΩ, the resistance at the halfway mark is 5 kΩ and the resistance at the one-quarter mark is 2.5 kΩ. They're called *linear taper* potentiometers because the resistance change is linear.
- **Logarithmic tapers:** Many potentiometers aren't linear. For example, potentiometers designed for audio applications usually have a *logarithmic taper,* which means that the resistance doesn't vary evenly as you move the dial.

- **Rheostats**: These variable resistors have only two terminals – one on an end of the resistor itself, the other attached to the wiper. Although properly called a rheostat, most people use the term *potentiometer* or *pot* to refer to both two- and three-terminal variable resistors.

Chapter 3

Working with Capacitors

In This Chapter

- Mining the mysteries of capacitors
- Measuring capacitors
- Calculating important capacitor numbers
- Building some basic capacitor circuits

In the film *Back to the Future,* Doc Brown's famous flux capacitor can convert 1.21 gigawatts of power into a distortion in the space-time continuum, which allows him to send his friend Marty McFly back to 1955 and change the course of history.

The flux capacitor is science fiction. But capacitors are real and they did change the course of history, because you find them in almost every electronic circuit, including those that you build.

What Is a Capacitor?

To begin understanding what a capacitor is, consider this question: what makes current flow within a conductor? In Book I, Chapter 2, we explain that opposite charges attract one another and like charges repel one another. The attraction or repulsion is caused by a force known as the *electromagnetic force.* All charged particles exhibit an electric field that's associated with this force. This field is what draws *electrons* (negative charges) towards *protons* (positive charges) and also what pushes electrons away from one another.

Within a conductor such as copper, the electric fields produced by individual electrons create a constant movement of electrons within the conductor. However, this movement is completely random. When voltage is applied across the two ends of a conductor, an electric field is created, which causes the movement of electrons to become organised. The electric field pushes electrons through the conductor in an orderly fashion from the negative side of the voltage to the positive side. The result is current.

Unlike current, an electric field doesn't need a conductor to travel. An electric field can reach right across an insulator (see Book 1, Chapter 1) and push away or tug on charges on the other side of the insulator.

This ability may look like magic, but you've encountered it before. Blow up a balloon, rub it quickly against your shirt and then hold it up to your hair. Rubbing the balloon creates a static charge in the balloon. When this charge comes close to your hair, its electric field tugs the charges in your hair, which is drawn up towards the balloon.

Catching up with capacitors

A *capacitor* (or sometimes simply a *cap*) is an electronic component that takes advantage of the apparently magical ability of electric fields to reach out across an insulator. It consists of two flat plates made from a conducting material such as silver or aluminium, separated by a thin insulating material such as Mylar or ceramic, as shown in Figure 3-1. The two conducting plates are connected to terminals so that a voltage can be applied across the plates.

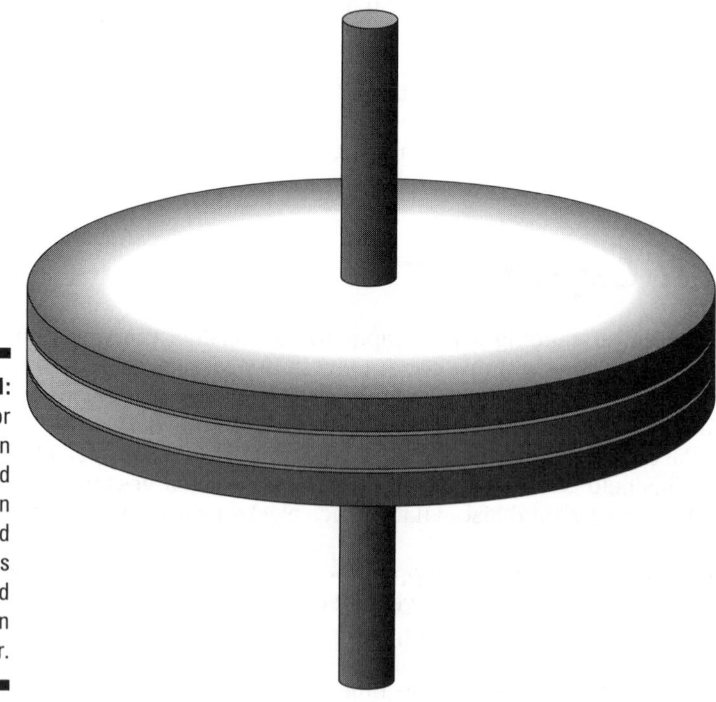

Figure 3-1: A capacitor creates an electric field between two charged plates separated by an insulator.

Note that because an insulator separates the two plates, a closed circuit *isn't* formed. Nevertheless, current flows – for a moment, anyway.

How can this be? When the voltage from a source such as a battery is connected, the negative side of the battery voltage immediately begins to push negative charges towards one of the plates. Simultaneously, the positive side of the battery voltage begins to pull electrons (negative charges) away from the second plate.

The electric field that quickly builds up between the two plates permits current to flow. As the plate on the negative side of the circuit fills with electrons, the electric field created by those electrons begins to push electrons away from the plate on the other side of the insulator, towards the positive side of the battery voltage.

As this current flows, the negative plate of the capacitor builds up an excess of electrons, whereas the positive side develops a corresponding deficiency of electrons. Thus, voltage is developed between the two plates of the capacitor. (Remember, the definition of *voltage* is the difference of charge between two points.)

The catch is that this current flows only for a brief time. As the electrons build up on the negative plate and are depleted from the positive plate, the voltage between the two plates increases because the difference in charge between the two plates increases. The voltage continues to increase until the capacitor voltage equals the battery voltage. When the voltages are the same, current stops flowing through the circuit, and the capacitor is said to be *charged.*

At this point the magic gets even better. After a capacitor has been charged, you can disconnect the battery from the capacitor and the charge remains in the capacitor. In other words, although the battery creates the voltage in the capacitor, this voltage isn't dependent on the battery for its continued existence. Disconnect the battery and the voltage remains across the two plates of the capacitor.

Therefore, capacitors have the ability to store charge – an ability known as *capacitance.*

If a charged capacitor is connected to a circuit, the voltage across the plates drives current through the circuit, which is called *discharging* the capacitor. Just as the current that charges a capacitor lasts for only a short time, the current that results when a capacitor discharges also lasts for only a short time. As the capacitor discharges, the charge difference between the two plates decreases and the electric field collapses. When the two plates reach equilibrium, the voltage across the plates reaches zero and no current flows.

Charged or not charged?

From a pure physics standpoint, saying that a capacitor has the ability to store charge isn't quite accurate. What a capacitor actually stores is energy in the form of an electric field that creates a voltage across the two plates. Strictly speaking, the capacitor doesn't have any more charge in it when fully 'charged' than when it isn't.

The difference is that when a capacitor isn't charged, the total negative and positive charges of the capacitor are divided evenly between the plates, and so no voltage exists across the plates. In contrast, when a capacitor is charged, the negative charges are concentrated on one plate and the positive charges on the other plate. The total amount of charge in the capacitor is the same either way.

By the way (for bonus points), the insulating material between the two conducting plates is called *dielectric.* This term refers to the ability of the insulating layer to become polarised by the electric field that exists between the two plates when they become charged.

Knowing the capacitor symbols

Here are some of the capacitor symbols you encounter in schematic diagrams:

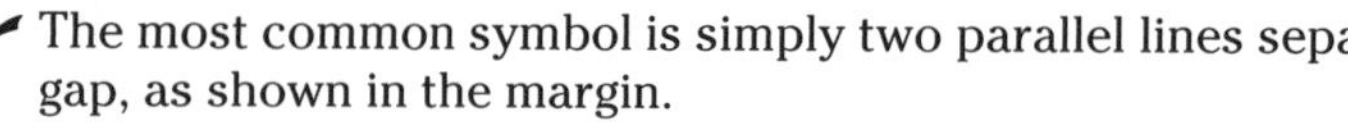

- The most common symbol is simply two parallel lines separated by a gap, as shown in the margin.

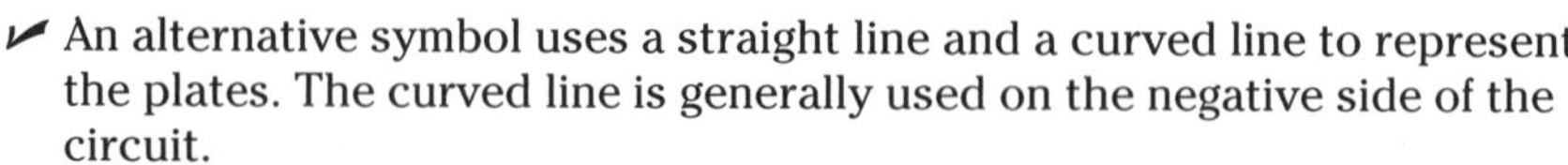

- An alternative symbol uses a straight line and a curved line to represent the plates. The curved line is generally used on the negative side of the circuit.

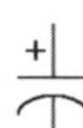

- Although some capacitors aren't sensitive to polarity, many others are. This sensitivity is connected to the choice of materials used to create the capacitors: with certain materials, connecting the voltage in the wrong direction can damage the capacitor. Capacitors with distinct positive and negative terminals are called *polarised capacitors.* A plus sign is used in the schematic diagram to indicate the polarity, as shown in the margin.

Counting Capacitance

Capacitance refers to the ability of a capacitor to store charge. It's also the measurement used to indicate how much energy a particular capacitor can store. The more capacitance a capacitor has, the more charge it can store.

Finding out about farads

Capacitance is measured in units called *farads* (abbreviated F). The definition of 1 farad is deceptively simple. A 1-farad capacitor holds a voltage across the plates of exactly 1 V when it's charged with exactly 1 ampere per second of current.

Note that '1 ampere per second of current' refers to the amount of charge present in the capacitor. No rule states that the current has to flow for a full second. It can be 1 ampere for 1 second, or 2 amperes for half a second or half an ampere for 2 seconds. Or it can be 100 mA for 10 seconds or 10 mA for 100 seconds.

One ampere per second corresponds to the standard unit for measuring electric charge, called the *coulomb.* So another way of stating the value of 1 farad is to say that it's the amount of capacitance that can store 1 coulomb with a voltage of 1 V across the plates.

One farad is a huge amount of capacitance, because 1 coulomb is a very large amount of charge. To put it into perspective, the total charge contained in an average lightning bolt is about 5 coulombs, and you need only five 1-farad capacitors to store the charge contained in a lightning strike. (Some lightning strikes are much more powerful, as much as 350 coulombs.)

Capacitors used in electronics are charged from *much* more modest sources than lightning bolts. In fact, the largest capacitors you're likely to use have capacitance measured in millionths of a farad, called *microfarads* and abbreviated *μF* (the μ is the Greek letter mu, a common abbreviation for micro). Smaller capacitors are measured in millionths of a microfarad, called a *picofarad* and abbreviated *pF.*

Values of 1,000 pF or more are commonly represented in μF rather than pF. For example, 1,000 pF is written as 0.001 μF and 22,000 pF is written as 0.022 μF.

Like resistors (which we discuss in Chapter 2 of this minibook), capacitors aren't manufactured to perfection. Instead, most capacitors have a margin of error, also called *tolerance.* In some cases, the margin of error may be as much as 80%. Fortunately, that degree of imprecision rarely has a noticeable effect on most circuits.

Reading capacitor values

If enough room is available on the capacitor, most manufacturers print the capacitance directly on the capacitor along with other information such as the working voltage and perhaps the tolerance. Small capacitors don't have enough room for all that, however, and so many manufacturers use a shorthand notation to indicate capacitance on small caps.

Historical note

Like many units of measure in electronics, the farad was named after one of the all-time great pioneers of electricity, an Englishman named Michael Faraday. He did the groundbreaking research into magnetism and its relationship with electric current.

As an example of how much Faraday is respected for his work, Albert Einstein kept portraits of three people on the wall of his study at Princeton. They were Sir Isaac Newton, considered by many to be the greatest physicist of all time; James Maxwell, who developed the theory of electromagnetism, perhaps as great an accomplishment as Newton's work with gravity; and Michael Faraday.

If you have a capacitor with nothing other than a three-digit number printed on it, the third digit represents the number of zeros to add to the end of the first two digits. The resulting number is the capacitance in pF. For example, *101* represents 100 pF: the digits 1 and 0 followed by one additional zero.

If only two digits are listed, the number is simply the capacitance in pF. Thus, the digits *22* indicate a 22 pF capacitor.

Table 3-1 lists how some common capacitor values are represented using this notation.

Table 3-1 Capacitance Markings

Marking	*Capacitance (pF)*	*Capacitance (μF)*
101	100	0.0001
221	220	0.00022
471	470	0.00047
102	1,000	0.001
222	2,200	0.0022
472	4,700	0.0047
103	10,000	0.01
223	22,000	0.022
473	47,000	0.047
104	100,000	0.1
224	220,000	0.22

Marking	Capacitance (pF)	Capacitance (μF)
474	470,000	0.47
105	1,000,000	1
225	2,200,000	2.2
475	4,700,000	4.7

You may also see a letter printed on the capacitor to indicate the tolerance. You can interpret the tolerance letter according to Table 3-2.

Table 3-2 Capacitor Tolerance Markings

Letter	Tolerance
A	±0.05 pF
B	±0.1 pF
C	±0.25 pF
D	±0.5 pF
E	±0.5%
F	±1%
G	±2%
H	±3%
J	±5 %
K	±10%
L	±15%
M	±20%
N	±30%
P	–0%, + 100%
S	–20%, + 50%
W	–0%, + 200%
X	–20%, + 40%
Z	–20%, + 80%

Note that the tolerances for codes P to Z are a little odd:

- **Codes P and W:** The manufacturer promises that the capacitance is no less than the stated value, but may be as much as 100% or 200% over the stated value. For example, if the marking is 101P, the actual capacitance is no less than 100 pF but may be as much as 200 pF.

- **Codes S, X and Z:** The actual capacitance may be as much as 20% below the stated value or as much as 50%, 40% or 80% over the stated value. So, if the marking is 101Z, the capacitance is between 80 pF and 180 pF.

Sizing Up Capacitors

Capacitors come in all sorts of shapes and sizes, influenced mostly by three things: the type of material used to create the plates, the type of material used for the dielectric (the insulating material between the two conducting plates) and the capacitance. Figure 3-2 shows some varieties of capacitors.

Figure 3-2: Capacitors are made in many different shapes and sizes.

The most common types of capacitors are as follows:

- **Ceramic disc:** The plates are made by coating both sides of a small ceramic or porcelain disc with silver solder. The ceramic or porcelain disc is the dielectric and the silver solder forms the plates. Leads are soldered to the plates, and the entire thing is dipped in resin.

  Ceramic disc capacitors are small and usually have low capacitance values, ranging from 1 pF to a few microfarads. This small size means

that their values are usually printed using the three-digit shorthand notation that we describe in the preceding section.

Ceramic disc capacitors aren't polarised and so you don't have to worry about polarity when using them.

- **Electrolytic:** One of the plates is made by coating a foil film with a highly conductive, semiliquid solution called *electrolyte.* The other plate is another foil film on which an extremely thin layer of oxide has been deposited; this thin layer serves as the dielectric. The two layers are then rolled up and enclosed in a metal can, epoxy or plastic.

  WARNING!

  Electrolytic capacitors are polarised, and so you have to be sure to connect voltage to them in the proper direction. If you apply voltage in the wrong direction, the capacitor may be damaged or even explode.

  You find two common types of electrolytic capacitors:

  - *Aluminium:* Can be quite large, with as much as a tenth of a farad or more (100,000 µF).
  - *Tantalum:* Smaller and range up to about 1,000 µF.

- **Film:** The dielectric is made from a thin film-like sheet of insulating material and the plates are made from film-like sheets of metal foil. In some cases, the plates and the dielectric are then tightly rolled together and enclosed in a metal or plastic can. In other cases, the layers are stacked and then dipped in epoxy.

  Depending on the materials used, capacitance for film capacitors can be as small as 1,000 pF or as large as 100 µF. Film capacitors aren't polarised.

- **Silver mica:** The dielectric is made from mica, and this capacitor is sometimes referred to as a *mica capacitor.* As with ceramic capacitors, the plates in a silver mica capacitor are made from silver. Electrodes are joined to the plates and then the capacitor is dipped in epoxy.

  Silver mica capacitors come in about the same capacitance range as ceramic disc capacitors. However, they can be made to much higher tolerances – as close as 1% in some cases. Like ceramic disc capacitors, silver mica capacitors aren't polarised.

  Although ceramic disc and mica capacitors are constructed in a similar way, they're easy to tell apart. Ceramic disc capacitors are thin, flat discs and are nearly always a yellowish-brown or orange colour. Silver mica capacitors are thicker, bulge at the ends where the leads are attached, and are shiny and sometimes colourful – red, blue, yellow and green are common colours for mica capacitors. (Interestingly enough, we've never seen a silver silver mica capacitor, if you get what we mean!)

- **Variable:** A capacitor whose capacitance can be adjusted by turning a knob. One common use for a variable capacitor is to tune a radio circuit to a specific frequency.

In the most common type of variable capacitor, air is used as the dielectric and the plates are made of rigid metal. As shown in Figure 3-3, several pairs of plates are typically used in an intermeshed arrangement. One set of plates is fixed (not moveable), but the other set is attached to a rotating knob. When you turn the knob, you change the amount of surface area on the plates that overlap. This, in turn, changes the capacitance of the device.

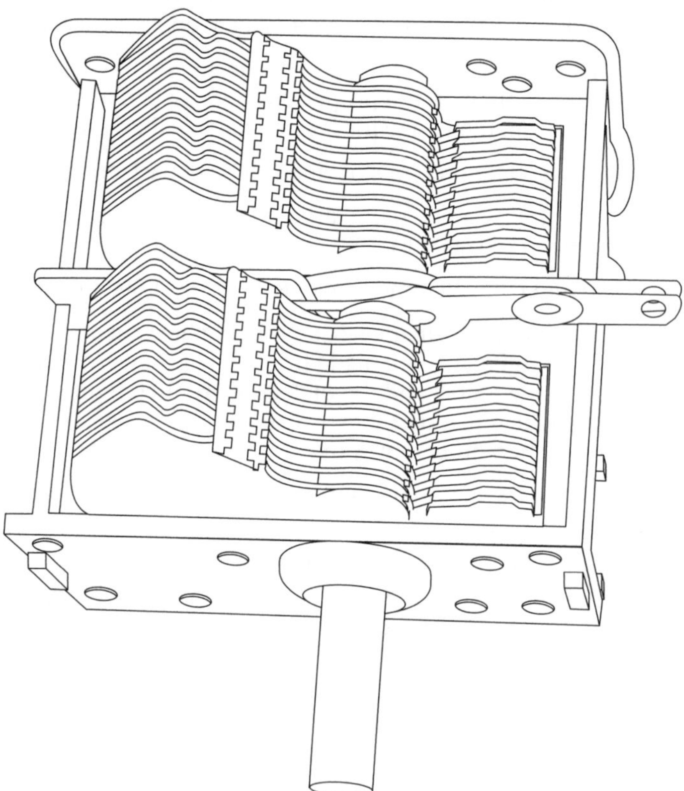

Figure 3-3: A variable capacitor.

The schematic symbol for a variable capacitor is shown in the margin.

Calculating Time Constants for Resistor/Capacitor Networks

As we mention earlier in the 'What Is a Capacitor' section, when you put a voltage across a capacitor, a bit of time is necessary for the capacitor to charge fully. During this period, current flows through the capacitor. Similarly, when you discharge a capacitor by placing a load across it, the capacitor takes time to discharge fully. Knowing exactly how much time charging a capacitor takes is one of the keys to using capacitors correctly in your circuits, and you can get that information by calculating the RC time constant.

Calculating the exact amount of time required to charge or discharge a capacitor requires more maths than we're willing to throw at you in this book; simple algebra is one thing, but we draw the line at calculus! Fortunately, you can skip the calculus and use simple algebra if you're willing to settle for close approximations. Given that most resistors and capacitors have a tolerance of plus or minus 5 or 10% anyway, using approximate calculations doesn't have any significant effect on your circuits.

Looking at the calculations

Before making the approximate calculations, we need to emphasise the importance of understanding the concepts behind the calculus – even if you don't do the calculus itself.

When a capacitor is charging, current flows from a voltage source through the capacitor. In most circuits, a resistor is working in series with the capacitor as well, as shown in Figure 3-4.

In fact, a circuit always has resistance even if you don't use a resistor, because the perfect conductor doesn't exist – even solid wire has some resistance. For the purposes of this discussion, assume that a resistor is in the circuit, but that its resistance may be 0 Ω.

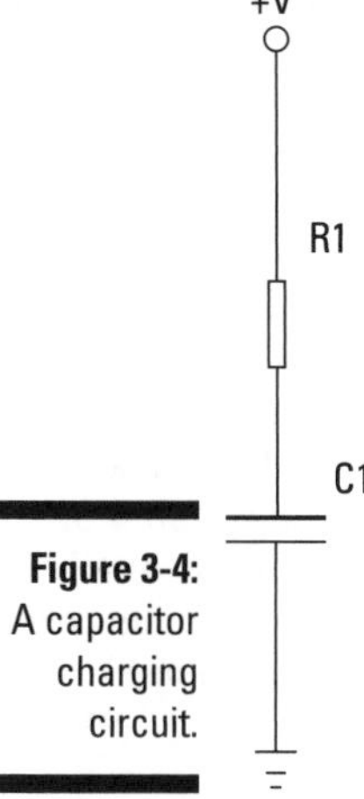

Figure 3-4: A capacitor charging circuit.

The rate at which the capacitor charges through a resistor is called the *RC time constant* (the *RC* stands for *resistor-capacitor*), which you can calculate by multiplying the resistance in ohms by the capacitance in farads. Here's the formula:

$$T = R \times C$$

For example, suppose the resistance is 10 kΩ and the capacitance is 100 μF. Before you do the multiplication, you must first convert the μF to farads. One μF is one-millionth of a farad, and so you can convert μF to farads by dividing the μF by one million. Therefore, 100 μF is equivalent to 0.0001 F. Multiplying 10 kΩ by 0.0001 F gives you a time constant of 1 second.

If you want to increase the RC time constant, you can increase the resistance or the capacitance, or both. Also, you can use an infinite number of combinations of resistance and capacitance values to reach a desired RC time constant. For example, all the following combinations of resistance and capacitance yield a time constant of 1 second:

Resistance	***Capacitance***	***RC Time Constant***
1 kΩ	1,000 μF	1 s
10 kΩ	100 μF	1 s
100 kΩ	10 μF	1 s
1 MΩ	1 μF	1 s

Appreciating the RC time constant

Importantly, in each interval of the RC time constant, the capacitor moves 63.2% closer to a full charge. For example, after the first interval, the capacitor voltage equals 63.2% of the battery voltage. So if the battery voltage is 9 V, the capacitor voltage is just under 6 V after the first interval, leaving it just over 3 V away from being fully charged.

In the second time interval, the capacitor picks up 63.2%, not of the full 9 V of battery voltage but 63.2% of the difference between the starting charge (just under 6 V) and the battery voltage (9 V). Thus, the capacitor charge picks up just over 2 additional volts, bringing it up to about 8 V.

This process keeps repeating: in each time interval, the capacitor picks up 63.2% of the difference between its starting voltage and the total voltage. In theory, the capacitor is never fully charged, because with the passing of each RC time constant it picks up only a percentage of the remaining available charge. But within just a few time constants, the capacitor becomes very close to fully charged (check out the nearby sidebar 'Full charge ahead!' for a little more detail).

Full charge ahead!

Although in theory a capacitor never becomes fully charged, in reality the capacitor does eventually become fully charged (say what!). The difference between theory and reality in this context is that in mathematics you can use as many digits beyond the decimal point as you want. In other words, no limit exists to how small a number can be. But a limit does exist to how small a charge can be: it can't be smaller than a single electron.

When enough time constants have elapsed, the difference between the battery voltage and the capacitor charge is a single electron. When that electron joins the charge, the capacitor is full.

The following minitable gives you a helpful approximation of the percentage of charge that a capacitor reaches after the first five time constants. For all practical purposes, you can consider the capacitor fully charged after five time constants have elapsed.

RC Time Constant Interval	*Percentage of Total Charge*
1	63.2%
2	86.5%
3	95.0%
4	98.2%
5	99.3%

Combining Capacitors

In Chapter 2 of this minibook, we explain how you can combine resistors in series or parallel networks to create any arbitrary resistance value you need. Well, you can do the same with capacitors. But as we discuss in this section, the formulas for calculating the total capacitance of a capacitor network are the reverse of the rules you follow to calculate resistor networks.

In other words, the formula you use for resistors in series applies to capacitors in parallel, and the formula you use for resistors in parallel applies to capacitors in series.

Combining capacitors in parallel

Calculating the total capacitance of two or more capacitors in parallel is simple: just add up the individual capacitor values to get the total capacitance.

For example, if you combine three 100 μF capacitors in parallel, the total capacitance of the circuit is 300 μF.

This rule makes sense because when you connect capacitors in parallel, you're essentially connecting the plates of the individual capacitors. Therefore, connecting two identical capacitors in parallel doubles the size of the plates, which effectively doubles the capacitance.

Figure 3-5 shows how parallel capacitors work. Here, the two circuits have identical capacitances. The right-hand circuit accomplishes the job with one capacitor and the left-hand one does it with three, but the circuits are equivalent.

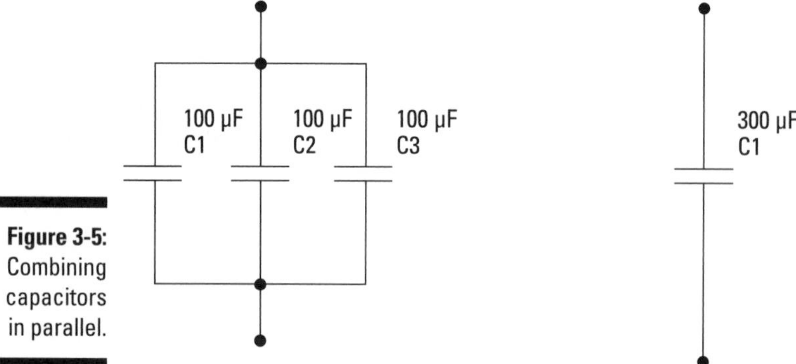

Figure 3-5: Combining capacitors in parallel.

Whenever you see two or more capacitors in parallel in a circuit, you can substitute a single capacitor whose value is the sum of the individual capacitors. Similarly, any time you see a single capacitor in a circuit, you can substitute two or more capacitors in parallel as long as their values add up to the original value.

The total capacitance of capacitors in parallel is always greater than the capacitance of any of the individual capacitors, because each capacitor adds its own capacitance to the total.

Connecting capacitors in series

You can also combine capacitors in series to create equivalent capacitances, as shown in Figure 3-6. When you do, however, the maths is a little more complicated; the calculations required for capacitors in series are the same as for calculating resistors in parallel.

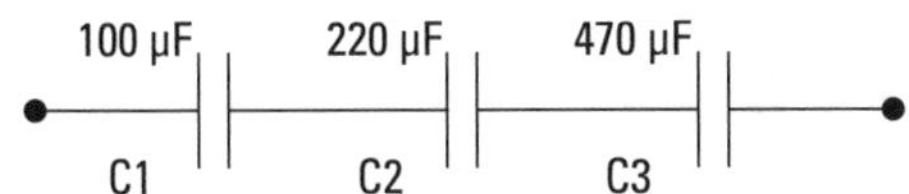

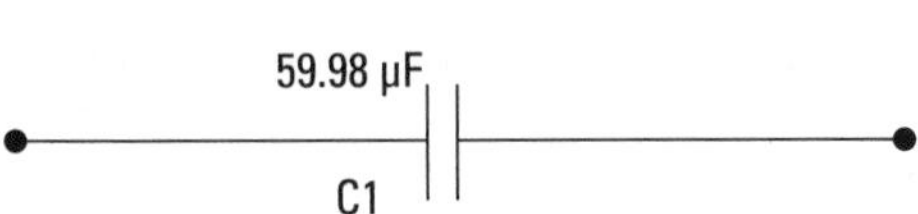

Figure 3-6: Combining capacitors in series.

Here are the rules for calculating capacitances in series:

- **Capacitors of equal value in series:** You're in luck! All you have to do is divide the value of one of the individual capacitors by the number of capacitors. For example, the total capacitance of two 100 µF capacitors is 50 µF.
- **Two capacitors of different values in series:** Use this calculation (*C*1 and *C*2 are the values of the two capacitors):

  $$C_{total} = \frac{C1 \times C2}{C1 + C2}$$

  Here's an example, based on a 220 µF and a 470 µF capacitor in series:

  $$C_{total} = \frac{C1 \times C2}{C1 + C2} = \frac{220\ \mu\text{F} \times 470\ \mu\text{F}}{220\ \mu\text{F} + 470\ \mu\text{F}} = 149.855\ \mu\text{F}$$

- **Three or more capacitors of different values in series:** The formula is as follows:

  $$C_{total} = \frac{1}{\frac{1}{C1} + \frac{1}{C2} + \frac{1}{C3} \cdots}$$

  Note that the ellipsis at the end of the expression indicates that you keep adding up the reciprocals of the capacitances for as many capacitors as you have.

  Here's an example for three capacitors whose values are 100 µF, 220 µF and 470 µF:

  $$C_{total} = \frac{1}{\frac{1}{100\ \mu\text{F}} + \frac{1}{220\ \mu\text{F}} + \frac{1}{470\ \mu\text{F}}} = \frac{1}{0.016673\ \mu\text{F}} = 59.977\ \mu\text{F}$$

  As you can see, the final result is 59.977 µF. Unless your middle name is Pedantic, you probably don't care about the answer being quite so precise, and so you can safely round it to an even 60 µF.

Putting Capacitors to Work

Capacitors can be incredibly useful in your circuits. Here are the most common ways to put capacitors to work:

- **Storing energy:** When charged, a capacitor has the ability to store a lot of energy and discharge it when needed (see Project 3-1 in the next section). Two of the most familiar uses of this ability are in a camera's flash and in devices such as alarm clocks that need to stay powered up even when household power is disrupted for a few moments.
- **Timing circuits:** Many circuits depend on capacitors along with resistors to provide timing intervals. For example, in Chapter 6 of this minibook you discover how to use a resistor and capacitor together along with a transistor to create a circuit that flashes an LED on and off.
- **Stabilising shaky direct current (DC):** Many electronic devices run on DC but derive their power from household alternating current (AC). These devices use a power-supply circuit that converts AC to DC. As we discuss in Book IV, Chapter 3, an important part of this circuit is a capacitor that creates steady DC voltage from the constantly changing voltage of an AC circuit.

- **Blocking DC while passing AC:** Sometimes you need to prevent DC from flowing while allowing AC to pass. A capacitor can do this job for you. The basic trick is to choose a capacitor that fully charges and discharges within the AC cycle. Remember that as the capacitor is charging and discharging, it allows current to flow. So as long as you can keep the capacitor charging and discharging, it passes the current. Flip to Project 3-2 in the later section 'Blocking DC while Passing AC'.
- **Filtering certain frequencies:** Capacitors are often at work in audio or radio circuits to select certain frequencies. For example, a variable capacitor (check out the earlier section 'Sizing Up Capacitors') is often used in a radio circuit to tune the circuit to a certain frequency. You see how that works in Book V, Chapter 2.

Charging and discharging a capacitor

One of the most common uses for capacitors is to store a charge that you can discharge when needed. Project 3-1 presents a simple construction project on a solderless breadboard that demonstrates how you can use a capacitor to this end. You connect an LED to a 3 V battery power supply and use a capacitor so that when you disconnect the battery from the circuit, the LED doesn't immediately go out. Instead, it continues to glow for a moment as the capacitor discharges.

Figure 3-7 shows the completed project.

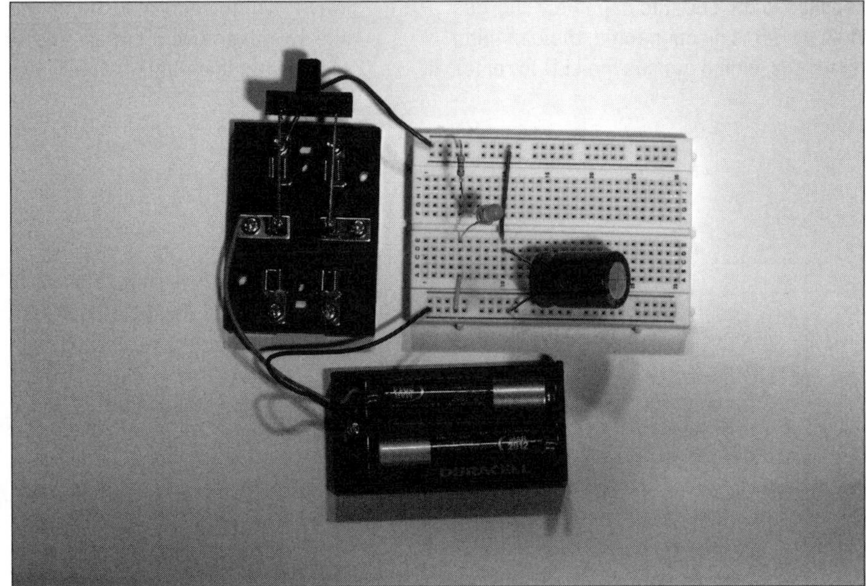

Figure 3-7: The capacitor discharge circuit.

Here are some additional points to consider or things you may want to try after you complete Project 3-1:

- Pull the LED out of the breadboard, and then close the switch to charge the capacitor. After a few seconds, open the switch. Then, use the voltmeter function of your multimeter to measure the voltage across the two leads of the capacitor. Notice that even though the battery is disconnected from the circuit, the capacitor reads 3 V.
- Try varying the size of the capacitor to see what happens. If you use a smaller capacitor, the LED extinguishes more quickly when you remove the power. If you use a larger capacitor, the LED fades more slowly.

- You may not be able to find a capacitor larger than 1,000 μF, but you can increase the capacitance by adding more capacitors in parallel with the 1,000 μF capacitor.
- Try adding a resistor in series with the capacitor. To do that, simply replace the jumper wire you connect in Step 6 of Project 3-1 with a resistor.

Project 3-1: Charging and Discharging a Capacitor

In this project, you build a circuit that places a capacitor in parallel with an LED. When power is applied to the LED, the capacitor charges. When power is disconnected, the capacitor discharges, which causes the LED to remain lit for a short while after the power has been disconnected. You need a Phillips-head screwdriver to secure the connections to the knife switch and wire cutters and strippers to cut and strip the jumper wires.

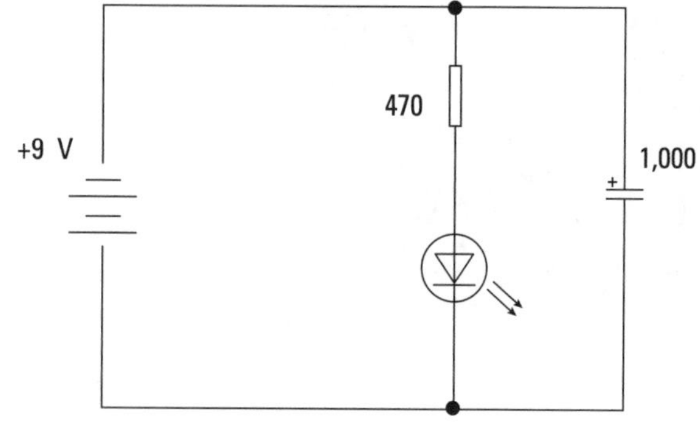

Parts List
1 Solderless breadboard
2 AA batteries
1 Battery holder
1 470 Ω resistor (yellow-violet-brown)
1 1,000 μF electrolytic capacitor
1 Red LED, 5 mm
1 DPDT knife switch
1 1–2 cm 22-AWG solid jumper wire
1 2–3 cm 22-AWG solid jumper wire
1 7–8 cm length 22-AWG solid jumper wire

Steps

1. **Connect the battery holder.**

   Insert the black lead in the first hole on the ground bus strip, located at the bottom of the solderless breadboard near hole A1.

   Connect the red lead of the battery holder to terminal 1X on the switch.

2. **Connect the switch to the breadboard.**

   Use the 7–8 cm solid jumper wire to connect terminal 1A on the switch to the first hole in the positive bus strip on the top of the solderless breadboard (near hole hole J1).

3. **Connect the resistor.**

   Insert the resistor into the breadboard so that one lead is on the positive voltage bus strip (the strip that's connected to the jumper leading to the switch) and the other lead is in hole 5G.

4. **Connect the LED.**

   Notice that one lead of the LED is shorter than the other. Insert the short lead into hole 5E and the long lead into hole 5F.

   Note that the circuit doesn't work if you insert the LED backwards. The short lead must be on the negative side of the circuit.

5. **Use the 1–2 cm jumper wire to connect the ground bus to the LED.**

   Insert one end of the jumper wire into a hole in the ground bus near row 5. Then, insert the other end row 5. Then, insert the other end into hole 5A.

6. **Use the 2–3 cm jumper wire to connect the capacitor to the positive bus.**

   Insert one end of the jumper wire in hole 10E and the other end in a hole in the positive bus strip near row 10.

7. **Insert the capacitor.**

   The negative terminal of the capacitor, marked by a series of dashes on the capacitor, goes into a hole in the ground strip near row 10. The other lead goes in hole 10D.

8. **Insert the batteries.**

9. **Close the switch.**

   Push the lever into the A position to close the circuit. The LED lights up. You can't see it, but the capacitor charges up in just a few seconds.

10. **After a few seconds, open the switch.**

    Watch how the LED goes out. Instead of turning off instantly, it fades over the course of a few seconds. That's because when the battery is disconnected from the circuit, the capacitor discharges through the LED. As the capacitor discharges, the LED goes gradually dim and then goes out altogether.

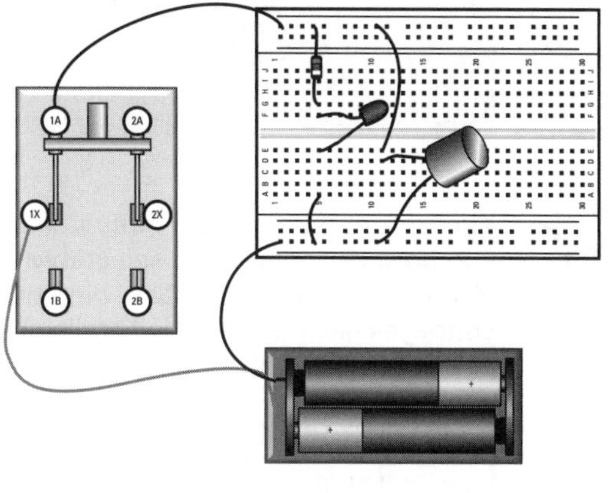

Blocking DC while passing AC

One of the important features of capacitors is their ability to block DC while allowing AC to pass. This works because of the way a capacitor allows current to flow while the plates are charging or discharging, but halts current in its tracks when the capacitor is fully charged. Thus, as long as the source voltage is constantly changing, the capacitor allows current to flow. When the source voltage is steady, the capacitor blocks current from flowing.

In Project 3-2, you build a simple circuit that demonstrates this principle in action. The circuit places a resistor and a LED in series with a capacitor. Then, it uses a knife switch so that you can switch the voltage source for the current between a 9 V battery (DC) and a 9 VAC power adapter (AC). Figure 3-8 shows the finished project.

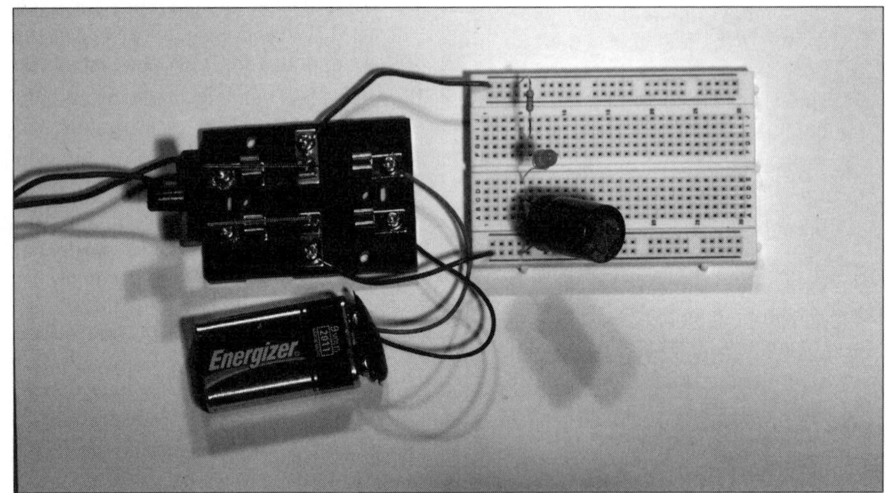

Figure 3-8: The finished AC/DC circuit.

The trick to building this circuit is finding a 9 VAC power adapter. You can find power supplies with selectable outputs and a choice of connectors at electric shops for under £20, but you may find one much cheaper if you keep your eyes peeled on market stalls or pound shops. The printing can be difficult to see but look for markings on the back that read something like this:

- INPUT: 100-230VAC 50-60Hz 0.15A
- OUTPUT: 5VAC 0.5A

The input must be compatible with the UK mains supply at 230-240 VAC, but for this project the output voltage must be listed at 9 VAC rather than DC.

Project 3-2: Blocking Direct Current

In this project, you build a circuit that uses a capacitor to allow alternating current to pass but blocks direct current. The circuit simply places a capacitor in series with an LED, and then uses a DPDT switch to connect the LED/capacitor circuit to 9 VDC or 9 VAC.

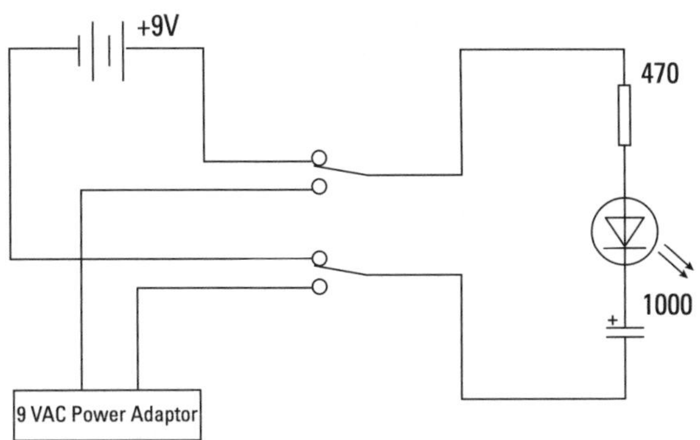

Parts List

- 1 Solderless breadboard
- 1 9 V battery
- 1 9 V battery snap holder
- 1 9 VAC power adapter
- 1 470 Ω, resistor (yellow-violet-brown)
- 1 1,000 μF electrolytic capacitor
- 1 Red LED, 5 mm
- 1 DPDT knife switch
- 2 7–8 cm 22-AWG solid jumper wire

Steps

1. **Connect the battery snap holder to the switch.**

   Connect the red lead to terminal 1A on the switch, and then connect the black lead to terminal 2A.

2. **Connect the 9 VAC adapter to the switch.**

   Connect the 9 VAC adapter to terminals 1B and 2B on the switch. The easiest way to connect the power adapter is to cut the plug off the end of the power adapter's output cable, and then strip about 1 cm of insulation from the ends of the two wires that make up the cable.

3. **Connect the switch to the breadboard.**

   Use one of the 7–8 cm solid jumper wires to connect terminal 1X on the switch to the first hole in the positive bus strip on the top of the solderless breadboard. Then, use the other 7–8 cm jumper wire to connect terminal 2X on the switch with the first hole in the ground bus strip at the bottom of the breadboard.

4. **Connect the resistor.**

   Insert the resistor into the breadboard so that one lead is on the positive voltage bus strip (the strip that's connected to the jumper leading to switch terminal 1X) and the other lead in hole 5I.

5. **Connect the LED.**

   Insert the short LED lead into hole 5E and the long lead into hole 5F. Note that the circuit doesn't work if you insert the LED backwards. The short lead must be on the negative side of the circuit.

6. **Insert the capacitor.**

   The negative terminal of the capacitor, marked by a series of dashes on the capacitor, goes into a hole in the ground strip near row 5. The other lead goes in hole 5C.

7. **Move the switch to the vertical position so that neither side of the switch is closed.**

8. **Insert the 9 V battery.**

9. **Plug the 9 VAC power adapter into a wall outlet.**

   The circuit is now finished and ready for testing.

10. **Close the switch toward the A contacts.**

    This connects the 9 V battery to the circuit. The LED initially turns on, but after a few moments it begins to fade. When it goes completely out, the capacitor is fully charged. When charged, the capacitor doesn't allow direct current to pass, and so the LED remains dark.

11. **Flip the switch to the B contacts.**

    This action disconnects the battery from the circuit and instead connects the AC power adapter. Now the LED comes on and stays on. As long as the AC power source is applied across the circuit, the capacitor allows the current to flow, and the LED lights up.

12. **Flip the switch back to the A contacts.**

    Because the capacitor is now fully charged, the LED goes out immediately.

13. **Open the switch, and then short out the capacitor leads.**

    With the switch in the neutral position so that neither power source is connected to the circuit, touch both leads of the capacitor with the blade of your screwdriver. This shorts out the leads, which causes the capacitor to quickly discharge itself.

14. **You're done!**

Chapter 4

Working with Inductors

In This Chapter

- Unravelling the mystery of induction
- Discovering reactance
- Combining inductors in series and parallel
- Using inductors in electronic circuits

Nearly a century ago, when radio was brand new and the invention of the transistor was another quarter of a century away, a book was published called *A Course in Electrical Engineering,* written by Chester L. Dawes. Technology moves so quickly that you'd think that an electrical engineering book written in 1920 would be completely obsolete today. But here's how Chapter 1 begins:

> Magnets and magnetism are involved in the operation of practically all electrical apparatus. Therefore an understanding of their underlying principles is essential to a clear conception of the operation of all such apparatus.

On the face of it magnets don't seem to have much to do with today's solid-state computing or LCD televisions. Yet electricity and magnetism are closely related and the relationship between electric current and magnetism still forms the core of many different types of essential circuits and everyday gadgets. For example, a power adapter that converts 230 VAC from a wall outlet to a safer level of 9 V uses a transformer that relies on magnetism to step down the voltage. An analogue multimeter (the kind with a needle that moves with voltage, current or resistance and which we describe in Book I, Chapter 8) relies on magnetism to deflect the needle. Plus, electric motors use magnetism to convert electric current into motion.

In this chapter, we look at a special class of component called *inductors,* which exploit the nature of magnetism and its relationship to electric current. Inductors are sort of like cousins to capacitors (which we discuss in Chapter 3 of this minibook), in that you use them to do similar things and they play by similar rules. For example, an important characteristic of a capacitor is its ability to oppose changes in voltage. Inductors have a similar ability to oppose changes, but in current not voltage.

This chapter is a bit different from the others in this minibook, because it contains no construction projects. The practical applications for inductors are in circuits that you may not be ready to build yet, such as radio tuners or power supplies. We explain the important concepts of how inductors work in this chapter so that later, when you come to use an inductor in a circuit, you know what it does. You can then use this chapter to refresh your memory as and when the need arises.

What Is Magnetism?

When Albert Einstein was 5-years-old and sick in bed, his father gave him a compass to play with. Young Albert saw that no matter how he spun the compass around, the needle always swung back to point north, and he was amazed. Years later, he wrote that this compass launched his lifelong interest in physics. He realised then that 'something [was] behind things, something deeply hidden'.

In the case of the compass, that deeply hidden thing that Einstein marvelled at was *magnetism*. Although everyone is familiar with magnetism, exactly what it is remains mysterious.

A *magnetic field* extends through space and attracts or repels certain materials. Materials that are strongly affected by magnetic fields are called *magnetic* and materials that create magnetic fields are called *magnets.*

Pointing to the north and south of magnetism

Magnetic fields and electric fields are distinct things, but they're closely related and have much in common. For example, magnetic fields are polarised in much the same way that electric fields are polarised. Electric fields exist between electric charges of opposite polarity (negative and positive) and magnetic fields exist between opposite magnetic poles, called *north* and *south.*

Just as opposite electric charges attract and like charges repel, opposite magnetic poles attract and like magnetic poles repel. That's why if you take two strong magnets and try to push the two north poles together or the two south poles together, the magnet fights back and you can't connect them. But if you turn one of the magnets around and try to hook up the north pole with the south pole, the magnets attract each other and you have trouble keeping them apart.

A magnetic field has a distinct shape that you can visualise by using a simple experiment that's still done in schools. All you need is a bar magnet, some iron filings and a piece of paper. Put the paper over the magnet and sprinkle the filings on the paper. The filings line up to reveal the shape of the magnetic field, as shown in Figure 4-1. As you can see, the filings align themselves from one pole of the magnet to the other.

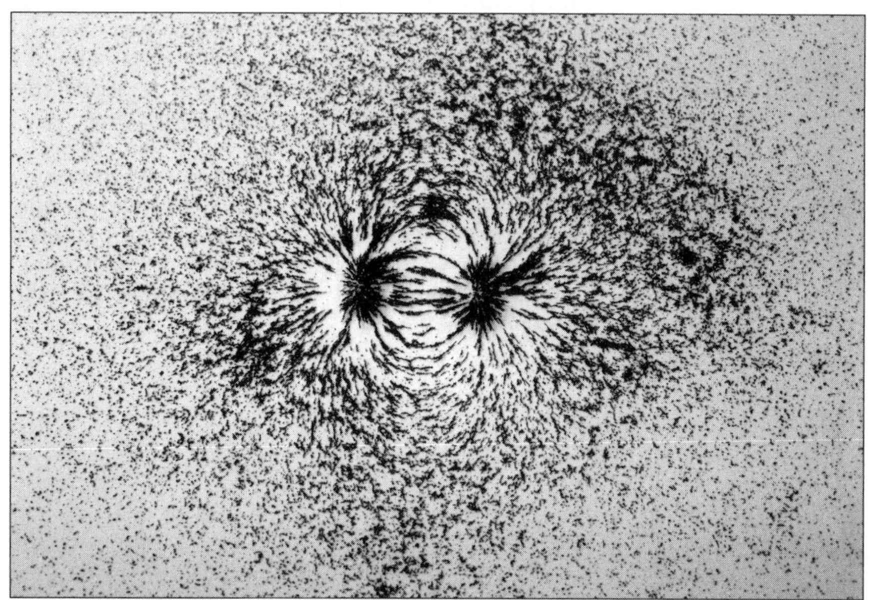

Figure 4-1: Iron filings revealing the shape of a magnetic field.

Pondering permanent magnets

A *permanent magnet* is a material that creates its own magnetic field. Some naturally occurring materials, such as lodestone, are inherently magnetic and produce magnetic fields on their own. But most permanent magnets are made from materials that aren't inherently magnetic, but become so when they're exposed to a powerful magnetic field.

You probably have several permanent magnets around your home, such as on your refrigerator holding up your children's pictures or your shopping list.

Examining Electromagnets

Whereas permanent magnets create a magnetic field all by themselves (see the preceding section), an *electromagnet* relies on the key relationship between electricity and magnetism to create a magnetic field. Specifically, whenever electric current flows, a magnetic field is created by the moving charges. In short, an electron in motion becomes a magnet.

As we say several times in this book, electrons are always in motion, and so, you may wonder, aren't little magnets everywhere? The answer is yes, in the same sense that electric current is everywhere. But when the motion of electrons within a material is random, the magnetic fields created by the electrons are oriented randomly, and so they end up just cancelling each other out. But when you give the electrons a nudge with voltage, they all start moving in the same direction. This strengthens and organises the magnetic fields so they can combine to form one large magnetic field.

The magnetic field created by current flowing through a single wire is measurable but small. If you coil the wire tightly, however, as shown in Figure 4-2, the magnetic fields are strengthened because of their proximity to one another. For example, you can create a simple electromagnet by wrapping a metre or so of small, insulated wire around a pencil, a ballpoint pen or any other rigid tube or cylinder.

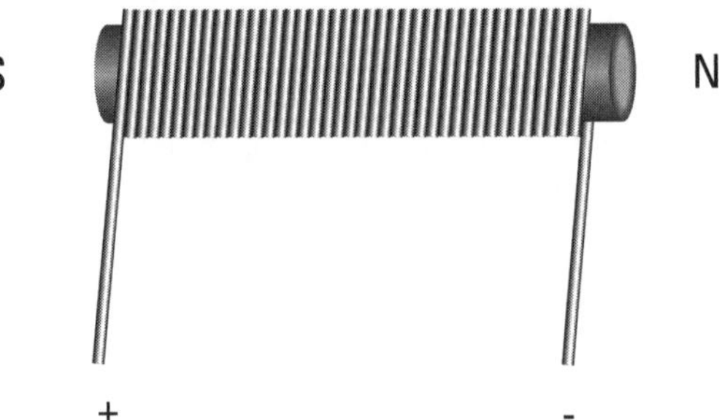

Figure 4-2: An electromagnet.

Michael Faraday strikes again

Michael Faraday discovered electromagnetic induction in 1831. This discovery is one of the most important in the history of electrical science, because it's at the heart of nearly all forms of electrical power generation. Coal-burning, hydroelectric and even nuclear power plants use moving water to spin turbines that are connected to generators. Those generators use the principle of electromagnetic induction to turn the spinning motion of magnetic fields into electric current.

The strength of the magnetic field of such a coil depends on several factors, the most important being:

- **The number of turns in the coil:** More turns makes for a stronger magnetic field.
- **The amount of current flowing through the electromagnet:** Increasing the current increases the strength of the electromagnet exponentially. For example, if you double the current the electromagnet is four times as strong.
- **The material used for the core that the electromagnet is wrapped around:** Any coil wrapped around an iron core is about ten times as strong as the same coil wrapped around an inert core such as air, plastic or wood.

The polarity of an electromagnet is easy to determine: the negative side of the coil is the magnet's north pole and the positive side is the south pole.

Inducing Current

Electromagnets are possible because a moving current creates a magnetic field. Interestingly enough, the reverse is also true: a moving magnetic field creates an electric current. In other words, if you wave a magnet over a wire, you create a current in the wire. This effect is called *electromagnetic induction,* or sometimes just *magnetic induction.*

You can increase the strength of the current induced in a wire by coiling the wire into turns so that you expose a greater length of wire to the magnetic field. Whether the magnet or the coil moves, a current is induced in the coil if the coil is moving relative to the magnet, provided of course that the coil passes through the magnet's magnetic field.

Whenever current is flowing, voltage exists. Thus, in addition to causing current to flow through the coil, electromagnetic induction creates a voltage across the coil.

Resisting change: Inductance

An *inductor* is a coil that's designed for use in electronic circuits. Inductors take advantage of an important characteristic of coils called *self-inductance,* also called just *inductance.* Inductors are simple devices, consisting of nothing more than a coil of wire, often wrapped around an iron core. But their ability to exploit the idea of self-inductance is a stroke of genius.

Self-inductance is similar to electromagnetic induction (which we describe in the previous section), but with a subtle twist. Whereas *electromagnetic induction* refers to the ability of a coil to generate a current when it moves across a magnetic field, *self-inductance* refers to the ability of a coil to create the very magnetic field that then induces the voltage. In other words, with self-inductance, the coil feeds back upon itself. A voltage applied across the coil causes current to flow, which creates a magnetic field, which in turn creates more voltage.

Inductance happens only when the current running through the coil changes, because only a *moving* magnetic field induces voltage in a coil. Whenever the current changes in a coil, the magnetic field created by the current grows or shrinks, depending on whether the current increases or decreases. When the magnetic field grows or shrinks, it's effectively moving, and so a voltage is inducted in the coil as a result of this movement. When the current stays steady, no inductance occurs.

Self-inductance is a tricky concept to get your mind around, and so don't panic if it doesn't make sense to you straight away. We took a while to get this concept clear at first. Therefore, we go over the idea in more detail, point by point, in the following list:

- **When voltage is applied across a coil, the voltage causes current to flow through the coil.** Remember, current always requires voltage, and voltage always results in a current when applied across a conductor.
- **The current flowing through the coil creates a magnetic field around the coil.** Keep in mind that the coil that creates the magnetic field is itself within the field and can therefore be influenced by it.
- **If the current flowing through the coil changes, the magnetic field created by the current also changes.** The magnetic field grows or shrinks, depending on whether the current increases or decreases. Either way, the changing magnetic field is in effect moving.
- **The magnetic field is moving, which means that voltage is induced in the coil.** This voltage is additional, on top of the voltage that's driving the main current through the coil.
- **The amount of voltage induced by the changing magnetic field depends on the speed in which the current changes.** The faster the current changes, the more the magnetic field moves and therefore the more voltage is induced.
- **The polarity of the induced voltage depends on whether the current is increasing or decreasing.** This is because the direction of movement in the magnetic field depends on whether the field is growing or shrinking,

and the voltage induced by a moving magnetic field depends on which direction the field is moving, according to the following rules:

  - *When the current increases,* the polarity of the induced voltage is opposite to the polarity of the voltage driving the coil. This inducted voltage is often called *back voltage,* because it has the opposite polarity to the supply voltage.
  - *When the current decreases,* the resulting self-induced voltage has the same polarity as the supply voltage.

- **The induced voltage creates a current in the coil that flows with or against the main coil current, depending on whether the coil current is decreasing or increasing, according to the following rules:**
  - *When the coil current is increasing,* the additional current flows against the main coil current. This has the effect of pushing back against the increasing main current, which effectively slows down the rate at which the current can change.
  - *When the coil current is decreasing,* the additional current flows with the main coil current, thus counteracting the decrease in coil current.
- **When the coil current stops changing, self-inductance stops.** Thus, when current is steady, an inductor is simply a straight conductor. (It's also an electromagnet (check out the earlier section 'Examining Electromagnets'), because the current travelling through it produces a magnetic field.)

Self-inductance means that an inductor is said to oppose changes in current. If the current increases, an opposite voltage is induced across the coil, which slows the rate at which the current can increase. If the current decreases, a forward voltage is induced across the coil, which slows the rate at which the current decreases. An inductor applies equal opposition to increases and decreases in current.

This ability to oppose changes in current is useful in electronic circuits, as we explain later in the section 'Putting Inductors to Work'.

Here are some additional important details concerning inductors:

- Inductors can't stop changes in current; they can only slow them down. Measuring how much an inductor can slow down a change in current is the topic of the next section.
- The magnetic field from one inductor can spill over into a nearby inductor and induce voltage in it as well. To prevent this from happening,

many inductors have special shielding around them to keep them magnetically isolated.

If you use unshielded inductors in your circuits, be sure to space them as far apart as possible.

Regarding henrys

Inductance is only a momentary occurrence. Exactly how momentary depends on the amount of inductance an inductor has.

Inductance is measured in units called *henrys.* One henry is the amount of inductance necessary to induce 1 V when the current in the coil changes at a rate of 1 ampere per second.

As you may guess, 1 henry is a fairly large inductor. Inductors in the single-digit henry range are often used when dealing with household current (230 VAC at 50 Hz). But for most electronics work, you use inductances measured in thousandths of a henry (*millihenrys,* abbreviated *mH*) or in millionths of a henry (*microhenrys,* abbreviated *μH*).

Here are a few factoids about inductors and henrys:

- **Abbreviation:** The letter *L* is often used to represent inductance in formulas. In addition, inductors in schematic diagrams are usually referenced by 'L'. For example, if a circuit calls for three inductors, they're identified L1, L2 and L3.
- **Name:** The henry is named after Joseph Henry, who discovered self-inductance and invented the inductor. Note that the plural of *henry* is *henrys*, not *henries.*
- **Schematics:** The symbol used to represent inductors in a schematic diagram is shown in the margin.

Calculating RL Time Constants

In Chapter 3 of this minibook, you discover how to calculate the RC time constant for a resistor-capacitor circuit. Well, you can do a similar calculation for inductors, except that instead of being called the RC time constant, it's known as the *RL time constant* (*L* is the symbol for inductance).

The RL time constant indicates the amount of time necessary to conduct 63.2% of the current that results from a voltage applied across an inductor. (If you're thinking that you've seen 63.2% before, you're right! It's the same percentage used to calculate time constants in resistor-capacitor networks. The value 63.2% derives from the calculus equations used to determine the exact time constants for both resistor-capacitor and resistor-inductor networks.)

Here's the formula for calculating an RL time constant:

$$T = \frac{L}{R}$$

In other words, the RL time constant in seconds is equal to the inductance in henrys divided by the resistance of the circuit in ohms.

For example, suppose that the resistance is 100 Ω and the capacitance is 100 mH. Before you do the multiplication, you first need to convert the 100 mH to henrys. Because 1 mH is 1 1-thousandth of a henry, 100 mH is equivalent to 0.1 H. Dividing 0.1H by 100Ω gives you a time constant of 0.001 second (s), or one millisecond (ms).

The following table gives you a helpful approximation of the percentage of current that an inductor passes after the first five time constants. For all practical purposes, you can consider the current fully flowing after five time constants have elapsed.

RL Time Constant Interval	*Percentage of Total Current Passed*
1	62.3%
2	86.5%
3	95.0%
4	98.2%
5	99.3%

Thus, in the example circuit in which the resistance is 100 Ω and the inductance is 0.1 H, you can expect current to be flowing at full capacity within 5 ms of when the voltage is applied.

Small can be beautiful

Five milliseconds is a very short amount of time, but then electronic circuits are often designed to respond within very short time intervals. For example, the sine wave of standard household alternating current swings from its peak positive voltage to its peak negative voltage is about 8 ms. Sound waves at the upper end of the human ear's ability to hear cycle in about 25 μs (microseconds) and the time interval for radio waves can be in small fractions of microseconds. Therefore, very small RL time constants can be extremely useful in certain types of electronic circuits, from audio electronics to smoothing power supplies.

Calculating Inductive Reactance

Although inductors oppose changes in current (as we discuss in the earlier section 'Resisting change: Inductance'), they don't oppose all changes equally.

Inductors present more opposition to fast current changes than they do to slower changes. Or, put another way, inductors oppose current changes in higher-frequency signals more than they do in lower-frequency signals.

The degree to which an inductor opposes current change at a particular frequency is called the inductor's *reactance*. Like resistance, inductive reactance is measured in ohms (check out Chapter 2 of this minibook for more details), and can be calculated with the following formula:

$$X_L = 2\pi \times f \times L$$

Here, the symbol X_L represents the inductive reactance in ohms, f represents the frequency of the signal in hertz (cycles per second) and L equals the inductance in henrys. Oh, and π is the mathematical constant you heard so much about at school, whose value is approximately 3.14.

For example, suppose you want to know the reactance of a 1 mH inductor to a 60 Hz sine wave. The maths looks like this:

$$X_L = 2\pi \times f \times L = 2 \times 3.14 \times 60 \times .001 = 0.3768\ \Omega$$

Thus, a 1 mH inductor has a reactance of about a third of an ohm at 60 Hz.

How much reactance does the same inductor have at 20 kHz? Much more:

$$X_L = 2 \times 3.14 \times 20{,}000 \times .001 = 125.6\ \Omega$$

Increase the frequency to 100 MHz and see how much resistance the inductor has:

$$X_L = 2 \times 3.14 \times 100{,}000 \times .001 = 628{,}000\ \Omega$$

At low frequencies, inductors are much more likely to let current pass than at high frequencies. As the next section explains, you can exploit this characteristic to create circuits that block frequencies above or below certain values.

Combining Inductors

Just like resistors and capacitors (the subjects of Chapters 2 and 3 of this minibook, respectively), you can combine inductors in series or parallel and use simple equations to calculate the total inductance of the circuit.

For the calculations to be valid, however, the inductors must be shielded. If they aren't shielded, they're not only affected by their own magnetic fields, but also by the magnetic fields of other inductors around them. In that case, all bets are off.

You calculate inductor combinations just like resistor combinations (from Chapter 2 of this minibook), using exactly the same formulas except substituting henrys for ohms. Here are the formulas:

- **Series inductors:** Just add up the value of each individual inductor.
- **Two or more identical parallel inductors:** Add them up and divide by the number of inductors.
- **Two parallel and unequal inductors:** Use this formula:

  $$L_{total} = \frac{L1 \times L2}{L1 + L2}$$

- **Three or more parallel and unequal inductors:** Use this formula:

  $$L_{total} = \frac{1}{\frac{1}{L1} + \frac{1}{L2} + \frac{1}{L3} \cdots}$$

Here's an example in which three inductors valued at 20 mH, 100 mH and 50 mH are connected in parallel:

$$L_{total} = \frac{1}{\frac{1}{L1} + \frac{1}{L2} + \frac{1}{L3} \cdots} = \frac{1}{\frac{1}{20} + \frac{1}{100} + \frac{1}{50}}$$
$$= \frac{1}{0.05 + 0.01 + 0.02} = \frac{1}{0.08} = 12.5 \text{ mH}$$

Therefore, the total inductance of this circuit is 12.5 mH.

Putting Inductors to Work

Here are a few common uses for inductors in real-life circuits:

- **Smoothing voltage in a power supply:** The final stage of a typical power-supply circuit that converts 230 VAC household current to a useable direct current is often a filter circuit that removes any residual irregularities in the voltage due to the fact that it was derived from a 50 Hz AC input.
- **Filters:** Select frequencies to be allowed to pass or to be blocked. You're probably familiar with the tone settings on a stereo system, which let you bump up the bass, tone down the midrange and ease up the upper range just for brightness. Three different types of filters exist:

- *High-pass filters* allow only frequencies above a certain value to pass.
- *Low-pass filters* allow only frequencies below a certain value to pass.
- *Band-pass filters* allow only frequencies that fall between an upper and a lower value to pass.

- **Radio-tuning circuits:** You can use coils to help a radio-tuning circuit tune to a particular frequency signal and hold it.
- **Transformers:** The simplest transformer consists of a pair of inductors placed next to each other. The transformer serves two functions: it can increase or decrease voltage and it can electrically isolate one part of a circuit from another. For more about transformers, flip to Book IV, Chapter 1.

Chapter 5

Working with Diodes and LEDs

In This Chapter

- Sussing out how semiconductors work
- Introducing the simple diode
- Changing AC to DC with diodes
- Lighting up with LEDs

From roughly 1900 to 1950, the world of electronics was dominated by a now all but obsolete technology: vacuum tubes. These glass tubes were large, expensive and fragile and required a lot of current. In the 1940s, however, researchers made a technological breakthrough, which led to new components that were a quantum-leap improvement over vacuum tubes. These components are called *semiconductors.*

This chapter introduces you to the most basic kind of semiconductor, called a diode. Although diodes look a little like resistors (for details, check out Chapter 2 of this minibook), they behave very differently. Diodes have one ability that sets them apart: they let current flow freely in one direction, but block current if it tries to flow in the other direction. In other words, a diode is like a turnstile gate that you can walk through in one direction but not the other. This characteristic turns out to be incredibly useful in electronic circuits.

What Is a Semiconductor?

As its name implies, a *semiconductor* is a material that conducts current, but only partly. The conductivity of a semiconductor is somewhere between that of an insulator, which has almost no conductivity, and a conductor, which has almost full conductivity. Most semiconductors are crystals made of certain materials, most commonly silicon.

To understand how semiconductors work, you need to know a little about how electrons are organised in an atom. As a result, this section is fairly difficult, because it delves just a little into the physics of how semiconductors work at the atomic level.

You don't have to understand the physics of diodes to use them in your circuits, and so you can skip over this section if you want, and go straight to 'Discovering Diodes', later in this chapter. However, we recommend perusing this section. Don't worry if you have difficulty understanding how semiconductors work, just note the key terms, such as p-type, n-type and p-n junction, and move on.

Examining elements and atoms

The electrons in an atom are organised in layers, kind of like the layers of an onion. These layers are called *shells.* The outermost shell is called the *valence* shell. The electrons in this shell are the ones that form bonds with neighbouring atoms. Such bonds are called *covalent bonds* because they share valence electrons, which are also the electrons that sometimes go wandering off in search of other atoms.

Most conductors, including copper and silver, have just one electron in the valence shell. Atoms with just one valence electron have a hard time keeping that electron, which is what makes copper and silver such good conductors. When valence electrons travel, they create moving electric fields that push other electrons out of their way. That's what causes current to flow.

Semiconductors, on the other hand, typically have four electrons in their valence shell. The best known elements with four valence electrons are carbon, silicon and germanium. Atoms with four valence electrons rarely lose one of them. However, they do like to share them with neighbouring atoms.

If all the neighbouring atoms are of the same type, all the valence electrons are able to bind with valence electrons from other atoms. When that happens, the atoms arrange themselves into neat and orderly structures called *crystals.* Semiconductors are made out of such crystals.

The most plentiful element with four valence electrons is carbon. However, carbon crystals are rarely used as semiconductors because they have other uses, such as in diamond rings. So instead, semiconductors are usually made from crystals of silicon and occasionally germanium.

Figure 5-1 shows the covalent bonds formed in a silicon crystal. Here, each circle represents a silicon atom and the lines between the atoms represent the shared electrons. Each of the four valence electrons in each silicon atom is shared with one neighbouring silicon atom. Thus, each silicon atom is bonded with four other silicon atoms.

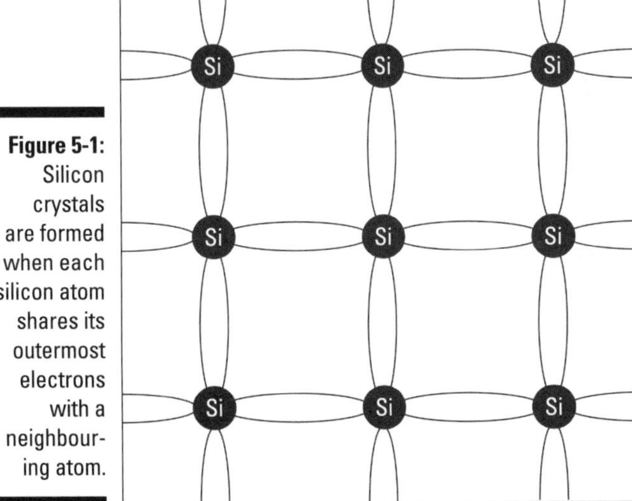

Figure 5-1: Silicon crystals are formed when each silicon atom shares its outermost electrons with a neighbouring atom.

Doping is really rather clever

By themselves, pure silicon crystals are pretty, but not all that useful. With all four valence electrons interlocked with neighbouring atoms, getting current to flow through a pure silicon crystal isn't easy. But things get interesting if you introduce small amounts of other elements into a crystal. Then the crystal starts to conduct in an interesting way.

The process of deliberately introducing other elements into a crystal is called *doping* and the element introduced by doping is called a *dopant*. By carefully controlling the doping process and the dopants that are used, silicon crystals can transform into one of two distinct types of conductors:

- **N-type semiconductor:** Created when the dopant is an element that has five electrons in its valence layer. Phosphorus is commonly used for this purpose. The phosphorus atoms join right in the crystal structure of the silicon, each one bonding with four adjacent silicon atoms just like a silicon atom would. Because the phosphorus atom has five electrons in its valence shell, but only four of them are bonded to adjacent atoms, the fifth valence electron is left hanging out with nothing to bond to.

The extra valence electrons in the phosphorus atoms start to behave like the single valence electrons in a regular conductor such as copper. They are free to move about, as shown in Figure 5-2. This type of semiconductor has extra electrons and is called an *n-type semiconductor.*

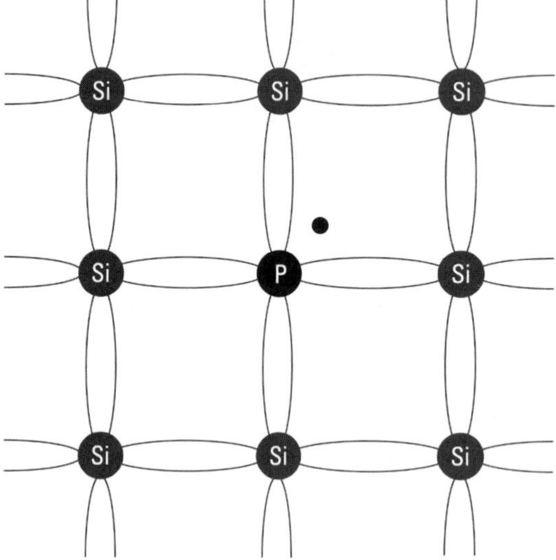

Figure 5-2: An n-type semiconductor has extra electrons.

- **P-type semiconductor:** Created when the dopant is an element (such as boron) that has only three electrons in the valence shell. When a small amount of boron is incorporated into the crystal, the boron atom is able to bond with four silicon atoms, but because it has only three electrons to offer, a gap is created where an electron would normally be, as shown in Figure 5-3. This electron gap is called a *hole.* The hole behaves like a positive charge, and so semiconductors doped in this way are called *p-type semiconductors.*

  Like a positive charge, holes attract electrons. But when an electron moves into a hole, the electron leaves a new hole at its previous location. Thus, in a p-type semiconductor, holes are constantly moving around within the crystal as electrons constantly try to fill them up. A p-type semiconductor is like crazed pieces of Swiss cheese in which you can't quite get a fix on the holes because they're always moving.

The physics behind the n- and p- designations is complex but you can think of the n- as standing for the negatively charged electronics that are free to move in an n-type semiconductor and the p- standing for the holes that behave like positive charges in a p-type semiconductor.

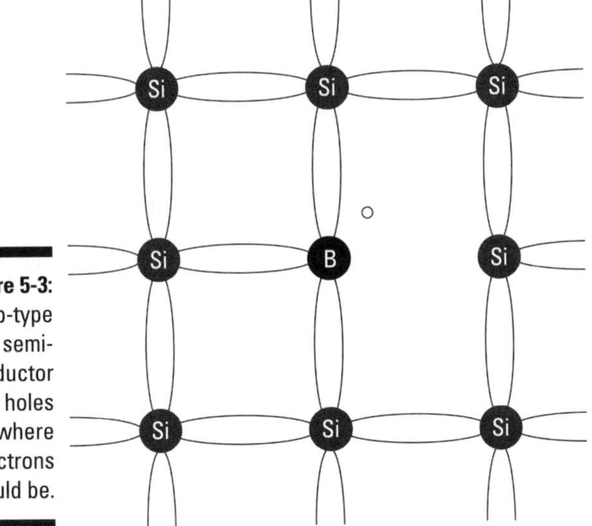

Figure 5-3: A p-type semi-conductor has holes where electrons should be.

When voltage is applied to an n-type or a p-type semiconductor, current flows, for the same reason that it flows in a regular conductor: the negative side of the voltage pushes electrons and the positive side pulls them. The result is that the random electron and hole movement that's always present in a semiconductor becomes organised in one direction, creating measurable electric current.

Combining types into one-way junctions

By themselves, p-type and n-type semiconductors are just conductors. But if you put them together, an interesting and very useful thing happens: current can flow through the resulting semiconductor, but only in one direction.

This *p-n junction*, as it's called, is like a one-way gate. If you put positive voltage on the p side of the junction and negative voltage on the n side, current flows through the junction. But if you reverse the voltage, putting negative voltage on the p side and positive voltage on the n side, current doesn't flow.

Picture a turnstile, such as the gates you have to go through to get into a sports stadium or an underground station: You can walk through the gate in one direction but not the other. That's essentially what a p-n junction does. It allows current to flow one way but not the other.

To understand why p-n junctions allow current to flow in only one direction, you have to understand what happens right at the boundary between the p-type material and the n-type material. Because opposite charges attract, the extra electrons on the n-type side of the junction are attracted to the holes on the p-type side. So they start to drift across to the other side.

When an electron leaves the n-type side to fill a hole in the p-type side, a hole is left in the n-type side where the electron was. The result is as if the electron and the hole trade places. The boundary of a p-n junction ends up being populated by defectors: electrons and holes have crossed the boundary and are now on the wrong side of the junction.

The region occupied by electrons and holes that have crossed over is called the *depletion zone*. Because one side of the depletion zone has electrons (negative charges) and the other side has holes (positive charges), a voltage exists between the two edges of the depletion zone. This voltage has an interesting effect on the defectors: it beckons them to turn around and come home. In other words, the holes that jumped to the negative side of the junction attract the electrons that jumped to the positive side.

Imagine being an electron that has jumped over the boundary and into the p-type side of the junction. Being negatively charged, you're attracted to move farther into the p side by the positively charged holes you see ahead of you. But you're also attracted by the positively charged holes that now lie behind you – the very same hole you traded places with now exerts a pull on you that discourages you from going any farther.

Unable to make up your mind, you decide to just stay put. That's exactly what happens to the electrons and holes that have crossed over to the other side. The depletion zone becomes stable – a state that's called *equilibrium*.

Now consider what happens when the equilibrium is disturbed by a voltage placed across the p-n junction. The effect depends on which direction the voltage is applied, as follows:

- **When you apply positive voltage to the p-type side and negative voltage to the n-type side, the depletion zone is pushed from both sides towards the centre, making it smaller.** Electrons in the n-type side of the junction are pushed by the voltage towards the depletion zone and eventually collapse it completely. When that happens, the p-n junction becomes a conductor and current flows.
- **When you apply voltage in the reverse direction, the depletion zone is pulled from both sides of the junction, and thus it expands.** The larger it gets, the more of an insulator the p-n junction becomes. Thus, when voltage is applied in the reverse direction, current doesn't flow through the junction.

Discovering Diodes

Diodes come in many shapes and sizes, but most of them resemble the ones shown in Figure 5-4. These diodes are about the size of resistors and are available from any electronics supply shop.

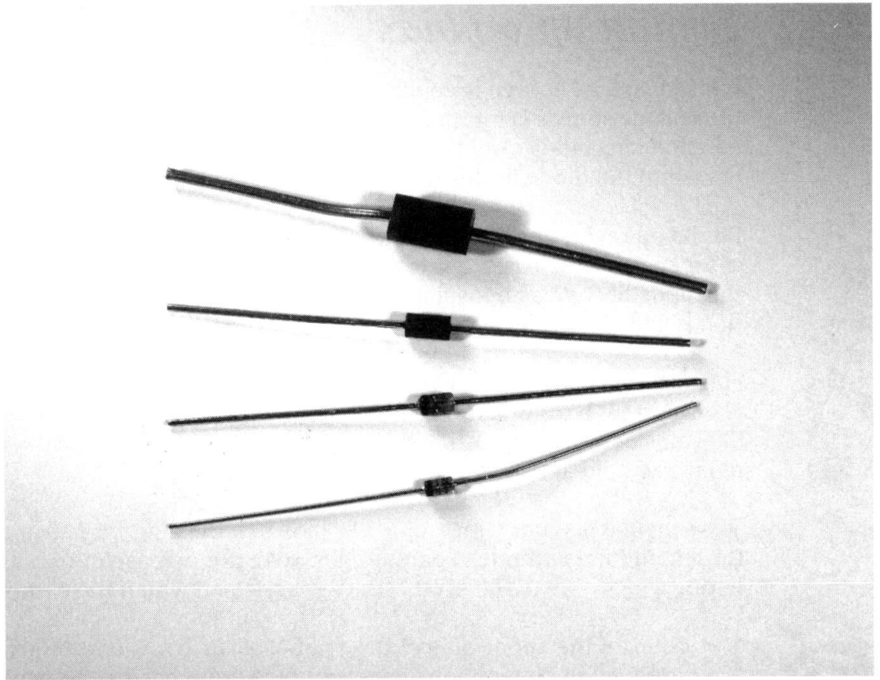

Figure 5-4: Several common varieties of diodes.

A *diode* is a device made from a single p-n junction (which we describe in the earlier section 'Combining types into one-way junction'). As Figure 5-5 shows, leads are attached to the two ends of the p-n junction, which is encased in an insulating package. These leads allow you to easily incorporate the diode into your circuits.

The lead attached to the n-type semiconductor is called the *cathode.* Thus, the cathode is the negative side of the diode. The positive side of the diode – that is, the lead attached to the p-type semiconductor – is called the *anode.*

Figure 5-5: A diode has a single p-n junction.

Cathode N-Type P-Type Anode

Boning up on bias

When a voltage source is connected to a diode such that the positive side of the voltage source is on the anode and the negative side is on the cathode, the diode becomes a conductor and allows current to flow. Voltage connected to the diode in this direction is called *forward bias.*

But if you reverse the voltage direction, applying the positive side to the cathode and the negative side to the anode, current doesn't flow. In effect, the diode becomes an insulator. Voltage connected to the diode in this direction is called *reverse bias.*

Forward bias allows current to flow through the diode. Reverse bias doesn't allow current to flow. (Up to a point at least; as we explain in the later section 'Moving voltage forward', limits apply to how much reverse bias voltage a diode can hold at bay.)

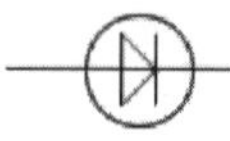

The schematic symbol for a diode is shown in the margin. The anode is on the left and the cathode is on the right. Here are two useful tricks for remembering which side of the symbol is the anode and which the cathode:

- Think of the anode side of the symbol as an arrow that indicates the direction of conventional current flow – from positive to negative. Thus, the diode allows current to flow in the direction of the arrow.
- Think of the vertical line on the cathode side as a giant minus sign, indicating which side of the diode is negative for forward bias.

Figure 5-6 illustrates forward and reverse bias with two very simple circuits that connect a lamp to a battery with diodes. In the circuit on the left the diode is forward biased, and so current flows through the circuit and the lamp lights up. In the circuit on the right the diode is reverse biased, and so current doesn't flow and the lamp remains dark.

Moving voltage forward – or in reverse

In a typical diode, a certain amount of forward – or reverse – voltage is required before any current will flow. This amount of voltage is usually very small: on most diodes, around half a volt. Up to this voltage, current doesn't flow. When the forward or reverse voltage is reached, however, current flows easily through the diode. Figure 5-6 shows both a diode wired for forward bias, in which the lamp will light, and reverse bias, in which the lamp won't light.

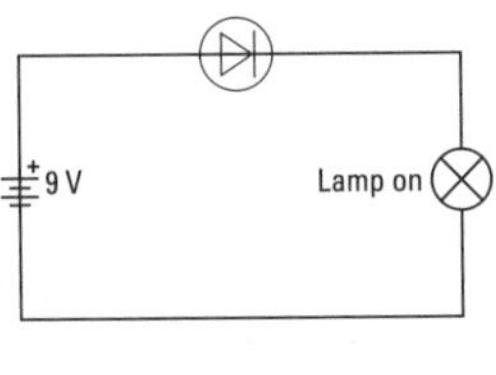

Forward bias

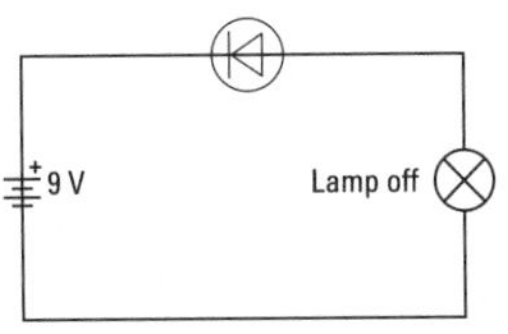

Reverse bias

Figure 5-6: Forward and reverse bias on a rectifier diode.

This minimum threshold of voltage in the forward direction is called the diode's *forward voltage drop,* because the circuit loses this voltage at the diode. For example, if you placed a voltmeter across the leads of the diode in the forward-biased circuit in Figure 5-6 (on the left), you'd read the forward voltage drop of the diode. Then, if you placed the voltmeter across the lamp terminals, the voltage would be the difference between the battery voltage (9 V) and the forward voltage drop of the diode.

For example, if the forward voltage drop of the diode is 0.7 V and the battery voltage exactly 9 V, the voltage across the lamp is 8.3 V.

Likewise, diodes also have a maximum reverse voltage they can withstand before they break down and allow current to flow backwards through the diode. This reverse voltage is called *peak inverse voltage* (PIV) (or sometimes *peak reverse voltage*).

PIV is an important specification for diodes you use in your circuits, because you need to ensure that your diodes aren't exposed to more than their PIV rating. In addition to the forward voltage drop and PIV, diodes are also rated for a maximum current rating. Exceed the current rating and the diode will probably be damaged beyond repair.

Meeting the many types of diodes

Diodes come in many different flavours. Some of them are exotic and rarely used by hobbyists, but many are commonly used. The following sections describe three types of diodes you're likely to encounter: rectifier, signal and Zener diodes. In the later section 'Introducing Light-Emitting Diodes', you also discover light-emitting diodes (LEDs).

Rectifier diodes

A *rectifier diode* is designed specifically for circuits that need to convert alternating current (AC) to direct current (DC). The most common rectifier diodes are identified by the model numbers 1N4001 to 1N4007. These diodes can pass currents of up to 1 A, and they have PIV ratings that range from 50 to 1,000 V.

Table 5-1 lists the PIVs for each of these common diodes. When choosing one of these diodes for your circuit, pick one that has a PIV that's at least double the voltage to which you expect to expose it. For most battery-power circuits, the 50 V PIV of the 1N4001 is more than sufficient.

Table 5-1 Peak Inverse Voltages for 1N400x Diodes

Model Number	*Diode Type*	*Peak Inverse Voltage*	*Current*
1N4001	Rectifier	50 V	1 A
1N4002	Rectifier	100 V	1 A
1N4003	Rectifier	200 V	1 A
1N4004	Rectifier	400 V	1 A
1N4005	Rectifier	600 V	1 A
1N4006	Rectifier	800 V	1 A
1N4007	Rectifier	1,000 V	1 A

Most rectifier diodes have a forward voltage drop of about 0.7 V and so need a minimum of 0.7 V to conduct current. For more details, check out the later section 'Putting Rectifiers to Work'.

Signal diodes

A *signal diode* is designed for much smaller current loads than a rectifier diode and can typically handle about 100 mA or 200 mA of current.

The most commonly used signal diode is the 1N4148. This diode has a close brother called 1N914, which you can use in its place if you can't find a 1N4148. This diode has a forward voltage drop of 0.7 and a PIV of 100 V, and can carry a maximum of 200 mA of current.

Here are a few other interesting points:

- **Signal diodes are noticeably smaller than rectifier diodes and their cases are often made of glass.** You have to look closely to see it, but a small black band marks the cathode end of a signal diode.
- **Signal diodes are better than rectifier diodes when dealing with high-frequency signals.** For this reason, they're often used in circuits that process audio or radio frequency signals. Because of their ability to respond quickly at high frequencies, signal diodes are sometimes called *high-speed diodes*. They're also sometimes called *switching diodes* because digital circuits (which you discover in Book VI) often use them as high-speed switches.

- **Some signal diodes are made of germanium rather than silicon.** (Germanium the crystal, not geranium the flower!) Germanium diodes have a much smaller forward voltage drop than silicon diodes – as low as 0.15 V. This makes them useful for radio applications, which often deal with very weak signals.

Freewheeling signal diodes

Signal diodes such as the 1N4148 variety often find themselves used with inductors – most commonly relays – to protect surrounding circuits from voltage spikes that can occur when the inductor suddenly shuts off. You discover more about relays in Book IV, Chapter 4, but put simply a relay is a mechanical switch that's operated not by a human finger but by a magnetic field created by a coil. Relays are incredibly useful because they allow you to use low-voltage circuits to control high-voltage circuits. For example, you can use a circuit powered by a 9 V battery to turn a mains powered lamp on or off.

The problem with relays is that they have coils, and as Joseph Henry discovered in the 19th century, coils have an interesting property called *self-inductance.* (If you want to know what self-inductance is and how it works, refer to the Chapter 4 in this minibook.)

Self-inductance can be a problem in circuits that include relays. When a current is applied to the relay, a magnetic field is formed around the coil. When the current is suddenly shut off, the magnetic field quickly collapses. This collapse induces a voltage in the coil that's opposite in polarity to the voltage that turned the coil on. If you don't do something to contain that voltage, it sends current travelling around your circuit like a drunk driver going down a motorway in the wrong direction.

Signal diodes are just the ticket for stopping this wrong-way voltage in its tracks. Simply place a signal diode across the relay, backward from the voltage applied to the relay coil, as shown in the following schematic:

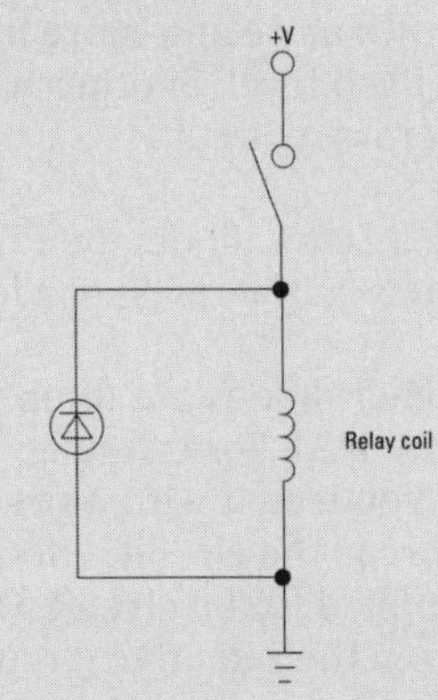

Consider now how this diode functions alongside the relay. When you close the switch, voltage is applied and current flows. All this current flows through the relay because the diode is reverse biased: the positive voltage is at the cathode end of the diode.

When you open the switch, the voltage across the relays is cut. Current stops flowing and the magnetic field around the coil suddenly collapses. This creates a voltage across the coil that's opposite to the voltage shown in the circuit. In other words, the induced voltage is negative at the top of the coil, positive at the bottom. Now the diode is forward biased, and so it becomes a conductor. The diode then creates a circuit through which current can flow, thus safely containing the self-induced voltage within the coil and the diode.

Diodes used in this way go by several names, including *freewheeling diodes, flyback diodes, catch diodes* and *snubber diodes.*

Zener diodes

In a normal diode, the PIV is usually pretty high – 50, 100 even 1,000 V. If the reverse voltage across the diode exceeds this number, current floods across the diode in the reverse direction in an avalanche, which usually results in the diode's demise. Normal diodes aren't designed to withstand a reverse avalanche of current, but *Zener diodes* are.

Zener diodes are specially designed to withstand current that flows when the PIV is reached or exceeded. And more than that, they're designed so that as the reverse voltage applied to them exceeds the threshold voltage, current flows more and more in a way that holds the voltage drop across the diode at a fixed level. In other words, Zener diodes can be used to regulate the voltage across a circuit.

In a Zener diode, the PIV is called the *Zener voltage*. This voltage can be quite low – in the range of a few volts – or it can be hundreds of volts.

Zener diodes are often used in circuits where a predictable voltage is required. For example, suppose that you have a circuit that will be damaged if you feed it with more than 5 V. In that case, you can place a 5 V Zener diode across the circuit, effectively limiting the circuit to 5 V. If more than 5 V is applied to the circuit, the Zener diode conducts the excess voltage away from the sensitive circuit.

Zener diodes have their own variation of the standard diode schematic symbol, as shown in the margin. You can find out more about working with Zener diodes in Book IV, Chapter 3.

Blocking reverse polarity with a diode

Project 5-1 presents a simple little construction project that demonstrates a diode's ability to conduct current in only one direction. For this project, you wire up a rectifier diode in series with a 3 V flashlight lamp and a pair of AA batteries. Use a DPDT knife switch between the battery and the diode/lamp circuit so that when the switch is changed from one position to the other, the polarity of the voltage across the diode and lamp is reversed. Thus, the lamp lights in only one of the two switch positions. Figure 5-7 shows the completed project.

Figure 5-7: The completed circuit for Project 5-1.

Project 5-1: Blocking Reverse Polarity

In this project, you build a simple circuit that uses a diode to allow current to flow through a lamp circuit in only one direction. The circuit uses a DPDT knife switch to reverse the polarity of the battery. When the polarity is reversed, the diode blocks the current so the LED doesn't light up.

You also need a small Philips-head screw-driver, wire cutters wire strippers, and a multimeter to complete this project.

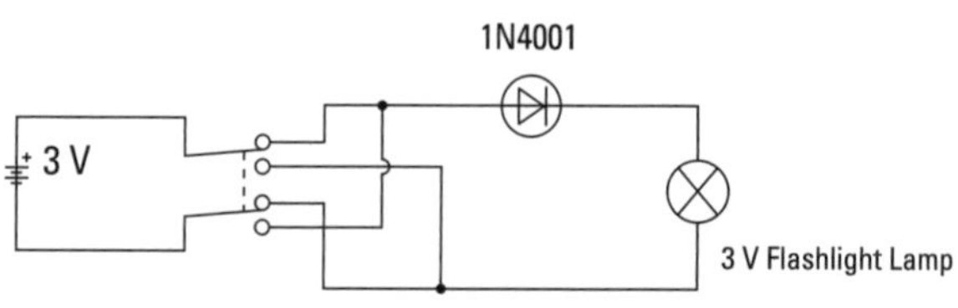

Parts List
2 AA batteries
1 Battery holder
1 Lamp holder
1 3 V lamp
1 DPDT knife switch
1 1N4001 rectifier diode
3 12-13 cm/22-AWG stranded wire

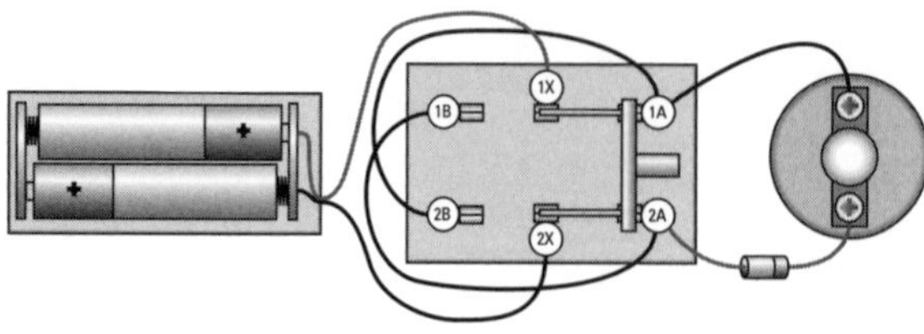

Steps

1. **Cut three, 12–13 cm lengths of wire and strip 1 cm of insulation from each end.**
2. **Open the switch.**

   Move the handles to their upright positions so none of the contacts are connected.
3. **Attach the red lead from the battery holder to terminal 1X of the switch.**
4. **Attach the black lead from the battery to terminal 2X of the switch.**
5. **Use one of the three 12–13 cm wires to connect terminal 1B of the switch to terminal 2A.**
6. **Use a second 12–13 cm wire to connect terminal 2B of the switch to terminal 1A.**
7. **Use the third 12–13 cm wire to connect terminal 1A of the switch to one terminal of the lamp socket.**
8. **Connect the diode to termnal 2A of the switch and the empty terminal of the lamp socket.**

   The cathode (the side with the stripe) should be connected to the lamp.
9. **Insert the lamp into the lamp holder.**
10. **Insert the batteries into the battery holder.**

    The circuit is now ready to test.
11. **Flip the switch to the B position.**

    The lamp lights because positive voltage is connected to the diode's anode, and negative voltage is applied to the cathode side of the circuit. This allows current to flow through the diode, and the lamp turns on.
12. **Flip the switch to the A position.**

    This time, the lamp goes out because voltage is reversed across the diode.
13. **Use your multimeter to measure the voltage across the battery, lamp and diode.**

    Flip the switch to the B position so the lamp lights. Set the multimeter to a DC volts range that will accomodate the 3 V battery voltage. Then, take the measurements indicated in the table 'Voltage Measurements' (below). The third measurement (the voltage across the diode) should be very close to 0.7 V. The second measurements (the voltage across the lamp) plus the voltage across the diode should add up to the first measurement (the voltage across the battery). Don't worry if they're slightly off, because the difference may be accounted for by inaccuracies in your voltmeter – especially if you have an analogue meter or an inexpensive digital meter.

Voltage Measurements

Black Lead	***Red Lead***	***Voltage***
Switch 2X	Switch 1X	
Lamp 1	Lamp 2	
Diode cathode	Diode anode	

Putting Rectifiers to Work

One of the most common uses for rectifier diodes is to convert household AC into DC that can be used as an alternative to batteries. We just concentrate on one part of a complete DC power supply – the rectifier circuit, which is typically made from a set of cleverly interlocked diodes. (A *rectifier* is a circuit that converts alternating current to direct current.)

Looking at rectifier circuits

In household mains supply or lower-voltage AC supplies derived from it, the voltage swings from positive to negative in cycles that repeat 50 times per second. If you place a diode in series with an AC voltage, you eliminate the negative side of the voltage cycle and so end up with just positive voltage, as shown in Figure 5-8.

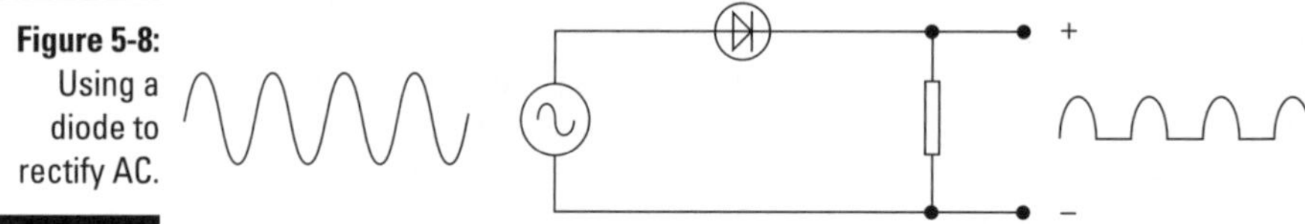

Figure 5-8: Using a diode to rectify AC.

If you look at the waveform of the voltage coming out of this rectifier diode, you see that it consists of intervals that alternate between a short increase of voltage and periods of no voltage at all. This is a form of DC because it consists entirely of positive voltage. However, it pulsates: first it's on, then it's off, then it's on again and so on.

Overall, voltage rectified by a single diode is off half of the time. So although the positive voltage reaches the same peak level as the input voltage, the average level of the rectified voltage is only half the level of the input voltage. This type of rectifier circuit is sometimes called a *half-wave rectifier,* because it passes along only half of the incoming AC waveform.

A better type of rectifier circuit uses four rectifier diodes, in a special circuit called a *bridge rectifier* (see Figure 5-9).

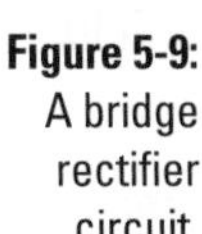

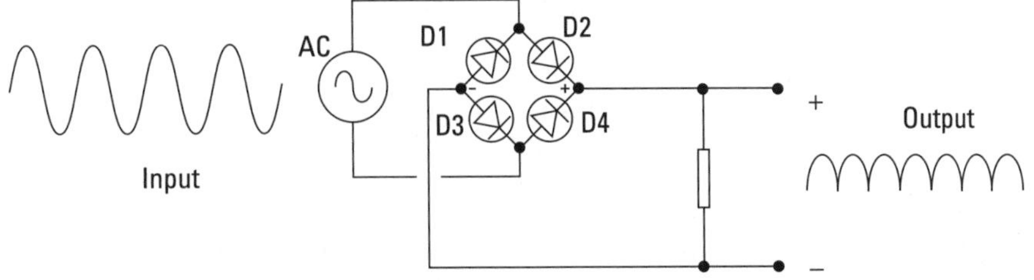

Figure 5-9: A bridge rectifier circuit.

Look at how this rectifier works on both sides of the AC input signal:

- In the first half of the AC cycle, D2 and D3 conduct because they're forward biased. Positive voltage is on the anode of D2 and negative voltage is on the cathode of D4. Thus, these two diodes work together to pass the first half of the signal through.
- In the second half of the AC cycle, D1 and D4 conduct because they're forward biased. Positive voltage is on the anode of D1 and negative voltage is on the cathode of D3.

The net effect of the bridge rectifier is that both halves of the AC sine wave are allowed to pass through, but the negative half of the wave is inverted so that it becomes positive.

The resulting DC signal doesn't drop to zero for half of the cycle, but still doesn't provide a steady DC voltage level. As we describe in Chapters 3 and 4 of this minibook, both capacitors and inductors can be used to slow down changes in current and voltage, and so they're often used in power supply circuits to improve the quality of DC voltage coming out of a rectifier circuit. Check out how that's done in Book IV, Chapter 3.

Building rectifier circuits

In Project 5-2, you build a simple half-wave circuit and a bridge rectifier circuit to see how diodes can convert AC to DC. You don't light any lamps or anything with this circuit; instead, just use a voltmeter to verify that the diodes are doing their job. And to keep things safe, use a 9 VAC power adapter as the source for the alternating current (see Book 2 Chapter 3 for further specification of a suitable adapter).

To make the voltage of the 9 VAC power adapter easier to work with, make sure that it's not plugged in to the mains, cut off the low-voltage power plug, separate the two low-voltage wire leads, strip a bit of insulation from the ends and attach crocodile clips to the leads. The crocodile clips make connecting the supply to your experimental circuits much easier.

Before you build the project, use the multimeter to measure the AC voltage created by your power adapter. Although it may be marked as 9 VAC, the actual voltage you measure is likely to be slightly more than 9 VAC.

You can find more information about rectifiers and how they're used in power supply circuits in Book IV, Chapter 3.

Project 5-2: Rectifier Circuits

In this project, you build a simple half-wave rectifier and a bridge rectifier. You connect the rectifiers to a 9 VAC power source, and then use your multimeter to measure the DC output to verify that the AC voltage has been converted to DC.

Parts List

- 1 9 VAC power adapter
- 2 Crocodile-clip leads (with crocodile clips on both ends)
- 1 Small solderless breadboard
- 4 1N4001 rectifier diode
- 1 Multimeter with AC and DC voltmeter functions
- 1 470 Ω, 0.25 W, 5% tolerance resistor

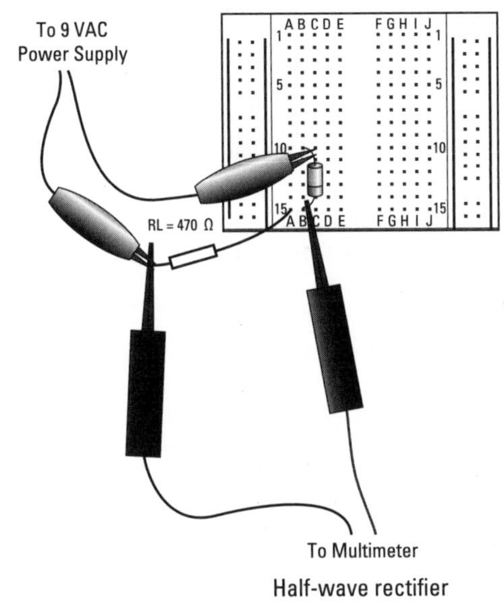

Half-wave rectifier

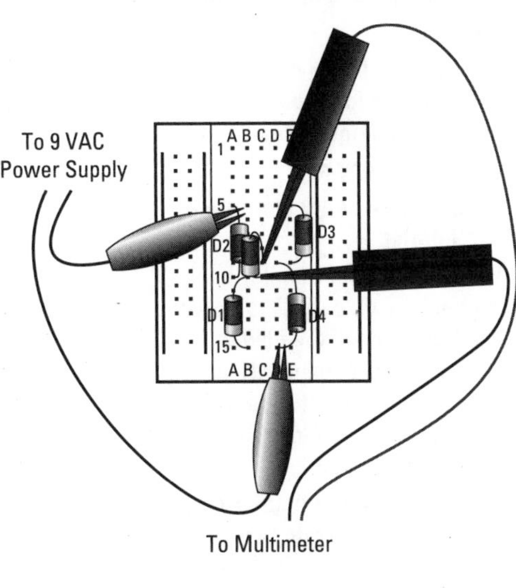

Bridge rectifier

Steps

1. **Start by building a simple half-wave rectifier: Insert one of the diodes in the solderless breadboard.**

   Insert the anode in hole B10 and the cathode in hole B15. (The cathode is the lead that has the stripe next to it.)

2. **Set the multimeter to a suitable VDC range.**

   10 VDC or 20 VDC should be fine.

3. **Connect one lead of the power adapter to the anode.**

   The anode is the lead inserted into hole B10.

4. **Insert one leg of the resistor into hole A15 and the other to the remaining lead of the power adapter.**

5. **Connect the black multimeter probe to the junction of the resistor and power lead.**

   Use a crocodile clip.

6. **Plug the power adapter into a wall outlet.**

   When the power adapter is plugged in, be careful to prevent the two leads of the power adapter from coming into contact with one another. The resulting short circuit could damage the power adapter.

7. **Touch the red multimeter probe to the diode cathode to measure the DC voltage rectified by the diode.**

   Note that the DC voltage is about half of the input AC voltage. That's because the single diode is allowing only half of the AC sine wave to pass.

8. **Unplug the power adapter.**

   That's it for the half-wave rectifier. Now build the bridge rectifier:

9. **Insert the three remaining diodes to complete the bridge rectifier.**

   When you're done, the four diodes should be inserted into the following holes.

Cathode (Striped End)	Anode
B15	B10
A5	A10
E9	E5
D9	D15

   Note that the first two diodes share a common lead in row 10, while the second two diodes share a common lead in row 9. Make sure you don't insert all four of these diodes into the same row. If you do, the bridge rectifier circuit doesn't work.

10. **Connect one lead of the AC power adapter to the diode lead in hole A5.**

11. **Connect the other lead of the AC power adapter to the diode lead in hole D15.**

12. **Plug in the 9 VAC power adapter.**

13. **Measure the voltage across the bridge rectifier.**

    Touch the black multimeter probe to the diode lead in hole C10 and the red probe to the diode lead in hole C9.

Introducing Light-Emitting Diodes

A *light-emitting diode* (LED) is a special type of diode that emits visible light when current passes through it. The most common type of LED emits red light, but you can also purchase LEDs that emit blue, green, yellow or white light.

You pronounce LED by spelling the letters out (*el-ee-dee*), not like the heavy metal 'lead' (or indeed Led Zeppelin!).

The schematic diagram symbol for an LED is shown in the margin, and Figure 5-10 shows an LED. The two leads protruding from the bottom of an LED aren't the same length: the shorter lead is the cathode and the anode is the longer lead.

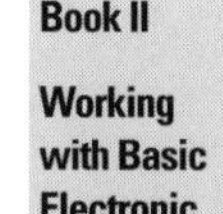

Figure 5-10: A typical LED.

You can obtain two LEDs, usually of different colours, combined into a single package: green and red are a common combination. In such cases, they're usually wired opposite of one another so that you can control which LED lights by changing the polarity of the voltage applied across the LED. In some cases, a third lead is used. This lead is connected to the cathode of both LEDs. The third lead enables you to light both LEDs, which yields a third colour. For example, when both LEDs are lit in a green and red combination LED, the resulting colour is yellow.

Providing the necessary resistance

Whenever you use an LED in a circuit, you need to provide some resistance in series with it, as shown in the schematic diagram in Figure 5-11. In this example, the LED is connected to a 9 V DC supply through a 470 Ω resistor.

Without resistance, the LED lights brightly for an instant and then burns itself out: it may even go bang!

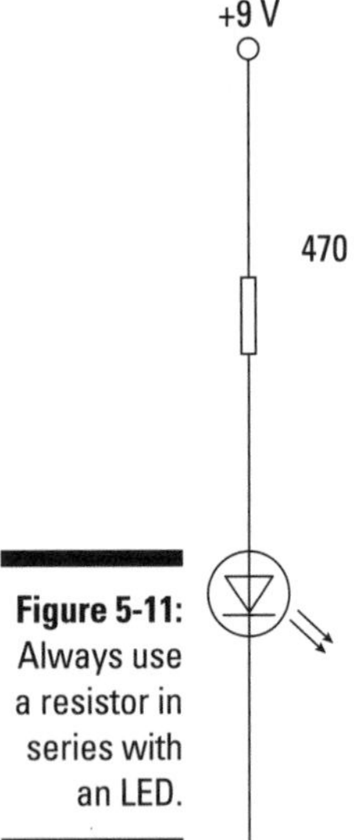

Figure 5-11: Always use a resistor in series with an LED.

To determine the value of the resistor you need to use, you need to know three bits of information:

- **Supply voltage:** For example, 9 V.
- **LED forward voltage drop:** For most red LEDs, the forward voltage drop is 2 V. For other LED types, the voltage drop may be different. Check the specifications on the package if you use other types of LEDs.
- **Desired current through the LED:** Usually, you need to keep the current flowing through the LED under 20 mA.

When you know these three things, you can use Ohm's law to calculate the desired resistance (check out Chapter 2 of this minibook). The calculation requires just four steps, as follows:

1. **Calculate the resistor voltage drop.**

   You do that by subtracting the voltage drop of the LED (typically 2 V) from the total supply voltage. For example, if the total supply voltage is 9 V and the LED drops 2 V, the voltage drop for the resistor is 7 V.

2. **Convert the desired current to amperes.**

   In Ohm's law, the current has to be expressed in amperes. You can convert milliamperes to amperes by dividing the milliamperes by 1,000. Thus, if your desired current through the LED is 20 mA, you have to use 0.02 in your Ohm's law calculation.

3. **Divide the resistor voltage drop by the current in amperes.**

   This gives you the desired resistance in ohms. For example, if the resistor voltage drop is 7 V and the desired current is 20 mA, you need a 350 Ω resistor.

4. **Round up to the nearest standard resistor value.**

   The next higher resistor value from 350 Ω is 390 Ω. If you can't find a 390 Ω resistor, a 470 Ω will do the trick.

   Note that the minor increases in resistances mean that slightly less current flows through the resistor, but the difference isn't noticeable. However, avoid going to a lower resistor value. Lowering the resistance increases the current, which can damage the LED.

If you're going to place more than one LED together in series, just add up the voltage drops to calculate the size of the resistor you need. For example, if you have a 9 V battery to supply voltage to three LEDs, each with a 2 V drop, the total voltage drop is 6 V, and so the voltage drop across the resistor is 3 V. Using Ohm's law, you can then calculate that the resistor needed to restrict the current flow to 20 mA is 150 Ω.

For 9 V and less, ¼ W resistors are more than adequate. If you're applying larger voltages to the LED circuit, you may need to use resistors that can handle more power. To calculate how much power in watts the resistor should be rated for, just multiply the voltage dropped across the resistor by the current in amps. For example, if the voltage drop on the resistor is 3 V and the current is 20 mA, the power dissipated by the resistor is 0.06 W, which is well under the limits of a ⅛ W resistor.

Detecting polarity with LEDs

In Project 5-3 you build a circuit that uses two LEDs to indicate the polarity of an input voltage. The voltage is provided by a 9 V battery connected to the circuit via a DPDT knife switch that's wired to reverse the battery polarity. The two LEDs and their corresponding resistors are mounted on a small solderless breadboard. Figure 5-12 shows the completed circuit.

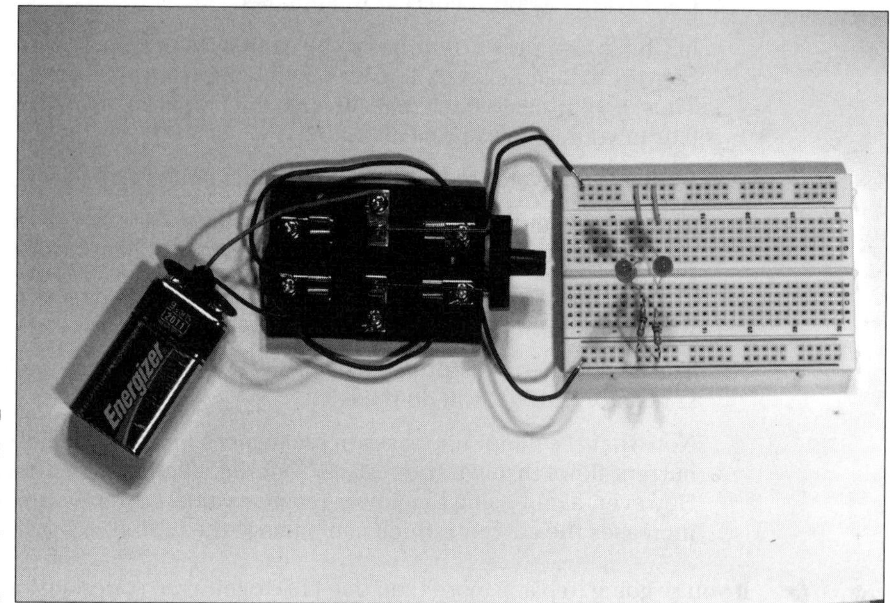

Figure 5-12: The completed LED polarity detector.

Project 5-3: An LED Polarity Detector

In this project, you build a polarity detector that uses LEDs to indicate the polarity of a power supply. You need a small Phillips-head screwdriver, wire cutters and wire strippers to assemble the circuit.

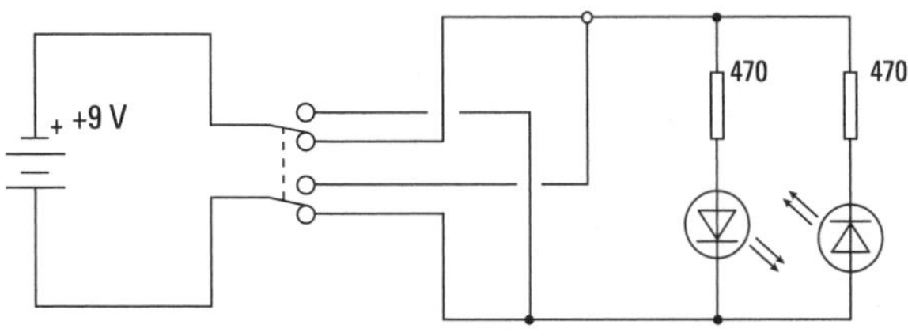

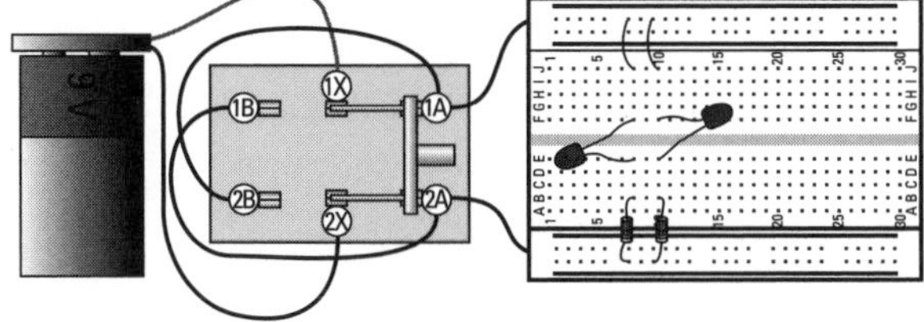

Parts List

1 9 V battery
1 9 V snap-on battery holder
1 DPDT knife switch
2 Red LEDs, 5 mm
2 470 Ω resistors
2 Short (under 2.5 cm) jumper wires
4 12–13 cm, 22-AWG stranded wire

Steps

1. **Cut four 12–13 cm lengths of wire and strip 1 cm of insulation from each end.**
2. **Open the switch.**

   Move the handles to their upright positions so none of the contacts are connected.
3. **Attach the red lead from the battery holder to terminal 1X of the switch.**
4. **Attach the black lead from the battery to terminal 2X of the switch.**
5. **Use a 12–13 cm wire to connect terminal 1B of the switch to terminal 2A.**
6. **Use a 12–13 cm wire to connect terminal 2B of the switch to terminal 1A.**
7. **Use a 12–13 cm wire to connect terminal 1A of the switch to any hole on the positive breadboard bus.**
8. **Use a 12–13 cm wire to connect terminal 2A of the switch to any hole on the negative breadboard bus.**
9. **Connect a short jumper wire on the breadboard to connect hole J8 to any nearby hole on the positive bus strip.**
10. **Use a short jumper wire to connect hole J10 to any nearby hole on the positive bus strip.**
11. **Insert one of the resistors in hole B8 and any nearby hole on the negative bus strip.**
12. **Insert the other resistor in hole B10 and any nearby hole in the negative bus strip.**
13. **Connect the two LEDs on the breadboard as follows:**

Cathode (short lead)	Anode
Hole E8	Hole F8
Hole F10	Hole E10

   Note that the cathode is the short lead.
14. **Insert the 9 V battery into the battery holder.**
15. **Flip the switch to the A position.**

   The LED in row 10 lights up because the voltage from the battery is forward biased on this LED.
16. **Flip the switch to the B position.**

   The LED in row 8 lights up because the voltage from the battery is forward biased on this LED.

Chapter 6

Working with Transistors

In This Chapter

- Understanding transistors
- Using a transistor as an amplifier
- Switching on and off with a transistor
- Creating a flashing circuit

Most people became aware of the transistor when transistor radios took off in the 1960s and 1970s. Before that, radios were more or less confined to the home – often as pieces of furniture in the living room that people gathered around. The transistor allowed the radio to go out into the street, the workplace and – to the older generation's annoyance – the beach and cafe.

That was just the first obvious sign of an extraordinary technology revolution that was to change everyone's lives. Today, transistors are in just about every piece of electronics you can think of, from your washing machine to satellites. They have led to new industries, such as computing, and revolutionised older ones, such as publishing. Without them your mobile phone wouldn't be mobile and your computer wouldn't fit in your office, let alone under your desk. The world wouldn't have laptops, MP3 players or even cassette or CD players, and people would still be playing music on vinyl records. Well, okay, we admit some of us still do, but that's from personal choice and not necessity.

In the half a century or more since the first transistor radio, little thimble-sized transistors have given way to transistors that are literally millions of times smaller. Nowadays, experts can put 100 million transistors on a single piece of silicon crystal about the size of your fingernail.

In this chapter, you take a look at what transistors are and how you can put them to use in your own circuits, as amplifiers, switches and LED flashers. Along the way, you build a few simple transistor circuits to discover how they work.

What's the Big Deal About Transistors?

When the transistor was invented in 1947, it didn't really do anything that hadn't already been done before. It did, however, do it in a radically different way (for the history, check out the later sidebar 'Why transistors were invented').

The basic idea behind a transistor is that it lets you control the flow of current through one channel by varying the intensity of a much smaller current that's flowing through a second channel.

Think of a transistor as an electronic lever. A lever is a device that allows you to lift a large load by exerting a small amount of effort. In essence, a lever amplifies your effort. Well, that's what a transistor does: it lets you use a small current to control a much larger current.

Figure 6-1 shows some of the many different kinds of transistors that are available today. As you can see, transistors come in a variety of different sizes and shapes. But all these transistors have one thing in common: they each have three leads.

Figure 6-1: Transistors come in many shapes and sizes.

Why transistors were invented

Devices that performed the function of transistors had been around for 30–40 years prior to the invention of the transistor. They were called *vacuum tubes.* A vacuum tube consisted of a vacuum chamber made from glass or metal, a heating element that heated the space inside the chamber and electrodes that protruded into the chamber. (We use past tense because although vacuum tubes still exist, they aren't used all that often.)

One specific type of vacuum tube was called a *triode*; it had three electrodes. In a triode, a large current flowing through two of the electrodes (called the anode and the cathode) could be regulated by placing a wire grid (called the *control grid*) between the cathode and the anode. Applying a small current to this grid slowed down the flow of electrons between the cathode and the anode.

Soon, people figured out that they were able to use a fluctuating signal such as a radio or audio wave on the control grid. When they did, the current on the anode followed the fluctuations of the control grid current but with much larger variations. Thus, the triode was an electronic lever: small variations in current at the control grid were amplified to create large variations in current at the anode.

The vacuum tube triode was patented in 1907 and was the key invention that enabled the development of radio, television and computers. But vacuum tubes had serious limitations: they were expensive to manufacture, large (the small ones were about the size of a thumb), required a lot of power to operate, generated a lot of heat and lasted only a few years before burning themselves out.

Transistors changed all that. They perform the same function as vacuum tube triodes but use semiconductor junctions instead of heated electrodes in a vacuum chamber. They had huge advantages over vacuum tubes: they were small, required very little power to operate, generated much less heat and lasted much longer.

Peering inside a transistor

The most basic kind of transistor is called a *bipolar transistor.* Bipolar transistors are the easiest to understand, and they're the ones you're most likely to work with as a hobbyist. As a result, most of this chapter focuses on bipolar transistors (though we do describe another type of transistor in the later sidebar 'Considering field-effect transistors'). Throughout this entire book, you can assume that whenever we use the term *transistor* by itself, we're referring to a bipolar transistor.

We now peer inside a transistor to see how it works.

In this minibook's Chapter 5, we explain that a *diode* is the simplest kind of semiconductor, made from a single p-n junction. The latter is simply a junction of two different types of semiconductors, one that's missing a few electrons to give it a positive charge (*p-type* semiconductor) and the other with a few extra electrons, which gives it a negative charge (*n-type* semiconductor).

By itself, a p-n junction works as a one-way gate for current. In other words, a p-n junction allows current to flow in one direction but not the other. A diode is simply a p-n junction with a lead attached to both ends.

A transistor is like a diode with a third layer of either p-type or n-type semiconductors on one end. Thus, a transistor has three regions rather than two. The interface between each of the regions forms a p-n junction.

You can think of a transistor as a semiconductor with two p-n junctions.

Figure 6-2 shows the structure of two common types of transistors along with their schematic diagram symbols. We explain the details of this figure in the following paragraphs.

You can make a bipolar transistor in two ways:

- **NPN transistor:** A p-type semiconductor sandwiched between two n-type semiconductors. This type of transistor has three regions: n-type, p-type and n-type (see the top part of Figure 6-2).
- **PNP transistor:** An n-type semiconductor sandwiched between two p-type semiconductors (in other words, the opposite way round). This type of transistor has three regions: p-type, n-type and p-type (check out the bottom part of Figure 6-2).

Each of the three regions of semiconductor material in a transistor has a lead attached to it, and each of these leads is given a name:

- **Collector:** This lead is attached to the largest of the semiconductor regions. Current flows through the collector to the emitter as controlled by the base.
- **Emitter:** This lead is attached to the second largest of the semiconductor regions. When the base voltage allows, current flows through the collector to the emitter.
- **Base:** This lead is attached to the middle semiconductor region. This region serves as the gatekeeper that determines how much current is

allowed to flow through the collector-emitter circuit. When voltage is applied to the base, current is allowed to flow.

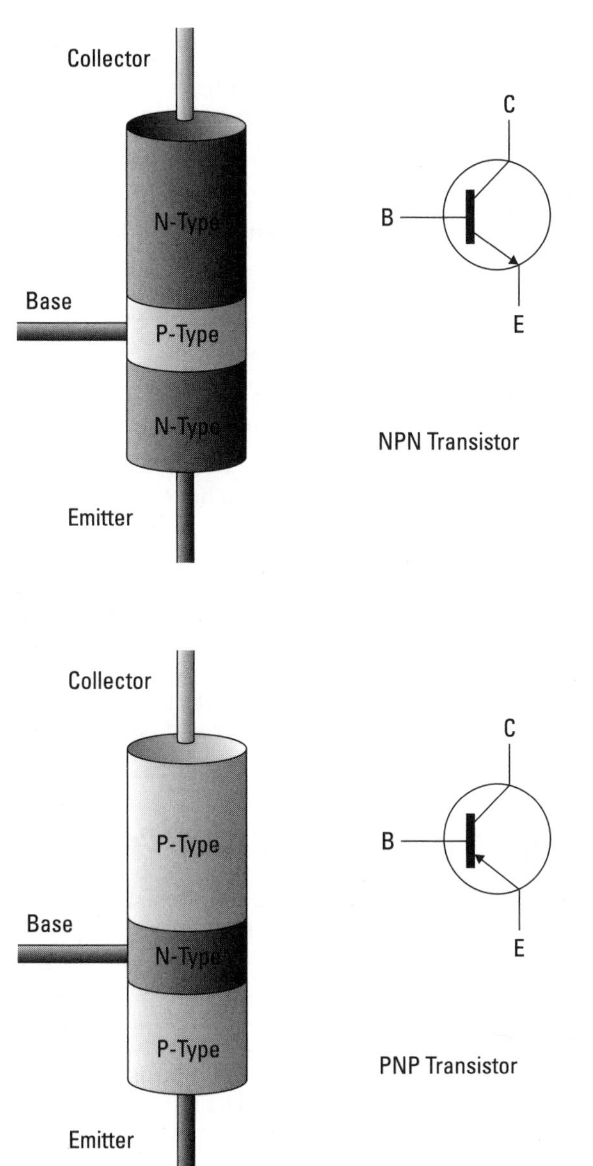

Figure 6-2: NPN and PNP transistors.

These two current paths are important in a transistor:

- **Collector-emitter:** The main current that flows through the transistor. Voltage placed across the collector and emitter is often referred to as V_{ce} and current flowing through the collector-emitter path is called I_{ce}.
- **Base-emitter:** The current path that controls the flow of current through the collector-emitter path. Voltage across the base-emitter path is referred to as V_{BE} and is also sometimes called *bias voltage.* Current through the base-emitter path is called I_{BE}.

Before we move onto the gory details in the next section, just note these transistor points:

- In an NPN transistor, the emitter is the negative side of the transistor. The collector and base are the positive sides.
- In a PNP transistor, the emitter is the positive side of the transistor. The collector and base are the negative sides.

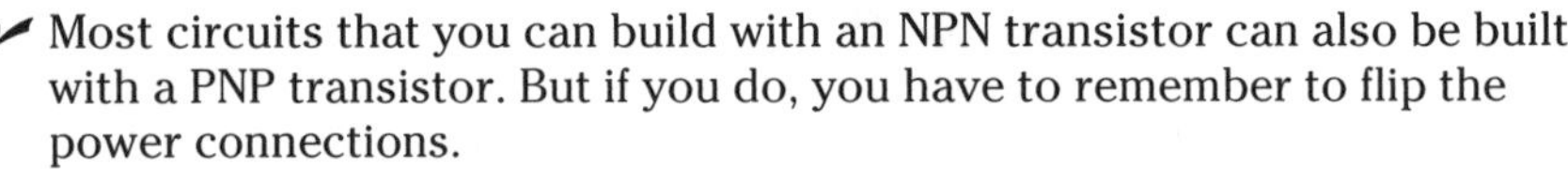

- Most circuits that you can build with an NPN transistor can also be built with a PNP transistor. But if you do, you have to remember to flip the power connections.

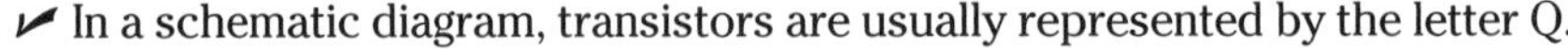

- In a schematic diagram, transistors are usually represented by the letter Q.
- Although good reasons exist why the terms collector, emitter and base were chosen for the three leads of a transistor, they have to do with the internal operation of the transistor at a deeper level than you need to go for this book. So please take our word for it: the inventors of the transistor didn't choose the terms collector, emitter and base just to confuse you!

Examining transistor specifications

Transistors are more complicated devices than resistors, capacitors, inductors and diodes (see Chapters 2 to 5 of this minibook, respectively). Whereas those components have just a few specifications to wrangle with, such as ohms of resistance and maximum watts of power dissipation, transistors have a whole range.

You can find the complete specifications for any transistor by looking up its *data sheet* on the Internet; just type the part number into your favourite search engine. The data sheet gives you dozens of interesting facts about the transistor you're interested in, with charts and graphs only a rocket scientist can love.

Considering field-effect transistors

Bipolar transistors aren't the only kids on the semiconductor block. Another variety of transistor, called a *field-effect transistor* (FET), has become extremely popular in recent years, especially as the building blocks for integrated circuits (the subject of Book III). Field-effect transistors can be made much smaller than bipolar transistors, and they use much less current.

Field-effect transistors behave much like bipolar transistors, but they have their own terms: instead of base, emitter and collector, the terminals in a field-effect transistor are called the *gate, drain* and *source.* Internally, a field-effect transistor is very different from a bipolar transistor. Instead of using a pair of p-n junctions, a field-effect transistor consists of a single piece of n- or p-type semiconductor with a special substance placed on it that can control the current flow through the semiconductor.

A dozen or so different types of field-effect transistors exist, but the most commonly used are called *MOSFET* (for metal-oxide-semiconductor field-effect transistor) and *JFET* (for junction field-effect transistor).

Be warned that field-effect transistors are quite susceptible to accidental static discharge. If you touch one and hear a little pop as static in your skin discharges through the transistor, you may as well throw it away. Always take precautions against static discharge whenever you handle a field-effect transistor or an integrated circuit that contains field-effect transistors.

If you happen to be a rocket scientist and you're thinking about using the transistor in a missile, by all means pay attention to every detail in the data sheet. But if you're just trying to do a little on-the-side circuit design, you need to pay attention only to the most important specifications, such as the following:

- **Collector current (I_C):** The maximum current that can flow through the collector-emitter path.

  Most circuits employ a resister to limit this current flow; use Ohm's law (from Chapter 2 of this minibook) to calculate the value of the resistor necessary to keep the collector current below the limit. If you exceed this limit for long, the transistor may be damaged.
- **Collector-base voltage (V_{CBO}):** The maximum voltage across the collector and the base. This is usually 50 V or more.
- **Collector-emitter voltage (V_{CEO}):** The maximum voltage across the collector and the emitter. This is usually 30 V or more, which is well above the voltage levels you work with in most hobby circuits.

- **Current gain (H_{FE}):** A measure of the amplifying ability of the transistor, which refers to the ratio of the base current to the collector current. Typical values range from 50 to 200. The higher this number, the more the transistor is able to amplify an incoming signal.
- **Emitter-base voltage (V_{EBO}):** The maximum voltage across the emitter and the base, which is usually a relatively small number such as 6 V. Most circuits are designed to apply only small voltages to the base, and so this limit isn't usually a concern.
- **Total power dissipation (P_D):** The total power that can be dissipated by the device. For most small transistors, the power rating is on the order of a few hundred milliwatts (mW).

You need to worry about these specifications only if you're designing your own circuits. If you're building a circuit from a book or the Internet, all you need to know is the transistor part number specified in the circuit's schematic.

Amplifying Current with a Transistor

The most common way to use a transistor as a current amplifier is shown in Figure 6-3. This type of circuit is sometimes called a *common-emitter* circuit because, as you can see in the figure, the emitter is connected to ground, which means that the input signal and the output signal share the emitter connection.

You can also use a transistor as an amplifier in two other ways: *common-base* and *common-collector*. As you may guess, they involve connecting the base and the collector to ground, respectively. Common-emitter circuits are used more often than common-base or common-collector, and so that's what we show you in this chapter.

The circuit in Figure 6-3 uses a pair of resistors as a voltage divider to control exactly how much voltage is placed across the base and emitter of the transistor. The AC signal from the input is then superimposed on this bias voltage to vary the bias current. Then, the amplified output is taken from the collector and emitter. Variations in the bias current are amplified in the output current.

As we describe in Chapter 2 of this minibook, a *voltage divider* is simply a pair of resistors. The voltage across both resistors equals the sum of the voltages across each resistor individually. You can divide the voltage any way you want by picking the correct values for the resistors. If the resistors are identical, the voltage divider cuts the voltage in half. Otherwise, you can use a simple formula to determine the ratio at which the voltage is divided. (To check out this formula, turn to Chapter 2 of this minibook.)

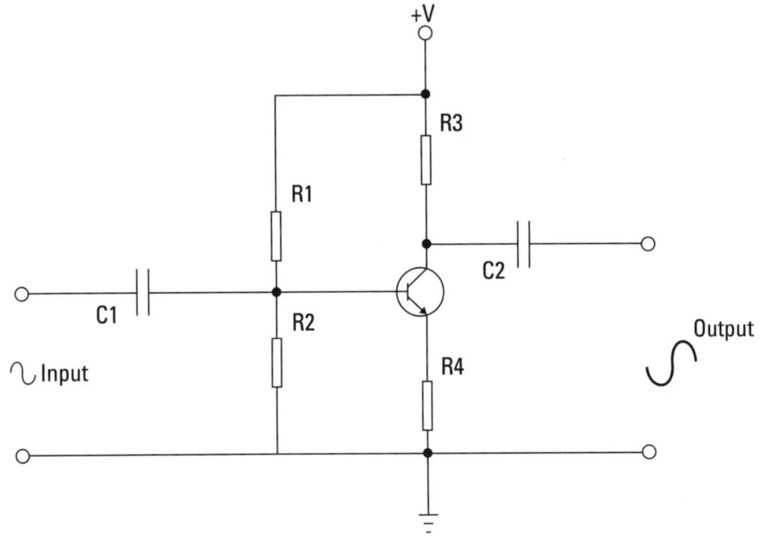

Figure 6-3: A basic transistor amplifier circuit.

If you look at the schematic diagram in Figure 6-3 and narrow your eyes just a bit, you can see that the circuit features two voltage dividers:

- The combination of resistors R1 and R2, which provide the bias voltage to the transistor's base.
- The combination of resistors R3 and R4, which provide the voltage for the output.

In reality, a third resistor exists in the output voltage divider: the collector-emitter path in the transistor itself. In fact, one common way to explain how a transistor works is to think of the collector-emitter path as a *potentiometer* (a variable resistor), whose knob is turned by the bias voltage. For a more detailed explanation, see the later sidebar 'The magic pot'.

The output voltage divider is a variable voltage divider: the ratio of the resistances changes based on the bias voltage, which means the voltage at the collector varies as well. The amplification occurs because very small variations in an input signal are reflected in much larger variations in the output signal.

We use the term *reflected* in the preceding paragraph, because in a common-emitter amplifier circuit, the amplified output is a reflection of the input signal. In other words, positive voltage variations in the input appear as negative variations in the output. So the output signal is *inverted* – which is just a fancy way of saying that it's upside down.

This idea is a little tricky and so now we look more closely at this circuit:

- The input arrives at the left side of the circuit in the form of a signal, which usually has a DC and an AC component. In other words, the voltage fluctuates but never goes negative.
- One side of the input is connected to ground, to which the battery's negative terminal is also connected. The transistor's emitter is also connected to ground (through a resistor), as is one side of the output.
- The purpose of C1 (see Figure 6-3) is to block the DC component of the input signal. Only pure AC gets past the capacitor. Without this capacitor, any DC voltage in the input signal would be added to the bias voltage applied to the transistor, which may spoil the transistor's ability to amplify faithfully the AC part of the input signal.
- R1 and R2 form a voltage divider that determines how much DC voltage is applied to the transistor base. The AC portion of the signal that gets past C1 is combined with this DC voltage, which causes the transistor's base current to vary with the voltage.
- R3, R4 and the variable resistance of the collector-emitter circuit form a voltage divider on the output side of the amplifier. Amplification occurs because the full power supply voltage is applied across the output circuit. The varying resistance of the collector-emitter path reflects the small AC input signal on the much larger output signal.
- C2 blocks the DC component of the output signal so that only pure AC is passed on to the next stage of the amplifier circuit.

The trick in designing transistor amplifiers is picking the right values for all the resistors and capacitors. That task involves more than a little bit of maths and engineering knowledge, however, and is beyond the scope of this book. Most hobbyists get along just fine with published circuits in kits or on the Internet. But if you really want to know how to calculate these values for yourself, you can find excellent tutorials on the subject on the Internet: just search for 'common emitter'.

The magic pot

A transistor functions like a 'magic potentiometer' ('magic pot' for short). Here's the basic idea. A transistor works like a combination of a diode and a variable resistor, also called a *potentiometer* (or pot). This magic pot's knob is mysteriously connected to the diode by invisible rays, kind of like this:

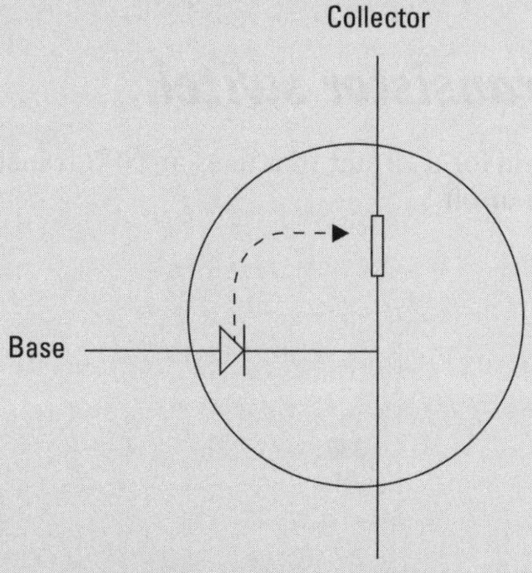

When forward voltage is applied to the diode, the knob of the magic pot turns much like the needle on a voltmeter. This changes the resistance of the potentiometer, which in turn changes the amount of current that can flow through the collector-emitter path.

Note that a magic potentiometer is wired so that when bias voltage increases, resistance decreases. When bias voltage decreases, resistance increases.

Besides being connected to the diode by invisible rays, the magic pot is magic in one more way: its maximum resistance is infinite. Real-world potentiometers have a finite maximum resistance, such as 10 kΩ or 1 MΩ, but the magic pot has infinite maximum resistance.

With this knowledge of the magic pot's properties in mind, you can visualise how a transistor works. The magic knob can be in one of three positions, which correspond to these three operating modes for a transistor:

- **Infinite resistance:** With no bias voltage, the magic pot's knob is spun all the way in one direction, providing infinite resistance, and so no current flows through the transistor. (Remember that the base of the transistor is like a diode, which means that a certain amount of forward voltage is required before current begins to flow through the base. The magic pot stays at its infinite setting until that voltage – usually about 0.7 V – is reached.) This state is called *cut-off* because current is cut off. No amps for you!
- **Some resistance:** As the bias voltage moves past 0.7 V, the diode begins to conduct, and the invisible rays start turning the knob on the magic pot. Thus current begins to flow. How much current flows depends on how far the bias voltage causes the knob to turn.
- **No resistance:** Eventually, the bias voltage turns the knob to its stopping point, and no resistance exists at all. Current flows unrestricted through the collector-emitter circuit. You can continue to increase the bias voltage, but you can't lower the resistance below zero! This state is called *saturation.*

Switching with a Transistor

One of the most common uses for transistors is as simple switches. In short, a transistor conducts current across the collector-emitter path only when a voltage is applied to the base. When no base voltage is present, the switch is off. When base voltage is present, the switch is on.

In an ideal switch, the transistor should be in only one of two states: off or on. The transistor is off when no bias voltage is present or when the bias voltage is less than 0.7 V. The switch is on when the base is saturated so that collector current can flow without restriction.

Exploring an NPN transistor switch

Figure 6-4 shows a schematic diagram for a circuit that uses an NPN transistor as a switch that turns an LED on or off.

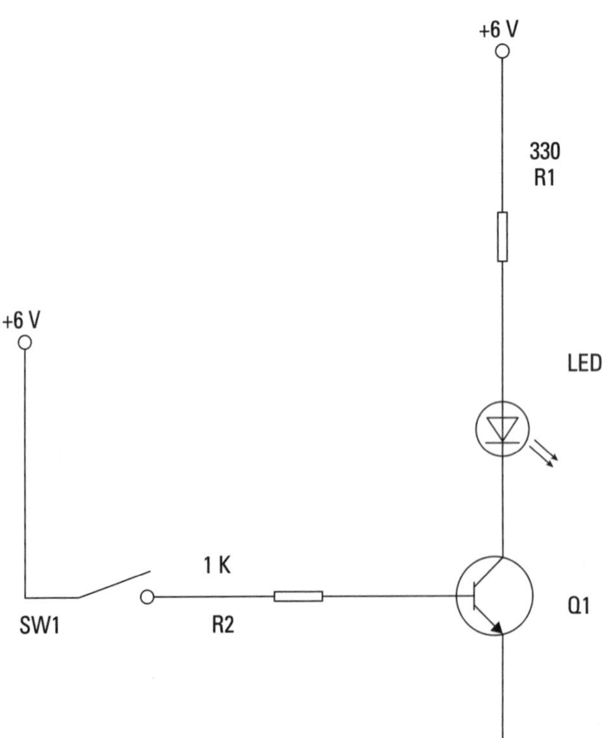

Figure 6-4: Switching an LED with an NPN transistor.

The following list looks at the circuit in Figure 6-4 component by component:

- **LED:** A standard 5 mm red LED. This type of LED has a voltage drop of 1.8 V and is rated at a maximum current of 20 mA.
- **R1:** This 330 Ω resistor limits the current through the LED to prevent the LED from burning out. You can use Ohm's law (for details, see this minibook's Chapter 2) to calculate the amount of current that the resistor allows to flow. Because the supply voltage is +6 V, and the LED drops 1.8 V, the voltage across R1 is 4.2 V (6 – 1.8). Dividing the voltage by the resistance gives you the current in amperes, approximately 0.127 A. Multiply by 1,000 to get the current in mA: the result is 12.7 mA, well below the 20 mA limit.
- **Q1:** A common NPN transistor. We use a 2N2222A transistor, but just about any NPN transistor works. R1 and the LED are connected to the collector and the emitter is connected to ground. When you turn on the transistor, current flows through the collector and emitter so lighting the LED. When you turn off the transistor, the transistor acts as an insulator and the LED doesn't light.
- **R2:** This 1 kΩ resistor limits the current flowing into the base of the transistor. You can use Ohm's law to calculate the current at the base. Because the base-emitter junction drops about 0.7 V (the same as a diode), the voltage across R2 is 5.3 V. Dividing 5.3 by 1,000 gives the current at 0.0053 A, or 5.3 mA. In this way the 12.7 mA collector current (I_{CE}) is controlled by a 5.3 mA base current (I_{BE}).
- **SW1:** This switch controls whether current is allowed to flow to the base. Closing this switch turns on the transistor, which causes current to flow through the LED. Thus, closing this switch turns on the LED even though the switch isn't placed directly within the LED circuit.

You may be wondering why you need to bother with a transistor in this circuit. After all, you can just put the switch in the LED circuit and do away with the transistor and the second resistor. But that would defeat the principle that this circuit illustrates: that a transistor allows you to use a small current to control a much larger one.

If the entire purpose of the circuit is to turn an LED on or off, by all means omit the transistor and the extra resistor. But when you're working with more advanced circuits in the future, you'll find plenty of cases when the output from one stage of a circuit is very small and you need that tiny amount of current to switch on a much larger current. In that case, the transistor circuit shown here is just what you need.

Building an LED driver circuit

In Project 6-1, you build a circuit similar to the one shown in the preceding section. The circuit uses a transistor to switch on an LED using a current that's much smaller than the LED current.

The only difference between this circuit and the one in Figure 6-4 is that the circuit in Project 6-1 adds an LED to the base circuit. When you close the switch, both LEDs light up. However, LED1 is brighter than LED2 because the collector current is larger than the base current.

Figure 6-5 shows the completed project.

Figure 6-5: The circuit for the LED driver circuit.

Project 6-1: A Transistor LED Driver

In this project, you build a circuit that uses a transistor to drive an LED. The transistor allows a small amount of current to control a larger amount of current that passes through an LED. The circuit includes two LEDs: one on the base and one on the emitter. Both LEDs light up when the knife switch is closed, but the LED on the emitter circuit glows brighter than the one on the base circuit, demonstrating the transistor's current gain.

To complete this project, you'll need wire cutters, wire strippers, and a multimeter.

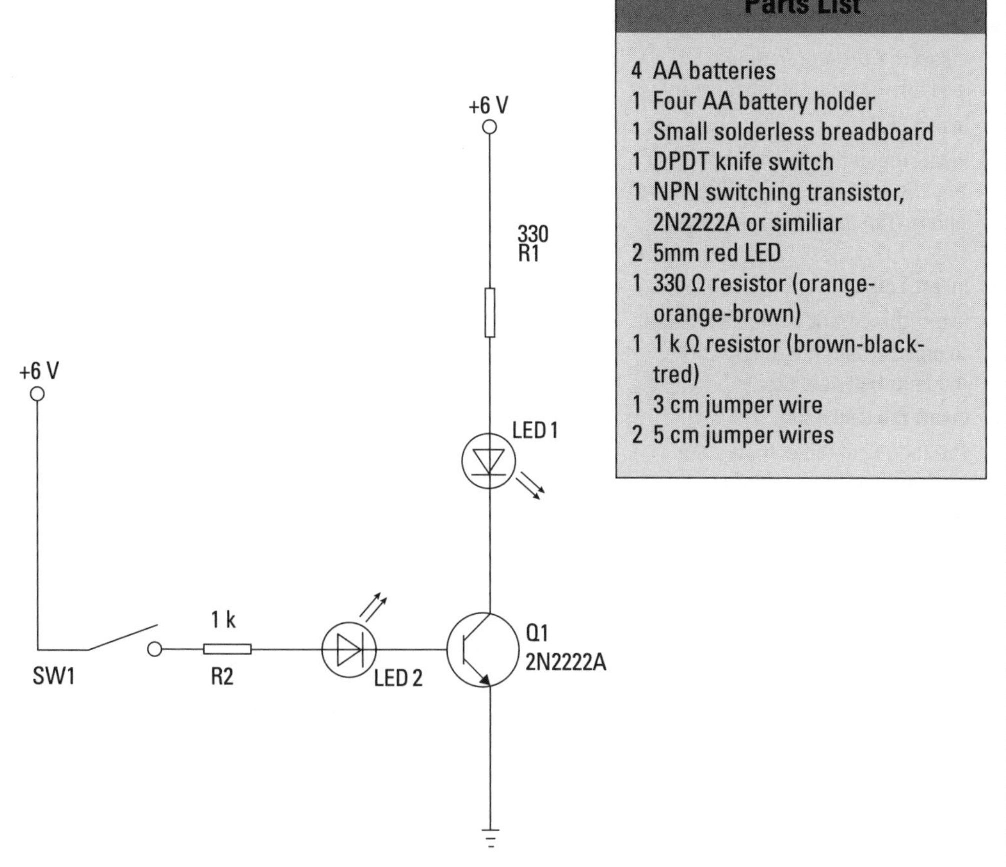

Parts List

- 4 AA batteries
- 1 Four AA battery holder
- 1 Small solderless breadboard
- 1 DPDT knife switch
- 1 NPN switching transistor, 2N2222A or similiar
- 2 5mm red LED
- 1 330 Ω resistor (orange-orange-brown)
- 1 1 k Ω resistor (brown-black-tred)
- 1 3 cm jumper wire
- 2 5 cm jumper wires

Steps

1. **Attach the red lead from the battery holder to any hole in the positive bus strip of the solderless breadboard.**
2. **Attach the black lead from the battery holder to any hole in the negative bus strip of the breadboard.**
3. **Insert the 1 KΩ resistor.**

   Insert the resistor in holes F1 and F6.
4. **Insert the 330 Ω resistor.**

   Insert the resistor in holes J15 and any hole on the positive bus.
5. **Insert LED1.**

   Insert the cathode (the shorter of the two leads) in hole H12 and the anode (the longer lead) in hole H15.
6. **Insert LED2.**

   Insert the cathode (the short lead) in hole I11 and the anode (the long lead) in hole I6.
7. **Insert the transistor.**

   The following table shows the connections for each of the three transistor leads:

Lead	Hole
Emitter	G10
Base	G11
Collector	G12

8. **Connect the transistor emitter to ground.**

   Use the 3 cm jumper wire. One end goes in hole F10; the other goes in any nearby hole in the negative bus on the left side of the breadboard.
9. **Use the two 5 cm jumper wires to connect the switch.**

   Connect 2X to J1 and 2A to any hole on the positive bus.
10. **Insert the four AA batteries into the battery holder.**
11. **Flip the switch to the A position.**

    The two LEDs light up. When you open the switch, the LEDs go out.

    Notice that LED1 is brighter than LED2. This is because more current is flowing through the collector than the base.

    Note: If you have trouble seeing the difference in brightness, look down on the LEDs from directly overhead.
12. **Congratulate yourself!**

    You've built your first transistor circuit!

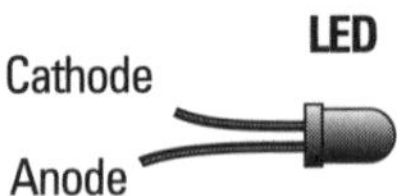

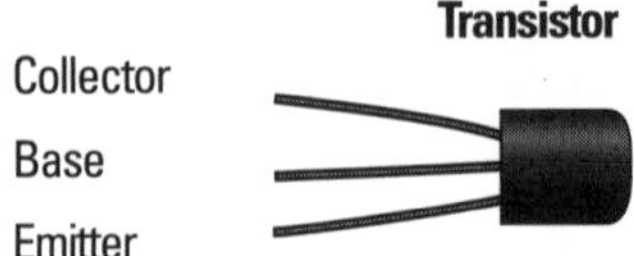

Walking Through a NOT Gate

A *gate* is a basic component of digital electronics, which you can read all about in Book VI. Gate circuits are built from transistor switches that are either on or off. A total of 16 different kinds of gates exist and you discover them all in Book VI, Chapter 2.

In this section, we introduce you to one of the simplest of all gate circuits, called a *NOT gate,* which simply takes an input that can be either on or off and converts it to an output that's the opposite of the input. In other words, if the input is on, the output is off. If the input is off, the output is ON.

Looking at a simple NOT gate circuit

Figure 6-6 shows the schematic diagram for a circuit that uses a single transistor to implement a NOT gate. Here's how the circuit works:

- The input is controlled by a single-pole switch. When the switch is closed, the input is ON. When the switch is open, the input is OFF.
- The input is sent through the R1 and LED1 to bias the transistor. Thus, when the input is ON, LED1 lights up and the transistor is turned on, which enables the collector-emitter path to conduct. When the input is OFF, LED2 is dark, the transistor turns off and no current flows through the collector.
- LED2 is connected directly between the +6 V power supply and ground, through a current-limiting resistor, of course, to keep the LED from burning itself out.
- The anode of LED2 is connected to the transistor's collector.
- When the transistor is off, current flows through R2 and LED2 and the LED lights up, indicating an ON output. But when the transistor turns on, a short circuit is created through the transistor's collector and emitter. This short circuit causes the current to bypass LED2, and so the LED goes dark to indicate an OFF output condition.

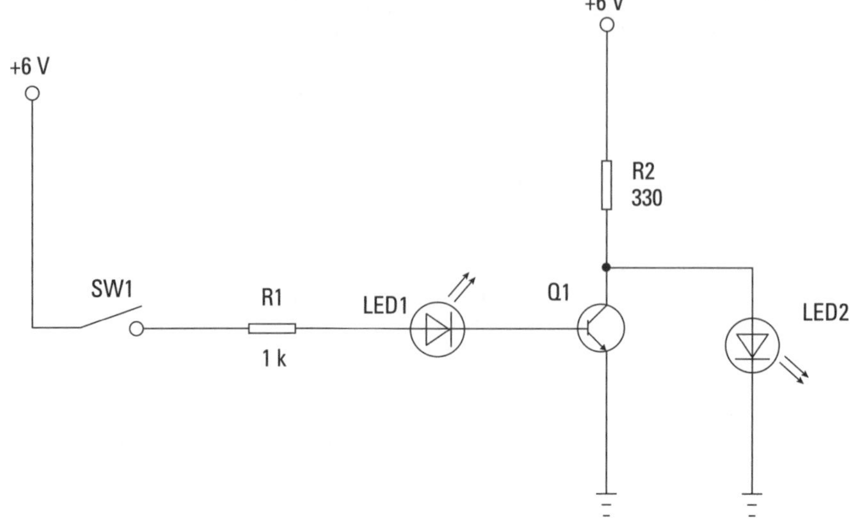

Figure 6-6: The schematic diagram for the NOT gate circuit.

Building a NOT Gate

Project 6-2 shows you how to build the one-transistor NOT gate circuit on a solderless breadboard. The completed project is shown in Figure 6-7.

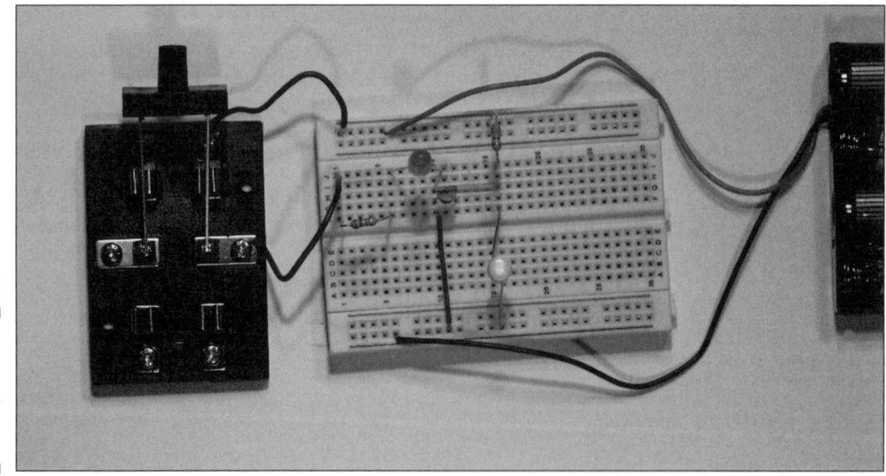

Figure 6-7: The transistor NOT gate.

When you complete this project, you can allow yourself a celebration that you've built your first digital logic circuit.

Project 6-2: A NOT Gate

In this project, you build a simple NOT gate circuit that uses a transistor to invert an input signal. That is, if the input signal is on, the output if off; if the input signal is off, the output is on. One LED is used to indicate the status of the input, and a second LED is used to indicate the status of the output.

You need just a few tools to complete this project: a small Phillips head screwdriver, wire cutters, and wire strippers.

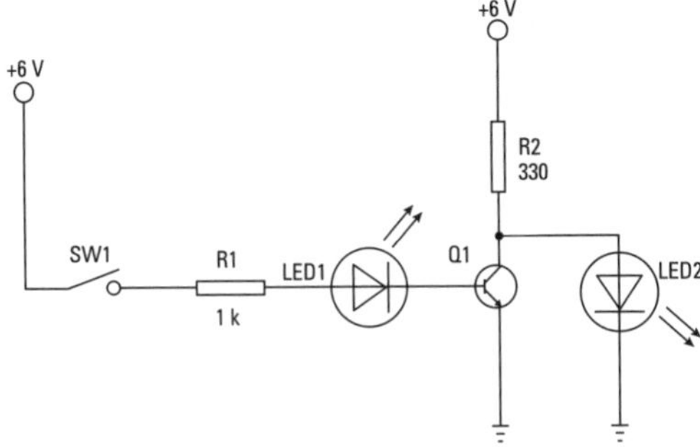

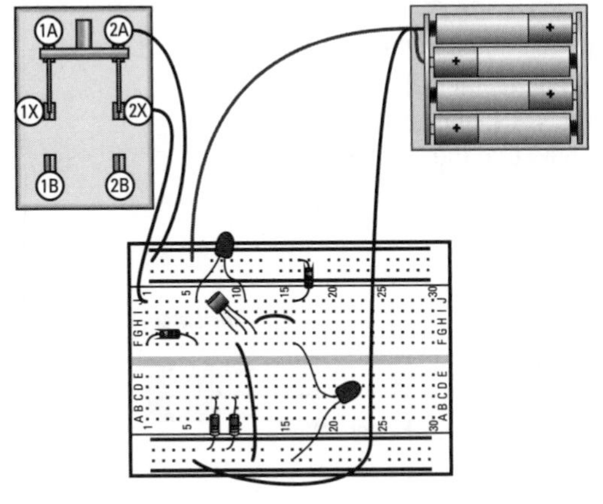

Parts List

- 4 AA batteries
- 1 Four AA battery holder
- 1 Small solderless breadboard
- 1 DPDT knife switch
- 1 NPN switching transistor, 2N2222A or similar
- 2 5 mm red LED
- 1 1 kΩ resistor (brown-black-red)
- 1 330 Ω resistor (orange-orange-brown)
- 1 3 cm jumper wire
- 1 1–2 cm jumper wire
- 2 5 cm jumper wires

Steps

1. **Attach the red lead from the battery holder to any hole in the positive bus strip of the solderless breadboard.**
2. **Attach the black lead from the battery holder to any hole in the negative bus strip of the breadboard.**
3. **Insert the resistors.**

   Insert the two resistors as follows:

Resistor	From	To
1 kΩ	F1	F6
330 Ω	J16	Any hole on the positive bus

4. **Insert LED1.**

   Insert the cathode (the shorter of the two leads) in hole J11 and the anode (the longer lead) in hole J6.
5. **Insert LED2.**

   Insert the anode (the long lead) in hole F16 and the cathode (the short lead) in any convenient hole in the negative bus on the left side of the board.
6. **Insert the transistor.**

   The following table shows the connections for each of the three transistor leads:

Lead	Hole
Emitter	G10
Base	G11
Collector	G12

7. **Connect the transistor emitter to ground.**

   Use the 3 cm jumper wire. One end goes in hole F10; the other goes in any nearby hole in the negative bus on the left side of the breadboard.
8. **Connect the collector to LED2's anode.**

   Use the 2–3 cm jumper wire. Insert one end in hole H12, the other in hole H16.
9. **Use the two 5 cm jumper wires to connect the switch.**

   Here are the connections:

From	To
Switch terminal 2X	Hole JI
Switch terminal 2A	Any hole on the positive bus on the right side of the breadboard

10. **Insert the four AAA batteries into the battery holder.**

    When the batteries are connected, LED2 lights up (assuming the switch is open).
11. **Flip the switch to the A position.**

    LED1 lights up to show that the input is ON, and LED2 goes dark to show that the input is OFF.

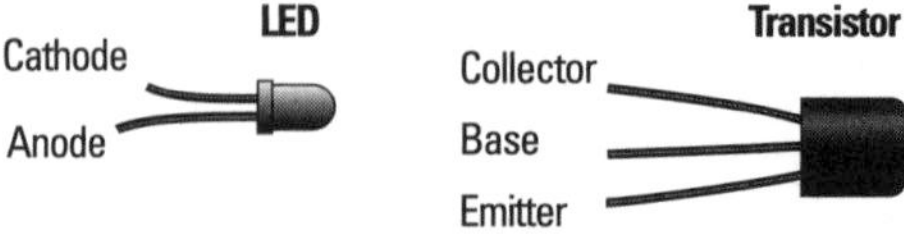

Oscillating with a Transistor

An *oscillator* is an electronic circuit that generates repeated waveforms. The exact waveform produced depends on the type of circuit used to create the oscillator. Some circuits generate sine waves, some generate square waves and others generate other types of waves. Oscillators are essential ingredients in many different types of electronic devices, including radios and computers. Flip to Book I, Chapter 9, for more about the different types of waves.

One of the most commonly used oscillator circuits is made from a pair of transistors rigged up to turn on and off alternately. This type of circuit is called a *multivibrator.* If the circuit is designed to cycle continuously between the two transistors, it's called an *astable multivibrator,* because the circuit never reaches a point of stability – that is, it never decides which of the two transistors should be on and so just keeps flipping back and forth between the two. Astable multivibrators are great for producing square waves, which are often used in digital circuits.

Inspecting an astable multivibrator

Figure 6-8 is a generalised schematic diagram for an astable multivibrator made from a pair of NPN transistors.

When you first power up this circuit, only one of the transistors turns on. You may think that they'd both turn on, because the bases of both transistors are connected to +V, but it doesn't happen that way: one of them goes first. For the sake of discussion, assume that Q1 is the lucky one.

When Q1 comes on, current flows through R1 into the collector and on through the transistor to ground. Meanwhile, C1 starts to charge through R2, developing a positive voltage on its right plate. Because this right plate is connected to the base of Q2, positive voltage also develops on the base of Q2.

When C1 is charged sufficiently, the voltage at the base of Q2 causes Q2 to start conducting. Now the current flows through the collector of Q2 via R4, and C2 starts charging through R3. Because the right-hand plate of C2 is bombarded with positive charge, the voltage on the left plate of C2 goes negative, which drops the voltage on the base of Q1 and causes it to turn off.

C1 discharges while C2 charges. Eventually, the voltage on the left plate of C2 reaches the point where Q1 turns back on, and the whole cycle repeats.

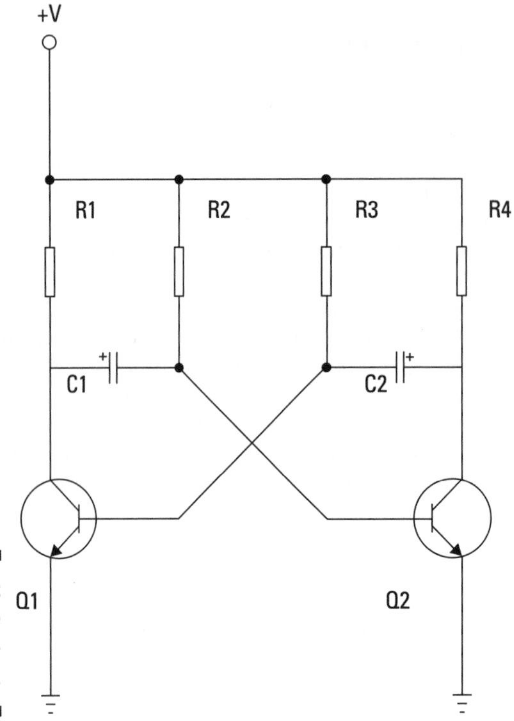

Figure 6-8: An astable multi-vibrator.

Don't worry if this process seems confusing: it is. If the details seem baffling, just focus on the big picture. The dueling capacitors alternately charge and discharge, turning the two transistors on and off, which in turn allows current to flow through their collector circuits. Back and forth it goes, like an amazing rally at Wimbledon that no one ever wins . . . the players just keep lobbing the ball back and forth forever, until their batteries run out.

Here are a few other interesting things about astable multivibrators:

- The time that each half of the multivibrator is on is determined by the RC time constant formed by the capacitor charging circuits. Therefore, you can vary the speed at which the circuit oscillates by adjusting the capacitor and resistor values. For more information about calculating resistor and capacitor time constants, refer to Chapter 3 of this minibook.
- You can create an astable multivibrator from PNP transistors simply by switching the ground with the +V voltage source.

- Output from the multivibrator circuit can be taken directly from the collector of either transistor. For example, you can place an LED or a speaker in series with R1 or R4 to see or hear the oscillator in action. For an example, check out the next section.
- You can use a third transistor to couple the multivibrator with an output load, as shown in Figure 6-9. Just connect the emitter of one of the multivibrator transistors to the base of the third transistor and connect the load to the collector, as shown in the figure.

  This arrangement has two advantages:

  - The load itself interferes with the multivibrator circuit if you take it directly from the collector of Q1 or Q2. By using a third transistor, you isolate the load from the multivibrator circuit.
  - The output is much closer to a true square wave when the coupling transistor is used. Without it, the output isn't a clean square wave because of the effects of the capacitor charging.

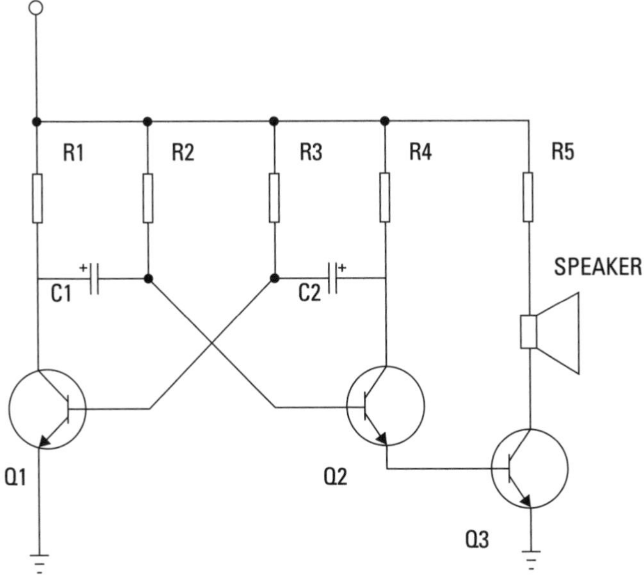

Figure 6-9: Using a transistor to couple an output load to an astable multivibrator.

Building an LED flashing circuit

In Project 6-3, you build a circuit that uses an astable multivibrator to flash two LEDs alternately. LED flasher circuits are a favourite of electronic hobbyists because flashing LEDs have all sorts of fun uses, from surprising Christmas cards to warning lights on your model railroad layout.

The LED flasher circuit is an astable multivibrator similar to the one shown in Figure 6-8. The only differences are that we add LEDs to the collector circuit of each transistor and fill in the resistor and capacitor values. With the values we select for this project, the lights alternate quickly, a bit faster than once per second.

If you feel like experimenting a bit after you complete this project, here are a few suggestions:

- Replace R2 and R3 with larger resistors, such as 220 kΩ and 470 kΩ, and see what effect this has on the flasher.
- Add a potentiometer in series with R2 or R3. Doing so allows you to vary the flash rate by turning the potentiometer knob.
- Change the capacitors.

Project 6-3: An LED Flasher

In this project, you build a circuit that uses a two-transistor astable multivibrator to flash two LEDs in sequence. You need a small Phillips-head screwdriver, wire cutters and wire strippers to build this project.

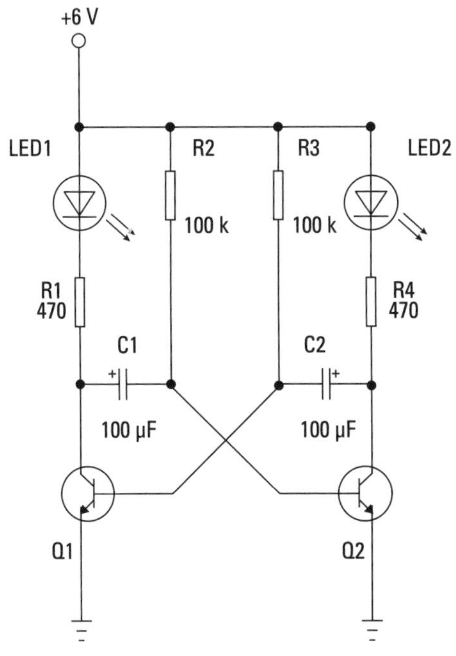

Parts List	
4	AAA batteries
1	Four AA battery holder
1	Small solderless breadboard
2	NPN switching transistors, 2N2222A or similar
2	5 mm red LED
2	100 µF electrolytic capacitors
2	100 kΩ resistor (brown-black-orange)
2	470 Ω resistor (yellow-violet-brown)
2	3 cm jumper wire
1	2–3 cm jumper wire
1	2 cm jumper wires

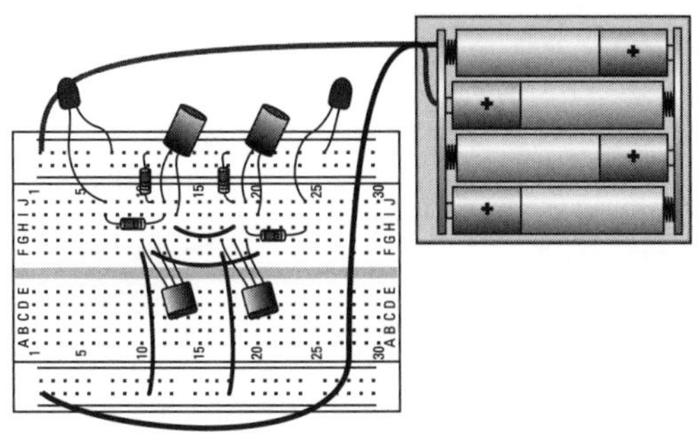

Steps

1. Insert the jumper wires.

Length	*From*	*To*
3 cm	F10	Any hole in the ground bus on the left side of the breadboard
3 cm	F17	Any hole in the ground bus on the left side of the breadboard
2–3 cm	F11	F20
2 cm	H13	H18

2. Insert the capacitors.

Insert the two electrolytic capacitors as follows:

Capacitor	–	+
C1	I13	J12
C2	I20	J19

Be sure to observe the correct polarity of the capacitors. The negative lead is marked with minus signs.

3. Insert the transistors.

The following table shows the connections for each of the three transistor leads of the two transistors:

Lead	Q1	Q2
Emitter	G10	G17
Base	G11	G18
Collector	G12	G19

4. Insert the LEDs.

LED	Cathode (Short)	Anode (Long)
LED1	J7	Any hole in the positive bus
LED2	J24	Any hole in the positive bus

5. Insert the resistors.

Resistor	*From*	*To*
R1 (470)	I7	I12
R2 (100K)	J11	Any hole in the positive bus
R3 (100K)	J18	Any hole in the positive bus
R4 (470)	H19	H24

6. Connect the battery holder to the breadboard.

Attach the red lead from the battery holder to any hole in the positive bus strip on the right side of the solderless breadboard.

Attach the black lead from the battery holder to any hole in the negative bus strip on the left side of the breadboard.

7. Insert the four AAA batteries into the battery holder.

8. Enjoy the show!

The LEDs flash alternately as long as you leave the batteries in.

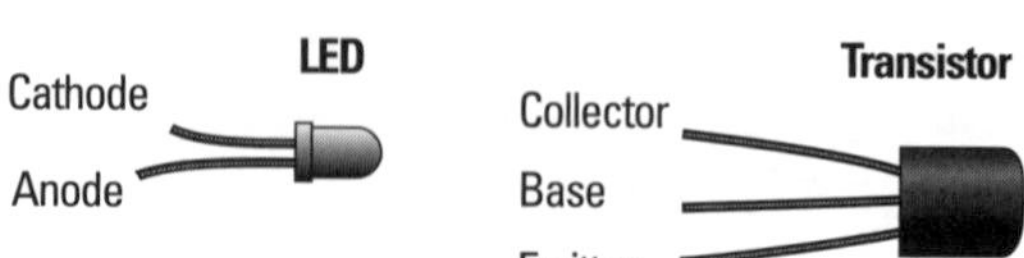

Book III

Working with Integrated Circuits

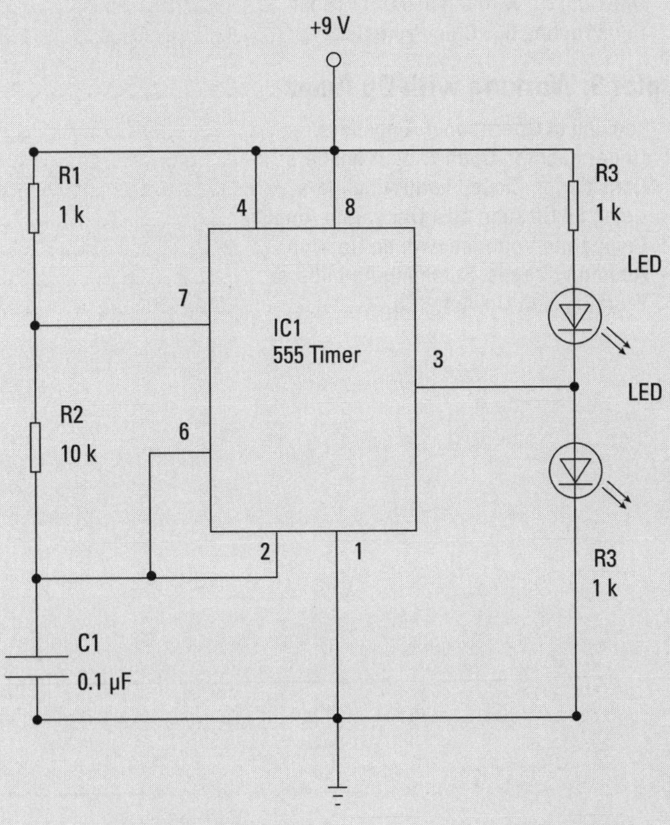

If you're interested in the nitty-gritty of how integrated circuits are made, you can find a great article about making nano-sized features for integrated circuits at `www.dummies.com/extras/electronicsaiouk`.

Contents at a Glance

Chapter 1

Introducing Integrated Circuits

In This Chapter

- Examining integrated circuits
- Protecting against damage
- Deciphering IC data sheets
- Meeting some of the IC family

On 25 April 1961, Robert Noyce (an engineer working in Palo Alto, California) received word that a patent application that he'd submitted a year before had been finally approved. The patent was for a new type of device that would soon become known as an integrated circuit. Exactly one month later, on 25 May, President John F. Kennedy announced to the world that the United States was going to the moon.

These two events have a lot in common, because without the integrated circuit NASA probably wouldn't have been able to pull off Kennedy's challenge.

In this chapter, you discover the device that put men on the moon and went on to change the world of electronics. You find out how integrated circuits are built, what they can do and why they keep getting smaller and cheaper. We also show you how to incorporate them into your own electronic projects.

'Fly me to the moon'

In May 1961, NASA had no idea how to get to the moon. But they did know that their moonship needed a computer like none the world had seen before. They'd have to figure out a way to shrink a computer that at the time filled an entire room into a box about the size of a picnic basket.

So the very first contract NASA awarded for the Apollo programme was for this computer. Many more contracts would follow: the command module, the lunar module, the launch vehicle, the space suit and thousands of other vital elements to make the moon landing possible. But NASA's first priority was to build the computer.

The computer contract went to the Massachusetts Institute of Technology, and the engineers there quickly decided that the only way to build the computer was to take advantage of the brand-new technology of integrated circuits. As a result, NASA became the first large-scale user of integrated circuits.

Investigating Integrated Circuits

An *integrated circuit* (also called an *IC, silicon chip* or just a *chip*) is an entire electronic circuit – consisting of multiple individual components such as transistors, diodes, resistors, capacitors and the conductive pathways that connect all the components – constructed from a single piece of silicon crystal.

An IC isn't a really small circuit board with components mounted on it. In an IC, the individual components are embedded directly into the silicon crystal. Previous circuit fabrication techniques relied on mounting smaller and smaller parts on smaller and smaller circuit boards, but an IC is all one piece.

Instead of just two or three p-n junctions, an IC has thousands of individual p-n junctions. In fact, many modern ICs have millions or even billions of them, all fashioned from a single piece of silicon.

The earliest ICs were simple transistor amplifier circuits with just a few transistors, resistors and capacitors. In fact, they weren't much more complicated than the circuits you breadboard in Book II, Chapter 6. But today's ICs are unbelievably complex. At the time we're writing, the most advanced Intel computer chip had 2.6 *billion* transistors. That number will probably have grown by the time this book reaches your hands.

Most of the ICs you're going to use for hobby projects are much more modest, having something in the order of a few dozen transistors. For example, the 555 timer IC, which we describe in Chapter 2 of this minibook, has 20 transistors, 2 diodes and 15 resistors and costs about a pound.

The miracle of Moore's law

In a nutshell, *Moore's law* predicts that the number of transistors that can be placed on a single IC doubles about every two years. Gordon Moore, one of the founders of Intel, first stated his prediction in 1965. Back then, the prediction was even more ambitious. Originally, Moore's law said that the transistor count would double every year, not every two years. In the mid-1970s, the pace slowed a bit and the prediction was scaled back.

The staggering reality of Moore's law is that the increase in complexity of electronic technology is exponential, not incremental as with most technologies. For example, consider the auto industry, where petrol mileage gets incrementally better every year. Gordon Moore said that if Moore's law applied to cars, a Rolls-Royce would get half a million miles per gallon, and it would be cheaper to buy a new one than pay to park the one you have.

Several times over the years, scientists feared that the end of Moore's law was on the horizon, as the chip manufacturing technology approached some physical limit that couldn't be exceeded, such as the wavelength of the light used to etch the circuits. But each time, a technological breakthrough enabled manufacturers to simply bypass the old limit. Moore's law has held true now for nearly 50 years and is expected to continue into the foreseeable future.

One possible explanation for the uncanny accuracy of Moore's law is that it has become a self-fulfilling prophesy. Integrated circuit manufacturers rely on Moore's law to set their own engineering goals, and they then work feverishly to achieve those goals. Thus, Moore's law has become the objective of the semiconductor industry.

Manufacturing integrated circuits

You don't have to know how ICs are made to use them, and so you can skip this section if you prefer, but the process is pretty interesting. It's complex, and varies depending on the type of chip being made, but here's the typical process:

1. A large, cylindrical piece of silicon crystal is shaved into thin wafers a fraction of a millimetre thick. Each of these wafers are used to create several hundred or thousand finished ICs.
2. A special photoresist solution is deposited on top of the wafer.
3. A mask is applied over the photoresist. The mask is an image of the actual circuit, with some areas transparent to allow light through and others opaque to block the light.
4. The wafer is exposed to intense ultraviolet light, which etches the wafer under the transparent portions of the mask but leaves the areas under the opaque parts of the mask untouched.

5. The mask is removed and any remaining photoresist is cleaned off.
6. The wafer is then exposed to a doping material, which creates n-type and p-type regions in the etched areas of the wafer. (For a review on doping and n-type and p-type semiconductors, take a look at Book II, Chapter 5.)
7. If the circuit design calls for multiple layers stacked on top of one another, the process is repeated for each layer until all the layers are created.
8. The individual ICs are then cut apart and mounted in their final packaging.

The manufacturing process for ICs takes place in a clean room, where workers wear masks and special suits nicknamed 'bunny suits' (they only have ears on April Fools' Day!). This extreme cleanliness is necessary, because even the smallest speck of dust is enormous on a tiny IC.

Each IC goes through a variety of complicated quality tests after the circuit is finished. The process is by no means perfect and so many ICs are discarded.

Packaging integrated circuits

Integrated circuits come in a variety of different package types, but nearly all the ICs you work with in hobby electronics come in a type of package called *dual inline package* (DIP). Figure 1-1 shows two ICs in DIP packages.

Yes, we know that the phrase 'DIP package' is redundant because the 'P' in DIP already stands for 'package', but the phrase is commonly used. Some people justify it by claiming that the 'P' stands for 'pin', removing the redundancy. Like it or not, DIP stands for 'dual inline package' and 'DIP package' is widely used and considered correct. Get used to it.

The phrase 'DIP chip' is also sometimes used to describe ICs in DIP packages. It has a nice ring to it and sounds like something you'd serve at a party.

A DIP package consists of a rectangular plastic or resin case that encloses the IC itself, with two rows of pins on the long sides of the rectangle. The pins on each side jut out a bit from the case and then turn straight down. This arrangement makes the package look like an insect.

The pins on each side of a DIP package are spaced exactly 0.1" (2.54 mm) apart, and the two rows of pins are usually spaced 0.3" (7.62 mm) apart, although some larger DIP packages have wider spacing. In any event, the standard tenth-of-an-inch spacing is perfect for use with solderless breadboards, which have holes spaced at this distance. In fact, the gap that runs down the centre of a solderless breadboard happens to be 0.3" too, which makes it easy to mount DIP chips so that they straddle the gap, as shown in Figure 1-2.

Figure 1-1: Most ICs come in DIP packaging.

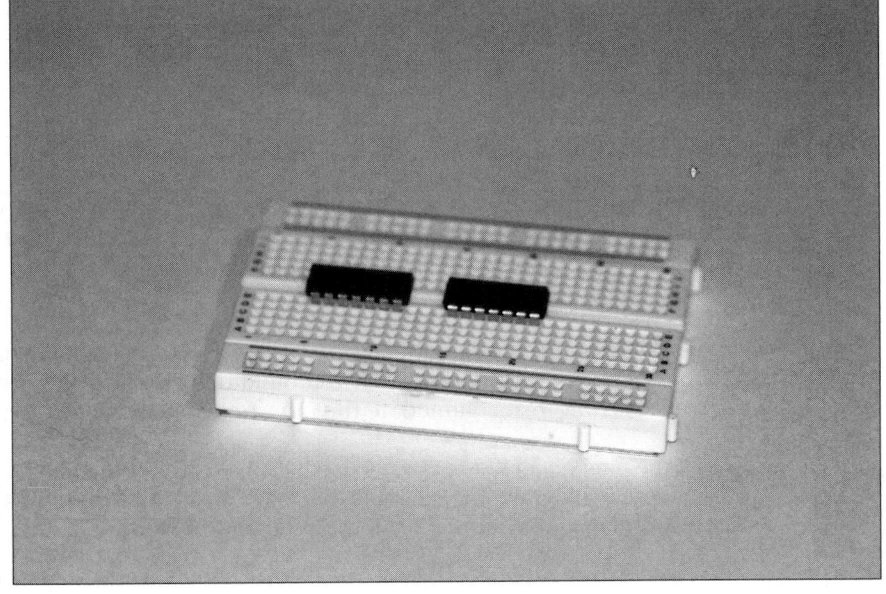

Figure 1-2: Solderless breadboards are designed with DIP chips in mind.

Each pin in a DIP package is numbered. Look down on the package from above and you can see an orientation mark, usually a notch, groove or dot. Orient the package so that this mark is on the top and pin 1 is immediately to the left of the mark. The pins are numbered counter-clockwise, working down the left side and then back up the right side until you get to the last pin, which is immediately to the right of the orientation mark (see Figure 1-3).

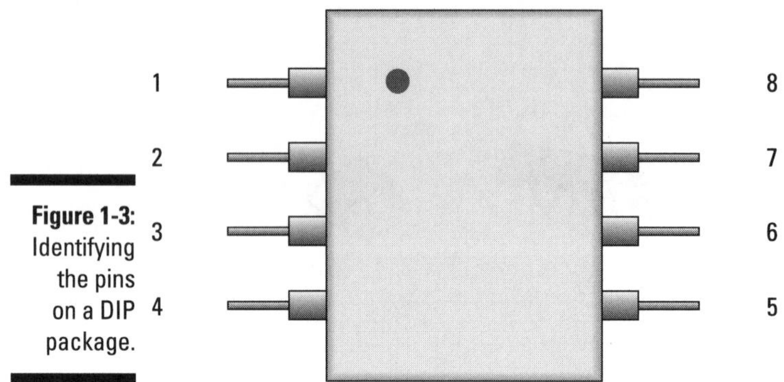

Figure 1-3: Identifying the pins on a DIP package.

The DIP package in Figure 1-3 is for an eight-pin DIP. Larger DIPs have more pins, but the numbering scheme is always the same: pin 1 is to the left of the orientation mark and the remaining pins are numbered counter-clockwise from it.

Depicting ICs in schematic diagrams

In a schematic diagram, an IC is usually represented simply as a rectangle with circuit connections placed conveniently around the rectangle without regard for the physical positioning of the pins. Each pin connection is labelled, as shown in Figure 1-4.

Notice that the pins in this schematic diagram aren't in the same order as in the actual DIP package. Thus, when you build this circuit, you have to adjust the wiring layout to accommodate the pin arrangement of the DIP package.

Not all the pins on an IC are always used. Unused pins are usually left out of the schematic diagram. For example, pin 5 isn't used in the circuit shown in Figure 1-4, and so it's omitted from the schematic.

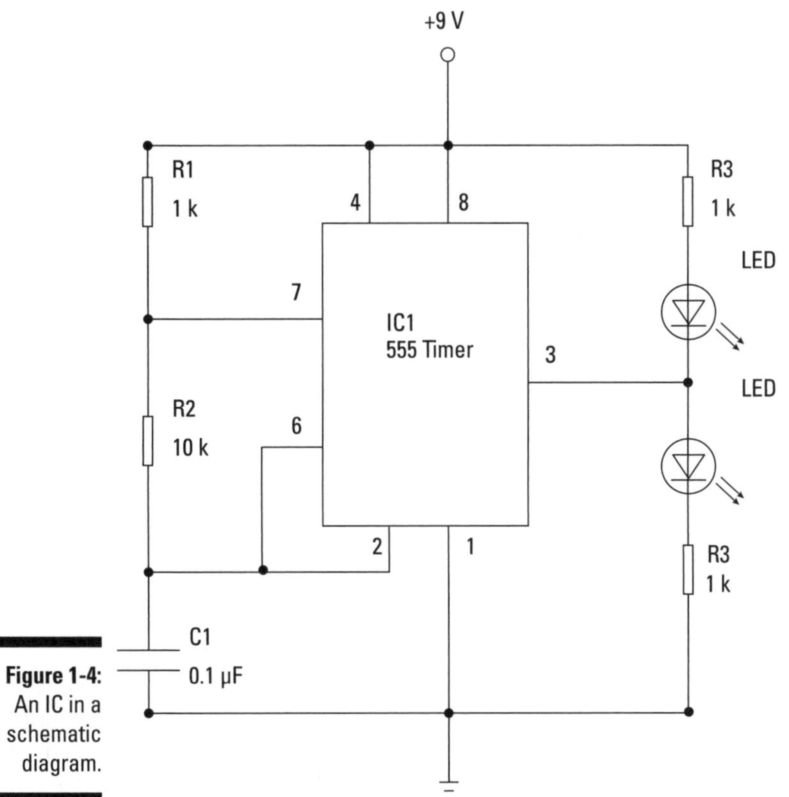

Figure 1-4: An IC in a schematic diagram.

Some ICs contain two or more independent circuits that share a common power supply. For example, the 556 dual timer chip contains two complete 555 timer circuits within a single 14-pin package. When chips like this one are used in a circuit, the schematic diagram may show them separately. For example, Figure 1-5 shows a circuit that calls for a 556 dual timer chip, but each section of the timer is listed separately in the schematic.

Please don't worry about the details of the circuits in Figures 1-4 and 1-5. We aren't trying to explain how these circuits work, but only to show you how the ICs are depicted in the schematic diagrams. You can discover more about how these circuits work in Chapter 2 of this minibook.

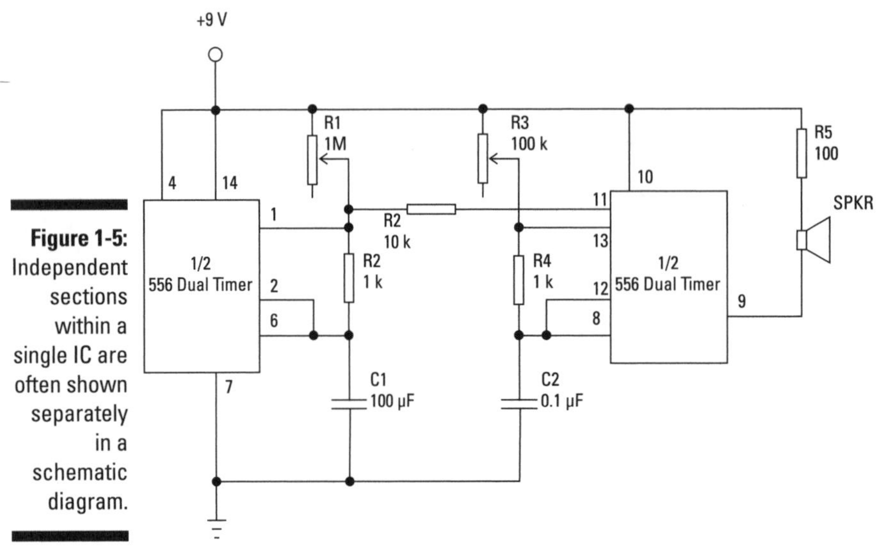

Figure 1-5: Independent sections within a single IC are often shown separately in a schematic diagram.

Powering ICs

In most DIP integrated circuits, two of the pins are used to provide power to the circuit. One of these is designated for positive voltage, typically identified with the symbol V_{CC}. The other is the ground pin. For example, the 555 timer chip (which you can read about in Chapter 3 of this minibook) requires a positive supply voltage between 4.5–15 V at pin 8, and pin 1 is connected to ground.

In the case of ICs that contain two or more separate circuits, the circuits usually share a common power supply. Thus, even though a 556 dual timer chip contains two separate 555 timer circuits, the chip has just one positive voltage pin and one ground pin.

Some ICs call for separate positive and negative supply voltages, not just a positive and a ground connection. You can create a power supply like that using the circuit shown in Figure 1-6.

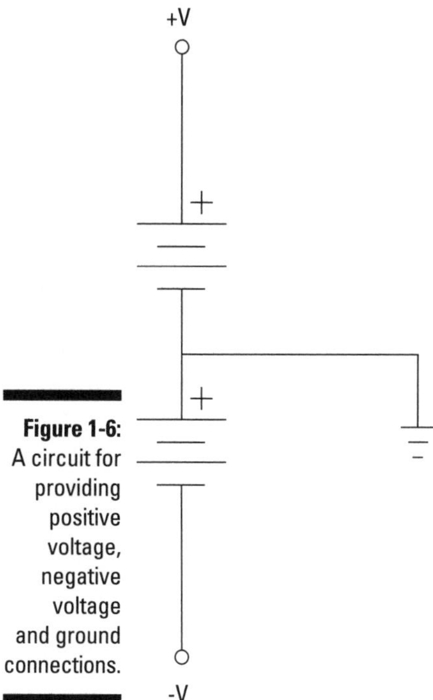

Figure 1-6: A circuit for providing positive voltage, negative voltage and ground connections.

Avoiding Static and Heat Damage

When you build a circuit board that contains one or more ICs, be careful that you don't damage the IC when you build your circuit. In particular, watch out for these two possible problems:

- **Heat damage:** Some ICs are sensitive to heat, and so take precautions whenever you solder an IC to a circuit board. If possible, attach a crocodile clip to the pin to act as a *heat sink* that helps to dissipate some of the heat from the IC itself.

  You can avoid soldering ICs altogether by using DIP sockets such as the ones shown in Figure 1-7. When you use these sockets, you solder the socket to your circuit board. Then, after the socket is safely soldered in place, you simply insert the IC into the socket.

- **Static discharge:** Many ICs can be damaged by static electricity discharged though your fingers when you handle the chips. Therefore, make sure that you discharge yourself by touching a grounded metal surface before handling an IC. You may also want to use an antistatic wristband when handling ICs.

Figure 1-7: DIP sockets let you avoid soldering delicate ICs.

Reading IC Data Sheets

Before you work with a specific type of IC, download a copy of its data sheet (available from many sources on the Internet, including from the vendors' websites). Search with the IC part number and 'data sheet' or 'datasheet'. For example, to find a data sheet for a 555 timer chip, search for '555 data sheet'.

In addition to basic info such as the manufacturer's name and the IC part number, the IC data sheet contains loads of useful details such as:

- A description of what the circuit does.
- Detailed pinout descriptions that tell you the purpose of each pin.
- A diagram of the internal circuitry of the chip. For simple circuits, you may get the entire detailed schematic. For more complicated chips, you get a conceptual diagram instead.
- Detailed electrical specifications, such as maximum voltage you can feed the circuit via the V_{CC} pin or the maximum current loads for output pins.
- Operating conditions such as maximum and minimum temperatures.
- Charts and graphs that illustrate the circuit's behaviour for different operating conditions.

- Formulas for calculating operating characteristics of the circuit. For example, if the operation of the circuit depends on an external RC (resistor/capacitor) circuit, you get formulas for calculating how these external components affect the operation of the circuit.
- Sample circuit diagrams.
- Mechanical descriptions including dimensions.

Meeting the Family: Popular Integrated Circuits

Thousands of different types of ICs are available, most designed for specific applications. However, many ICs are designed for general-purpose use and can be used in a wide variety of circuits.

In this section, we briefly describe some of the most popular general-purpose ICs. These ICs have been around for decades, but their multipurpose design, wide availability and low cost give them enduring popularity.

555 timer

The 555 timer chip was invented in 1971 but remains one of the most popular ICs in use. By some estimates, more than a billion are made and sold every year.

As its name implies, the 555 is a timer circuit. The timing interval is controlled by an external RC network. In other words, by carefully choosing the values for the resistors and capacitors, you can vary the timing duration.

The 555 can be configured in several different ways. In one configuration (called *monostable*), it works like an egg timer: you set it and it goes off after a certain period of time elapses. In a different configuration (called *astable*), the 555 works like a metronome, triggering pulses at regular intervals.

Besides the basic 555 chip, which comes in an 8-pin DIP package, you can also get a 556 dual timer, which contains two independent 555 timers in a single 14-pin DIP package. Many common circuits call for two 555 timers working together, and so the 556 package is very popular.

You can read about the 555 and 556 timers in Chapter 2 of this minibook.

741 and LM324 Op-Amp

An *operational amplifier* (or *op-amp* for short) is a special type of amplifier circuit that has many applications throughout electronics. Although many different types of op-amp circuits exist, the 741 and LM324 are the most common.

The 741 is a single op-amp circuit in an eight-pin DIP package. It was first introduced in 1968 and is still one of the most widely used ICs ever made. The 741 is one of those ICs that require positive and negative voltage, as described earlier in the section 'Powering ICs'.

The LM324 was introduced in 1972 and consists of four separate op-amp circuits in a single 14-pin DIP package. Unlike the 741, the LM324 doesn't require separate negative and positive voltage supplies.

You can discover more about op-amps in Chapter 3 of this minibook.

78xx voltage regulator

The 78xx is a family of simple voltage regulator ICs. A *voltage regulator* is a circuit that accepts an input voltage that can vary within a certain range and produces an output voltage that's a constant value, regardless of fluctuations in the input voltage.

The 'xx' represents the actual voltage regulated by the chip. For example, a 7805 produces a 5 V output. The input voltage must be at least a couple of volts over the output voltage, and can be as high as 35 V.

74xx logic family

One of the primary uses for ICs is in digital electronics, and the 74xx is one of the oldest and still most widely used families of digital ICs. The 74xx family includes a variety of chips that provide basic building blocks for digital circuits. Thus, you don't find complete microprocessors in the 74xx family, but you do find circuits such as logic gates, flip-flops, counters, buffers and so on.

Flip to Book VI, Chapter 3 to read about this useful family of ICs.

Chapter 2

The Fabulous 555 Timer Chip

In This Chapter

- Discovering the 555 timer chip
- Configuring a 555 timer chip
- Putting the 556 dual-timer chip to work
- Building basic 555 timer chip circuits

The 555 timer chip, developed in 1970, is probably the most popular integrated circuit (IC) ever made. By some estimates, more than a billion are manufactured every year. Its popularity is well deserved.

The 555 is a single-chip version of a commonly used circuit called a multivibrator, which is useful in a wide variety of electronic circuits (see Book II, Chapter 6, for more). You can use the 555 chip for basic timing functions, such as turning a light on for a certain length of time, creating a warning light that flashes on and off, producing musical notes of a particular frequency or controlling the position of a servo device (read more about servos in Book VII, Chapter 4). The list goes on and on.

In this chapter, you find out how to use this versatile chip in a variety of circuits. We explain how the 555 works and what each of its pins does, and you also view and build a variety of common 555 circuits.

Examining how the 555 Works

Think of the 555 as a hybrid of an analogue and digital circuit. The output produced by a 555 is purely digital: it's either off (0 V) or on (with positive voltage of at least 2.5 V). The timing mechanism within the 555 determines how long the output is on and how long it's off.

The analogue part of the circuit lies in how you control the length of time that the output signal is on and off. You do that by creating an RC network using a resistor and a capacitor. The values you choose for the resistor and capacitor determine the timing interval. We give you all the information you need to choose the correct values for the resistor and capacitor later in this chapter in the 'Understanding 555 Modes' section, but to check how RC networks work, flip to Book II, Chapter 3.

You can also control whether the 555 does its timing job just once, like an egg timer, or whether it cycles the output on and off repeatedly, like a metronome that keeps ticking over and over again. You do so by connecting the pins of the 555 chip in various ways.

Figure 2-1 shows the arrangement of the eight pins in a standard 555 IC. As you can see, the 555 comes in an 8-pin DIP package.

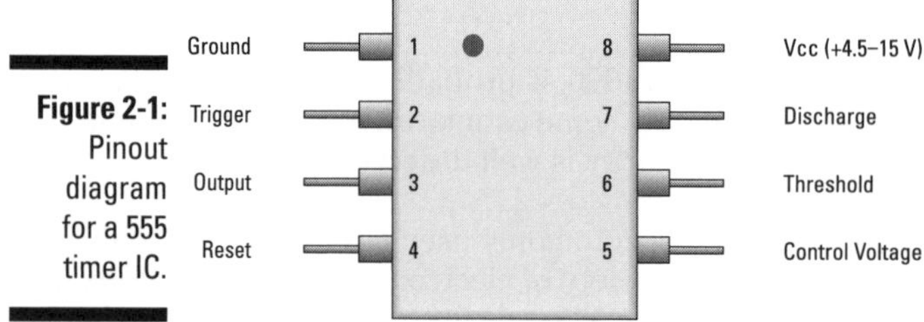

Figure 2-1: Pinout diagram for a 555 timer IC.

The following list describes the function of each of the eight pins (not in numerical order):

- **Ground:** Pin 1 is connected to ground.
- **V_{CC}:** Pin 8 is connected to the positive supply voltage. This voltage must be at least 4.5 V and no greater than 15 V. 555 circuits are commonly run using four AA or AAA batteries, providing 6 V, or a single 9 V battery.
- **Output:** Pin 3 is the output pin. The output is either low (very close to 0 V) or high (close to the supply voltage that's placed on pin 8). (On some 555 models, the output voltage may be as much as 2 V below the supply voltage.) The exact shape of the output – that is, how long it's high and how long it's low – depends on the connections to the remaining five pins.

  For more information about using the output from pin 3, see the 'Working with the 555 Timer Output' section later in this chapter.
- **Trigger:** Pin 2 is the *trigger,* which works like a starter's pistol to get the 555 timer running. The trigger is an *active low* trigger, which means that

the timer starts when voltage on pin 2 drops to below one-third of the supply voltage. When the 555 is triggered via pin 2, the output on pin 3 goes high.

For example, if you want to start the timer by pushing a button, you connect pin 2 to the supply voltage via a resistor. Then, to trigger the timer, you simply interrupt this supply voltage. You can do that in several ways, but the most common is to connect a normally open pushbutton between pin 2 and ground. When the button is pushed, the supply voltage is short-circuited to ground; the voltage at pin 2 drops to zero and the timer is triggered. You see an example of this type of triggering later in the section 'Using the 555 in monostable (one-shot) mode'.

- **Discharge:** Pin 7 is called the *discharge* and is used to discharge an external capacitor that works in conjunction with a resistor to control the timing interval. In most circuits, pin 7 is connected to the supply voltage through a resistor and to ground through a capacitor.
- **Threshold:** Pin 6 is called the *threshold.* The purpose of this pin is to monitor the voltage across the capacitor that's discharged by pin 7. When this voltage reaches two-thirds of the supply voltage (Vcc), the timing cycle ends, and the output on pin 3 goes low.
- **Control:** Pin 5 is the *control* pin. In most 555 circuits, this pin is simply connected to ground, usually through a small 0.01 µF capacitor. (The purpose of the capacitor is to level out any fluctuations in the supply voltage that may affect the operation of the timer.)

  In some circuits, a resistor is used between the control pin and Vcc to apply a small voltage to pin 5. This voltage alters the threshold voltage, which in turn changes the timing interval. Most circuits don't use this capability, however. In this chapter, all the 555 circuits simply connect pin 5 to ground through a 0.01 µF capacitor.
- **Reset:** Pin 4 is the reset pin, which can be used to restart the 555's timing operation. Like the trigger input, reset is an active low input. Thus, pin 4 must be connected to the supply voltage for the 555 timer to operate. If pin 4 is momentarily grounded, the 555 timer's operation is interrupted and won't start again until it's triggered again via pin 2.

When used in a schematic diagram, the pins of a 555 timer chip are almost always shown in the arrangement depicted in Figure 2-2. As you can see, this arrangement follows the customary order of elements in a schematic diagram: supply voltage is at the top, ground is at the bottom, inputs are at the left and outputs are at the right. This arrangement makes determining the operation of the circuit from the schematic diagram easy.

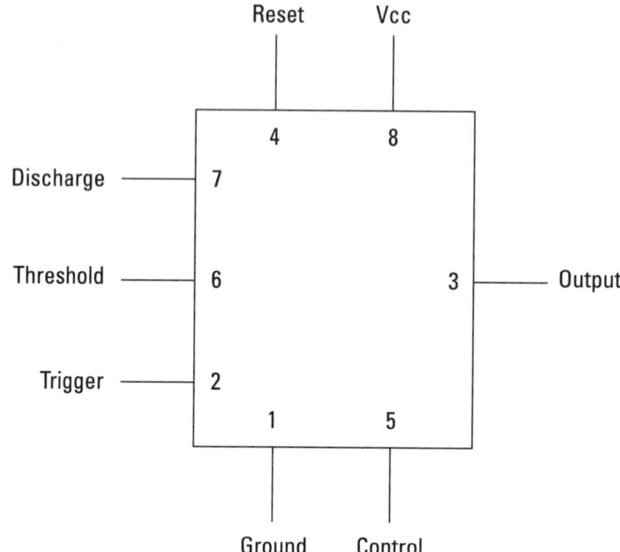

Figure 2-2: How a 555 timer chip is usually shown in a schematic diagram.

Understanding 555 Modes

555 timer ICs can be wired up in three basic ways, called *modes:*

- **Monostable mode:** Works like an egg timer. When you start it, the timer turns on the output, waits for the time interval to elapse and then turns the output off and stops.
- **Astable mode:** Works like a metronome; it keeps running until you turn it off.
- **Bistable mode:** Isn't really a timer mode. Instead, it uses the trigger input to turn the output alternately on and off. This type of circuit is often called a *flip-flop* and is very commonly used in digital electronics.

The following sections explain how each of these three operating modes work.

Using the 555 in monostable (one-shot) mode

Monostable mode lets you use the 555 timer chip as a single-event timer. This mode is called *monostable* because when wired this way, the 555 has just one stable mode, with the output at pin 3 off. When the 555 is sent a trigger pulse, this stable state is temporarily interrupted for an interval that's determined by the value of a resistor and a capacitor. During this interval, the output at pin 3 goes high, but after the time interval has passed, the 555 returns to its stable state, with pin 3 going low.

Monostable mode is sometimes called one-shot mode, which seems a little more descriptive to us. *One-shot mode* conveys the idea that when triggered, the 555 gives one and only one output pulse. When the time interval is reached, the output pulse stops and the circuit goes quiet until another trigger pulse is detected. Each trigger pulse results in a single output pulse.

Typical 555 monostable circuit

Figure 2-3 shows the typical wiring for a 555 timer used in monostable mode.

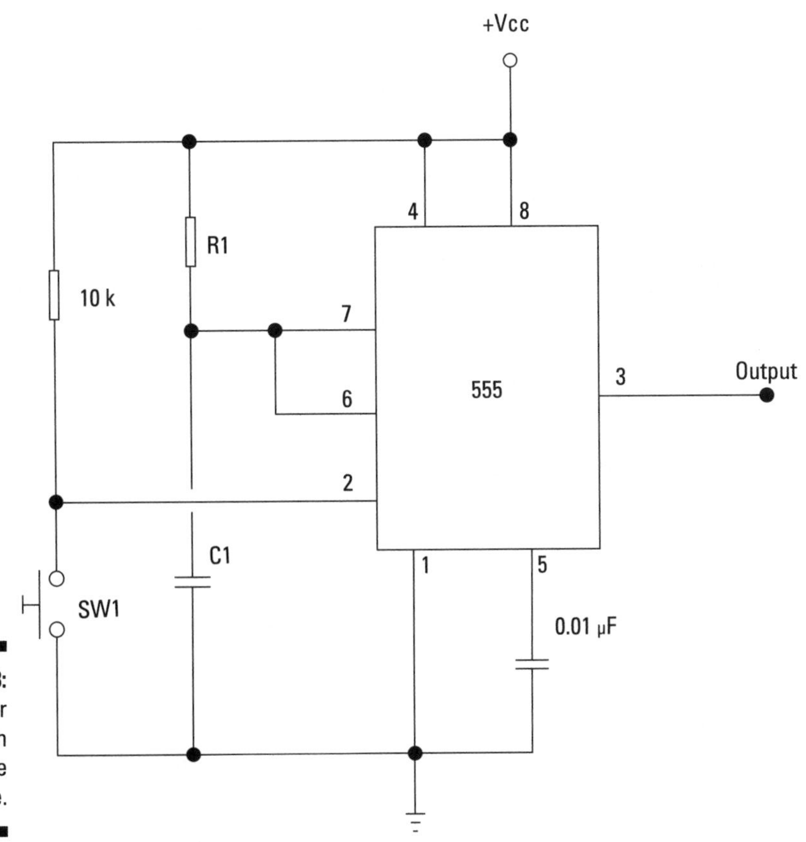

Figure 2-3: A 555 timer chip in monostable mode.

Look at the way the 10 kΩ resistor and the switch are wired to pin 2, the trigger input. The switch is a normally open pushbutton. When the button isn't depressed, the 10 kΩ resistor provides a voltage input to pin 2, which keeps the trigger input high. With the trigger input high, the output voltage at pin 3 is near zero.

When the button is pressed, pin 2 is connected to ground. This causes the voltage at pin 2 to drop to zero, triggering the timer. When the timer is triggered, the output voltage at pin 3 goes high and the timing interval begins.

Resistor-capacitor circuit in a monostable timer

Now look at how the RC circuit (R1 and C1) functions. The resistor and capacitor work together to determine how long the output remains high. In a nutshell, when the circuit is triggered C1 begins to charge.

Pins 6 and 7 – the threshold and discharge pins – are tied together in a monostable 555 circuit. Pin 6 watches the voltage across the capacitor. As the capacitor charges, this voltage increases. When the capacitor voltage reaches two-thirds of the Vcc supply voltage, the timing cycle ends and the output at pin 3 goes low.

The discharge pin (pin 7), charges and discharges the capacitor. To understand how pin 7 works, visualise the internal workings of pin 7 using the model in Figure 2-4. Here, pin 7 is connected to an imaginary switch that's controlled by the status of the output at pin 3. When the output is high, the switch is open; when the output is low, the switch is closed. When the switch is closed, a small 10 Ω resistor within the 555 connects pin 7 to ground. (Although a switch isn't really inside the 555, the model helps you understand how pin 7 functions.)

When the output on pin 3 is low, the imaginary switch inside the 555 is closed and pin 7 is connected to ground through the 10 Ω resistor, which allows the voltage on C1 to discharge through the 555.

But when the output on pin 3 goes high, the imaginary switch inside the 555 is opened. This change forces the current flowing through R1 to go through C1, which in turn causes the capacitor to charge at a rate that depends on the values of R1 and the capacitor.

While the capacitor is charging, pin 6 monitors the voltage that builds up across the capacitor. When this voltage reaches two-thirds of the supply voltage, pin 6 signals the 555 that the timing interval is ended, and the output goes low. This action, in turn, closes the imaginary switch inside the 555, which allows the capacitor to discharge.

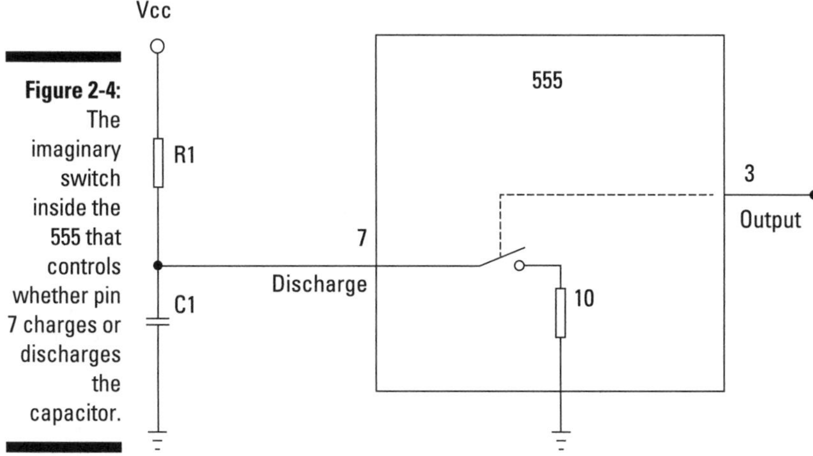

Figure 2-4: The imaginary switch inside the 555 that controls whether pin 7 charges or discharges the capacitor.

Time interval for a monostable circuit

The time interval for a 555 monostable circuit is a measure of how long the output stays high when it's triggered. To calculate the time interval, use this formula:

$$T = 1.1 \times R \times C$$

Here, T is the time interval in seconds, R is the resistance of R1 in ohms and C is the capacitance of C1 in farads.

For example, suppose R1 is 500 kΩ, and C1 is 10 μF. Then, you calculate the time interval like this:

$$T = 1.1 \times 500{,}000\ \Omega \times 0.00001\ F$$

$$T = 5.5\ s$$

The circuit stays on for 5.5 seconds after it's triggered.

When you make this calculation, make sure that you use the correct number of zeros for the resistance and the capacitance. Use Table 2-1 as a guide.

Table 2-1 Conversion of Resistance and Capacitance Values

Resistance	*Capacitance*
1 kΩ = 1,000 Ω	0.01 μF = 0.00000001F
10 kΩ = 10,000 Ω	0.1 μF = 0.0000001F
100 kΩ = 100,000 Ω	1 μF = 0.000001F
1M Ω = 1,000,000 Ω	10 μF = 0.00001F
10M Ω = 10,000,000 Ω	100 μF = 0.0001F
100M Ω = 100,000,000 Ω	1,000 μF = 0.001F

If you don't want to do these calculations yourself, you can find 555 timer calculators on the Internet. Just go to any search engine and type in '555 timer calculator'.

Using the 555 in astable (oscillator) mode

A common way to use a 555 timer is in *astable mode*. The term *astable* simply means that the 555 has no stable state: just as it gets settled into one state (say, the output at pin 3 high), it switches to the opposite state (output low). Then it switches back to the first state, and so on, forever.

This mode is also called *oscillator mode,* because it uses the 555 as an oscillator, which creates a square wave signal. (We discuss the different types of waveforms in Book I, Chapter 9.)

Typical astable circuit

Figure 2-5 shows the basic circuit for a 555 in astable mode.

Notice that the trigger pin (pin 2) is connected directly to C1. In a monostable circuit (see the earlier section 'Typical 555 monostable circuit'), the timer is triggered by a switch that short-circuits the voltage applied to pin 2. In an astable circuit, however, the timer is triggered when the capacitor discharges – when the voltage across the capacitor drops to one-third of the supply voltage, pin 2 triggers the timer to start another cycle.

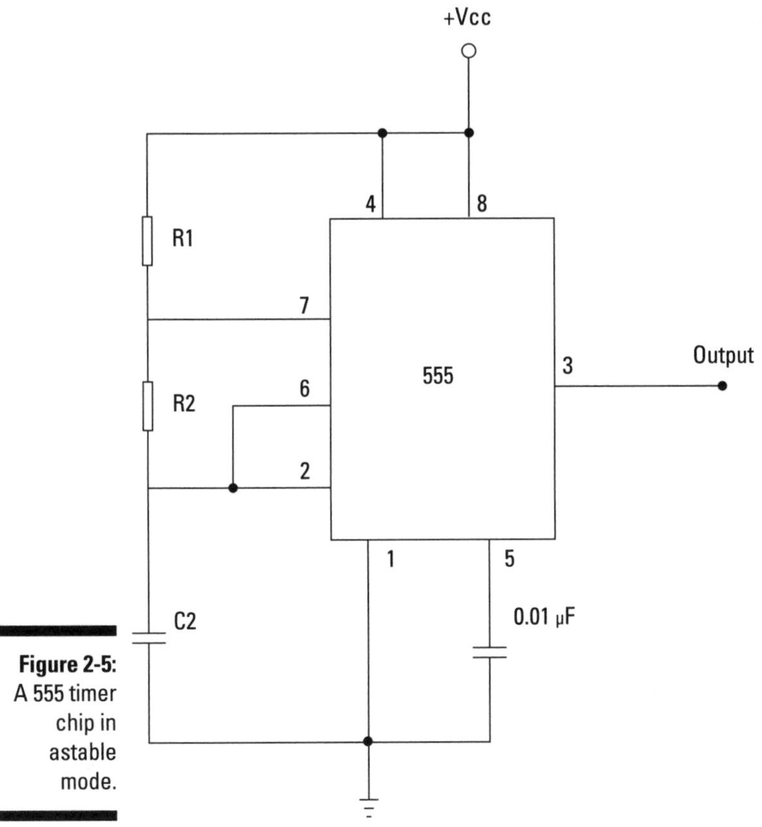

Figure 2-5: A 555 timer chip in astable mode.

Here we examine how this timing cycle works, step by step, starting with the output at pin 3 in the high condition:

1. With the output high, the discharge pin (pin 7) is open, forcing current through resistors R1 and R2 and capacitor C1. This causes the capacitor to charge at a rate that depends on the combined value of R1 and R2 and the value of C1.
2. As the capacitor charges, the voltage at pins 2 and 6 increases.
3. When the voltage at pin 6 (the threshold pin) reaches two-thirds of the supply voltage, the threshold circuitry within the 555 causes the output voltage at pin 3 to go low.
4. When the output at pin 3 goes low, the discharge pin (pin 7) is connected to ground within the 555, which allows C1 to discharge. This discharge occurs through R2, and so the value of R2 as well as the value of the capacitor determines the rate at which the capacitor discharges.

5. As the capacitor discharges, the voltage at pins 2 and 6 decreases.
6. When the voltage at pin 2 (the trigger pin) drops to one-third of the supply voltage, the trigger circuitry inside the 555 causes the output at pin 3 to go high.
7. When the output at pin 3 goes high, the discharge pin (pin 7) is opened, and the cycle starts over again.

Time intervals in an astable 555 circuit

The output of a 555 circuit in astable mode is a square wave, as depicted in Figure 2-6. Three important time measurements apply to a square wave:

- **T:** The total duration of the wave, measured from the start of one high pulse to the start of the next high pulse.
- **T_{high}:** The length of the high portion of the cycle.
- **T_{low}:** The length of the low portion of the cycle.

Naturally, the total time T is the sum of T_{high} and T_{low}.

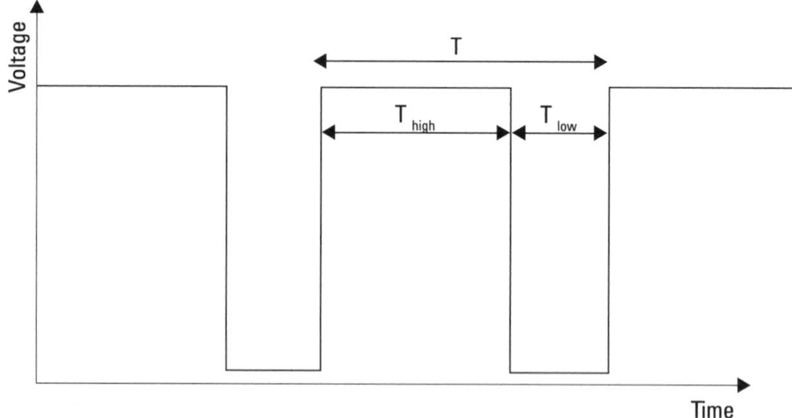

Figure 2-6: Timing the output wave created by an astable 555 timer circuit.

The values of these time constants depend on the values for the two resistors (R1 and R2) and the C1.

Here are the formulas for calculating each of these time constants:

$$T = 0.7 \times (R1 + 2R2) \times C1$$

$$T_{high} = 0.7 \times (R1 + R2) \times C1$$

$$T_{low} = 0.7 \times R2 \times C1$$

An interesting fact is buried in these calculations: C1 charges through both R1 and R2, but it discharges only through R2. That's why you have to add the two resistor values for the T_{high} calculation, but you use only R2 for the T_{low} calculation. It's also why you have to double R2 but not R1 for the total time (T) calculation.

Now, plug in some real numbers to see how the equations work out. Suppose both resistors are 100 kΩ and the capacitor is 10 μF. The total length of the cycle is calculated like this:

$$T = 0.7 \times (100{,}000\ \Omega + (2 \times 100{,}000\ \Omega)) \times 0.00001\ F$$
$$T = 21\ s$$

$$T_{high} = 0.7 \times (100{,}000\ \Omega + 100{,}000\ \Omega) \times 0.00001\ F$$
$$T_{high} = 14\ s$$

$$T_{low} = 0.7 \times 100{,}000\ \Omega - 0.00001\ F$$
$$T_{low} = 7\ s$$

The total cycle time is 21 s, with the output high for 14 s and low for 7 s.

If you want, you can also calculate the frequency of the output signal by dividing the total cycle time into 1. So, for the above calculations, the frequency is 0.47619 Hz.

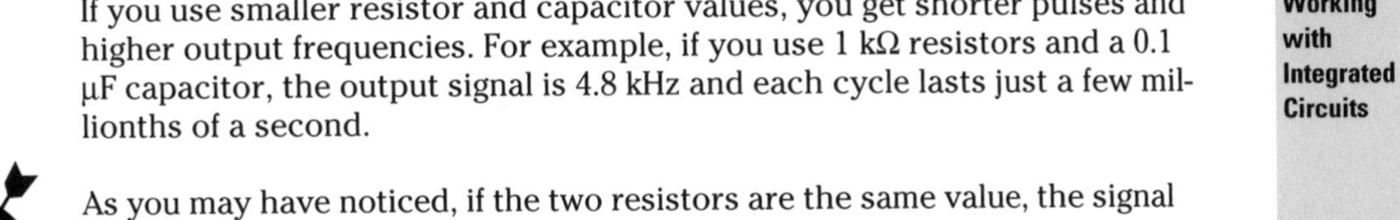

If you use smaller resistor and capacitor values, you get shorter pulses and higher output frequencies. For example, if you use 1 kΩ resistors and a 0.1 μF capacitor, the output signal is 4.8 kHz and each cycle lasts just a few millionths of a second.

As you may have noticed, if the two resistors are the same value, the signal is high for twice as long as it's off. By using different resistor values, you can vary the difference between the high and low portions of the signal.

Duty cycle

The *duty cycle* in a 555 circuit is the percentage of time that the output is high for each cycle of the square wave. For example, if the total cycle time is 1 s and the output is high for the first 0.4 s of each cycle, the duty cycle is 40%.

With an astable circuit such as the one shown in the earlier Figure 2-5, the duty cycle must always be greater than 50%. In other words, the duration for which the output is high must always be more than the duration during which the output is low.

The explanation for this requirement is pretty simple. For the duty cycle to be 50%, the capacitor would have to charge and discharge through the same resistance. The only way to accomplish that would be to omit R1 altogether, so that the capacitor charged and discharged through R2 only. But then you'd end up connecting pin 7 directly to Vcc. With no resistance between pin 7 and the voltage source, the current flowing through pin 7 would exceed the maximum that the circuitry inside the 555 can handle, and the chip would be damaged.

You can get around this limitation in a clever way: place a diode across R2, as shown in Figure 2-7. This diode bypasses R2 when the capacitor is charged. That way, the capacitor charges through R1 and discharges through R2.

When a diode is used in this way, you have complete control over the duration of the charge and discharge time. If R1 and R2 have the same value, the capacitor takes the same amount of time to charge as it does to discharge, and so the duty cycle is 50%. If R2 is smaller than R1, the duty cycle is less than 50%, because the capacitor discharges faster than it charges.

If you use the diode as depicted in Figure 2-7, you have to adjust the formulas for calculating the time intervals (from the preceding section) as follows:

$$T = 0.7 \times (R_1 + R_2) \times C_1$$

$$T_{high} = 0.7 \times R_1 \times C_1$$

$$T_{low} = 0.7 \times R_2 \times C_1$$

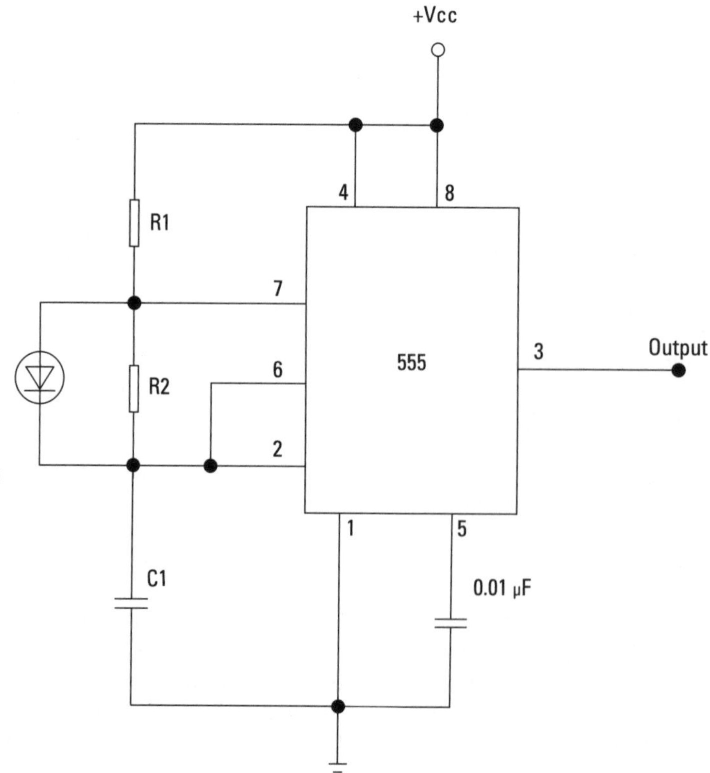

Figure 2-7: Using a diode to control separately the high and low part of the output signal.

Using the 555 in bistable (flip-flop) mode

A *flip-flop* is a circuit that alternates between two output states. In a flip-flop, a short pulse on the trigger causes the output to go high and stay high, even after the trigger pulse ends. The output stays high until a reset pulse is received, at which time the output goes low.

This type of circuit is called *bistable* because the circuit has two stable states: high and low. The circuit stays low until it is triggered and then stays high until it's reset. This type of circuit is used extensively in computers and other digital circuits.

For computer applications, the 555 is a poor choice for use as a flip-flop. That's because its output doesn't change fast enough in response to trigger or reset pulses in computer circuits that are driven by high-speed clock pulses. For computer applications, better flip-flop chips are readily available (see Book VI, Chapter 5).

The 555 is often used in bistable mode, however, for noncomputer applications where high-speed response isn't necessary. For example, imagine a simple robot that drives itself forward until it bumps into something in front of it, and then drives backward until it bumps into something behind it. You'd equip the robot with contact switches on the front and rear connected to the trigger and reset inputs of a 555 in bistable mode. You'd connect the robot's drive motor to the output such that when the output is low, the motor runs forward, and when the output is high, the motor runs backward. The bistable 555 causes the robot to drive back and forth between two obstacles.

Figure 2-8 shows the schematic for a 555 used in bistable mode. As you can see, this circuit doesn't require a capacitor, because in bistable mode the 555 isn't used as a timer. The highs and lows of the output signal are controlled by the trigger and reset inputs, not by the charging and discharging of a capacitor.

Both the trigger (pin 2) and the reset (pin 4) inputs are connected to Vcc through a 10 kΩ resistor. When the set switch is depressed, pin 2 is shorted to ground. This causes the voltage to bypass pin 2, resulting in a momentary low pulse, which triggers the 555. When triggered, the output pin goes high.

In astable or monostable mode, the output pin would remain high until the voltage at the threshold pin (pin 6) reaches two-thirds of the supply voltage. However, because pin 6 isn't connected to anything in this circuit, no voltage is ever present on pin 6. Thus, the threshold is never reached, and so the output remains high indefinitely until the 555 is reset by a low pulse on the reset pin (pin 4).

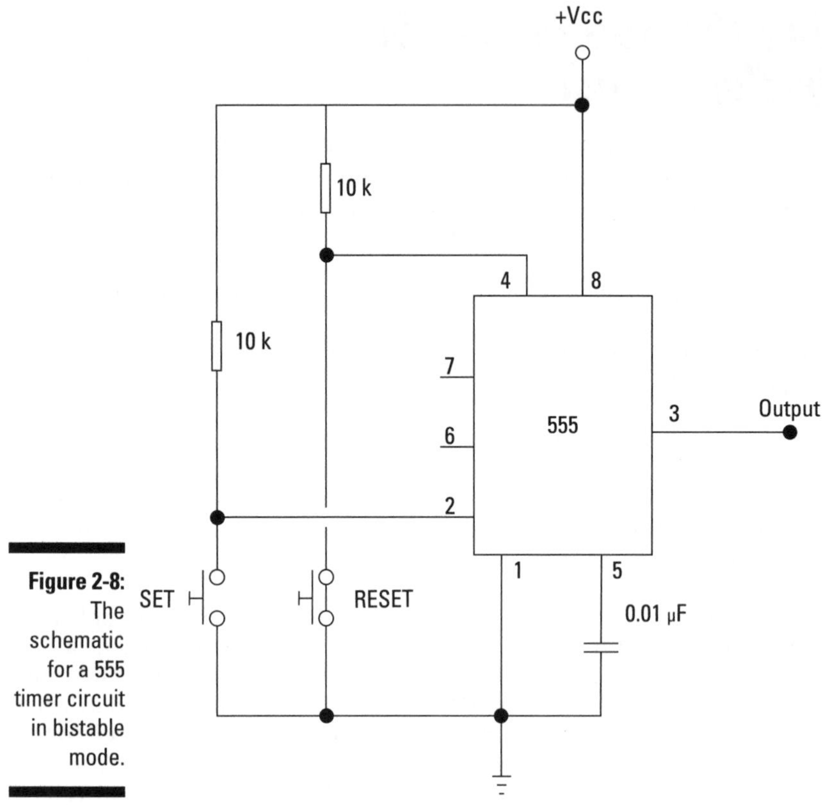

Figure 2-8: The schematic for a 555 timer circuit in bistable mode.

The reset input (pin 4) is connected to Vcc in the same manner as the trigger input. As a result, when the reset switch is pressed, pin 4 is short-circuited to ground, creating a low pulse, which resets the 555 and brings the output back to high.

Working with the 555 Timer Output

The output pin (pin 3) of a 555 can be in one of two states:

- **High:** The voltage at the pin is close to the supply voltage.
- **Low:** The voltage at the pin is 0 V.

You can connect output components to the output pin in two ways. Figure 2-9 illustrates these two configurations, using an LED as the output device. A resistor is also included in the circuit to limit the current flow.

Without the resistor, current flows through the circuit unimpeded, which quickly burns out the LED and probably ruins the 555 as well.

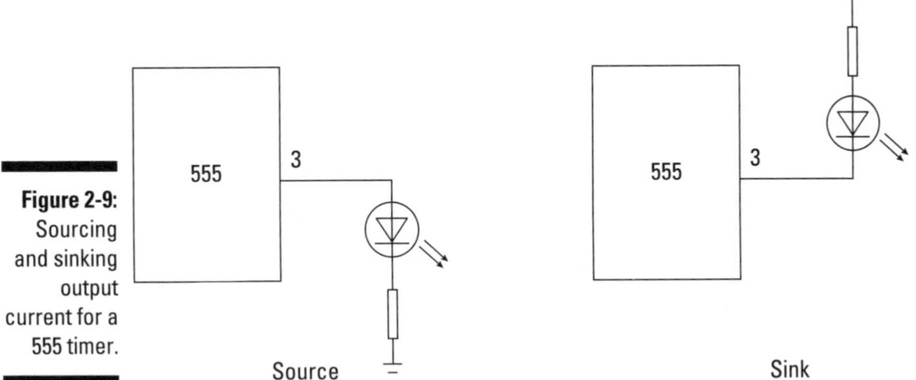

Figure 2-9: Sourcing and sinking output current for a 555 timer.

In the circuit on the left of Figure 2-9, current flows through the LED circuit when the output is high. The current flows from the output pin through the LED and resistor to ground. This output configuration is called *sourcing* because the 555 is the source of the current that drives the output.

In the circuit on the right, current flows through the LED circuit when the output is low. The current flows from the Vcc supply, through the LED and resistor, and into the 555, where it's internally routed to ground through pin 1. This output configuration is called *sinking* because the current is sent into the 555.

Whether you source or sink your output circuit depends on whether you want your output circuit to turn on when the output is high or low.

Figure 2-10 shows that you can combine sourcing and sinking in a single circuit. Here, two LEDs are connected to the output pin. One is sourced; the other is sunk. In this circuit, the LEDs flash alternately as the output switches from high to low. LED1 lights when the output is low, LED 2 when the output is high.

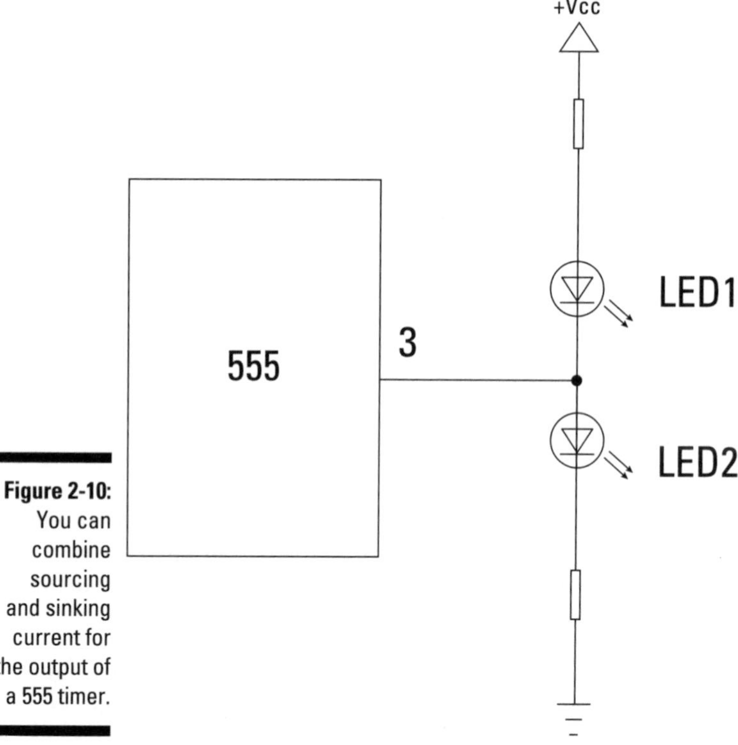

Figure 2-10: You can combine sourcing and sinking current for the output of a 555 timer.

The output circuit of a 555 timer can handle as much as 200 mA of current, which is much more current than most ICs can source or sink. If you need to drive a device that requires more than 200 mA, you can isolate the output device from the 555 by using a transistor, as shown in Figure 2-11. For more information about working with transistors, please refer to Book II, Chapter 6.

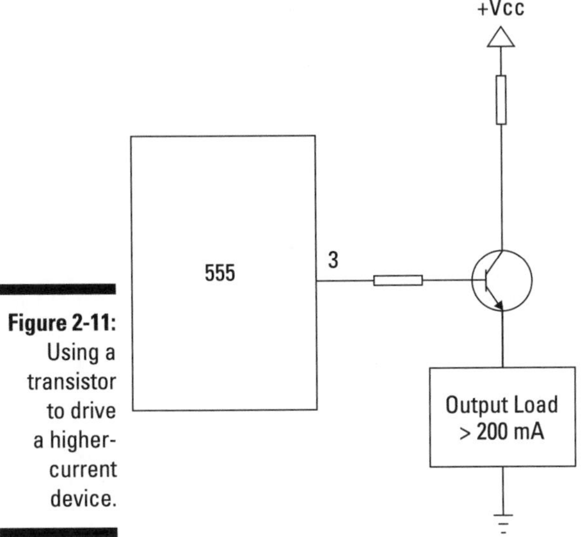

Figure 2-11: Using a transistor to drive a higher-current device.

Doubling Up with the 556 Dual Timer

If one 555 timer chip is good, two are even better! In fact, two (or more) 555 timers in a single circuit have so many uses that you can get two 555 timers in a single chip, called the 556 dual-timer chip.

The 556 dual-timer chip comes in a 14-pin DIP package. The two 555 timers share a common supply and ground pin. The remaining 12 pins are allocated to the inputs and outputs of the individual 555 timers. Table 2-2 lists the pin connections for each of the 555 timers in a 556 dual-timer chip. As an added bonus (no charge!), we also list the pinouts for a standard 555 timer chip.

Table 2-2 Pinouts for the 555 Timer and 556 Dual-Timer Chips

Function	*555 Timer*	*556 First Timer*	*556 Second Timer*
Ground	1	7	7
Trigger	2	6	8
Output	3	5	9
Reset	4	4	10
Control	5	3	11
Threshold	6	2	12
Discharge	7	1	13
Vcc	8	14	14

One common way to use a 556 dual timer is to connect both 555 circuits in monostable (one-shot) mode, with the output pin from the first 555 timer connected to the trigger pin of the second 555 timer. Then, when the output of the first timer goes low, the second timer is triggered. You can connect as many 555 timers as you want in this way, with each timer's output connected to the next timer's trigger so that the timers work in sequence, one after the other.

For example, Figure 2-12 shows a cascaded timer circuit that uses two separate 555 timer chips. In this circuit, both the 555 timer chips are configured in monostable mode, much like the circuit in Figure 2-3. The time interval for the first 555 is controlled by R1 and C1. For the second 555, the interval is controlled by R2 and C2. You can choose whatever values you want for these components to achieve whatever time intervals suit your fancy.

The first 555 chip is triggered when SW1 is depressed, taking pin 2 to ground. This action takes the output on pin 3 to high, which lights LED1. Notice, however, that pin 3 of the first 555 is connected through a small capacitor to the trigger input of the second 555. As soon as the time interval expires on the first 555, its output goes low, which turns off LED1 and at the same time triggers the second 555, which in turn lights up LED2. LED2 stays lit until C2 charges, and then it goes out. The circuit then waits to be triggered again by a press of the switch.

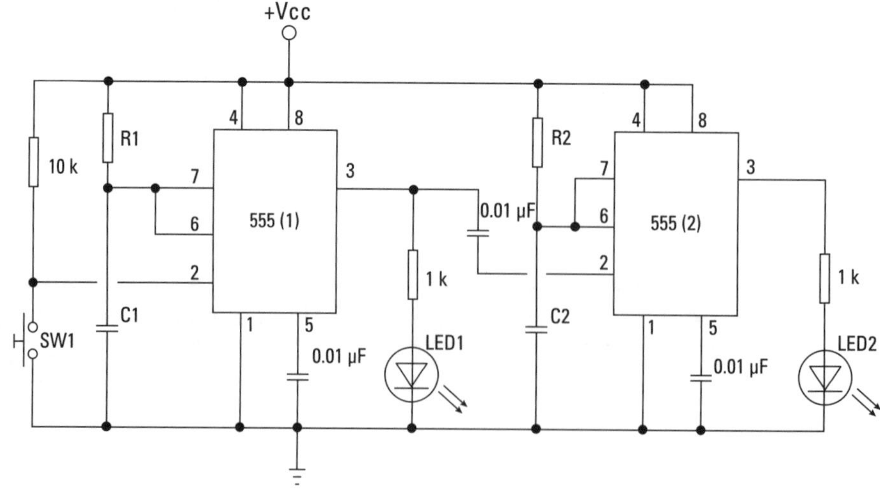

Figure 2-12: You can cascade 555 timers.

Figure 2-13 shows how you can implement this same circuit using a single 556 dual-timer chip. This schematic is nearly identical to the schematic shown in Figure 2-13, but with a few important differences:

- The two 555 timer circuits are designated as *556 (1)* and *556 (2)* to indicate that these timer circuits are part of a single 556 dual-timer chip.
- The pin numbers indicate the pin assignments for the two timer circuits of a 556 instead of the pin assignments for a 555.
- The second timer circuit doesn't show a supply or ground connection. That's because the two timer circuits share a common supply and ground connection, which is shown connected to the first timer.

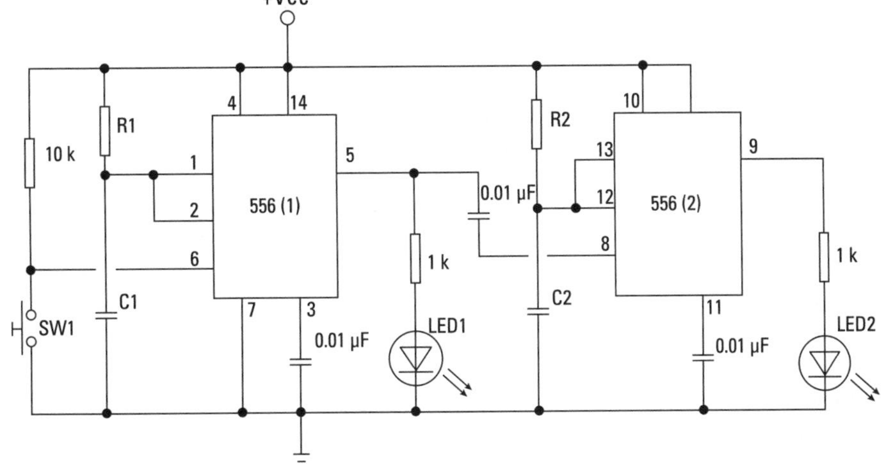

Figure 2-13: You can cascade the two halves of a 556 dual-timer circuit.

Although showing the two halves of a 556 dual timer as separate components in a schematic diagram is convenient, you can show the 556 as a single component if you prefer.

Figure 2-14 illustrates how you can draw the schematic for the cascaded timer circuit using a single component for the 556 dual timer. This circuit is nearly identical to the one shown in Figure 2-13; the only difference is the way the schematic depicts the two sections of the 556 dual-timer chip.

When you draw a 556 as a single component, place the connections for the two timers on opposite sides of the component. In Figure 2-14, we draw the connections for the first timer on the left and the second timer on the right, with the exception of the trigger input for the second timer (pin 8). Drawing that connection on the right side of the component would complicate the diagram, and so we place it on the left side just beneath the output pin for the first timer.

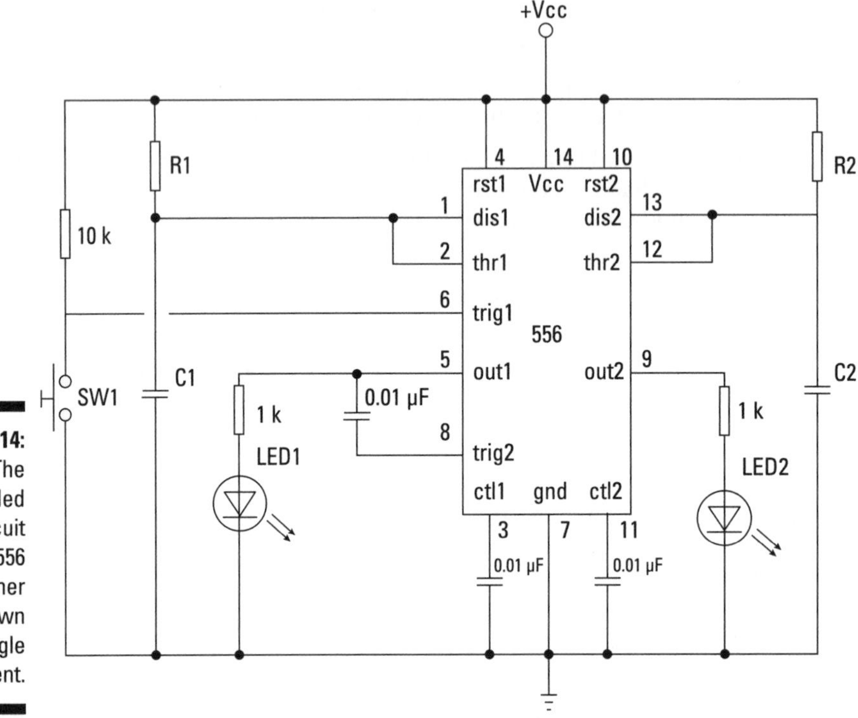

Figure 2-14: The cascaded timer circuit with the 556 dual-timer chip drawn as a single component.

Keeping track of which pin is which in a 556 is difficult, and so labelling the function of each pin helpful (as we do in Figure 2-14).

Constructing 555 Chip Projects

Now you get a chance to put this chapter's information into practice with four projects.

Making a one-shot timer

In this section, you build a circuit that uses a 555 timer chip in monostable mode. When a trigger switch is pressed, an LED lights and stays lit for approximately five seconds. Then, the LED goes dark until the button is pressed again.

Project 2-1 provides all the information you need to assemble this circuit, which is based on the schematic for the monostable circuit in the earlier Figure 2-3. The only differences are that an LED is added to the output pin (pin 3) and the resistor, and capacitor values are included in the capacitor charging circuit.

This project uses a pushbutton switch that isn't designed to be inserted in a solderless breadboard. To make the pushbutton usable with the solderless breadboard, solder a small length of 18- or 20-guage solid wire to the end of each lead of the test strip. You can then insert these leads directly into holes on the breadboard.

When you're finished, the circuit looks like Figure 2-15. To test the circuit, press the pushbutton. The LED should light, stay lit for just over five seconds and then go back off. It should light again only when you press the pushbutton again.

If the circuit doesn't work, here are a few things to check:

- Make sure that the battery is good. (Test it with a voltmeter.)
- Double-check carefully all the jumper wires and other components to ensure that they're connected properly.
- Make sure that the LED isn't inserted backwards. As a test, pull it out and insert it with the leads reversed.
- Make sure that the electrolytic capacitor is inserted with the negative end on the ground side of the circuit.
- Make sure that the solder connections to the pushbutton are solid.

Figure 2-15: The finished one-shot timer project.

Project 2-1: A One-Shot 555 Timer Circuit

In this project, you build a circuit that uses a 555 timer chip in monostable mode to create a one-shot timer. The circuit includes a pushbutton and an LED. When you press the pushbutton, the LED turns on and stays on for about five seconds. Then, the LED turns off and stays off until you press the pushbutton again.

The only tools you need to complete this project are wire cutters and wire strippers.

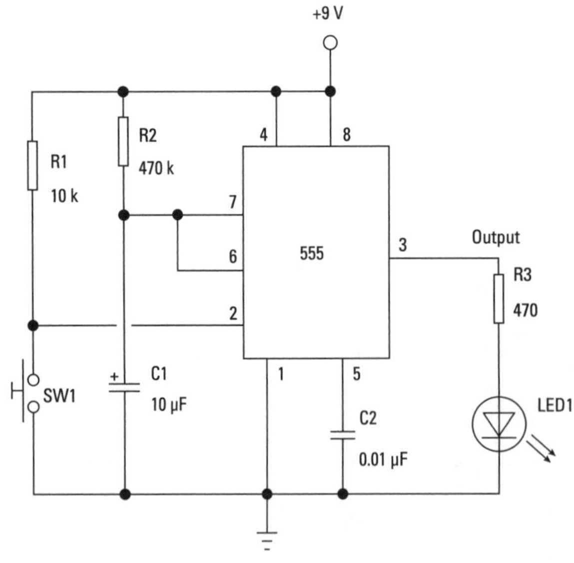

Parts List

- 1 9 V battery
- 1 9 V battery snap holder
- 1 Small solderless breadboard
- 1 Normally open pushbutton switch
- 1 555 timer chip
- 1 5 mm red LED
- 1 10 kΩ resistor (brown-black-orange)
- 1 470 Ω resistor (yellow-violet-brown)
- 1 470 kΩ resistor (yellow-violet-yellow)
- 1 10 µF electrolytic capacitor
- 1 0.01 µF ceramic disk capacitor
- 1 Normally open momentary contact pushbutton
- 8 Jumper wires (various lengths)

Steps

Throughout these steps, use the bus strip on the bottom of the board for the ground bus and the bus strip on the top of the board as the positive voltage bus.

1. **Insert the 555 timer chip.**

   Insert the chip so that it straddles the gap in the centre of the solderless breadboard, with pin 1 in hole E5 and pin 8 in hole F5.

2. **Insert the jumper wires.**

   If you're using precut jumper wires, choose an appropriate length for each segment. Otherwise, cut your own jumper wires as needed. You need a total of eight jumper wires, inserted into the solderless breadboard as follows:

From	*To*
A5	Any hole in the ground bus
D3	D6
D8	G5 (note that this wire crosses over the top of the 555)
F9	Any hole in the ground bus
B11	Any hole in the ground bus
G7	G11
H6	H11
J5	Any hole in the positive voltage bus

3. **Insert the resistors.**

Resistor	*From*	*To*
10 kΩ	E3	Any hole in the positive bus
470 kΩ	J11	Any hole on the positive bus
470 Ω	C7	C13

4. **Insert the capacitors.**

   Insert the two capacitors as follows:

Capacitor	*From*	*To*
0.01 μF	J8	J9
10 μF	E11	F11

   Make sure that the negative lead of the electrolytic capacitor is in hole E11.

5. **Insert the LED.**

   Insert the anode (the long lead) in hole A13 and the cathode (the short lead) in any hole in the ground bus.

6. **Insert the pushbutton.**

   Insert one of the leads you soldered to the pushbutton in hole A3 and the other one in the close available hole in the ground bus.

7. **Connect the battery.**

   Plug the 9 V battery into the battery snap connector, and then connect the red lead to the positive voltage bus and the black lead to the ground bus.

8. **You're finished!**

Brightening up with an LED flasher

Project 2-2 shows you how to build a LED flasher circuit. For this circuit, the 555 is configured in astable mode, and the resistor values are chosen so that they cause the high and low timings to be very close to one another, about one-tenth of a second each.

The schematic for this project is similar to the one in the earlier Figure 2-5, but adds a pair of LEDs to the output circuit. LED1 lights when the output is high and LED2 lights when the output is low. Note that this circuit uses both sinking and sourcing of the output current (which we describe earlier in the 'Working with the 555 Timer Output' section).

Figure 2-16 shows the completed project.

Figure 2-16: The finished LED flasher.

If the circuit doesn't work, take a look at the following:

- Check the battery voltage.
- Double-check all the jumper wires and resistors to make sure that they're inserted in the proper holes.
- Ensure that the diodes and the electrolytic capacitor are inserted correctly. For the LEDs, the anodes must be on the positive side of the circuit and the cathodes on the negative side. For the electrolytic capacitor, the negative side must be in the ground bus.
- For the duty cycle to be exactly 50-50, the circuit would need a diode added across the resistor R2 at the top end of C1, and R1 would need to have the same value as R2 (as in Figure 2-7, earlier in this chapter).

Project 2-2: An LED Flasher

In this project, you build a circuit that alternately flashes a pair of LEDs. The circuit uses a 555 timer configured in astable mode to flash the LEDs. The resistor and capacitor values are chosen so that the duty cycle is close to 50% and each LED stays on for about 1/10th of a second.

The only tools you'll need to complete this project are wire cutters and wire strippers.

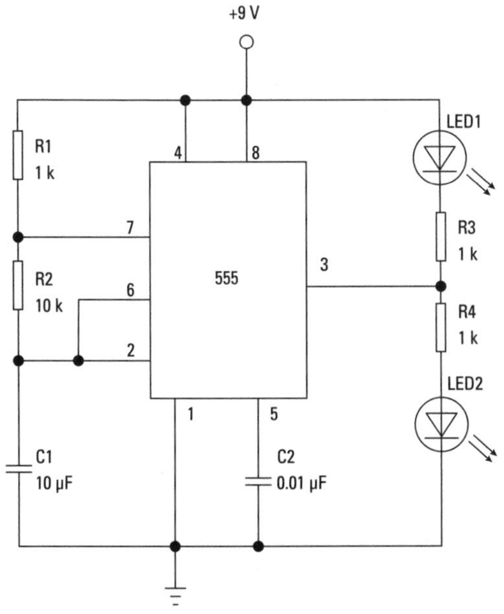

Parts List

- 1 9 V battery
- 1 9 V battery snap holder
- 1 Small solderless breadboard
- 1 555 timer chip
- 2 5 mm red LED
- 1 10 kΩ resistor (brown-black-orange)
- 3 1 kΩ resistor (brown-black-red)
- 1 10 µF electrolytic capacitor
- 1 0.01 µF ceramic disk capacitor
- 8 Jumper wires (various lengths)

Steps

1. **Insert the 555 timer chip.**

   Insert the chip so that it straddles the gap in the centre of the solderless breadboard, with pin 1 in hole E5 and pin 8 in hole F5.

2. **Insert the jumper wires.**

   If you're using precut jumper wires, choose an appropriate length for each segment.

   Otherwise, cut your own jumper wires as needed. You need a total of eight jumper wires, inserted in the solderless breadboard as follows:

From	*To*
A5	Any hole in the ground bus
C3	C6
C7	C10
D6	G7 (this wire crosses over the top of the 555)
D8	G5 (this wire also crosses over the top of the 555)
H3	H6
F9	Any hole in the ground bus
J5	Any hole in the positive voltage bus

3. **Insert the resistors.**

   Insert the four resistors as follows:

Resistor	*From*	*To*
10 kΩ	D3	G3
1 kΩ	J3	Any hole on the positive bus
1 kΩ	B10	B13
1 kΩ	D10	F13

4. **Insert the capacitors.**

   Insert the two capacitors as follows:

Capacitor	*From*	*To*
0.01 μF	G8	G9
10 μF	A3	Striped end into ground bus

5. **Insert the LEDs.**

   Insert the two LEDs as follows:

LED	*Cathode (Short Lead)*	*Anode (Long Lead)*
LED1	J13	Any hole in the positive bus
LED2	Any hole in the ground bus	A13

6. **Connect the battery.**

   Plug the 9 V battery into the battery snap connector, and then connect the red lead to the positive voltage bus and the black lead to the ground bus. The LEDs should begin flashing immediately when the battery is connected.

7. You're finished!

After you finish this project, leave it assembled if you intend to build Project 2-3.

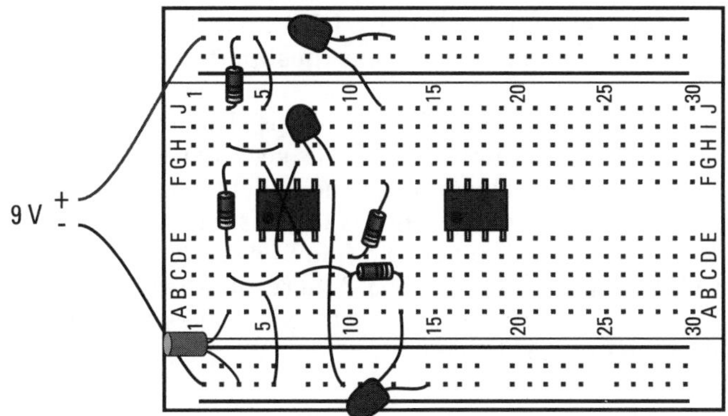

After you finish playing with the completed project, you may want to leave it assembled. The next section (Project 2-3) adds a second 555 timer to this circuit to include additional functionality.

Employing a set/reset switch

In this section, you modify the circuit that you built in Project 2-2 so that the circuit is controlled by two pushbuttons that function as a set/reset switch. When you connect the power to this circuit, LED1 turns on and stays on. When you press the set pushbutton, the two LEDs start flashing alternately and continue to flash until you press the reset pushbutton.

The circuit for this project uses two 555 timer chips. The first is configured in bistable mode with two pushbuttons acting as set and reset switches. The second is configured in astable mode, almost identical to the 555 timer that you used in Project 2-2. The difference is that instead of connecting the astable 555 Timer chip's supply voltage pin (pin 8) directly to the battery, you connect it to the output of the first 555. As a result, the first 555 controls the power to the second 555, and so the second 555 flashes the LEDs only when the output of the first 555 is high.

This project requires that you use two pushbutton switches that are not designed to be inserted into a solderless breadboard. To make the switches easier to use with the breadboard, solder short lengths of 20-AWG solid wire to the switch terminals.

Project 2-3 shows you how to build this circuit, and the completed project is pictured in Figure 2-17.

Figure 2-17: The completed circuit for Project 2-3.

Project 2-3: An LED Flasher with a Set/Reset Switch

In this project, you expand the circuit you built in Project 2-2 to add a set/reset switch that controls the flashing of the LEDs. You need to have built Project 2-2 before attempting this project.

The only tools you'll need to complete this project are wire cutters and wire strippers.

Parts List

You need all the parts from Project 2-2, plus the following:

1 555 timer chip
2 10 kΩ resistor (brown-black-orange)
1 0.01 µF ceramic disk capacitor
2 Normally open momentary contact pushbutton
10 Jumper wires (various lengths)

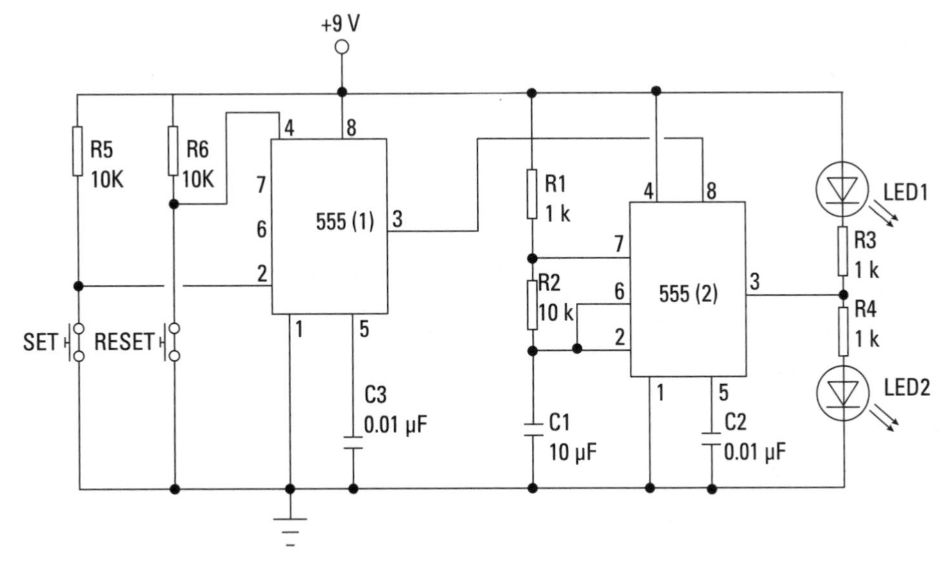

Steps

1. **If you haven't already done so, build the LED flasher circuit described in Project 2-2.**

   The 555 timer chip in that project is the one designated as 555 (2) for this project.

2. **Remove the jumper wire you inserted from hole J5 to the positive bus.**

   This step disconnects the Vcc supply (pin 8) of 555 (2). Voltage for 555 (2) is provided by the output of 555 (1).

3. **Insert 555 (1).**

   Insert the chip so that it straddles the gap in the centre of the solderless breadboard, with pin 1 in hole E17 and pin 8 in hole F17.

4. **Insert the additional jumper wires.**

   If you're using precut jumper wires, choose an appropriate length for each segment. Otherwise, cut your own jumper wires as needed. You need a total of ten jumper wires, inserted into the solderless breadboard as follows:

From	*To*
A17	Any hole in the ground bus
I5	I15
E15	F15
D15	D19
C18	C23
D20	D27
F21	Any hole in the ground bus
E23	F23
E27	F27
J17	Any hole in the positive voltage bus

5. **Insert the resistors.**

   Insert the two resistors as follows:

Resistor	*From*	*To*
10 kΩ	J23	Any hole on the positive bus
10 kΩ	J27	Any hole on the positive bus

6. **Insert the capacitor.**

   Insert the 0.01 μF capacitor in holes G20 and G21.

7. **Insert the pushbuttons.**

   Insert the two pushbutton switches as follows:

Button	*From*	*To*
Set	A23	Any hole in the ground bus
Reset	A27	Any hole in the ground bus

8. **Connect the battery.**

   Plug the 9 V battery into the battery snap connector, and then connect the red lead to the positive voltage bus and the black lead to the ground bus. LED1 should immediately light up. (If it doesn't, double-check all your connections and make sure that the battery isn't dead.)

9. **You're done!**

   Press the set button to start the LEDs flashing. Let them flash for awhile, and then press the reset button to make the flashing stop.

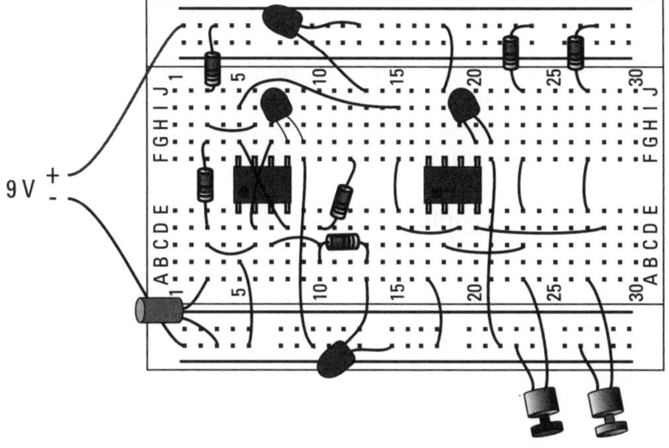

Building a beeper

In this section, you use two 555 timer chips to build an audible beeper, with both timers configured in astable mode. One timer generates an audible square-wave tone of approximately 476 Hz; the output of this 555 is sent to a speaker so that you can hear the tone. The other has a much lower frequency of about 1.5 Hz. You connect its output to the reset of the 476 Hz timer to turn the tone on and off, which creates the beeping effect.

Project 2-4 explains how to build this circuit. Before you start, have a look for a moment at the project's schematic diagram. For the first timer chip – designated 555 (1) in the schematic – the RC network uses resistors of 1 kΩ and 470 kΩ along with a 1 μF capacitor to produce the 1.5 Hz output. The second timer – 555 (2) – uses two 100 kΩ resistors and a 0.01 μF capacitor to create the 476 Hz output. The output of the first timer is sent to the reset of the second timer, and the output of the second timer is sent through a 22 μF capacitor to an 8 Ω speaker.

You can use any 8 Ω speaker in this circuit. If you have an old unamplified computer speaker, you can use that. We recommend you solder 5–8 cm lengths of 20-AWG solid wire to the speaker terminals so that you can connect the speaker easily to the breadboard.

Figure 2-18 shows the completed project.

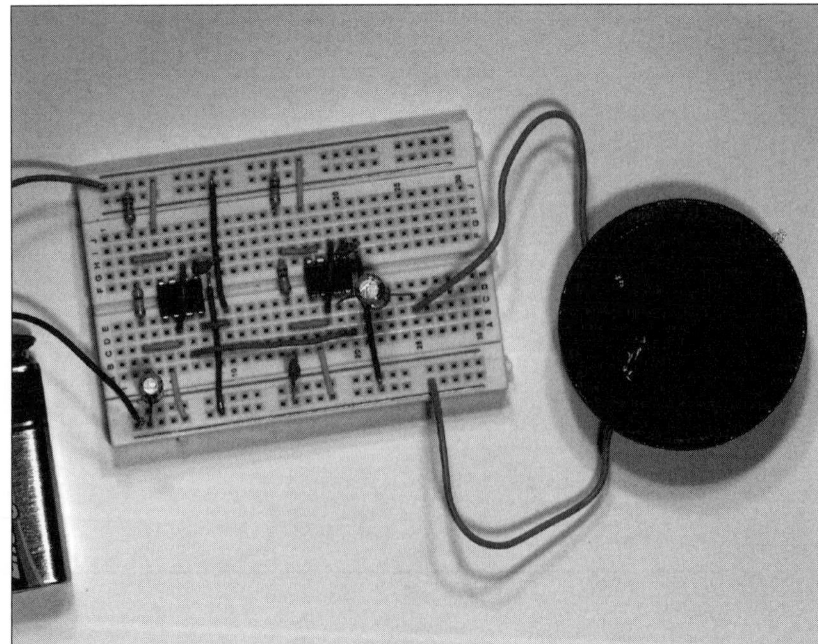

Figure 2-18: The completed beeper project.

Project 2-4: An Audible Beeper

In this project, you use a pair of 555 timer ICs to build a circuit that produces an audible beep on a small speaker. Both of the 555 timer ICs are configured in astable mode. The first 555 provides the beeping interval, which is about one beep every 7/10th second. The second 555 generates a 476 Hz frequency that's fed to the speaker to produce the audible output.

To build this project, you'll need wire cutters and wire strippers in addition to the parts listed below.

Parts List

1 9 V battery
1 9 V battery snap holder
1 Small solderless breadboard
2 555 timer chip
1 1 kΩ resistor (brown-black-red)
1 470 kΩ resistor (yellow-violet-yellow)
2 100 kΩ resistor (brown-black-yellow)
1 1 μF electrolytic capacitor
1 22 μF electrolytic capacitor
3 0.01 μF ceramic disk capacitor
15 Jumper wires (various lengths)
1 8Ω speaker

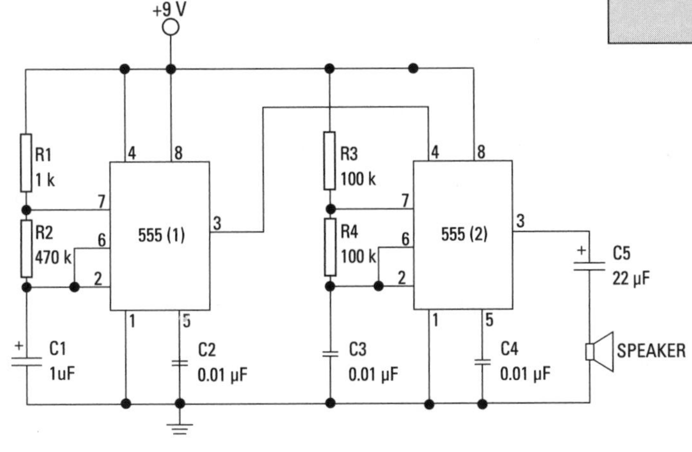

Steps

1. **Insert the 555 timer chips.**

   Insert the chips so that they straddle the gap in the centre of the solderless breadboard. For the first chip, insert pin 1 in hole E5 and pin 8 in hole F5. For the second chip, insert pin 1 in hole E17 and pin 8 in hole F17.

2. **Insert the jumper wires.**

   If you're using precut jumper wires, choose an appropriate length for each segment. Otherwise, cut your own jumper wires as needed. You need a total of 15 jumper wires, inserted into the solderless breadboard as follows:

From	*To*
A5	Any hole in the ground bus
A17	Any hole in the ground bus
J5	Any hole in the positive bus
J17	Any hole in the positive bus
C3	C6
H3	H6
B7	B20
C15	C18
G15	G18
F9	Any hole in the ground bus
E10	Any hole in the positive bus
D8	D10
D6	G7 (this wire crosses over the top of the first 555 chip)
D18	G19 (this wire crosses over the top of the second 555 chip)
F21	Any hole in the ground bus

3. **Insert the resistors.**

   Insert the four resistors as follows:

Designation	*Resistor*	*From*	*To*
R1	1 KΩ	J3	Any hole on the positive bus
R2	470 KΩ	E3	F3
R3	100 KΩ	J15	Any hole in the positive bus
R4	100 KΩ	E15	F15

4. **Insert the capacitors.**

   Insert the five capacitors as follows:

Designation	*Resistor*	*From*	*To*
C1	1 μF Electrolytic	A3	Any hole in the ground bus
C2	0.01 μF Ceramic Disk	G8	G9
C3	0.01 μF Ceramic Disk	G20	G21
C4	0.01 μF Ceramic Disk	A15	Any hole in the ground bus
C5	22 μF Electrolytic	D19	D25

   Note that for C1, the negative lead should be inserted into the ground bus. For C5, the negative lead should be in D25.

5. **Connect the speaker.**
   Insert one lead from the speaker in hole C25. You can insert the other in any hole in the ground bus.
   Note that speakers aren't sensitive to polarity, and so don't worry which lead goes in the ground bus and which goes in A25.
6. **Connect the battery.**
   Plug the 9 V battery into the battery snap connector, and then connect the red lead to the positive voltage bus and the black lead to the ground bus. You should immediately hear the speaker beeping.
7. **You're finished!**
   Remove the battery when you can no longer stand the beeping!

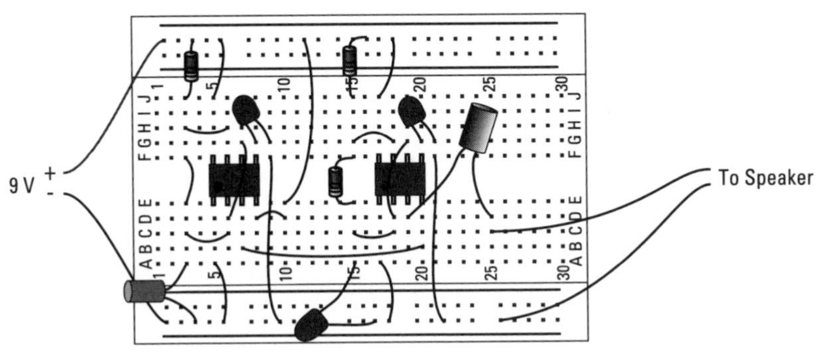

Here are some extra things to try with this project, in case you want to experiment with the circuit a bit:

- Build the circuit with a single 556 dual-timer chip instead of two 555 chips. Remember to adjust the pin designations on the schematic accordingly.
- Replace the 100 kΩ R3 with a 1 kΩ resistor, and then add a 1 MΩ potentiometer in series with the resistor. As you turn the potentiometer, the tone changes.
- Add a 1 MΩ potentiometer in series with R1. Then, as you turn the potentiometer, the beeping rate changes.

Chapter 3

Working with Op Amps

In This Chapter

- Familiarising yourself with op amps
- Exploring how feedback circuits work with op amps
- Looking at voltage comparators and summing amplifiers
- Considering popular op-amp packages

Have you ever played *Operation,* the game in which you use electrified tweezers to remove plastic body parts from little holes in a body? The edges of the holes are metal conductors, and so if you touch the edge of the hole with the tweezers while trying to remove the plastic piece inside, a buzzer sounds and the patient's nose (a red light bulb) lights up.

An operational amplifier (op amp for short) is a little like this game, in that the slightest variation in the input (your hand holding the tweezers) is amplified into a huge variation in the output (the flashing red nose and jarring buzzer).

Op amps are among the most common types of integrated circuits (ICs) – probably second in popularity only to the 555 timer chip we describe in Chapter 2 of this minibook. In this chapter, you find out what an op amp is and how you can use one in different circuits. So scrub up and get ready to operate!

Looking at Operational Amplifiers

An *op amp* is a super-sensitive amplifier circuit that's designed to amplify the difference of two input voltages. Thus, an op amp has two inputs and one output. The output voltage is often tens or even hundreds of thousands of times greater than the difference in the input voltages. Therefore, a very small difference in the two input voltages – perhaps a few hundredths or even a few thousandths of a volt – can result in a large output voltage.

Although an op amp is a type of IC, op amps were invented long before ICs. Op-amp circuits are a natural for ICs, however, and so not long after the introduction of the first ICs, IC versions of op amps became available. Today, op amps are among the most popular types of ICs.

The name *operational amplifier* may be a bit confusing. Originally, the op-amp circuit was created for use as an amplifier in telephone distribution systems. Later, computer engineers discovered that they were able to adapt it easily to do mathematical operations such as addition, subtraction, multiplication and division. The term *operational amplifier* was coined around that time, because the circuits are amplifiers that can perform (mathematical) operations. (For more information, see the nearby sidebar 'How the op amp came to be'.)

Internally, the simplest op amps consist of several dozen transistors, and more complicated varieties have many more. In this chapter, we ignore completely the internal circuitry of an op amp and treat it as what it is: a handy device you can use without understanding how it works. You can thank the engineers who, many decades ago, worked out all the internal details that make an op amp do its magic.

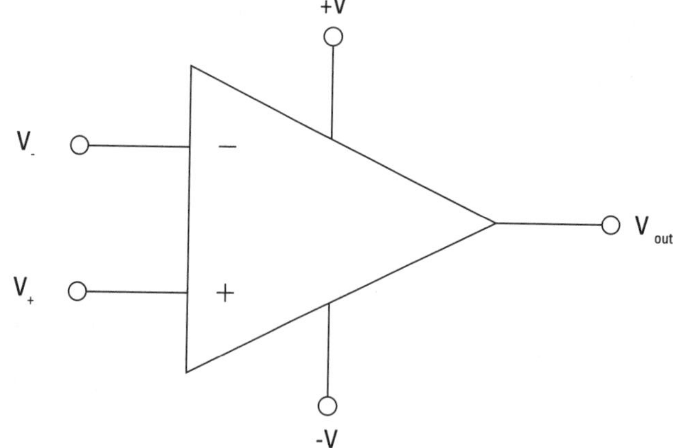

Figure 3-1: Schematic symbol for an op amp.

Many types of op-amp chips are manufactured today, but all have the five connections shown in Figure 3-1. The following list describes the function of each of these connections:

- **+V and –V:** The power supply for an op amp is provided via two pins usually labelled +V and –V. (These pins may be labelled V_{s+} and V_{s-} instead, but their functions are the same.) Most op amps require a positive and a negative voltage power supply, with voltages usually ranging from ±6 V to ±18 V. This type of power supply is called a *split supply*. The

± symbol indicates that both positive voltage and negative voltage are required: ±6 V, for example, means that +6 V *and* –6 V are required.

You can build a split supply easily by using two batteries connected end to end, as shown in Figure 3-2. Here, two 9 V batteries are connected to create a ±9 V supply. Note that the +9 V and –9 V are measured relative to ground, which is accessed between the two batteries.

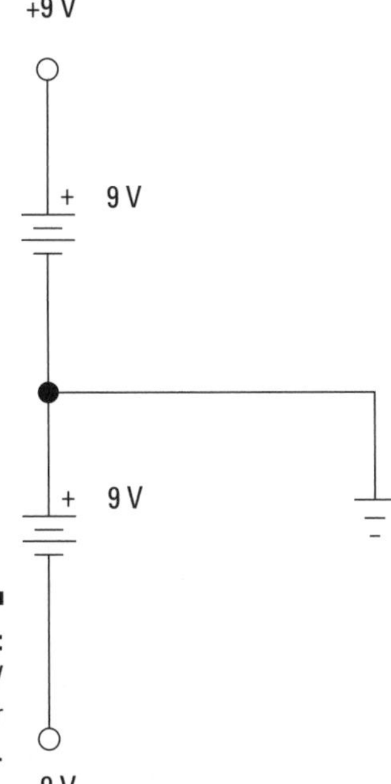

Figure 3-2: A split ±9V supply for an op amp.

Some op amps don't require split-voltage power supplies. Op amps that use single power supplies have a ground terminal instead of a –V terminal.

- **V_{out}:** The output of the op amp is taken from the V_{out} terminal. The voltage at the output terminal can be positive or negative, depending on the voltage difference between the two input terminals. The maximum voltage is usually a few volts less than the supply voltage at the +V and –V terminals. Thus, if the power supply for an op amp is ±9 V, the maximum output is around ±7 V or 8 V.

  Most op amps can handle only a small amount of current through the output terminal – usually, in the neighbourhood of 25 mA or less. As Figure 3-3 shows, the output is passed through an external resistance, designated R_L. The other end of this resistance is connected to ground. Thus, the output current that flows through the op amp must eventually end up at ground.

The load resistance isn't necessarily in the form of a simple resistor; any other circuit can provide load resistance, such as the base-emitter circuit of a transistor or even the input of another op amp.

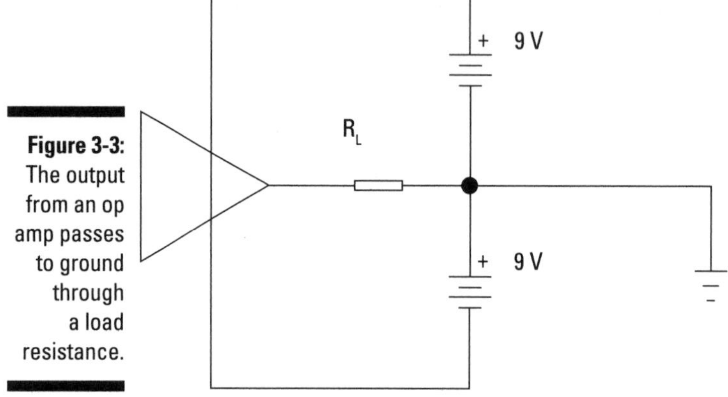

Figure 3-3: The output from an op amp passes to ground through a load resistance.

- **V_+ and V_-:** The two inputs of an op amp are the V_+ and V_- terminals. These terminals are sometimes identified by + and – signs inside the triangle. The inputs are called *differential inputs* because the output voltage, which appears on the V_{out} terminal, depends on the difference between the voltage of the + and – terminals.

  For most op amps, the maximum allowable input voltage is a bit less than the maximum power supply voltage: ±12 V is a typical limit. Remember, though, that an op amp amplifies the *difference* between the two input voltages. In many cases, the two input voltages are very close, and so the difference is very small.

  This fact is worth emphasising: the polarity of the op-amp output depends on the polarity of the difference between the V_+ and V_- inputs. Thus, if V_+ is greater than V_-, the output is a positive voltage, but if V_+ is less than V_-, the output is a negative voltage.

  In many op-amp circuits, one input is connected to ground. If the V_+ input is grounded, the output polarity is always the opposite of the polarity of the input voltage on the V_- terminal. In other words, negative voltage on V_- gives positive voltage on V_{out} and positive voltage on V_- gives negative voltage on V_{out}. For this reason, the V_- input is often called the *inverting input,* because its polarity is inverted in the output.

  If, on the other hand, the V_- input is connected to ground, the polarity of the output is the same as the polarity of the input voltage applied to V_+. Thus, if V_+ is positive, V_{out} is positive; if V_+ is negative, V_{out} is negative. For this reason, the V_+ input is called the *noninverting input,* because its polarity is the same in the output – that is, the V_- input voltage is *not* inverted.

How the op amp came to be

The modern operational amplifier dates back to the early 1930s, when Bell Telephone was just starting to run telephone cables throughout the US. In the early days of the telephone, engineers had a difficult problem with phone lines that ran more than a few thousand feet. Long phone lines needed amplifiers to give their signals a boost, but the amplifiers available at the time were very unreliable – too sensitive to weather (temperature and humidity) and unable to work consistently over the range of voltages used in early phone lines.

A Bell engineer named Harry Black was working on the amplifier problem in 1934 when an idea struck him as he was riding the ferry home from work. His insight was a stroke of genius that seems obvious decades later. Instead of trying to design an amplifier that would have the exact amplitude gain needed for the job, Black's idea was to use an amplifier that had far more gain than was needed – thousands of times more gain, in fact – and then feed some of the output back into the input through a resistor. This feedback circuit would reduce the overall gain of the amplifier based on the amount of resistance in the circuit.

The circuit didn't gain the name *operational amplifier* until the computer age began a decade or so later, and computer researchers figured out how to use the amplifier's unique characteristics to perform basic mathematical operations on the input voltages. Eventually, digital computers replaced the analogue computers built from op amps. Op amps still play an important role in computers today, however, primarily to provide an interface with real-world input measuring devices such as voltage sensors and moisture detectors.

The original op-amp circuits were built with vacuum tubes. They were large, required several hundred volts to operate and generated substantial heat. When transistors replaced vacuum tubes in the 1950s, op amps became smaller, and when ICs were invented in the 1960s, op amps were among the first chips to be designed.

Understanding Open Loop Amplifiers

As its name suggests, one of the most basic uses of an op amp is as an amplifier. If you connect an input source to one of the input terminals and ground the other input terminal, an amplified version of the input signal appears on the out terminal.

An important concept in op-amp circuits is *voltage gain,* which simply represents the amount by which the difference between the two input voltages is multiplied to produce the output voltage. If the input voltage difference is 2 V and the output voltage is 12 V, for example, the voltage gain of the amplifier is 6.

If you simply apply an input signal to one of the input terminals of an op amp, as shown in Figure 3-4, the circuit is called an *open loop amplifier.* We reveal the reason for this name in the next section's discussion on closed loop amplifiers, but for now just realise that this type of circuit goes by the name *open loop*.

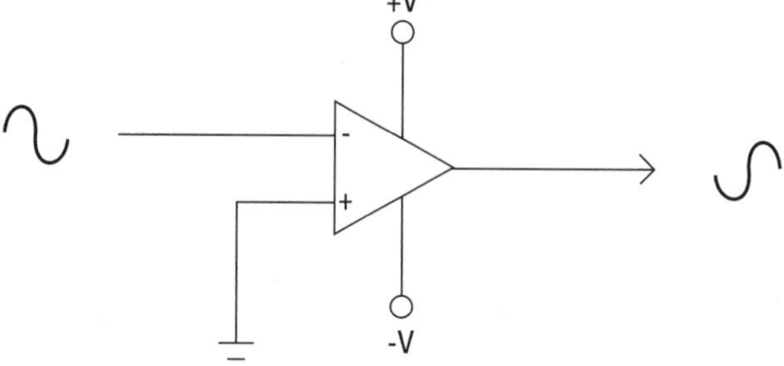

Figure 3-4: An op amp configured as an open loop amplifier.

In the open loop op-amp circuit, the V_+ input is connected to ground, and an input signal is placed on the V_- input. In this arrangement, the voltage to be amplified is the same as the voltage of the V_- input. Although Figure 3-4 shows the input as alternating current (AC), the open loop op-amp circuit works for direct current (DC) as well.

The voltage gain in an open loop op-amp circuit is extraordinarily high – in the order of tens or even hundreds of thousands. Suppose that you're using an op amp whose open loop voltage gain is 200,000 and that the power supply is ±9 V. In that case, an input voltage of +0.000025 V results in an output voltage of +5 V. An input voltage of +0.00004 V gives you an output voltage of 8 V.

The output voltage can never exceed the power supply voltage. In fact, the maximum output voltage usually is about 1 V less than the power supply voltage. So if you're using a pair of 9 V batteries to provide a ±9 V power supply, the maximum output voltage is ±8 V. As a result, the most that an open loop op-amp circuit with an open loop gain of 200,000 can reliably amplify is 0.00004 V. If the input voltage difference is any larger than 0.00004 V, the op amp is said to be *saturated,* and the output voltage goes to the maximum.

No matter how much money you invest in a top-quality voltmeter, it isn't sensitive enough to measure voltages that small. Particle physicists at CERN may worry about smaller voltages, but for all practical purposes 0.00004 V is the same as 0 V.

As a result, one of the basic features of an open loop op-amp circuit is that if the input voltage difference is anything other than zero, the op amp is saturated and the output voltage is the same as its maximum. So if the maximum output voltage is ±8 V, the output is one of only three voltages: +8 V, 0 V or –8 V.

The ideal op amp

While reading about op amps, you'll undoubtedly come across the term ideal op amp. An *ideal op amp* is a hypothetical op amp with certain characteristics that real op amps strive to achieve. Real op amps come very close to the ideal op amp, but no op amp in existence achieves the perfection of an ideal op amp. So you see, in many ways, op amps are almost human.

Depending on which list you read, an ideal op amp has anywhere between two and seven characteristics, the most important of which are:

- **Infinite open loop gain:** Several times in this chapter, we mention that the open loop gain in an op amp is very large – tens or even hundreds of thousands. In an ideal op amp, the open loop gain is infinite, which means that *any* voltage differential on the two input terminals results in an infinite voltage on the output. In real op amps, the output voltage is limited by the power supply voltage. Because the output voltage can't be infinite, the gain can't be infinite.
- **Infinite input impedance:** *Impedance* represents a circuit's opposition to current flow, whether the current is alternating or direct. In an ideal op amp, the impedance of the two input terminals is infinite, which means that no current enters the op amp from the inputs. The inputs are able to see and react to the voltage, but that voltage is unable to push any current into the op amp. In practice that means that the op amp has no effect on the input voltage. In a real op amp, a small amount of current – usually, a few milliamps or less – does leak into the op amp's input circuits.
- **Zero output impedance:** In an ideal op amp, the output circuitry has zero internal impedance, which means that the voltage provided from the output is the same regardless of the amount of load placed on it by the circuit to which the output is connected. In reality, most op amps have an output impedance of a few ohms: the actual voltage provided by the output terminal varies a small amount depending on the load connected to the output.
- **Zero offset voltage:** The *offset voltage* is the amount of voltage at the output terminal when the two inputs are exactly the same. If you connect both inputs to ground, for example, ideally you want exactly 0 V at the output. Real-world op amps have a very small voltage on the output even when both inputs are grounded, connected to each other or not connected to anything at all. For most op amps, this offset voltage is just a few millivolts.
- **Infinite bandwidth:** The term *bandwidth* refers to the range of AC frequencies within which an op amp can accurately amplify. In an ideal op amp, the frequency of the input signal has no effect on how the op amp behaves. In real-world op amps, the op amp doesn't perform well above a certain frequency – typically, a few megahertz (millions of cycles per second).

These characteristics are often summed up with the following two golden rules of op amps:

1. **The output attempts to do whatever is necessary to make the voltage difference between the inputs zero.**

Book III Working with Integrated Circuits

(continued)

(continued)

This rule, which applies only to closed-loop amplifier circuits, means that the feedback sent from the output to the input causes the two input voltages to become the same.

2. **The input draws no current.**

   This rule means that the input terminals look at the voltage placed across them but don't allow any current to flow into the op amp.

Although no op amp is able to live up to the standards of the ideal op amp, most come pretty close. Close enough, in fact, that you can safely design an op amp circuit as if the op amps were ideal. In particular, the two golden rules apply: the feedback equalises the input voltages and the op amp draws no current from the input.

Open loop op-amp circuits may not sound very useful, but they have many useful applications. You see one example in 'Comparing Voltages with an Op Amp', later in this chapter.

Considering Closed Loop Amplifiers

As we explain in the preceding section, open loop op-amp circuits aren't very useful as amplifiers because they're so easily saturated.

To make an op amp useful as an amplifier, you have to use it in a *feedback circuit,* which reduces the gain to a more manageable amount so that input voltages that are usable (and even measurable!) can be amplified reliably.

No doubt, you're familiar with the concept of feedback. You've probably been sitting in an auditorium listening to someone talk into a public-address system when suddenly a piercing screech comes out of the speakers. That screech is feedback. In the case of the public-address system, the microphone picks up some of the output from the speakers and sends it back through the amplifier again. The result is the annoying high-pitched squeal.

Not all feedback is bad, though. In an op-amp amplifier circuit, negative feedback is used to reduce the enormous open loop amplification gain to a more manageable gain, such as 10. To do so, a portion of the output signal is fed back into the input via the V_- terminal through a feedback resistor. This type of circuit is called a *closed loop amplifier,* because a closed circuit path exists between the output and the input. (Now you understand why an op-amp circuit without the feedback loop is called an *open loop amplifier.*)

Investigating inverting amplifiers

The most common op-amp configuration is called an *inverting amplifier,* because the voltage of the output is opposite the voltage of the input. Figure 3-5 shows a basic inverting amplifier circuit.

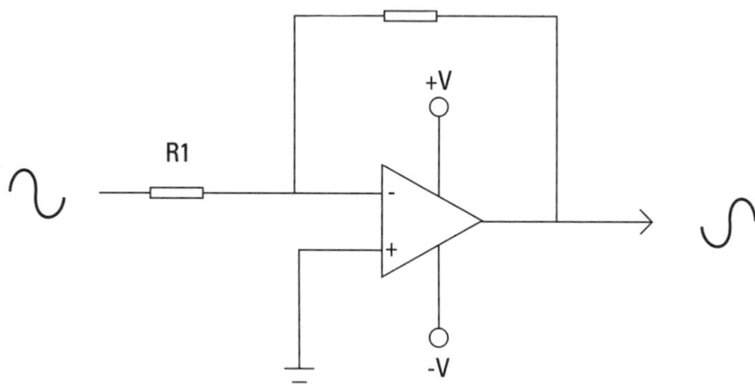

Figure 3-5: An op amp configured as an inverting amplifier.

In an inverting amplifier circuit, the input signal works its way through a resistor on its way to the V_ input, and the output is looped back into the V_ input through a second resistor. In Figure 3-5, these resistors are designated R1 and R2. You can easily calculate the overall voltage gain of the circuit by using this formula:

$$A_{CL} = -\frac{R2}{R1}$$

Here, the gain is designated A_{CL} (CL stands for *closed loop*).

If R1 is 1 kΩ and R2 is 10 kΩ, the voltage gain of the circuit is –10. Then if the input voltage is +0.5 V, the output voltage is –5 V (0.5 × –10).

The negative sign is required because Figure 3-5 is an inverting amplifier circuit, and so positive inputs give negative outputs, and vice versa.

Reversing inputs: Noninverting amplifiers

A closed loop amplifier can also be designed as a *noninverting amplifier* in which the output voltage isn't reversed. To do that, you simply reverse the inputs, as shown in Figure 3-6. Instead of connecting the input voltage to

V_- through a resistor and grounding V_+, you ground V_- through a resistor and connect the input voltage to V_+. The feedback circuit is the same; the output is connected to the V_- input through resistor R2.

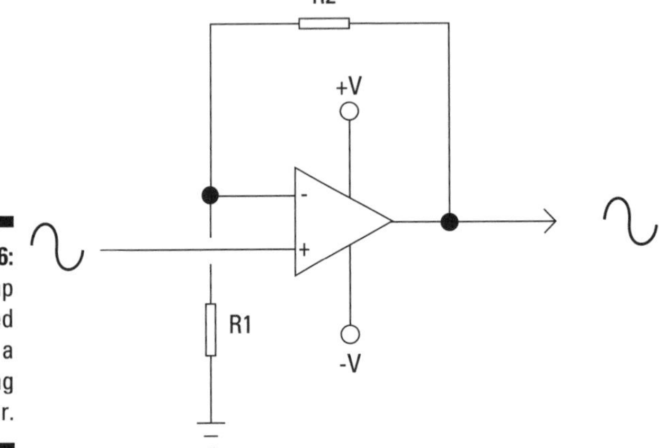

Figure 3-6: An op amp configured as a noninverting amplifier.

Delving deeper into feedback

If you're interested in understanding how the feedback circuit works, consider the rule that we describe in the earlier section 'Understanding Open Loop Amplifiers': if the input voltage is anything other than zero, the op amp is saturated and the output voltage is the maximum allowed. The purpose of the feedback circuit is to return some of the output voltage to the inverting input, which results in the input-voltage difference's being driven towards zero. As the voltage approaches zero, the op amp's gain begins to drop to a useful range.

Suppose that the input voltage difference in an inverting op amp circuit is +0.5 V. This difference results in the op amp becoming saturated, and so –8 V appears at the output. A portion of the negative voltage that depends on the voltage divider created by R1 and R2 is returned to the V_- input, which has the effect of reducing the input voltage. That results in a smaller voltage difference, but not small enough to prevent the op amp from still being saturated.

Remember, however, that the feedback loop is just that: a loop. As more and more of the saturated output voltage gets fed back through the loop, the input voltage difference gets closer to zero. When it gets oh-so-close to zero, the output voltage drops to a range between zero and the maximum voltage.

The best feature of the closed loop amplifier circuit is that the two resistors, which are outside the op-amp IC, give you precise control of the amount of gain that the circuit ultimately has. All you have to do to get any gain you want (within the limits of the op amp) is choose the right resistor values.

The formula for calculating the gain for a noninverting amplifier is a little different from the formula for an inverting amplifier:

$$A_{CL} = 1 + \frac{R2}{R1}$$

If R1 is 1 kΩ and R2 is 10 kΩ, the gain is 11. Thus, an input voltage of +0.5 V results in an output voltage of +5.5 V.

Using an Op Amp as a Unity Gain Amplifier

A *unity gain amplifier* is an amplifier circuit that doesn't amplify. In other words, it has a gain of 1. The output voltage in a unity gain amplifier is the same as the input voltage.

You may be thinking that such a circuit is worthless. After all, isn't a simple piece of wire a unity gain circuit? Yes, but a unity gain amplifier provides one important benefit: it doesn't take any current from the input source (that's one of the golden rules of the ideal op amp; see the earlier sidebar 'The ideal op amp'). Therefore, it completely isolates the input side of the circuit from the output side of the circuit. Op amps are often used as unity gain amplifiers to isolate stages of a circuit from one another.

Unity gain amplifiers come in two types:

- **Voltage followers:** Circuits in which the output is exactly the same voltage as the input.
- **Voltage inverters:** Circuits in which the output is the same voltage level as the input but with the opposite polarity.

If you think about it for a moment, you may be able to come up with the circuit for unity gain followers and inverters on your own. As we explain in the earlier section 'Considering Closed Loop Amplifiers', the formula for calculating the gain of both an inverting amplifier and a noninverting amplifier requires you to divide R2 by R1, and so all you have to do is choose resistor values that results in a gain of 1.

The following sections explain how to create unity followers and unity inverters.

Configuring a voltage follower

A unity gain voltage follower is simply a noninverting amplifier with a gain of 1. Recall that the formula for calculating the value of a noninverting amplifier is:

$$A_{CL} = 1 + \frac{R2}{R1}$$

To create a unity gain voltage follower, you just omit R2 and connect the output directly to the inverting input, as shown in Figure 3-7. R2 is zero and so the value of R1 doesn't matter, because zero divided by anything equals zero. So R1 is usually omitted as well, and the V_ input isn't connected to ground.

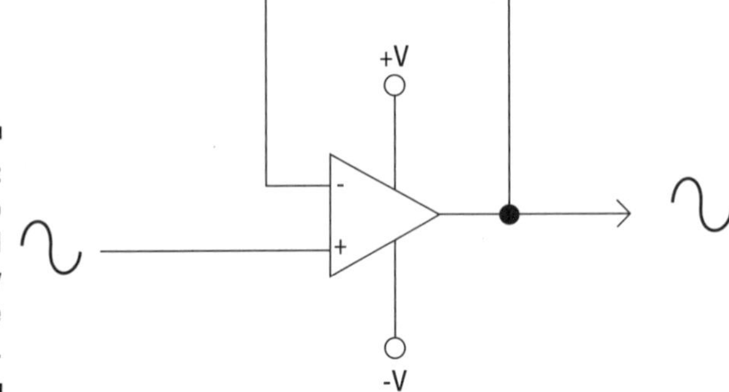

Figure 3-7: An op amp configured as a unity gain voltage follower.

Configuring a unity inverter

The formula for calculating gain for an inverting amplifier is:

$$A_{CL} = -\frac{R2}{R1}$$

In this case, all you have to do is use identical values for R1 and R2 to make the amplifier gain equal to 1. Figure 3-8 shows a unity gain inverter circuit using 1 kΩ resistors.

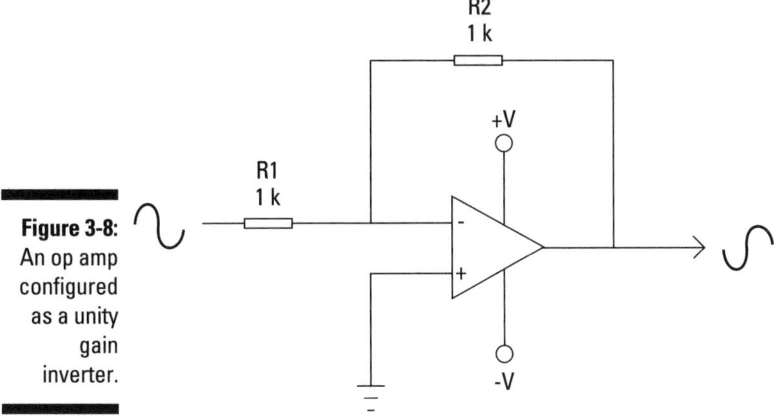

Figure 3-8: An op amp configured as a unity gain inverter.

Comparing Voltages with an Op Amp

A *voltage comparator* is a circuit that compares two input voltages and lets you know which of the two is greater. Suppose that you have a photocell that generates 0.5 V when it's exposed to full sunlight, and you want to use this photocell as a sensor to determine when it's daylight. You can use a voltage comparator to compare the voltage from the photocell with a 0.5 V reference voltage to determine whether or not the sun is shining.

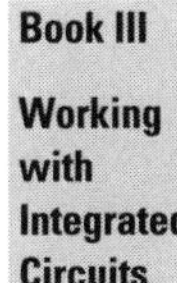

Creating a voltage comparator from an op amp is easy, because the polarity of the op-amp's output circuit depends on the polarity of the difference between the two input voltages. Figure 3-9 shows the basic circuit for an op-amp voltage comparator.

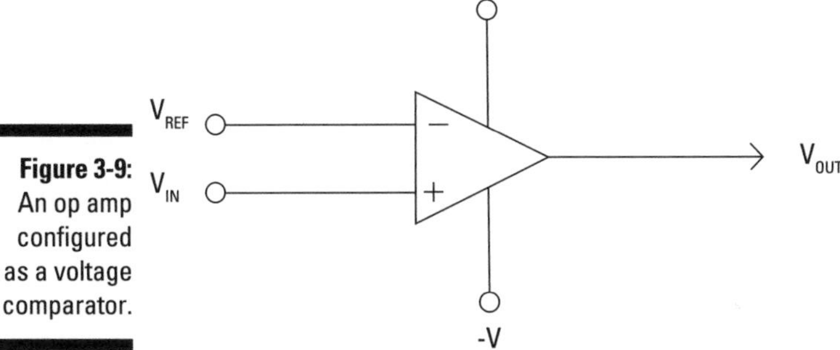

Figure 3-9: An op amp configured as a voltage comparator.

In the voltage-comparator circuit, a reference voltage is applied to the inverting input (V_-) and then the voltage to be compared with the reference voltage is applied to the noninverting input. The output voltage depends on the value of the input voltage relative to the reference voltage, as follows:

Input Voltage	***Output Voltage***
Less than reference voltage	Negative
Equal to reference voltage	Zero, in theory
Greater than reference voltage	Positive

The voltage level for the positive and negative output voltages is about 1 V less than the power supply. Thus, if the op-amp power supply is ±9 V, the output voltage is: +8 V if the input voltage is greater than the reference voltage, 0 V if the input voltage is equal to the reference voltage and –8 V if the input voltage is less than the reference voltage.

You can modify the circuit to eliminate the negative voltage if the input is less than the reference by sending the output through a diode, as shown in Figure 3-10. In this circuit, a positive voltage appears at the output if the input voltage is greater than the reference voltage; otherwise, no output voltage exists.

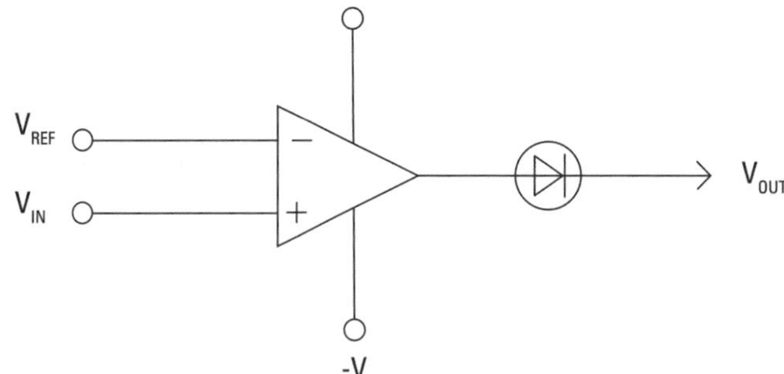

Figure 3-10: Using a diode in a voltage-comparator circuit.

To create a voltage comparator that creates a positive voltage output if the input voltage is *less than* a reference voltage, use the circuit shown in Figure 3-11. Here the reference voltage is applied to the inverting (V_-) input and the input voltage is applied to the noninverting (V_+) input.

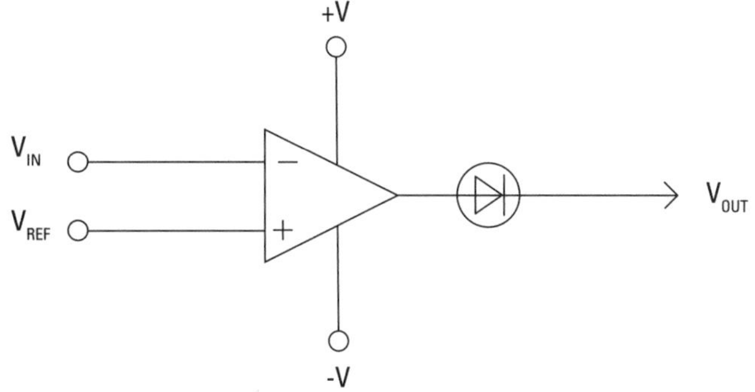

Figure 3-11: A voltage comparator that tests for a voltage that's less than a reference voltage.

The final voltage-comparator circuit you need to know about is the *window comparator,* which lets you know whether the input voltage falls within a given range. A window comparator requires three inputs: a low reference voltage, a high reference voltage and an input voltage. The output of the window comparator is a positive voltage only if the input voltage is greater than the high reference voltage or less than the high reference voltage. If the input voltage is between the two thresholds, then the output is zero.

You need two op amps to create a window comparator, as shown in Figure 3-12. As you can see in the figure, one op amp is configured to produce positive output voltage only if the input is greater than the high reference voltage ($V_{REF(HIGH)}$). The other op amp is configured to produce positive output voltage only if the input is less than the low reference voltage ($V_{REF(LOW)}$).

The input voltage is connected to both op amps; the output voltage is sent through diodes to allow only positive voltage before being combined. The resulting output has positive voltage only when the input voltage falls outside the low and high reference voltages.

Notice in Figure 3-12 that the power supply connections aren't shown separately for each op amp in the circuit. Omitting the power supply connections is common practice when multiple op amps are used in a single circuit. If the power supply connections were shown for all the op amps, the power supply connections would complicate the schematic unnecessarily. And no one needs unnecessary complications in life.

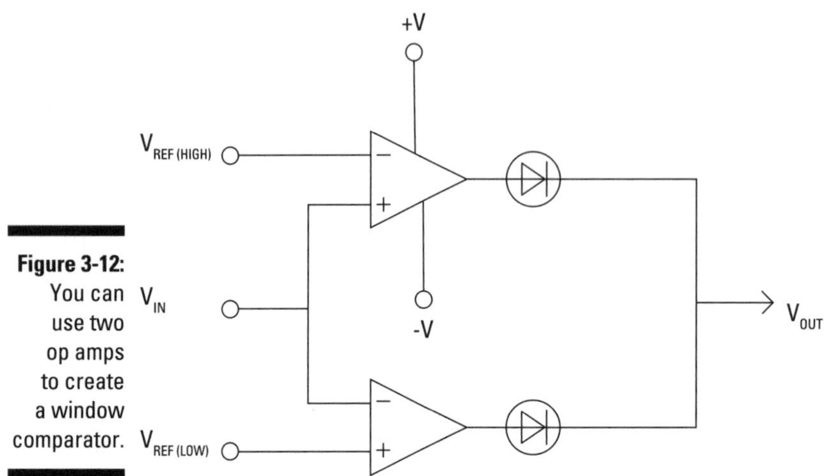

Figure 3-12: You can use two op amps to create a window comparator.

Adding Voltages: Summing Amplifiers

You can use an op amp to add or subtract two or more voltages. A circuit that adds voltages is called a *summing amplifier.* A summing amplifier has two inputs and an output whose voltage is the sum of the two input voltages but with the opposite polarity. If one of the inputs is +1.5 V and the other is +1.0 V, for example, the output voltage is –2.5 V.

Figure 3-13 shows a basic circuit for a summing amplifier. For the summing amplifier to work, resistors R1, R2 and R3 need to all be the same value.

If all the resistors in a summing amplifier are the same, the output voltage is the sum of the input voltages. This is the usual way to configure a summing amplifier, though you can vary the resistor values if you want.

If the resistors have different values, each of the input voltages is weighted according to the value of the resistor on its input circuit, which has the effect of multiplying each input voltage by a certain value before the voltages are summed. The exact value by which each input is multiplied depends on the mix of resistors you use.

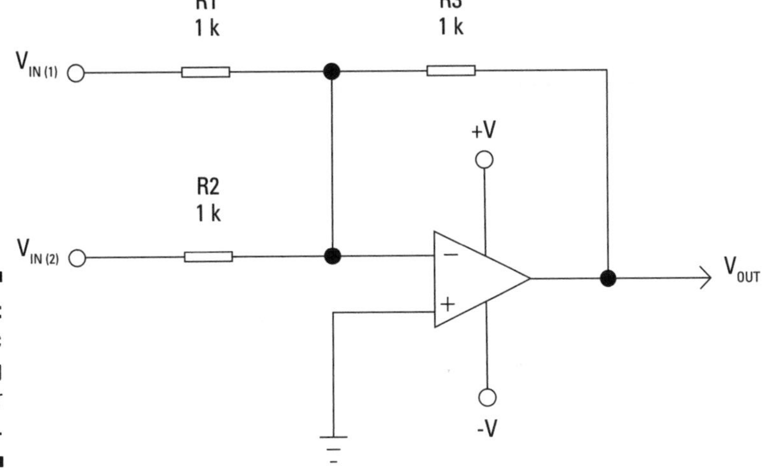

Figure 3-13: A basic summing amplifier circuit.

If R1 is 1 kΩ and both R2 and R3 are 10 kΩ, for example, the input voltage applied through the 1 kΩ resistor is multiplied by 10 before being added to the voltage applied through the 10 kΩ resistor. Thus, if the input at R1 is +1 V, and the input at R2 is +2 V, the output voltage is –12 V.

The formula for calculating the output voltage based on the input voltages and the resistor values is:

$$V_{out} = -R3\left(\frac{V1}{R1} + \frac{V2}{R2}\right)$$

We leave you to work out the maths for various combinations of resistor values and input voltages. Here, though, are a few examples to give you an idea of how the circuit behaves when R1 is 1 kΩ and both R2 and R3 are 10 kΩ:

$V_{IN(1)}$	$V_{IN(2)}$	V_{OUT}
+1 V	+1 V	–11 V
+1 V	+5 V	–15 V
0 V	+5 V	–5 V
+2 V	–5 V	–15 V
–1 V	–5 V	+15 V

One drawback of the summing amplifier is that it inverts the polarity of the input, but you can easily feed the output of a summing amplifier into the input of a unity gain inverter, as shown in Figure 3-14. Here, the second op amp inverts the polarity of the output from the summing amplifier, which has the effect of returning the output voltage polarity to the polarity of the original inputs. (For more information about the voltage-inverter portion of this circuit, refer to 'Using an Op Amp as a Unity Gain Amplifier,' earlier in this chapter.)

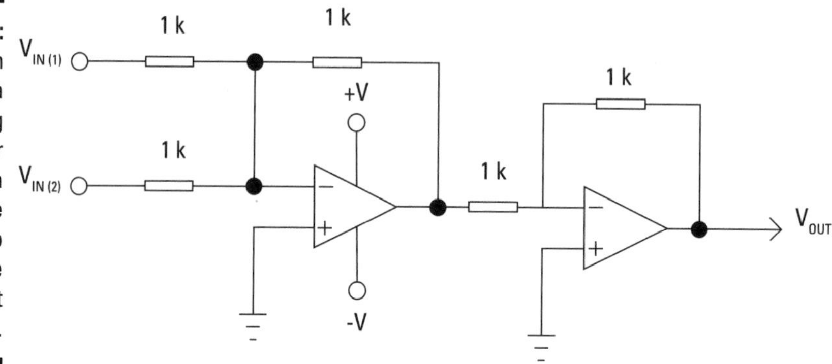

Figure 3-14: You can combine a summing amplifier with a voltage inverter to preserve the input polarity.

A common use for a summing amplifier circuit is as an audio mixer. When this type of circuit is used as an audio mixer, each input is connected to a microphone. The summing amplifier combines all the microphone inputs by adding the voltages from each microphone, and the resulting output is sent on to another amplifier stage.

The resistors in each input circuit are often potentiometers, which allows you to vary the signal level from each input source. When you increase the resistance on one of the input circuits, less of that input is represented in the output mix – especially useful if one of your singers is a bit off-key.

You can extend a summing amplifier circuit with additional inputs. Figure 3-15 shows a circuit with four inputs that uses potentiometers to control the level of each input. You can add as many inputs as you want, but you need to ensure that the total combined voltage from all inputs doesn't exceed the power supply voltage (minus a volt or two).

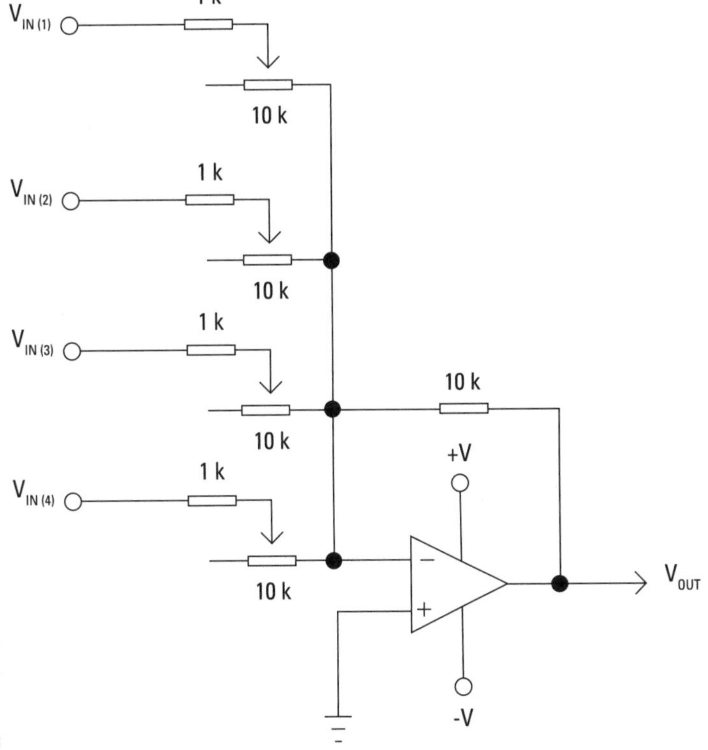

Figure 3-15: A simple audio mixer with four inputs.

Working with Op-Amp ICs

All the examples in this chapter assume that you're using a generic op amp in your circuits, but when you start building a real-world op-amp circuit you need to use a real op amp. Fortunately, op-amp ICs are plentiful and nearly all electronic component shops sell several types of inexpensive op-amp ICs.

The most popular op-amp IC is the LM741, which comes in a standard eight-pin DIP package. Figure 3-16 shows the pin connections for this op amp.

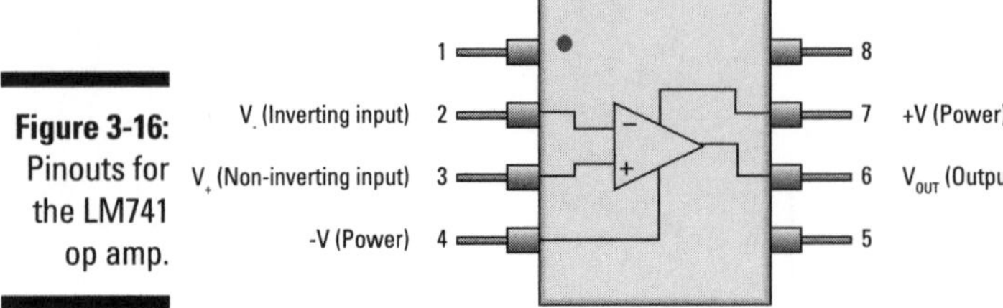

Figure 3-16: Pinouts for the LM741 op amp.

You can also get ICs that contain two or more op amps in a single package. One of the most common is the LM324 quad op amp, which contains four op amps in a single 14-pin DIP package. Unlike the LM741, the LM324 uses single-power-supply op amps. Thus, instead of a split + and – voltage supply, you provide just a positive voltage power supply and a ground.

Figure 3-17 shows the pinouts for an LM324. As you can see, the first op amp is accessed via pins 1–3, the second via pins 5–7, the third via pins 8–10 and the fourth via pins 12–14. The positive voltage power supply is connected to pin 4 and pin 11 is connected to ground.

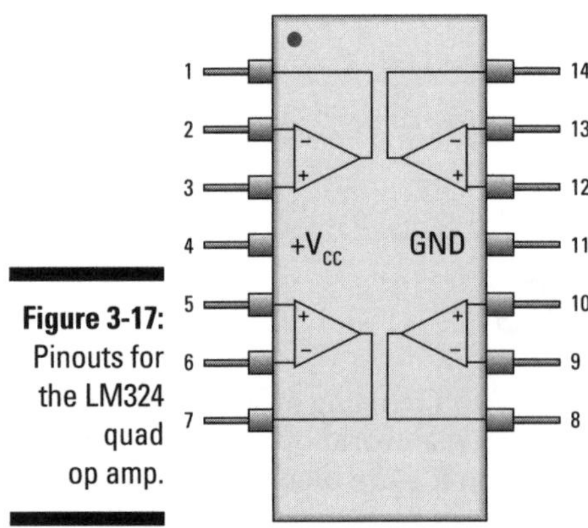

Figure 3-17: Pinouts for the LM324 quad op amp.

Book IV

Getting into Alternating Current

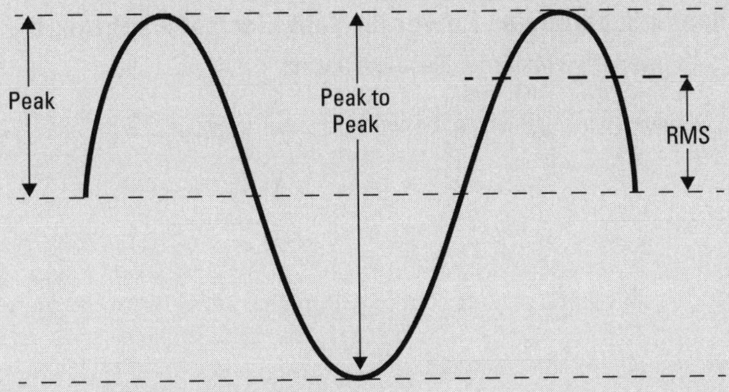

To get a rundown of the various ways in which we can harness renewable energy sources to generate power, head online and take a look at `www.dummies.com/extras/electronicsaiouk`.

Contents at a Glance

Chapter 1

Understanding Alternating Current

In This Chapter

- Discovering and using alternating current
- Converting mechanical power to alternating current
- Working with electric motors
- Stepping up and down with transformers

When you paddle in a river as a child, you can feel the cool water flowing between your toes and the current flowing steadily in one direction. Water doesn't flow uphill and so a river always flows one way, towards the sea.

Direct current (DC), as with the current you get from a battery, is like that – it flows only one way around the circuit and carries on doing so until you switch it off or turn the battery around so that the current goes the other way. For example, in Book III we focus mostly on working with DC and electric current flowing in one direction only.

Paddling in the sea, however, feels quite different to a river. You still sense the cool water around your toes, but it doesn't just go one way. The waves sweep in over your feet, and then you feel the water pulling the sand from under you as the water recedes. Another wave comes in, and then the water recedes and so on. The water forever ebbs and flows, back and forth, first one way then the other.

In *alternating current* (AC), the current flows in both directions – forward and backward – just like the waves breaking on a shore. Alternating current is of vital importance in electronics for the simple reason that the electric current you access by plugging a circuit into a wall outlet happens to be AC. So if you want to free your circuits from the tyranny of batteries, which eventually die, you need to discover how to make your circuits work from an AC power supply.

In this chapter, you take a look at the nature of AC and how it delivers reliable voltage. You also examine three fundamental AC devices: alternators, which generate AC from a source of motion such as a steam turbine or windmill; motors, which turn AC into motion; and transformers, which can transfer AC from one circuit to another without any physical connection between the circuits.

What Is Alternating Current?

In a DC circuit the electric current flows continuously in a single direction, caused by electrons that tend to move in one direction. Within a wire carrying DC, a given electron that starts its trek at one end of the wire eventually ends up at the other end of the wire. But in AC, the electrons don't tend to move in only one direction. Instead, they just move back and forth.

When the electrons in AC switch direction, the direction of current and the voltage of the circuit reverses itself. In public power-distribution systems in the UK, the voltage reverses itself 50 times per second. In some countries, such as the United States, the voltage reverses itself 60 times per second.

The rate at which AC reverses direction is called its *frequency,* which is expressed in hertz (Hz). Thus, standard household current in the UK is 50 Hz.

In an AC circuit, the voltage and therefore the current is always changing. The voltage doesn't, however, instantly reverse polarity; it increases steadily from 0 volts (V) until it reaches a maximum voltage, which is called the *peak voltage.* The voltage then begins to decrease again to 0 V, before reversing polarity and dropping below zero, again heading for the peak voltage but of negative polarity. When it reaches the peak negative voltage, it begins climbing back again until it gets to zero. Then the cycle repeats.

The swinging change of voltage is important because of the basic relationship between magnetic fields and electric currents. When a conductor (such as a wire) moves through a magnetic field, the magnetic field induces a current in the wire. But if the conductor is stationary relative to the magnetic field, no current is induced.

Physical movement isn't necessary to create this effect. If the conductor stays in a fixed position but the intensity of the magnetic field increases or decreases (that is, if the magnetic field expands or contracts), a current is induced in the conductor just as if the magnetic field were fixed and the conductor was physically moving across the field.

Therefore, the voltage in an AC circuit is always increasing or decreasing as the polarity swings from positive to negative and back again, which means that the magnetic field that surrounds the current is always collapsing or expanding. So, if you place a conductor within this expanding and collapsing magnetic field, AC is induced in the conductor.

As if by magic, with AC the current in one wire can induce current in an adjacent wire, even though no physical contact exists between the wires.

The bottom line is that AC can be used to create a changing magnetic field and changing magnetic fields can be used to create AC. This relationship between AC and magnetic fields makes three important devices possible:

- **Alternator:** A device that generates AC from a source of rotating motion, such as a turbine powered by flowing water, steam or a windmill. Alternators work by using the rotating motion to spin a magnet that's placed within a coil of wire. As the magnet rotates, its magnetic field moves, which induces an AC in the coiled wire. (Coils of wires are used instead of straight wires simply because coiling up the wire allows a greater length of wire to be exposed to the changing magnetic field.) For more details, flip to the later section 'Understanding Alternators'.
- **Motor:** The opposite of an alternator. A motor converts AC to rotating motion. In its simplest form, a motor is simply an alternator that's connected backwards. A magnet is mounted on a shaft that can rotate; the magnet is placed within the turns of a coil of wire. When AC is applied to the coil, the rising and falling magnetic field created by the current causes the magnet to spin, which turns the shaft. Check out 'Meeting up with Motors' later in this chapter to find out more.
- **Transformer:** Consists of two coils of wire placed within close proximity. When an AC is placed on one of the coils, the collapsing and expanding magnetic field induces an AC in the other coil. We discuss transformers later in the 'Thinking about Transformers' section.

Measuring Alternating Current

With DC, determining the voltage that's present between two points is easy: you simply measure the voltage with a voltmeter (see Book I, Chapter 8). But with AC, measuring the voltage isn't so simple, because the voltage in an AC circuit is constantly changing. So, for example, when people say that the voltage at a wall socket is 230 VAC, what does that mean in practice?

You can measure voltage in an AC circuit in three ways, as Figure 1-1 illustrates:

- **Peak voltage:** A measurement of the largest voltage present between 0 V and the highest point on the AC cycle (check out the preceding section for a little more). Peak voltage is the maximum voltage that the AC voltage attains.
- **Peak-to-peak voltage:** The difference between the highest and lowest peaks of the AC voltage. In most AC voltages, the peak-to-peak voltage is double the peak voltage.

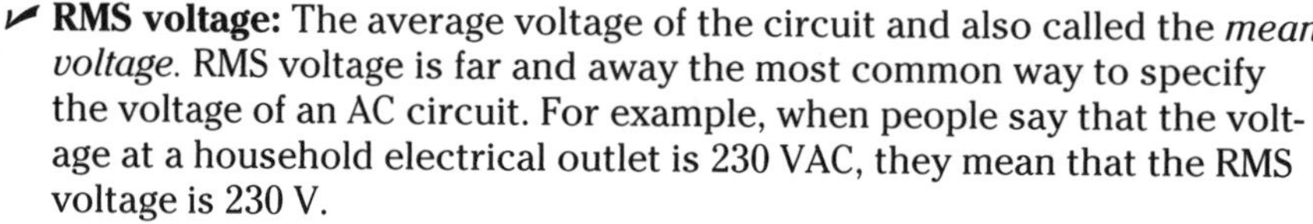

- **RMS voltage:** The average voltage of the circuit and also called the *mean voltage.* RMS voltage is far and away the most common way to specify the voltage of an AC circuit. For example, when people say that the voltage at a household electrical outlet is 230 VAC, they mean that the RMS voltage is 230 V.

  If the AC voltage follows a true sine wave (which we define in Book I, Chapter 9), the RMS voltage is equal to 0.707 multiplied by peak voltage. Or to turn it around, the peak voltage is equal to about 1.4 multiplied by the RMS voltage. Thus, the actual peak voltage at a household electrical outlet is about 322 V.

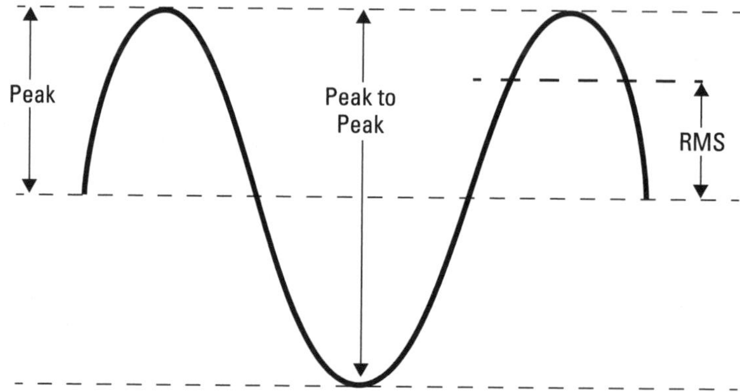

Figure 1-1: Three ways to measure alternating current.

RMS stands for *root mean square,* but that's important only if you're studying for an exam. The true RMS voltage is a bit tricky to work out, because it involves some fairly complicated maths. You calculate RMS by sampling the actual voltage in very small time increments. Then, you square the sample voltages, add up the squares of the voltages and calculate the average of all the squared values. Finally, you calculate the square root of the average to obtain the actual RMS value.

For a true sine wave, the preceding calculation turns out to be very close to multiplying the peak voltage by 0.707. For AC voltages that aren't true sine waves, however, the precise RMS value can be different from what the 'multiply by 0.707' shortcut indicates.

The current wars

Today, AC is the worldwide standard for power distribution, but this wasn't always the case. When electricity was first being put to practical use, DC was the normal way to distribute electricity. The biggest champion of DC was none other than Thomas Edison, the great inventor who's credited with inventing just about everything from the light bulb to the phonograph to the motion picture.

In 1880, Edison patented a system of electrical power distribution based on DC and, in 1882, he started providing electricity to customers in London and New York. By 1890, he had more than 100 power plants across America.

Edison's biggest rival was a fellow named George Westinghouse, who advocated the use of AC for power distribution and promoted a system developed by the brilliant but eccentric inventor Nicola Tesla. Westinghouse promoted the benefits of AC over DC – primarily its ability to transmit power efficiently over much larger distances. Edison's DC system required that power plants be located within a few miles of customers, but Tesla's AC system was able to deliver power hundreds of miles away from the power plants.

Edison responded to the negative publicity of DC like any true-blooded entrepreneur: he launched a smear campaign! In 1887, a man was accidentally killed when he touched bare power lines. Edison had one of his employees develop a method of *intentionally* killing people with AC electricity and the electric chair was born.

The electric chair used AC to electrocute victims. Edison launched a nationwide publicity campaign to convince the public that AC was so dangerous that it was used in prisons to kill condemned murderers. He even went so far as to conduct public executions of stray dogs and, in one case, an elephant. The message was clear: you don't want this dangerous stuff in your house.

Fortunately, the smear campaign didn't work and the benefits of AC eventually won out. The turning point came when the AC generators at Niagara Falls began operating in 1896, delivering power 20 miles away to the city of Buffalo. By the early 20th century, nearly all power distribution worldwide was done with AC.

Direct current distribution lasted much longer than you may think, however. Con Edison – one of the largest electric companies in the world and the direct descendant of Edison's original electric company – didn't convert its last few holdout customers to AC until 2007.

Nearly all AC voltmeters report the RMS voltage, but only more expensive AC voltmeters calculate the actual RMS by sampling the input voltage and doing the sum-of-the-squares thing. Inexpensive voltmeters simply measure the peak voltage and multiply it by 0.707. Fortunately, this result is close enough for most purposes.

WARNING!

Don't try to measure a household mains voltage, not even with a voltmeter whose scale seems to cover 230 VAC. Voltmeters and other electrical measurement meters often lack the required isolation, probes and other safety measures needed to measure such a dangerous voltage.

Understanding Alternators

One good way to get your mind around how AC works is to look at the device that's most often used to generate it: the alternator. As we describe in the earlier section 'What Is Alternating Current?', an *alternator* is a device that converts rotary motion into electric current. By its very nature, an alternator creates AC.

Figure 1-2 shows a simplified diagram of how an alternator works. Essentially, a large magnet is placed within a set of stationary wire coils. The magnet is mounted on a rotating shaft that's connected to a turbine or windmill. When water or steam flows through the turbine or when wind turns the windmill, the magnet rotates.

As the magnet rotates, its magnetic field moves across the coils of wire. The phenomenon of electromagnetic induction means that the moving magnetic field induces an electric current within the wire coils. The strength and direction of this electric current depends on the position and direction of the rotating magnet.

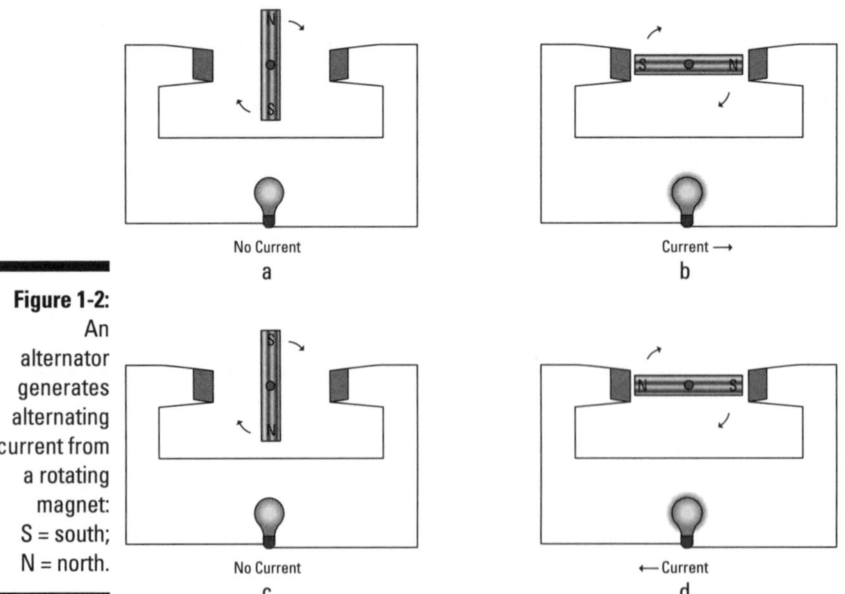

Figure 1-2: An alternator generates alternating current from a rotating magnet: S = south; N = north.

In Figure 1-2, you can see how the current is induced in the wire at four different positions of the magnet's rotation. Here's the cycle:

1. **Position a shows the magnet at its farthest point away from the coils and oriented in the same direction as the coils.** At this moment, the magnetic field doesn't induce any electric current at all. Thus, the light bulb is dark.

   But as the magnet begins to rotate clockwise, the magnet comes closer to the coils, thus exposing more of its magnetic field to the coils. The moving magnetic field induces a current that gets stronger as the magnet continues to rotate closer to the coils, which causes the light bulb to glow.

2. **Position b shows where the magnet reaches its closest point to the coils.** At this point, the current and the voltage are at their maximum and the light bulb glows at its brightest.

   As the magnet continues to rotate clockwise, it begins to move away from the coil. The moving electric field continues to induce current in the coil, but the current (and the voltage) decreases as the magnet retreats farther away from the coils.

3. **Position c shows the magnet reaching its farthest point from the coils, when the current stops and the light bulb goes dark.** As the magnet continues to rotate, it gets closer again to the coils.

4. **Position d shows that this time the polarity of the magnet is reversed.** Thus, the electric current induced in the wire by the moving magnetic field is in the opposite direction. Again, the light bulb glows as the current passing through it increases.

And so on. With each revolution of the magnet, voltage starts at zero and rises steadily to its maximum point, and then falls until it reaches zero again. Then the process is reversed, with the current flowing in the opposite direction.

Here are a few other interesting facts about alternators:

- The term *generator* refers to any device that converts mechanical energy into electrical energy. An alternator is a specific type of generator, and so people commonly – and quite correctly – refer to an alternator as a generator.
- You can generate DC from rotating magnetic fields. A DC generator is more complex than an alternator, however, and contains additional components that can wear out over time.
- The rate at which the magnet rotates dictates the frequency of the AC generated by an alternator: the faster the magnet rotates, the higher the frequency of the resulting AC.

If you place two sets of coils spaced evenly around the magnet, each forming its own complete circuit, AC is induced in each set of coils. However, the polarities of the two voltages are mirror images of one another. In other words, when the voltage is positive in one of the circuits, it's negative in the other. The relationship between the polarity of the circuits is called a *phase* and a power-generating system with two circuits arranged in this way is called a *two-phase system.* The two circuits are said to be 180 degrees out of phase with one another.

If you use three sets of coils, the system is called a *three-phase system* and the three circuits are 120 degrees out of phase. Most public power-generation systems are three-phase systems, because that results in the most efficient generation of power from the rotating magnetic fields.

Meeting up with Motors

An electric *motor* converts electrical energy in the form of electric current to rotating mechanical energy. The simplest type of electric motor is essentially the same thing as an alternator (see the preceding section). The difference is that instead of using some other mechanical force such as water or steam to turn the magnet – which in turn induces electric current in the coils – electric current is applied to the coils, which in turn causes the magnet to rotate.

We don't need to show you a diagram depicting the operation of a motor, because it looks pretty much like the alternator diagram in the earlier Figure 1-2. The only difference is that the light bulb is replaced by an AC power source. The same phenomenon that causes an electric current to be induced in a coil when the coil passes through a moving magnetic field causes a magnetic field to be created when a current is passed through a coil. The magnetic field, in turn, causes the magnet to rotate. This rotation is transferred to the shaft to which the magnet is attached.

As with alternators, you can create motors that work with DC circuits instead of AC, but DC motors are more complicated than AC motors. In a DC motor, the polarity of the coils must be reversed every half-revolution of the magnet to keep the magnet moving in complete rotations. Usually, metal brushes are used to accomplish this task. In an AC motor, the brushes aren't necessary because the AC reverses polarity on its own.

Thinking about Transformers

In Book II, Chapter 4, you discover the basic principles of magnetism and inductance, as follows:

- **Magnetism:** A changing current passing through a wire creates a moving magnetic field around the wire.
- **Inductance:** A changing current is induced in a wire that's exposed to a moving magnetic field.

A transformer is a device that exploits and combines these two principles by placing two coils of wire in close proximity to one another, as shown in Figure 1-3. When a source of AC is connected to one of the coils, that coil creates a magnetic field that expands and collapses in concert with the changing voltage of the AC. In other words, as the voltage increases across the coil, the coil creates an expanding magnetic field. When the voltage reaches its peak and begins to decrease, the magnetic field created around the coil begins to collapse.

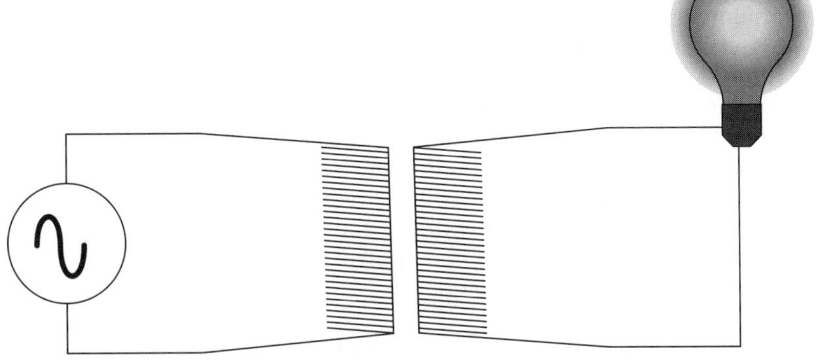

Figure 1-3: A transformer uses magnetic induction to pass current from one circuit to another.

The second coil is located within the magnetic field created by the first coil. As the magnetic field expands, it induces current in the second coil. The voltage across the second coil increases as long as the magnetic field expands. When the magnetic field begins to collapse, the voltage across the second coil begins to decrease.

Thus, the current induced in the second coil mirrors the current that's passed through the first coil. A small amount of energy is lost in the process, but if the transformer is well constructed the strength of the current induced in the second coil is very close to the strength of the current passed through the first coil.

Coiling up to create energy

All transformers have a primary and a secondary coil:

- **Primary coil:** The first coil in a transformer – the one that's connected to the AC voltage.
- **Secondary coil:** The second coil – the one in which an AC voltage is induced.

One of the most useful characteristics of a transformer is that the voltage induced in the secondary coil is equal to the voltage applied to the primary coil multiplied by the ratio of the number of turns in the primary coil and the second coil.

In the simplest case, where the primary and secondary coils have the same number of turns, the voltage induced in the secondary coil is the same as the voltage applied to the primary coil.

If the primary coil has more turns than the secondary coil, the voltage induced in the secondary coil is less than the voltage applied to the primary. How much less depends on the ratio of the turns in the primary and secondary coils. If the secondary coil has half as many turns as the primary coil, the voltage induced in the secondary coil is half the voltage applied to the primary coil. For example, if you apply 230 VAC to the primary coil, 115 VAC is induced in the secondary coil.

If the secondary coil has more turns than the primary coil, the induced voltage is more than the voltage applied to the primary coil. For example, suppose that the primary coil has 1,000 turns and the secondary coil has 2,000 turns. In this case, if you apply 115 V to the primary coil, 230 V is induced in the secondary coil.

A transformer whose primary coil has more turns than its secondary coil is called a *step-down transformer* because it reduces voltage – that is, the voltage at the secondary coil is less than the voltage at the primary coil. Similarly, a transformer that has more turns in the secondary than in the primary is called a *step-up transformer* because it increases voltage.

Although the voltage increases in a step-up transformer, the current is reduced proportionately. For example, if the primary coil has half as many turns as the secondary coil, the voltage induced in the secondary coil is twice the voltage that's applied to the primary coil, but the current that flows through the secondary coil is half the current flowing through the primary coil.

Similarly, when the voltage decreases in a step-down transformer, the current increases proportionately. Thus, if the voltage is cut in half, the current doubles.

This effect makes perfect sense, because a transformer can't just conjure up power out of thin air (otherwise the planet's energy problems would have been solved long ago). Unfortunately, free energy simply doesn't exist.

The basic formula for calculating electric power is power in watts equals voltage multiplied by current in amperes (as we discuss in Book I, Chapter 2):

$$P = V \times I$$

A transformer transfers power from the primary coil to the secondary coil. The power must stay the same, and so if the voltage increases, the current has to decrease. Likewise, if the voltage decreases, the current has to increase.

Producing huge power efficiently with AC

Transformers are the main reason why people use AC instead of DC in large power-distribution systems. When you send large amounts of power over a long distance, transmitting that power in the form of high voltage and low current is much more efficient – which is why overhead power-transmission lines often carry voltages as high as 400,000 VAC. Such high voltages allow the electrical power to be transmitted using much smaller wires than would be required if the same amount of power were transmitted at 230 VAC.

Power-distribution systems use large step-up transformers to increase voltages generated at power plants to hundreds of thousands of volts. Then, as the power gets closer to its final destination (such as your house), a series of step-down transformers drops the voltage down to more manageable levels, until the voltage is dropped to its final level (230 VAC) before it enters your house.

Transformers work only with AC, because the *change* of the magnetic field created by the primary coil is what induces voltage in the secondary coil. To create a changing magnetic field, the voltage applied to the primary coil must be constantly changing. But DC is a steady, fixed voltage and so creates a fixed magnetic field that doesn't induce voltage in the secondary coil.

Chapter 2

Working with Mains Voltage

In This Chapter

- Examining how mains-voltage wiring works
- Making safe connections with mains-voltage circuits

Mains voltage is the voltage that's available in standard residential or commercial wall outlets. In the UK, this voltage is supposed to be around 230 VAC (volts, alternating current), since harmonisaiton with other European countries. Mains voltage used to be 240 VAC in the UK and, in some places, it still is. In fact, mains supply voltages do vary a bit in practice but they will generally be around 230-240 VAC, and we refer to it as 230 VAC in this book. In other parts of the world, the voltage is often lower or occasionally higher.

In the US, mains voltage is often called *line voltage.*

Unlike the voltage available from household batteries, mains voltage is dangerous: if you're not careful, it can kill you. So, avoid connecting your circuits directly to the mains voltage and use an adaptor to step down voltages for your circuits instead.

In this chapter, you learn about ways to use the mains supply safely so that neither you nor anyone else gets hurt.

Adapting the Mains Supply for Use in Your Projects

Although batteries are a convenient source of power for your circuits, the fact is that they wear out and have to be replaced. For certain real-world projects, however, you require a power source that lasts indefinitely. In these cases you need to use mains voltage and an adaptor, which allows you to plug the project in and not have to worry about changing batteries.

Of course, most electronic components require direct current (DC) rather than AC, and at much lower voltage levels than those that mains voltage supplies. Thus, for your project to use mains voltage as its source of power, you need to provide the project with a *power supply* that converts the 230 VAC mains voltage to something more useful, such as 5 VDC.

You can fulfil this requirement in two basic ways:

- **Use a power adapter:** The easiest way is to use an external *power adapter* (often called a *wall wart* or a *power brick*). Figure 2-1 shows a typical external power adapter. You can purchase power adapters from just about any shop with a consumer electronics department. Just get one that provides the right level of DC voltage and use it instead of batteries.

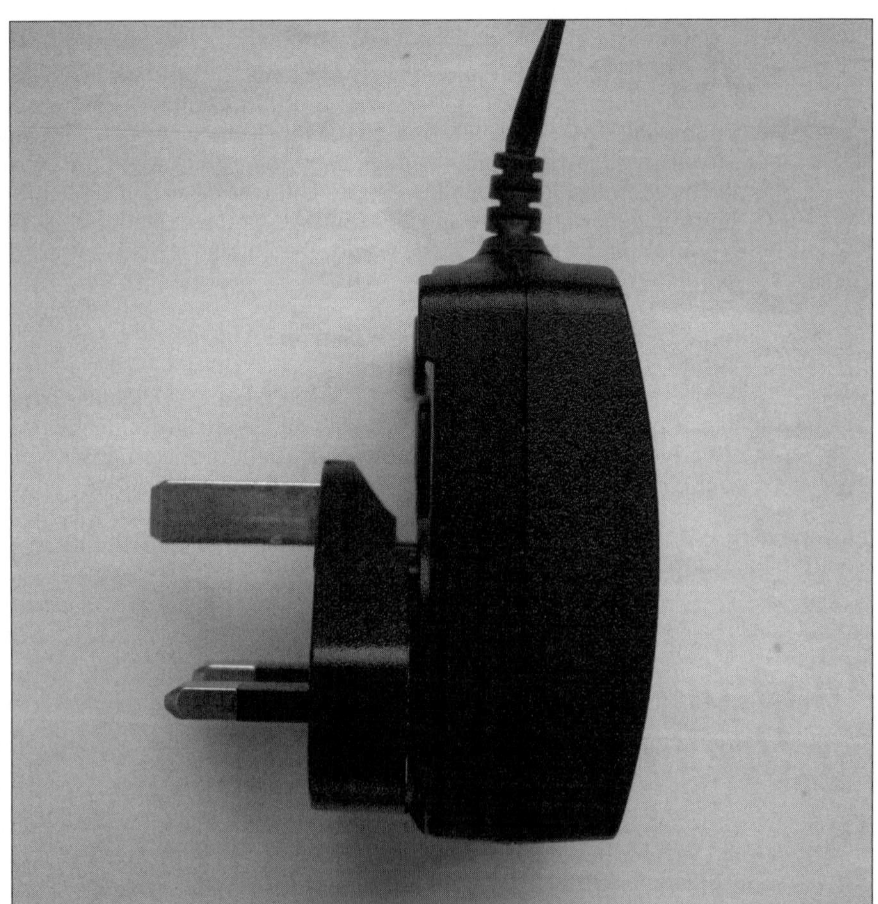

Figure 2-1: An external power adapter.

- **Buy a mains power supply unit (PSU) module:** The alternative to purchasing a power adapter is to buy a ready-made PSU module that goes in the box with your project's circuit rather than sticking out of the wall socket. You connect it to the mains in a similar way to wiring a plug - follow the instructions that come with it and be careful, because even wiring a plug requires caution to stay safe. This circuit does two things:

  It steps down the voltage from 230 VAC to whatever voltage your circuit requires.

  It converts the AC voltage to DC voltage.

You learn how a power-supply circuit works in Chapter 3 of this minibook, but buying a ready-made power adaptor is easier, much safer and usually cheaper.

Another common reason for using mains voltage in a project is if it needs to control an external device that runs on mains voltage, such as a flood lamp. In that case, your project has to be able to turn the mains voltage on and off.

The most common way to turn a mains voltage device on and off from an electronic circuit is to use a device called a *relay,* which is basically an electronic switch that uses a low-voltage input to control a high-voltage output. For example, a relay can let you use a 12 VDC circuit to control a separate mains-voltage circuit. We describe how to use a relay for this purpose later in this chapter in 'Controlling mains-voltage circuits using relays'.

Staying safe with mains voltage

Whenever you build an electronics project that uses mains voltage, you must take extra precautions to ensure your safety and the safety of anyone who may come in contact with your project. Mains voltage is potentially deadly, and so these precautions are absolutely mandatory.

Mains voltage of 230 VAC is more than enough voltage to kill, given the right conditions. In fact, treat any voltage above 50 V as potentially lethal.

When you work with mains voltage, be sure to take the following precautions:

- *Never* let children work with mains, and if you're under 16 years of age, this means you!
- *Never* work on the circuit when the power plug is plugged in. This includes probing the circuit or making measurements with a multimeter or oscilloscope. These devices aren't always designed with the levels of isolation from the mains voltage needed to keep you safe. And it's all too easy to slip when probing a circuit and accidentally touch the live circuit.

- *Never* leave exposed mains-voltage connections anywhere that you or anyone who comes into contact with your project may accidentally touch. *All* mains-voltage connections must be completely insulated or contained within an insulated project box.
- *Always* enclose projects that use mains voltage in a sealed project box so that stray hands can't accidentally come in contact with bare wires or other components.
- *Always* use 3-core grounded power cords if your project is contained in a metal box, and *always* connect the metal box itself to the power cord's earth ground lead. Also, always use 3-core wire if an insulated project box passes mains onto other devices outside the box.
- *Always* use the correct gauge of wire for the amount of current your circuit is carrying. For more information, see the later section, 'Working Safely with Mains Voltage'.
- *Always* ensure that all mains-voltage connections are tight and secure. When using stranded wire, always check for stray strands at your connections. Do not use solid wire as it can fracture and cause a short in places where it has been stripped back to connect to terminals.
- *Always* interconnect mains wiring using terminal blocks rated for UK mains use. Never be tempted to twist and/or solder the wires together.
- *Always* provide some form of strain-relief for wires that carry mains voltage. The most common way to do so is to pass the wire through a grommet-protected hole in the project box. The wire cord must firmly secured to stop it pushing further inside the box and shorting to something it shouldn't, as well as coming loose and pulling out of the box.
- *Always* incorporate a fuse in the primary side of your mains-voltage circuit. The fuse automatically detects when too much current is flowing and immediately breaks the circuit. (For more on fuses, see the section 'Protecting mains-voltage circuits with fuses', later in this chapter.)
- *Always* use stranded mains wire and label the wires with the standard brown for live, blue for neutral and green/yellow for earth colour scheme to make it obvious which wire is which.
- *Never* use a fuse that's rated for more than the maximum current your circuit is designed to bear. For example, if you're using a relay that can switch 5 amps (A) of current, use a fuse rated for 5 A or less.
- *Never* arrange the wiring for your project in a way that causes wires to move or rub against one another. The rubbing eventually wears off the insulation and creates a shock hazard. Use cable ties to secure the wiring inside the enclosure.
- *Always* be aware of heat sinks (pieces of metal that help dissipate heat) that may be hot.
- *Never* become complacent. Always stay as careful as the day you started.

Understanding live, neutral and earth

This section provides you with an introduction to how most residential and commercial buildings are wired. It does not provide enough information for you to attempt your own wiring, however, and you should always use a qualified professional electrician for this, anyway.

The following description applies only to the UK; if you're in a different country, make sure that you determine the standards for your country's wiring.

Standard mains-voltage wiring in the UK is done with plastic-sheathed cables, which usually have three conductors, as shown in Figure 2-2. This type of cable is called *T&E cable,* short for Twin and Earth cable and sometimes called *Flat Twin and Earth.*

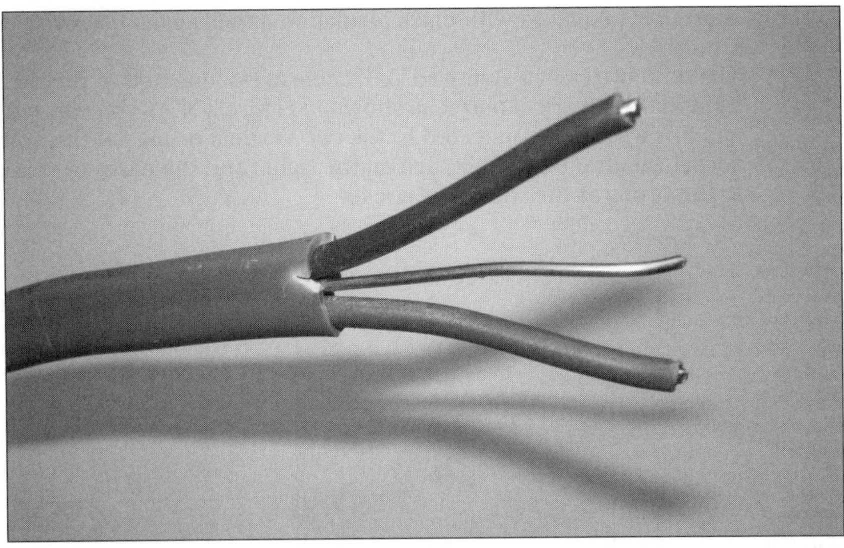

Figure 2-2: T&E cabling.

Two of the conductors in T&E cable are covered with plastic insulation. In new installations one wire is blue (neutral) and the other brown (live), though most houses still have the old black and red wiring colours in use before the UK harmonised with Europe in 2006. The third conductor is bare copper within T&E cable. The conductors are designated as follows:

- **Live:** The brown wire in new installations is the *live wire,* which provides a 230 VAC source. In the older colour scheme in use before 2006, the live wire was red.
- **Neutral:** The blue wire is called the *neutral wire.* It provides the return path for the current provided by the live wire. The neutral wire is connected to an earth ground. In the older colour scheme before 2006,

the neutral wire was black. This does not mean the neutral wire is the same potential as the earth wire. And although it is grounded, you should treat it with as much respect as the live wire. The two could also be swapped and you wouldn't necessarily know.

- **Earth:** The bare wire in T&E cable is called the *earth wire.* Like the neutral wire, the earth wire is also connected to an earth ground. However, the neutral and earth wires serve two distinct purposes. The neutral wire forms part of the live circuit along with the live wire, whereas the earth wire is connected to any metal parts in an appliance such as a microwave oven or coffee pot. This safety feature ensures that if the live or neutral wires somehow come in contact with metal parts, the fact that the metal parts are connected to earth ground eliminates the shock hazard in the event of a short circuit. Outside the T&E cable, the earth wire is sometimes sheathed with green and yellow bi-coloured plastic.

Note that some circuits require a fourth conductor. When a fourth conductor is used, it's covered with black insulation and is also a live wire.

The three wires in a standard T&E cable are connected to the three prongs of a standard electrical outlet as shown in Figure 2-3. As you can see, the neutral and live wires are connected to the two vertical prongs at the bottom of the socket (neutral on the left, live on the right) and the earth wire is connected to the prong at the top of the socket.

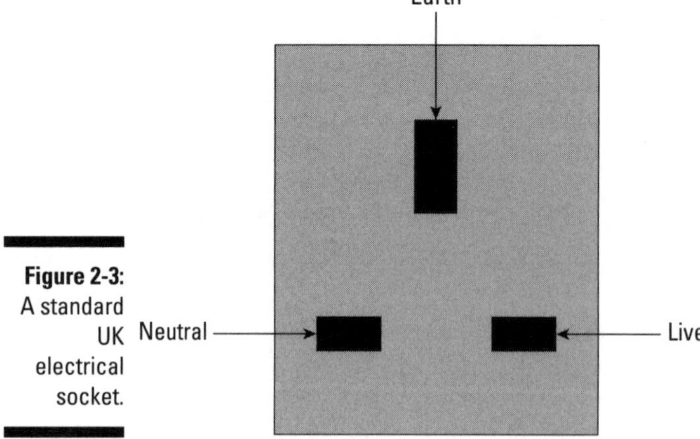

Figure 2-3: A standard UK electrical socket.

Always place switches or fuses on the live wire rather than on the neutral wire. That way, if the switch is open or the fuse blows, the current in the live wire is prevented from proceeding beyond the switch or fuse into your circuit. This approach minimises any risk of shock that may occur if a wire comes loose within your project.

Working Safely with Mains Voltage

In this section, we discuss safe wiring and connections, and describe the roles that fuses and relays play in mains-voltage circuits.

Wiring and connecting

When working with mains voltage, you must always use stranded wire (with standard brown, blue and green/yellow insulation colours) that's designed specifically to handle mains-voltage currents.

When choosing wire, make sure that you select the right gauge for the current your circuit is going to be carrying. For circuits that are designed to carry up to 13 A, 1.3 mm diameter wire is sufficient. For less than 10 A, 1 mm diameter wire is sufficient.

Always buy your electrical parts from a reputable and reliable dealer. Many unsafe electrical components exist on the black market in the UK. They may well be improperly constructed, have not passed the required British standards or fail to meet UK wiring regulations. They may be cheaper, but they're also illegal and potentially dangerous. Make sure legal parts are also suitable for UK mains usage too - component distributors sometimes stock others.

Make sure that all connections you make with wires carrying mains voltage are secure. The easiest way to connect the wires is to use a *terminal block* (also known as a *barrier strip*). Terminal blocks come in various sizes and shapes.

To use a terminal block, simply strip away a short length of insulation from the end of the wires you want to connect, insert them into each end of the terminal block and tighten the screws on top to clamp the wires in place. If you're using stranded wire, make sure that the screw is holding all the strands, because loose strands can cause short circuits.

Protecting mains-voltage circuits with fuses

A *fuse* is an inexpensive device that can carry only a certain amount of current. If the current exceeds the rated level, the fuse melts (blows), thus breaking the circuit and preventing the excessive current from flowing. Fuses are an essential component of any electrical system that uses mains voltage and has the possibility of short-circuiting or overheating and causing a fire.

The most common type of fuse is the *cartridge fuse,* which consists of a cylindrical body that's usually made of glass, plastic or ceramic and has two metal ends (see Figure 2-4). These metal ends are the two terminals of the fuse. Inside the body is a thin wire conductor that's designed to melt away if the current exceeds the rated threshold. As long as the current stays below the maximum level, the conductor passes the current from one metal end to the other. But when the current exceeds the rated maximum, the conductor melts and the circuit is broken.

Maplin (`www.maplin.co.uk/components/fuses`) sells fuses and sets of fuses with various current ratings. Use fast blow types or standard mains plug fuses.

Figure 2-4: A 2 A fuse.

Always connect the fuse to the live wire and place it before any other component in the circuit. In most projects, the fuse should be the first thing the live wire connects to after it enters your project enclosure. Figure 2-5 shows how a fuse is depicted in a schematic diagram. Here, the fuse is placed on the live wire before the lamp.

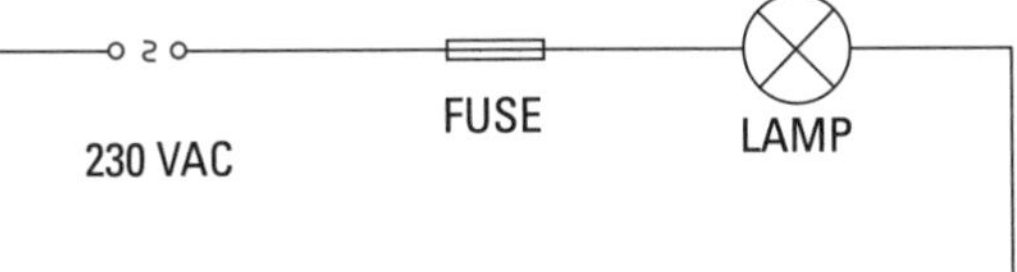

Figure 2-5: A fuse in a schematic diagram.

If you plan on using a fuse in your circuit, you need to purchase a fuse holder to contain it. If the fuse is to be mounted inside your project's enclosure, you can use a chassis-type fuse holder. If you want the fuse to be accessible from outside the project's enclosure, choose a panel-mount holder instead. Figure 2-6 shows a panel-mount holder. Component suppliers such as Maplin sell parts that combine externally accessible fuse holders with power-input sockets and on/off switches.

Figure 2-6: A panel-mount holder.

Controlling mains-voltage circuits using relays

In many projects, you need to turn mains-voltage circuits on and off using circuits that use low-voltage DC power supplies. For example, suppose that you want to flash a 230 VAC flood lamp on and off at regular intervals. You can build a circuit to provide the necessary timing using a 555 timer IC (as we describe in Book III, Chapter 2), but this integrated circuit requires just a small DC power supply, in the range of 5 to 15 V. Plus the output current can't exceed 200 milliamperes (mA), which isn't nearly enough to light a flood lamp.

But never fear, because in this case it's relays to the rescue!

A *relay* is an electromechanical device that uses an electromagnet to open or close a switch. The circuit that powers the electromagnet's coil is completely separate from the circuit that's switched on or off by the relay's switch, and so you can use a relay whose coil requires just a few volts to turn a mains-voltage circuit on or off.

Figure 2-7 shows a typical relay. For this one, the coil requires just 12 VDC to operate and pulls just 75 mA, well under the current limit that can be sourced by the 555 timer IC's output pin. But the switch part of this relay can handle up to 10 A of current at 230 VAC, which is more than enough to illuminate a flood lamp.

Figure 2-7: A relay is a switch that's controlled by an electromagnet.

The switch part of a relay is available in different configurations just like manual switches. The most common switch configuration is double pole, double throw (DPDT), which means that the relay controls two separate switches that operate together and each switch has normally open *and* normally closed contacts. However, the circuits in Figures 2-8 and 2-9 use simple single pole, double throw (SPDT) relays to switch simple powered circuits on or off. To find out more about switches, check out Book II, Chapter 1.

Figure 2-8 shows a schematic diagram for a simple circuit that uses a 9 VDC circuit with a handheld pushbutton to turn a 230 VAC lamp on and off. The relay in the circuit has a coil rated for 9 VDC and a switch rating of 5 A at 230 VAC. Thus, only 9 VDC passes through the pushbutton. If you're holding the switch and decide to take it apart, you aren't exposed to dangerous voltage.

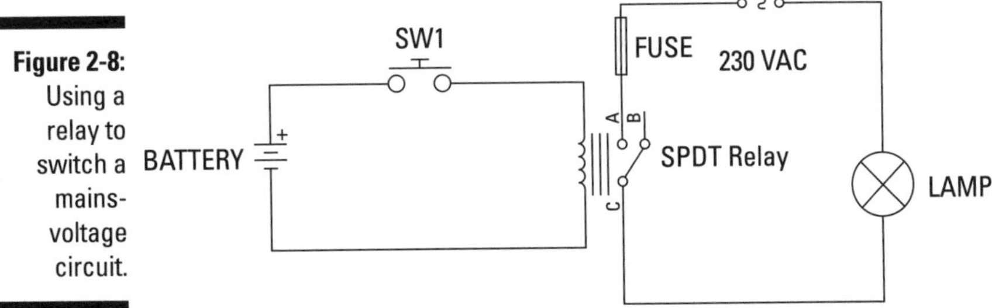

Figure 2-8: Using a relay to switch a mains-voltage circuit.

Figure 2-9 shows a more complicated circuit, in which a 555 timer IC controls a flood lamp via a relay. One end of the relay coil is connected to the 555 timer IC's output pin (pin 3) and the other end is connected to ground. When the 555's output switches on, the relay closes and the flood lamp circuit is completed.

Note the diode that's placed across the relay coil in this circuit. This diode is required to protect the 555 timer IC from any back-current that may be created within the relay's coil when the coil is energised. Because of electromagnetic induction, relay coils are prone to this problem.

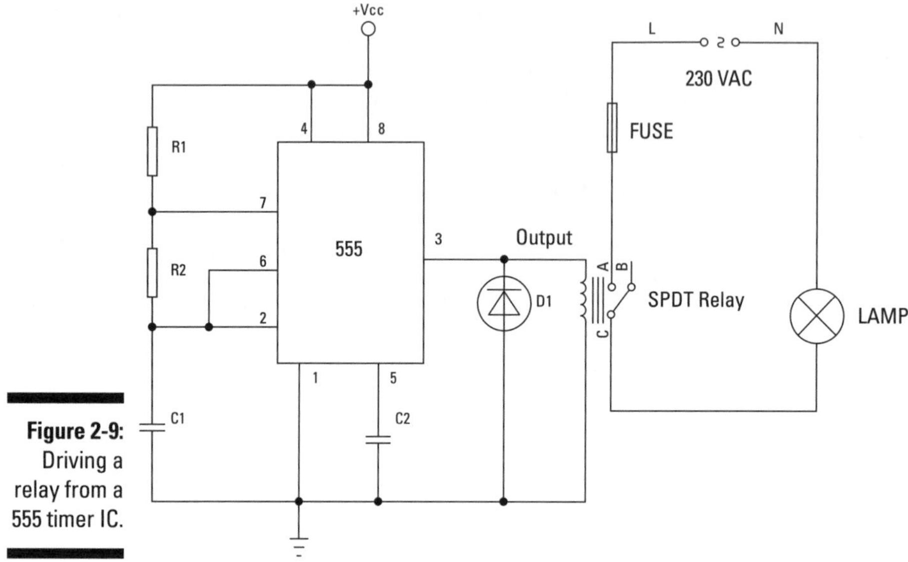

Figure 2-9: Driving a relay from a 555 timer IC.

When the coil is energised, it creates a magnetic field that causes the relay's switch contacts to move. However, this magnetic field has a subtle side effect. In the instant that the voltage on the coil goes from zero to the Vss supply voltage, the magnetic field surrounding the coil expands from nothing to its maximum strength. During this expansion, the magnetic field is moving relative to the coil itself. The principle of induction means that this moving magnetic field induces a current in the coil itself, in the opposite direction to the current that's energising the coil.

Depending on the circumstances, this back-current can be powerful enough to overwhelm the output current coming from the 555 timer IC and possibly powerful enough to send current into the output pin, which can damage or destroy the 555 chip. The diode D1 prevents this from happening by providing the equivalent of a short circuit across the coil for current flowing back towards the output pin.

Whenever you drive a relay from a circuit that has delicate components such as integrated circuits or transistors, always include a diode across the relay coil to prevent the relay from damaging your circuits.

Chapter 3

Supplying Power for Your Electronics Projects

In This Chapter

- Looking at how power supplies work
- Buying preassembled power adaptors
- Understanding what happens inside a power supply

With very few exceptions, electronic circuits require a power supply of some sort. Although some projects run off solar power or more exotic power sources, such as wind turbines, fuel cells or nuclear reactors, most of the projects you build are going to get their power from one of two sources: batteries or an electrical outlet. (After a tiny incident, we've been asked not to build another nuclear reactor in our garage!)

Electrical outlets have the compelling advantage over batteries of not dying on you at an inopportune moment (barring a power cut), although unless you use really long extension cords, you can't take your project very far from the outlet.

We show you how to employ power from an electrical outlet, and describe using power adaptors and how power supplies work.

Powering up from Your Electrical Outlet

Most electronic circuits require a relatively low direct current (DC) voltage, typically in the range of 3 to 12 volts (V). Getting that range of voltage out of batteries is easy; because each battery provides about 1.5 V, you just team up two or more batteries to get the right voltage. For example, if your circuit needs 6 V, you use four batteries connected in series.

Powering a project from an electrical outlet is a little more challenging than using a battery for a number of reasons:

- The 230 V provided at the electrical outlet is much more than most circuits require, and so you have to step-down the voltage to a more appropriate level.
- Circuits that run directly on 230 VAC are inherently more dangerous than circuits that run on lower voltages because of the shock danger that accompanies higher voltages.
- Electronic circuits usually require DC and the wall outlet provides alternating current (AC). Therefore, you have to convert the AC to DC.

The circuit that converts 230 VAC to DC at a lower voltage is called a *power supply.* In the later section 'Understanding the Power Supply', we describe the basics of power supplies. Given the low cost of off-the-shelf power supplies and the dangers inherent in working with mains voltages, however, always buy an adapter rather than ever attempting to build your own.

Using Power Adapters

You can purchase a preassembled *power adapter* that provides the voltage you need for just a few pounds. A power adapter, also called an adaptor or a *wall wart,* is a self-contained power-supply circuit that plugs into a wall outlet and provides a specified level of AC or DC voltage as its output. An adapter with a selectable output is sometimes referred to as a power-supply unit (PSU).

As long as the power adapter supplies the correct voltage, you can use it instead of batteries in just about any circuit.

When you buy a power adapter, check the specifications carefully to ensure that you're purchasing the correct one. The spec is usually printed on the adapter itself. Look for the following important details, commonly listed in this order:

- **AC or DC:** Not all power adapters supply DC; some are made to power low-voltage AC devices. So make sure that you get an adapter that provides DC output.
- **Voltage level:** Check the output voltage. Some power adapters have a switch that lets you choose from among several output voltages. If you use such an adapter, ensure that you set the switch to the correct output level for your circuit.

- **Current capacity:** Most power adapters have a maximum current rating expressed in milliamps (mA). Smaller adapters can handle a few hundred mA, whereas larger adapters may be able to handle an ampere or more. Make sure that the adapter you use can handle the current requirements of your project. (Although some power adapters can handle several amps, few can handle more than that.)
- **Polarity:** Most power adapters use a *barrel connector* to plug the power adapter into the circuit. In nearly all modern power adapters, the centre connection of the barrel connector is positive and the outer connection is negative. However, some power adapters are wired the opposite way round, with negative in the centre and positive on the outside. The polarity of the connector is usually printed on the adapter along with the voltage and current specifications.
- **Connector size:** Unfortunately, far too many different sizes and styles of connectors are used for power adapters. When you've purchased a power adapter, you can go to your local electronics shop and buy a jack that's compatible with the connector on the power adapter. Then, you can use the jack to connect the power adapter to your circuit. (You can buy power adapters equipped with different interchangeable plugs in different sizes from electronics shops.)

Using a power adapter instead of building your own power supply makes your project safer to build and use, because the potentially dangerous part of your project – the part that works directly with 230 VAC mains voltage – is fully contained inside the preassembled power adapter.

You soon discover that you get what you pay for as regards power supplies. Inexpensive power adapters convert AC to DC and step-down the voltage, but most don't provide power that's very clean (that is, a pure level of DC) or stable (that is, with a predictable voltage). Thus, even if you use a power adapter to power your project, you may still need to add circuitry that improves the quality of the DC supplied by the adapter.

Understanding the Power Supply

To convert 230 VAC mains voltage to a DC voltage that's suitable for your circuit, a power-supply circuit has to perform at least three distinct functions:

- **Voltage transformation:** Reduces the 230 VAC mains voltage to what your circuit needs, as we explain in the following section 'Transforming voltage'.
- **Rectification:** Converts the reduced AC voltage to DC voltage.

The DC voltage produced by a rectifier circuit is technically DC, but it isn't steady DC. Instead, a rectifier produces pulsating DC in which the voltage fluctuates in sync with the 50 hertz (Hz) AC that's fed into it from the transformation stage (assuming that the household current is 50 Hz, as it is in the UK). Check out the later section 'Turning AC into DC' for more details.

- **Filtering:** Smoothes out the ripples in the DC voltage produced by the rectification stage. Read the 'Filtering rectified current' section later in this chapter.

Never try to build your own power supply or to convert mains voltages. Always use a mains adaptor or power supply module instead.

Transforming voltage

As we discuss in Chapter 1 of this minibook, a transformer is a device that uses the principle of electromagnetic induction to transfer voltage and current from one circuit to another. The transformer uses a primary coil that's connected to mains voltage and a secondary coil that provides the output voltage.

In most power supplies, the transformer reduces the voltage. The amount of the voltage reduction depends on the ratio of the number of turns in the primary coil versus the number of turns in the secondary coil. For example, if the secondary coil has half as many turns as the primary coil, the primary coil voltage is cut in half at the secondary coil. In other words, if 230 VAC is applied to the primary coil, 115 VAC is available at the secondary coil.

Common secondary voltages for transformers used in low-voltage power supplies range from 6 to 24 VAC. Note that because some voltage is lost in the rectifier and filtering stages, the secondary coil voltage has to be a few volts higher than the final DC voltage your circuit requires.

The precise DC voltage level used for most circuits, however, isn't all that critical. A power supply for a circuit that calls for 6 VDC would use a transformer that provides 6 VAC in its secondary coil, the output from the power supply after it's rectified to DC voltage is going to be closer to 5 VDC. Most likely, 5 VDC is close enough for the circuit to work just fine.

Many transformers have more than one tap in the secondary coil. A *tap* is simply a wire connected somewhere in the middle of a coil, effectively dividing a single coil into two smaller coils. Multiple taps let you access several different voltages in the secondary coil. The most common arrangement is a *centre-tapped* transformer, which provides two voltages as shown in Figure 3-1.

In a centre-tapped transformer, the voltage measured across the two outer taps is double the voltage measured from the centre tap to either one of the two outer taps. Thus, if the voltage across the two outer taps is 24 VAC, the voltage across the centre tap and either of the outer taps is 12 VAC.

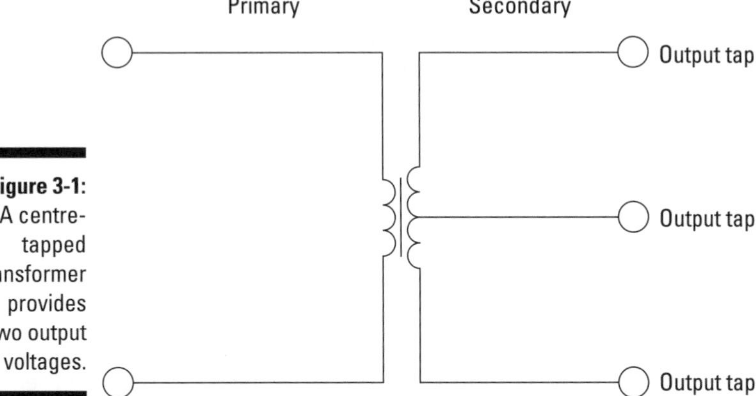

Figure 3-1: A centre-tapped transformer provides two output voltages.

When a transformer reduces voltage, it increases current. So if a transformer cuts the voltage in half, the current doubles. As a result, the overall power in the system (defined as the voltage multiplied by the current) remains the same.

A transformer is strictly an AC device, which means that:

- Transformers work only when AC is applied to the primary coil. If you apply DC to the primary coil, no voltage appears across the secondary coil.

  Strictly speaking, a brief spike of voltage does appear across the secondary coil the moment voltage is applied to the primary coil, but in most circuits this fleeting voltage is insignificant.

- A step-down transformer reduces the voltage from the primary to the secondary coils but *doesn't* convert AC to DC. The voltage at the secondary coil is always AC.
- A transformer isolates the circuit attached to the secondary coil from the circuit connected to the primary coil.

Turning AC into DC

The task of turning AC into DC is called *rectification* and the circuit that does the job is called a *rectifier.* The most common way to convert AC into DC is to use one or more diodes, those handy electronic components that allow current to pass in one direction but not the other. We cover diodes in detail in Book II, Chapter 5, and so check out that chapter if the concept doesn't sound familiar.

Although a rectifier converts AC to DC, the resulting DC isn't a steady voltage; more accurately it's 'pulsating DC'. Although the pulsating DC always moves in the same direction, the voltage level has a distinct ripple to it, rising and falling a bit in sync with the waveform of the AC voltage that's fed into the rectifier. For many DC circuits, a significant amount of ripple in the power supply can cause the circuit to malfunction. Therefore, additional filtering is required to 'flatten' the pulsating DC that comes from a rectifier to eliminate the ripple. (For more on filtering, see the section, 'Filtering rectified current', later in this chapter.)

Rectifier circuits come in three distinct types: half-wave, full-wave and bridge. The following sections describe these three rectifier circuits.

Handling half-wave rectifiers

The simplest type of rectifier is made from a single diode, as Figure 3-2 shows. This kind is called a *half-wave rectifier* because it passes just half of the AC input voltage to the output. When the AC voltage is positive on the cathode side of the diode, the diode allows the current to pass through to the output. But when the AC reverses direction and becomes negative on the cathode side of the diode, the diode blocks the current so that no voltage appears at the output.

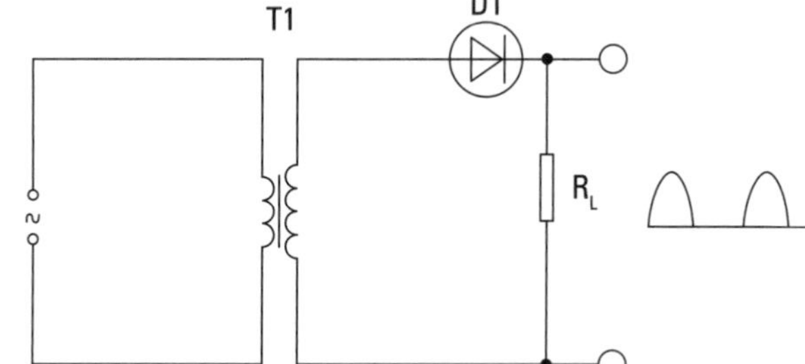

Figure 3-2: A half-wave rectifier uses just one diode.

Half-wave rectifiers are simple but aren't very efficient, because they block the entire negative cycle of the AC input. As a result, output voltage is zero half of the time, which causes the average voltage at the output to be half of the input voltage.

Note the resistor marked R_L in Figure 3-2: it isn't part of the rectifier circuit. Instead, it represents the resistance imposed by the load that's going to be placed ultimately on the circuit when the power supply is put to use.

Finding out about full-wave rectifiers

A *full-wave rectifier* uses two diodes, which enables it to pass the positive *and* the negative side of the AC input. The diodes are connected to the transformer, as Figure 3-3 illustrates.

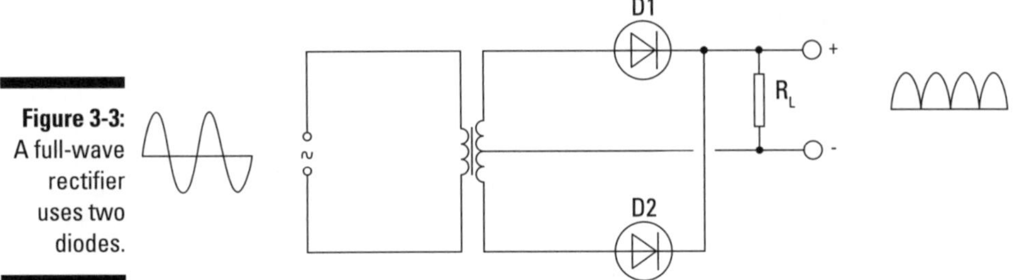

Figure 3-3: A full-wave rectifier uses two diodes.

The full-wave rectifier requires a centre-tapped transformer (see the earlier section 'Transforming voltage'). The diodes are connected to the two outer taps and the centre tap is used as a common ground for the rectified DC voltage. The full-wave rectifier converts both halves of the AC sine wave to positive-voltage DC. The result is DC voltage that pulses at twice the frequency of the input AC voltage. In other words, assuming that the input to the primary side of the transformer is 50 Hz household current, the output from the rectifier on the secondary side of the transformer is going to be DC pulsing at 100 Hz.

Understanding bridge rectifiers

The problem with a full-wave rectifier is that it requires a centre-tapped transformer, and so it produces DC that's just half of the total output voltage of the transformer.

A *bridge rectifier* (see Figure 3-4) overcomes this limitation by using four diodes instead of two. The diodes are arranged in a diamond pattern so that, on each half phase of the AC sine wave, two of the diodes pass the current to the positive and negative sides of the output, and the other two diodes block current. A bridge rectifier doesn't require a centre-tapped transformer.

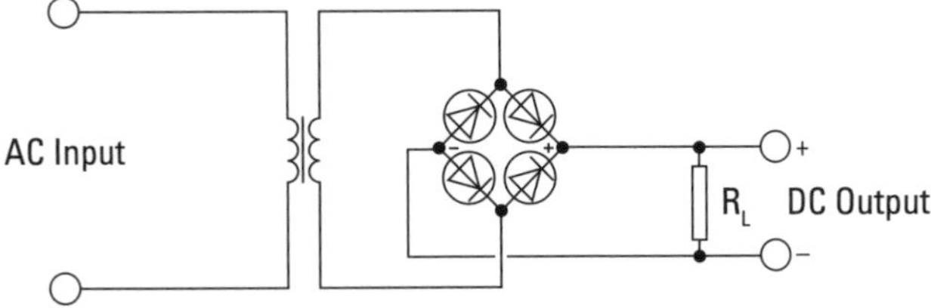

Figure 3-4: A bridge rectifier uses four diodes.

The output from a bridge rectifier is pulsed DC, just like the output from a full-wave rectifier. However, the full voltage of the transformer's secondary coil is used.

Filtering rectified current

Although the output from a rectifier circuit is technically DC, because all the current flows in the same direction, it isn't stable enough for most purposes. Even full-wave and bridge rectifiers (which we describe in the two preceding sections) produce DC that pulses in rhythm with the 50 Hz AC sine wave that originates with the 230 VAC that's applied to the transformer primary side.

That pulsing current isn't suitable for most electronic circuits, of course, which is where filtering comes in. The filtering stage of a power-supply circuit smoothes out the ripples in the rectified DC to produce a smooth DC that's suitable for even the most sensitive of circuits.

Filtering is usually accomplished by a capacitor being brought into the power-supply circuit, as shown in Figure 3-5. Here, the capacitor is simply placed across the DC output.

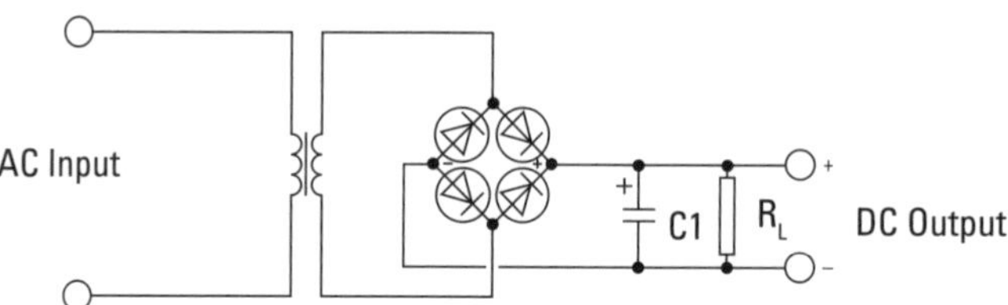

Figure 3-5: A capacitor can be used to filter the output from the rectifier.

As you discover in Book II, Chapter 3, a capacitor has the useful characteristic of resisting changes in voltage. It accomplishes this feat by building up a charge across its plates when the input voltage is increasing. When the input voltage decreases, the voltage across the capacitor's plates decreases as well, but more slowly than the input voltage decreases. This has the effect of levelling out the voltage ripple, as shown in Figure 3-6.

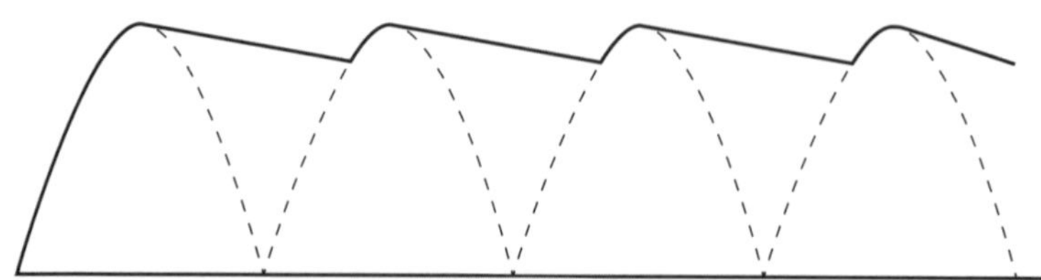

Figure 3-6: A filter circuit smoothes the output voltage.

The difference between the minimum DC voltage and the maximum DC voltage in the filtering stage is called the *voltage ripple* (or just *ripple*), which is usually measured as a percentage of the average voltage. For example, a 10% ripple in a 5 V power supply means that the actual output voltage varies by 0.5 V.

The filter capacitor usually has to be large to provide an acceptable level of filtering. For a typical 5 V power supply, a 2,200 µF electrolytic capacitor (which we describe in Book II, Chapter 3) does the job: the bigger the capacitor, the lower the resulting ripple voltage. With an electrolytic capacitor, the positive side must be connected to the positive voltage output from the rectifier and the negative side must be connected to ground.

Two capacitors in combination with a resistor, as shown in Figure 3-7, improves a filter circuit. In this circuit, the first capacitor acts like the capacitor in Figure 3-6, eliminating a large portion of the ripple voltage. The resistor and second capacitor work as a resistor-capacitor low pass filter network that eliminates the ripple voltage even further.

The advantages of this circuit are that the resulting DC has a smaller ripple voltage and the capacitors can be smaller. The disadvantage is that the resistor drops the DC output voltage: by how much depends on the amount of current the load draws. For example, with a 100 Ω resistor and the load drawing 100 mA, the resistor drops 10 V (100 × 0.1). Thus, to provide a final output of 5 V, the rectifier circuit needs to supply 15 V because of the 10 V drop introduced by the resistor.

Figure 3-7: Two capacitors and a resistor cut ripple voltage but also reduce the DC output voltage.

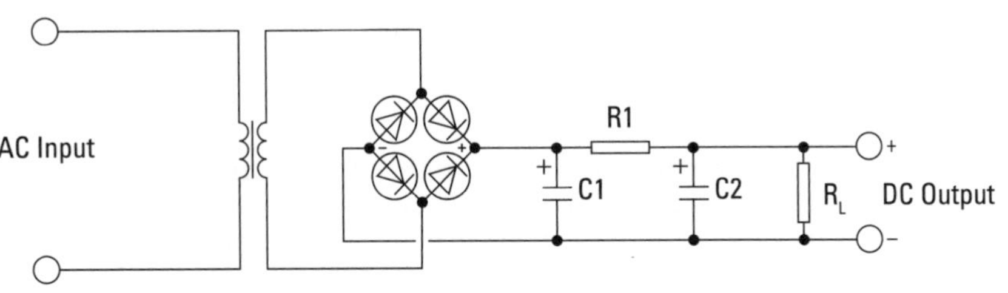

A filter circuit can also use an inductor, as shown in Figure 3-8. Unlike a resistor-capacitor filter, an inductor-capacitor filter doesn't significantly reduce the DC output voltage. Although inductor-capacitor filter circuits create the smallest ripple voltage, inductors in the required typical range of 10 henrys (a unit of measuring inductance that we discuss in Book II, Chapter 4) are large and relatively expensive. Thus, most filter circuits use a single capacitor or a pair of capacitors coupled with a resistor.

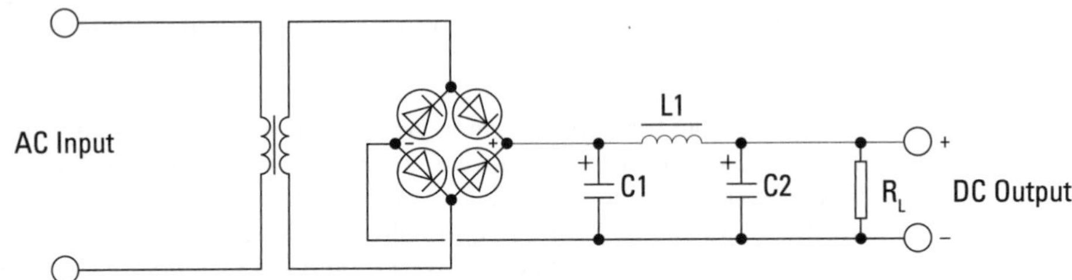

Figure 3-8: You can use an inductor in a filter circuit to minimise DC voltage loss.

Regulating voltage

The purpose of a power supply is to provide power for an electronic circuit, and as we explain in Book I, Chapter 2, a basic formula is available for calculating the amount of power that a circuit uses:

$$P = V \times I$$

So, power is equal to voltage multiplied by current.

If you have any two of these three elements for a circuit, you can easily calculate the third. For example, if you know that the current is 0.5 A and the voltage is 10 V, you can calculate that the circuit consumes 5 watts of power by multiplying 0.5 by 10.

For a given amount of power, an inverse relationship exists between voltage and current. Whenever current increases, voltage must decrease, and whenever current decreases, voltage must increase. This simple fact, unfortunately, has an adverse effect on power-supply circuits.

When a voltmeter is connected to the output terminals of a power supply, the meter itself draws an almost insignificant amount of current. As a result, it reads very close to the voltage you expect to obtain from the power supply. But when a circuit is connected that draws significant current from the power supply, the voltage from the power supply drops in proportion to the current. Depending on the nature of the circuit connected to the power supply, this voltage drop may or may not be important. Some circuits designed for 12 VDC work fine if given only 9 VDC. But other circuits are sensitive to the input voltage, and so the power supply needs to work harder to make sure that it delivers the desired voltage.

To maintain a steady voltage level regardless of the amount of current drawn from a power supply, you can incorporate a *voltage regulator* circuit into the power supply. This voltage regulator monitors the current drawn by the load and increases or decreases the voltage accordingly to keep the voltage level constant.

A power supply that incorporates a voltage regulator is called a *regulated power supply.*

Voltage regulator circuits can be designed using a couple of transistors, some resistors and a Zener diode, but buying one of the many available IC voltage regulators is far easier. Voltage regulator ICs are inexpensive (they cost a few pounds) and, with just three pins to connect, they're easy to incorporate into circuits.

The most popular type of voltage regulator IC is the *78XX* series, sometimes called the *LM78XX* series. These voltage regulators combine 17 transistors, three Zener diodes and a handful of resistors into one handy package with three pins and a heat sink that helps dissipate the excess power consumed by the regulator as it compensates for increases or decreases in current draw to keep the voltage at a constant level.

The last two digits of the 78XX ID number indicate the output voltage regulated by the IC. The most popular models are:

Model	***Voltage***
7805	5
7806	6
7809	9
7810	10
7812	12
7815	15
7818	18
7824	24

Of these, the most common are the 7805 (5 V) and 7812 (12 V), which are available from most electronics component distributors.

To use a 78XX voltage regulator, you just insert it in series on the positive side of the power-supply circuit and connect the ground lead to the negative side, as we show in Figure 3-9. As you can see, placing a small capacitor (typically 1 μF) after the regulator is a good idea.

You have to supply a voltage regulator with about 3 V more than the regulated output voltage. Thus, for a 7805 regulator, you have to give it at least 8 V. The maximum input voltage for a 7805 is 35 V.

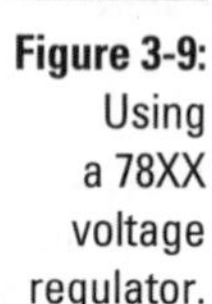

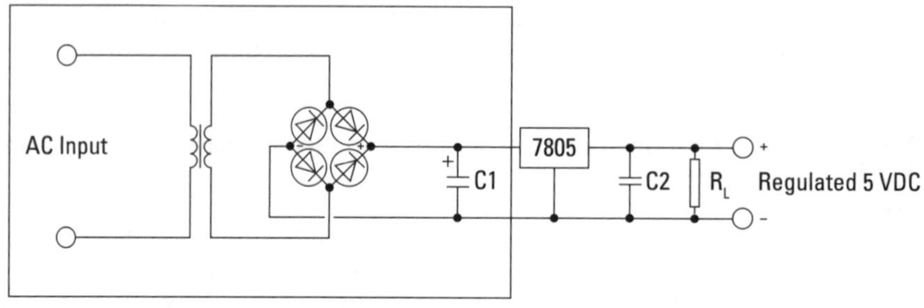

Figure 3-9: Using a 78XX voltage regulator.

The four diodes in a bridge rectifier (see the earlier section 'Understanding bridge rectifiers' for details) each drop about 3 V from the transformer output, and so a transformer whose secondary coil delivers at least 11 V is required to produce 5 V of regulated output.

The Mains to DC PSU section can be bought as a ready-made and relatively safe mains adaptor or PSU module; then you can add the DC-DC regulator electronics to get the output you want.

Eleven-volt transformers are rarer than shy reality-TV stars, but 12 V transformers are readily available. Thus, a 5 V regulated power supply starts with a 12 VAC transformer that delivers 12 V to the bridge rectifier, which converts the AC to DC and drops the voltage down to about 9 V and then delivers the voltage to the filter circuit. This filter smoothes out the ripples and passes the voltage on to the 7805 voltage regulator, which holds the output voltage at 5 V.

Another popular voltage regulator IC is the LM317, which is an adjustable voltage regulator. The LM317 regulator works much like a 78XX regulator, except that instead of connecting the middle lead directly to ground, it connects to a voltage divider built from a pair of resistors, as shown in Figure 3-10. The value of the resistors determines the regulated voltage. In Figure 3-10, we use a potentiometer so that the user can vary the output voltage (flip to Book II, Chapter 2, for more about using potentiometers).

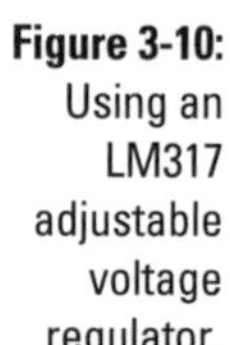

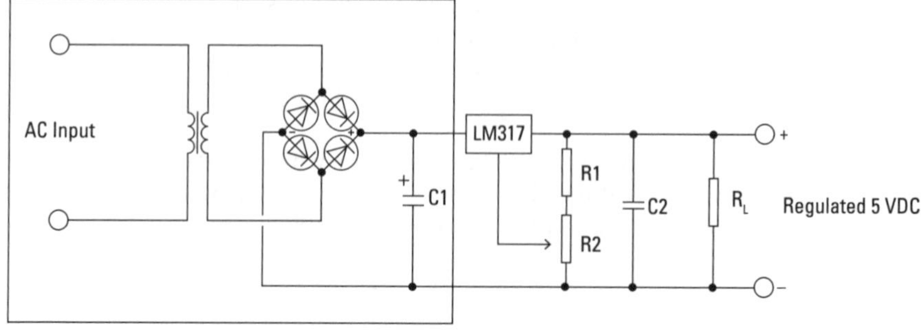

Figure 3-10: Using an LM317 adjustable voltage regulator.

Book V

Working with Radio and Infrared

Ever heard of RFID? If not, why not head online to www.dummies.com/extras/electronicsaiouk for a great introduction to radio frequency identification.

Contents at a Glance

Chapter 1

Tuning in to Radio

In This Chapter

- Realising how radio waves work
- Using transmitters and receivers
- Discovering the differences between AM and FM

The American cable TV channel Music Television, better known as MTV, launched on 1 August 1981. Appropriately, the first music video it played, at one minute past midnight, was 'Video Killed the Radio Star' by the Buggles. The song laments how the golden age of radio would be lost to the rise of television. Ironically, most teenagers today get their music-video fix by searching YouTube instead of switching on the television in the hope of catching their favourite stars, whereas radio is still doing fine. Video didn't kill the radio star after all.

In the 1930s and 1940s, people had only one use for radio: broadcast audio signals. Today, audio broadcast over radio is as commonplace as ever, but the list of other types of information being broadcast by radio technology has skyrocketed:

- **Broadcast television:** Nothing more than the combination of audio and video broadcast over radio.
- **Mobile phones:** Extends the telephone networks to places that phone cables can't reach.
- **Wireless networking:** Replaces bulky computer network cables with data transmitted over radio.
- **Traditional audio radio and television:** Received a technology overhaul with digital television and digital radio.
- **Other popular uses for radio technology:** Includes radar, GPS navigation systems and wireless Bluetooth devices.

In this chapter, you discover the basic concepts of radio: what it is, how it works and how it was discovered. This chapter provides the important foundation for this minibook's Chapters 2 and 3, in which you find out how to build circuits that receive and play radio signals broadcast in the AM radio band.

Throughout this chapter, we sometimes refer to radio as if it's only for the broadcast of sound. Just keep in mind that radio is also used for broadcasting video and digital data as well as other types of information.

Rolling with Radio Waves

Most people think of radio as the wireless broadcast of sound, often music and speech. But the term is much broader than that; the broadcast of sound is just one application of the extremely useful electrical phenomenon called radio.

Radio takes advantage of one of the most interesting of all electrical phenomena: *electromagnetic radiation* (often abbreviated EMR): a type of energy that moves in waves at the speed of light, through the air and even in the vacuum of space.

EMR waves can oscillate at any imaginable frequency. The rate of the oscillation is measured in cycles per second, also known as *hertz* (abbreviated Hz). The term hertz doesn't refer to the car rental company! Instead, it honours the great German physicist Heinrich Hertz, who was the first person to build a device that was able to create and detect radio waves.

Radio is simply a specific range of frequencies of EMR waves. The low end of this range is just a few cycles per second and the upper end is about 300 billion cycles per second (also known as *gigahertz,* abbreviated GHz). That's a pretty big range, but EMR waves with much higher frequencies exist as well, and are in fact commonplace. EMR waves with frequencies higher than radio waves go by various names, including infrared (see Chapter 3 of this minibook), ultraviolet, X-rays, gamma rays and – most importantly – visible light.

That's right; what people call light is exactly the same thing as what they call radio, but at higher frequencies. The frequency of visible light is measured in billions of hertz, also called *terahertz* and abbreviated THz. The low end of visible light (red) is around 405 THz and the upper end (violet) is around 790 THz.

Disputed paternity: Who's the Daddy?

The history of radio technology is plagued by controversy over the question of who actually invented it. The answer most often given is Italian inventor Guglielmo Marconi, but many others made important discoveries that give them good claim to the title.

Here's a rundown of the contenders for the title of The Father of Radio:

- **Marconi:** He was the first person to demonstrate radio successfully and exploit it commercially. In 1901, Marconi sent a message via radio across the Atlantic from England to Canada. The message was faint, consisted of nothing but the letter 'S' and reception of the message wasn't independently confirmed. Nevertheless, Marconi's accomplishment was astonishing, and he made many important contributions to the technology and business of radio.
- **Tesla:** In 1943, the US Supreme Court ruled that many of Marconi's important radio patents were invalid because Nikola Tesla had already described the devices covered by Marconi's patents. Tesla was a brilliant engineer who's best known for being the champion of alternating current over direct current for power distribution. He publicly demonstrated wireless communication devices as early as 1893. Tesla believed that wireless technology would be used not only for communication, but also for power distribution.
- **Lodge:** In England, Sir Oliver Lodge was building wireless telegraph systems in the mid-1890s.
- **Popov:** In Russia, Alexander Stepanovich Popov was demonstrating wireless telegraph transmissions around the same time as Lodge.
- **Bose:** In India, Sir Jagadish Chandra Bose was also demonstrating wireless telegraph transmissions in the early 1890s. Whether these demonstrations occurred before, after or at the same time as other demonstrations by Lodge, Popov and others is under dispute.
- **Many others:** The list of names of others who did important research and made important discoveries in the last decades of the nineteenth century is long: Heinrich Hertz, Edouard Branly, Roberto de Moura, Ernest Rutherford, Ferdinand Braun, Julio Baviera, and Reginald Fessenden are just a few of the many individuals who made important contributions.

Therefore no one person has a clear-cut claim to being the first to invent radio. Work was being done all around the world and discoveries were being made almost every day. Perhaps the best answer is that no one 'invented' radio, because it's a natural phenomenon. Radio was discovered, not invented.

What was invented were ways to exploit the phenomenon of radio by building devices to generate radio waves and modulate them to add information, as well as devices that could receive radio waves and extract the information that was added.

So here's an interesting thought. Radio stations broadcast on a specific frequency. For example, London's Capital radio station has broadcast on 95.8 MHz for many years. Plenty of other radio stations operate in the area, but only Capital broadcasts at 95.8 MHz – or only Capital is supposed to. Pirate stations sometimes cause interference in certain areas. The term *channel* is often used to refer to a radio station broadcasting at a particular frequency.

Purple is the colour you perceive when you see light whose frequency is right around 680 THz. Many other colours exist, of course, but only the colour purple is at 680 THz. So in a way, colour is the same thing as channel. If EMR waves are vibrating at 95.8 MHz, they're Capital radio. If those same EMR waves vibrate millions of times faster, at 680 THz, they're the colour purple.

An important concept that's related to frequency is the idea of wavelength. The term *wavelength* refers to the distance between the crests of each cycle of EMR at a particular frequency. Because EMR waves travel at the speed of light, you can calculate the wavelength of a given frequency by dividing the distance that light travels in a single second by the number of cycles per second.

Light is pretty fast: it scoots along at 186,282 miles per second – or, to be precise about it, 299,792,458 metres per second (m/s). Thus, the wavelength of an EMR wave oscillating at 100 kHz is about 1.86 miles: 186,282 divided by 100,000 (or 2997924.58 m/s).

The higher the frequency, the shorter the wavelength. The wavelength of most AM broadcast radio stations is a small fraction of a mile. The wavelength of visible light is a very small fraction of a centimetre.

Sound waves aren't a type of EMR. Sound waves are created when particles of matter bump against each other. Thus, you need matter – such as air or water – to transmit sound waves. Radio waves don't require particles of matter to travel. In fact, radio waves travel best in the vacuum of space, where no matter is present to get in the way.

Transmitting and Receiving Radio Waves

Many natural sources of radio waves exist. But in the later part of the 19th century, scientists figured out how to generate radio waves using electric currents. In a nutshell, if you pass an alternating current (AC) into a length of wire, radio waves at the same frequency as the AC are generated.

Radio communication requires two components:

- **Transmitters** generate radio waves.
- **Receivers** detect radio waves.

The following two sections describe the basic operation of radio transmitters and receivers.

Making waves with radio transmitters

A radio transmitter consists of several elements that work together to generate radio waves that contain useful information such as audio, video or digital data. Figure 1-1 shows these components, as follows:

- **Power supply:** Provides the necessary electrical power to operate the transmitter.
- **Oscillator:** Creates AC at the frequency on which the transmitter is going to transmit. The oscillator usually generates a sine wave, which is referred to as a *carrier wave.*
- **Modulator:** Adds useful information to the carrier wave. This information can be added in two main ways:
    - *Amplitude Modulation* (AM) makes slight increases or decreases to the *intensity* of the carrier wave.
    - *Frequency Modulation* (FM) makes slight increases or decreases to the *frequency* of the carrier wave.

  For more information about AM and FM, see the later sections 'Approaching AM Radio' and 'Finding out about FM Radio'.

  A third method of adding information to a radio signal is simply turning the signal on and off in a pattern that represents the information. For example, radio signals can send Morse code in this way.
- **Amplifier:** Amplifies the modulated carrier wave to increase its power. The more powerful the amplifier, the more powerful the broadcast.
- **Antenna:** Converts the amplified signal to radio waves.

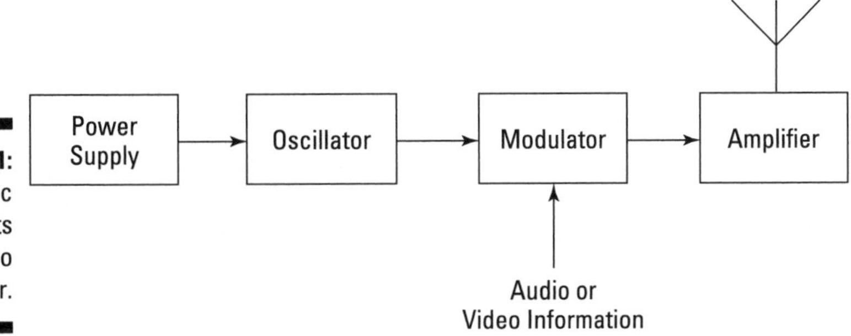

Figure 1-1: The basic components of a radio transmitter.

Catching the waves: Radio receivers

A radio receiver is the opposite of a radio transmitter. It uses an antenna to capture radio waves, processes those waves to extract only those waves that are at the desired frequency, extracts the audio signals that were added to those waves, amplifies the audio signals and finally plays them on a speaker. Figure 1-2 shows these components. Here's how each one works:

- **Antenna:** Captures the radio waves. Typically, the antenna is simply a length of wire. When this wire is exposed to radio waves, the waves induce a very small AC in the antenna.
- **RF amplifier:** A sensitive amplifier that amplifies the very weak radio frequency (RF) signal from the antenna so that the signal can be processed by the tuner.
- **Tuner:** A circuit that can extract signals of a particular frequency from a mix of signals of different frequencies. On its own, the antenna captures radio waves of all frequencies and sends them to the RF amplifier, which dutifully amplifies them all. Unless you want to listen to every radio channel at the same time, you need a circuit that can pick out just the signals for the channel you want to hear. That's the role of the tuner.

  The tuner usually employs the combination of an inductor (for example, a coil) and a capacitor to form a circuit that resonates at a particular frequency. This frequency, called the *resonant frequency,* is determined by the values chosen for the coil and the capacitor. This type of circuit tends to block any AC signals at a frequency above or below the resonant frequency.

  You can adjust the resonant frequency by varying the amount of inductance in the coil or the capacitance of the capacitor. In simple radio receiver circuits (such as the one you read about in Chapter 2 of this minibook), the tuning is adjusted by varying the number of turns of wire in the coil. More sophisticated tuners use a variable capacitor (also called a *tuning capacitor*) to vary the frequency.
- **Detector:** Responsible for separating the audio information from the carrier wave. For AM signals, you can do this separation with a diode that rectifies the AC signal and a low pass filter. What's left is the audio AC signal with a DC offset that you can feed to an audio amplifier circuit through a series capacitor to remove the DC signal component. For FM signals, the detector circuit is more complicated.
- **Audio amplifier:** Amplifies the weak signal that comes from the detector so that you can hear it. You can achieve this amplification by using a simple transistor amplifier circuit as we describe in Book II, Chapter 6. You can also use an op-amp IC, which is the subject of Book III, Chapter 3.

Looking at the radio spectrum

The term *spectrum* simply means a range of frequencies. Radio is generally considered to be frequencies between 3 Hz and 300 GHz. That broad range of frequencies is carved up into smaller pieces that are used for specific types of radio:

Frequency	***Abbreviation***	***Description***
3–30 Hz	ELF	Extremely low frequency, used for communications with submarines.
30–300 Hz	SLF	Super low frequency, also used for communications with submarines.
300 Hz–3 kHz	ULF	Ultra low frequency, used for underground communications within mines.
3–30 kHz	VLF	Very low frequency, also for submarine communications and a few other unusual applications.
30–300 kHz	LF	Low frequency, used for navigation, Radio Frequency Identification (RFID) and a few other applications.
300–3,000 kHz	MF	Medium frequency, used for AM radio.
3–30 MHz	HF	High frequency, used for shortwave and Citizens' Band (CB) radio ('that's a big 10-4, little buddy!').
30–300 MHz	VHF	Very high frequency, used for FM radio and television.
300–3,000 MHz	UHF	Ultra high frequency, used for television, mobile phones, wireless networking, Bluetooth and so on.
3–30 GHz	SHF	Super high frequency, used for high-speed wireless networking, radar and communication satellites.
30–300 GHz	EHF	Extreme high frequency, used for microwave communications.
300–3,000 THz	THF	Tremendously high frequency (no we didn't make that up), used for exotic applications that border on science fiction.
Above 3,000	GHz	You're off the edge of the spectrum map, mate. Here be dragons!

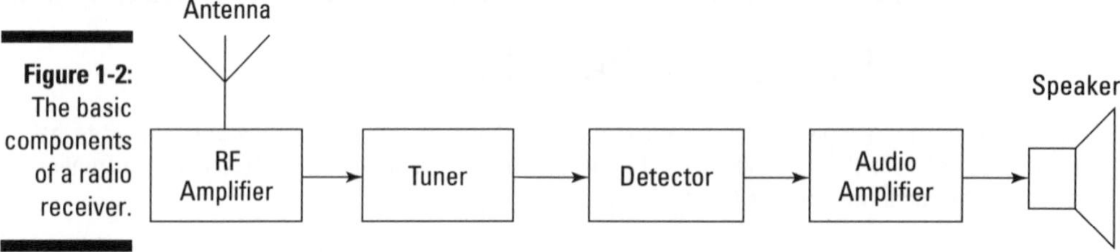

Figure 1-2: The basic components of a radio receiver.

Of course, loads of variations exist to this basic radio receiver design. Many receivers include additional filtering and tuning circuits to better lock on to the intended frequency – or to produce better-quality audio output – and exclude other signals. Still, most receiver circuits contain these basic elements.

Approaching AM Radio

The original method of encoding sound information on radio waves is called *amplitude modulation* (AM), which was developed in the first few decades of the 20th century. AM is a relatively simple way of adding audio information to a carrier wave so that sounds can be transmitted.

One of the straightforward forms of AM modulators simply runs the power supply for an oscillator circuit through an audio transformer that's coupled to a microphone or other sound source. Figure 1-3 shows this arrangement.

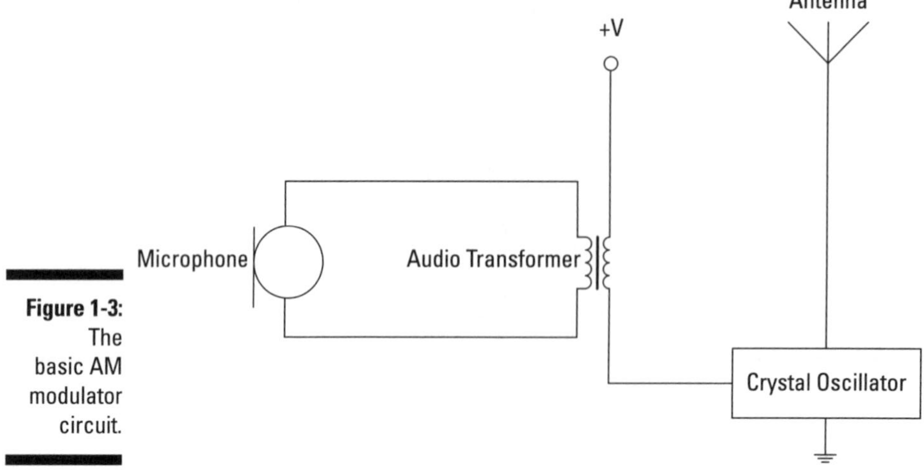

Figure 1-3: The basic AM modulator circuit.

The circuit in Figure 1-3 uses a 1 MHz *crystal oscillator,* which is often used to generate the clock frequencies for microprocessor circuits. 1 MHz is perfect for a simple AM transmitter circuit because 1 MHz falls right in the middle of the band that's used for AM radio transmissions.

Although you can't buy a crystal oscillator very easily at the shops, you can get one on the Internet easily enough. Search for '1 MHz crystal oscillator' and you can find several component distributors to sell you one for a few pounds.

The crystal oscillator is contained in a metal can that has three pins, one each for:

- Ground
- Supply voltage (typically 9 VDC)
- Oscillator output

By running the supply voltage through the secondary coil of a transformer whose primary coil is connected to an audio input source such as a microphone, the actual voltage supplied to the oscillator fluctuates based on the variations in the input signal. Because crystal oscillators are very stable, these voltage variations don't affect the frequency generated by the oscillator, but they do affect the voltage of the oscillator output. Thus, the audio input signal is reflected as voltage changes in the oscillator's output signal.

A better AM modulation circuit uses a transistor as shown in Figure 1-4. In this circuit, the carrier-wave generated by an oscillator that isn't shown in the circuit is applied to the base of a transistor. Then, the audio input is applied to the transistor's emitter through a transformer. The AM signal is taken from the transistor's collector.

This circuit works by the transistor amplifying the input from the oscillator through the emitter-collector circuit. As the audio input varies, however, it induces a small current in the secondary coil of the transformer. This, in turn, affects the amount of current that flows through the collector-emitter circuit. In this way, the intensity of the output varies with the audio input.

Figure 1-5 shows how a carrier wave is combined with an audio signal to produce an AM radio waveform. As you can see, the carrier wave is a constant frequency and amplitude. In other words, each cycle of the sine wave is of the same intensity. The current of the audio wave varies, however. When the modulator circuit combines the two, the result is a signal with a steady frequency, but the intensity of each cycle of the sine wave varies depending on the intensity of the audio signal.

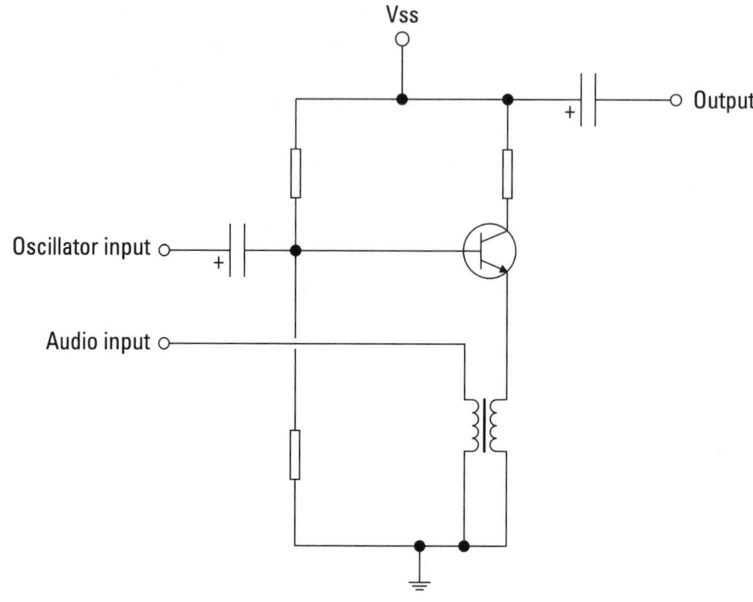

Figure 1-4: Using a transistor for amplitude modulation.

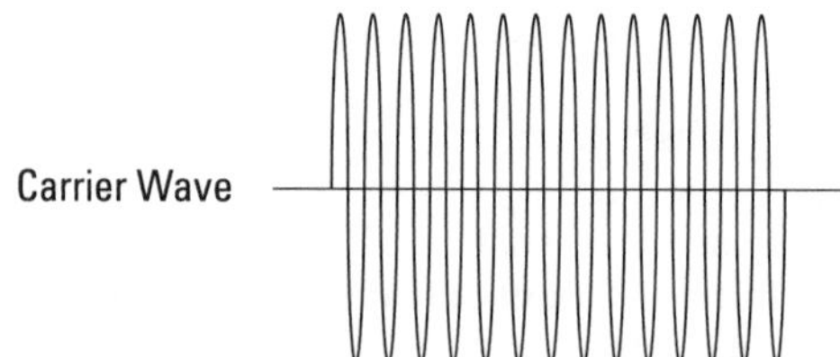

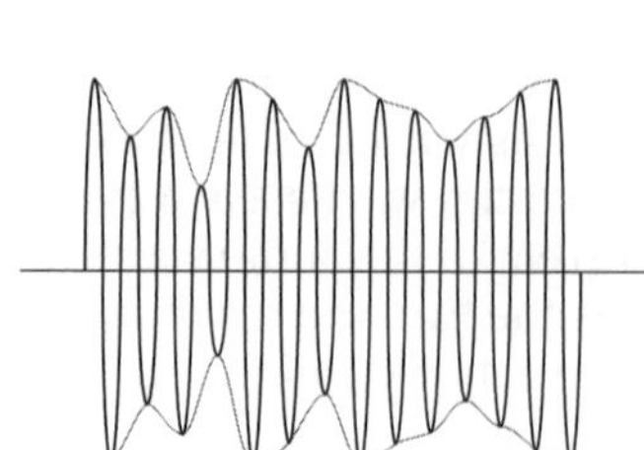

Figure 1-5: How the carrier wave and the audio signal are combined to produce an AM waveform.

Finding out about FM Radio

The AM radio system that we describe in the preceding section is relatively simple but has several weaknesses. The main drawback is that getting an AM radio receiver to distinguish between a signal broadcast by a radio transmitter and spurious signals at the same frequency (generated by other sources) is difficult, if not impossible.

The most obvious example is lightning. When lightning strikes, it generates a brief but powerful burst of electromagnetic radiation with a very large spectrum of frequencies. The noise generated by a lightning strike includes just about the entire range of frequencies used by AM radio. If you're listening to an AM radio station when the lightning strikes, the sudden burst of radio energy on the frequency you're listening to is interpreted as sound. Thus, when lightning strikes, you can hear a crackle on the radio.

Signals that interfere with an intentional broadcast are called *static,* which is the main drawback of AM radio. To counteract it, a better method of superimposing information on a radio wave, called *frequency modulation* (FM), was developed in 1933. (See the sidebar 'The tragic genius behind FM radio' for the fascinating and sad story about the inventor of FM radio.)

In FM, the intensity of the carrier wave isn't varied. Instead, the exact frequency of the carrier wave is varied in sync with the audio signal. When the audio signal is higher, the frequency of the broadcast signal goes up a little. When the audio signal is lower, the frequency slows down a bit.

Figure 1-6 shows how this variation appears in a graph:

- At the top of the figure is the carrier wave that clocks the specific frequency of the broadcast station.
- In the middle is the audio signal that's to be superimposed on the carrier wave.
- At the bottom is the resulting modulated signal.

As you can see from the bunching of the waveforms, the frequency decreases when the input signal gets lower and increases when the input signal is higher.

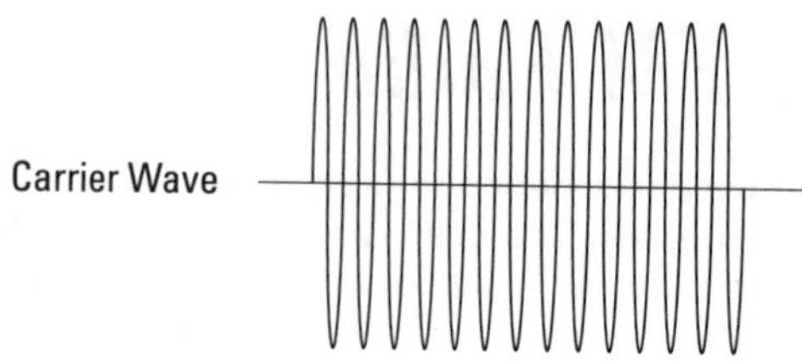

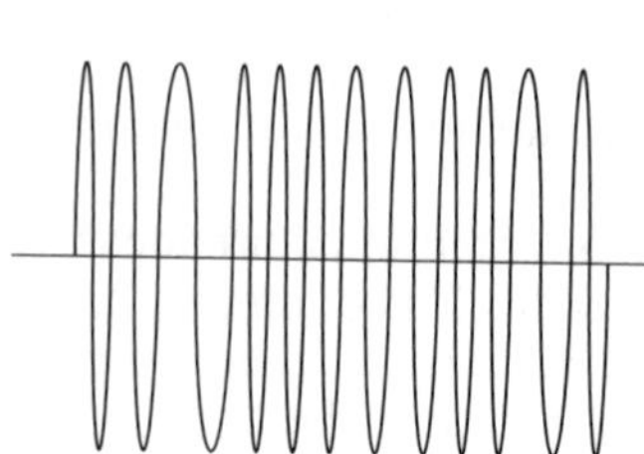

Figure 1-6: How the carrier wave and the audio signal are combined to produce an FM waveform.

The frequency variations in a frequency-modulated signal are all within a small proportion of the carrier-wave frequency. Typically, the frequency stays within 100 kHz of the base frequency.

FM modulators usually employ a type of electronic component called a *varactor,* which is a type of diode that has an unusual characteristic: it has capacitance like a capacitor, and its capacitance increases when voltage is applied across the diode. In essence, a varactor is a voltage-controlled variable capacitor.

The schematic symbol for a varactor, shown in the margin, looks like a cross between a diode and a capacitor.

You can use varactors in oscillator circuits to create an oscillator that vibrates faster when voltage increases. This ability makes them ideal as FM radio modulators. As the voltage of the audio input increases, the capacitance of the varactor increases and thus the frequency of the oscillator increases. When the voltage decreases, the capacitance of the varactor decreases and so does the oscillator's frequency. Figure 1-7 shows a sample of an FM modulator circuit that uses a varactor.

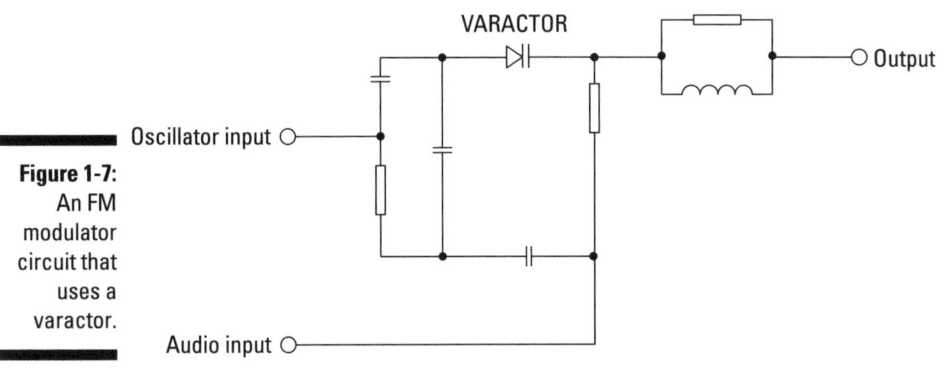

Figure 1-7: An FM modulator circuit that uses a varactor.

The tragic genius behind FM radio

One of the great inventors in the history of radio was a brilliant engineer named Edwin H. Armstrong. Born in 1890, he was fascinated with electrical technology from a very young age. At the age of 14, he started experimenting with wireless radio circuits, building an antenna in his family's backyard that was more than 30 m tall.

He made his first major contribution to radio technology while at Columbia University in 1912. His invention was a circuit that amplified incoming radio signals by feeding them back though the amplifier tube in what came to be called a *regenerative circuit.* This important early breakthrough in radio technology allowed for the first time radio to be heard through a speaker rather than with headphones.

During World War I, Armstrong invented another type of radio receiver, which he called the *superheterodyne circuit.* The basic principal of the superheterodyne circuit is that a radio signal broadcasting at a high frequency – say 1,500 kHz – can be combined with a nearby frequency from an oscillator – say, 1,560 kHz – so that the original signal can also be detected at 60 kHz – the difference between the original signal's frequency (1,560 kHz) and the oscillator's frequency (1,500 kHz). The superheterodyne circuit may be one of the most important electronic circuits ever invented and is still used in nearly all radio receivers to this day.

Armstrong's third great invention was in 1933, when he created a method for transmitting radio signals that wasn't subject to interference from atmospheric disturbances such as lightning. His new system was called *frequency modulation* (FM radio).

Armstrong patented his inventions, but his patents were challenged or ignored by the titans of radio. He lost his lawsuit to protect his patent for his regenerative circuit in 1934 because the justices of the Supreme Court didn't understand how the circuit worked, and the industry challenged his FM radio patents and used his technology freely throughout the 1940s and 1950s.

Finally, in 1954, ill and broke from his legal battles, Armstrong committed suicide by jumping from his high-rise apartment window.

Eventually his widow, Marion, won a series of patent lawsuits and was awarded damages of US$10 million.

Chapter 2

Building a Crystal Radio

In This Chapter

- Understanding how crystal radios work
- Building your own crystal radio
- Listening to stations on your radio

In this chapter, we guide you step by step through how to build one of the simplest of all useful electronic circuits: a crystal radio. On this radio receiver you can pick up AM broadcasts.

Now, this radio is pretty basic: it isn't a particularly good receiver, only one person can listen to it at a time (it uses an earphone instead of a speaker) and its tuning isn't very sensitive. You're lucky if you can receive two or three different stations even if dozens are broadcasting in your area. But this crystal radio is unique in that, unlike every other electronic circuit we describe in this book, it has no obvious source of power: no batteries or other power supply. The only source of power a crystal radio uses is from the radio waves themselves.

If you prefer, you can buy a kit to build your own crystal radio. Amazon sells some pretty good ones priced from $8 to 20. Although you can probably round up the parts separately for less than the cost of a kit, a few of the parts are relatively hard to find otherwise. (One option to consider is buying a kit to obtain these few hard-to-get parts, but build the radio itself using the instructions we provide in this chapter.)

'Here's one I made earlier'

Crystal radios have been around since the very beginning of radio broadcasting. In the 1920s, people often built crystal radio receivers for themselves. Newspapers and magazines published articles telling readers how to construct crystal radios using mostly household items such as scraps of wood and metal and empty oatmeal boxes, plus a few speciality items including a crystal and a telephone headset.

The total cost of a 1920's crystal radio was around £6.50. That sounds cheap, but adjusted for inflation it's more like £80 today. Fortunately, the total cost for a crystal radio today is still around £6.50.

Looking at a Simple Crystal Radio Circuit

Figure 2-1 shows a basic crystal radio receiver circuit. As you can see, this circuit consists of just a few basic components: an antenna and a ground connection, a coil, a variable capacitor, a diode and an earphone.

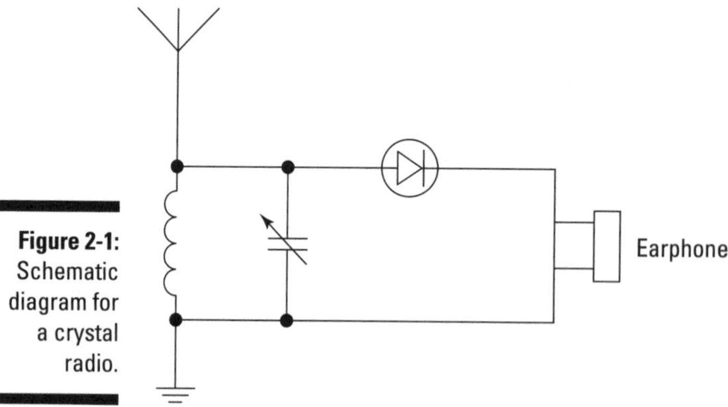

Figure 2-1: Schematic diagram for a crystal radio.

The antenna, of course, captures the radio waves travelling through the air and converts them into alternating current (AC). In order for current to flow, a complete circuit is required. The ground connection is what completes the circuit, allowing current to flow.

The combination of the coil and the capacitor form the tuning circuit. The inductance of the coil combines with the capacitance of the variable capacitor to create a circuit that resonates at a particular frequency, allowing that frequency to pass but blocking others. In a basic crystal radio such as the one shown in Figure 2-1, the tuning circuit isn't very precise. As a result, you may well hear several stations simultaneously. You can, however, build more sensitive tuning circuits that can hone in on individual stations.

The diode forms the detector part of the circuit. It simply converts the amplitude modulated RF signal that comes from the antenna and tuning circuit to an audio frequency AC signal combined with the RF carrier wave. The characteristics of the sensitive piezoelectric earphone eliminate the RF component and any DC offset without the need for a low pass filter or AC decoupling capacitor (see Chapter 1 of this book). The remaining audio frequency AC signal is extremely small, but it's enough for the earphone to convert the current to sound.

So that, in a nutshell, is how a crystal radio works. The rest of this chapter shows you how to build a crystal radio of your own.

Figure 2-2 shows the finished crystal radio that you build. Note that you can construct a crystal radio in many different ways, and so the instructions in this chapter are by no means definitive. Use your imagination when looking for materials to build your radio. You could use an old tin, the case of an old gadget or a really beautiful wooden box, for example.

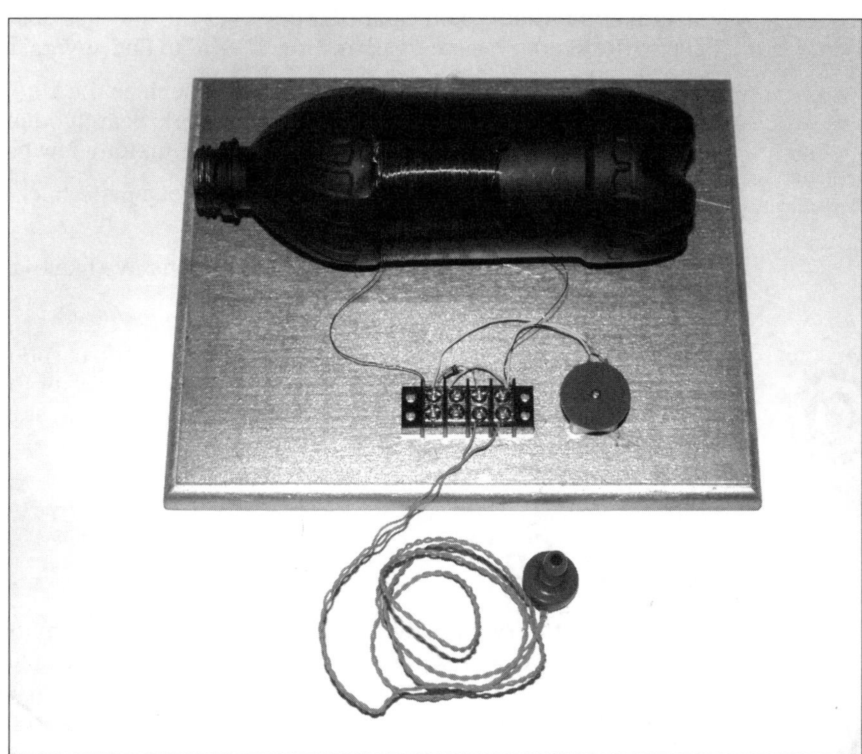

Figure 2-2: A finished crystal radio.

Gathering Together Your Parts

You need a handful of parts to build your crystal radio. The following is a recommended list:

- **At least 15 m of antenna wire:** You can use almost any wire for the antenna. A roll of 18-AWG or 1 mm diameter solid core wire does just fine.
- **A few metres of hook-up wire:** To connect the radio to a ground connection.
- **At least 15 m of 30-AWG or 0.3 mm diameter, enamel-coated transformer wire for the coil:** Plus, don't forget something to wrap the coil on – we use an empty drinks bottle.
- **Variable capacitor (also called a *tuning capacitor*):** These are becoming hard to obtain, though you can purchase them online. Alternatively, you can easily harvest one out of an old radio that no longer works.

  The variable capacitor is an optional component. If you can't find one, you can still build your crystal set; you just can't tune out competing stations.
- **Germanium diode:** You can order one over the Internet. Just use your favourite search engine to search for '1N34A' to find several suppliers.
- **Piezoelectric earphone:** Regular earphones such as the kind you use with an MP3 player or mobile phone don't work. Search online for 'piezoelectric earphone' and you can pick one up for a few pounds.
- **Board to mount the radio on:** About the size of a paperback book (15 by 23 cm) is sufficient.
- **Something to make your electrical connections:** We like to use a four-pole terminal block.

The germanium diode, variable capacitor and piezoelectric earphone are the three parts that can be a bit tricky to track down. You may want to purchase a crystal radio kit from a hobby or school-supply shop and harvest those three parts from the kit.

Creating the Coil

When you look at a crystal radio, the first thing you're likely to notice is the large coil. This coil usually consists of 100 turns or more of small-gauge transformer wire wrapped around a non-conductive tube anywhere from 3 to 13 cm in diameter. The coil is an essential part of the radio's tuning circuit.

You can use many different types of materials to wrap the coil around. Here are a few ideas:

- A small empty drinks bottle.
- An empty toilet-paper roll.
- An empty bottle of contact-lens fluid or another similarly sized plastic bottle.
- A short length of PVC sprinkler pipe.
- A short length of broom handle.
- A short length of a cardboard tube.

In short, you can use any sturdy cylindrical object made of an insulating material as the core of your coil. As long as it's cylindrical and not made of metal, go right ahead.

We use an empty drinks bottle (see Figure 2-3). To make the coil look better, you can spray-paint the bottle with black paint (but don't use metallic paint!).

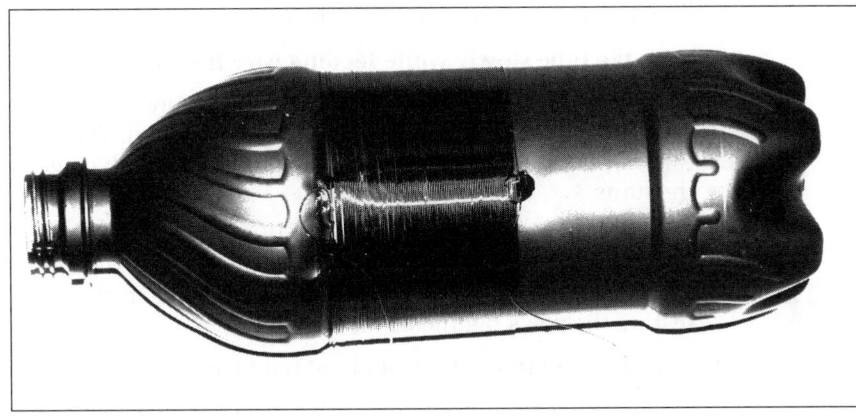

Figure 2-3: A coil wound on an empty drinks bottle.

As for the choice of wire for the coil, the most common is transformer wire, which is coated with thin enamel insulation rather than encased in plastic insulation. Wire wrapped with plastic insulation works, but enamel insulation is thinner and so allows the turns to be spaced closer together.

The number of turns in the coil and the diameter of the cylinder you wrap the coil around determine how much wire you need, but you want to wrap at least 100 turns.

To determine how much wire you need for each turn, multiply the diameter of the cylinder by 3.14. Then, multiply the result by the number of turns to determine how much wire you need.

For example, suppose you're wrapping the coil around a 5 cm wide cylinder and you want to wrap 100 turns. In this case, each turn requires 15.7 cm of wire (5×3.14) and so you need around 16 m of wire (15.7×100). Remember to allow another 50 cm or so of extra wire at each end of the coil to connect the coil to the radio circuit, meaning that you need about 17 m of wire in total for the coil.

Here's the best way we've found of winding the coil:

1. **Place the tube you're winding the coil around on a screwdriver blade or other long narrow object so that the tube spins freely.**

   That way, you can turn the tube and slowly feed wire from its spool onto the tube. This method keeps the wire from becoming twisted as you wind the coil. If you have a vice on your workbench, you can clamp the screwdriver horizontally in the vice and then slide the tube onto the screwdriver so that the tube spins easily.

2. **Attach one end of the transformer wire to one end of the tube.**

   You can do so with a dab of strong glue or you can punch a hole through the tube and feed the wire through. Either way, be sure to leave 15 cm or more of wire free. This gives you plenty of wire to connect the coil to the circuit when you finish winding the coil.

3. **Turn the tube slowly while feeding wire from the spool onto the tube.**

   Each half turn or so, use your fingers to carefully scoot the wire you've just fed up against the turns you've already wound. The goal is for each turn of wire to be adjacent to the previous turn, with no gaps between the turns.

   You need to keep a bit of tension on the wire as you feed it onto the tube in order to keep the windings nice and tight. If you slip and let go of the tension, several turns may unravel and you have to untangle them to restore the coil's tightness.

4. **Wrap the coil in sections of about ten turns each.**

   When you finish each section, dab a little hot glue on it to hold it in place.

5. **Use a little hot glue to secure the last turn of the coil, or cut a slit in the tube and slide the wire through it, when you reach the end of the tube (or run out of wire).**

   Be sure to leave about 15 cm of free wire after the last turn.

The coil needs to have a nice, tight appearance with no major gaps between the turns and you require about 15 cm of free wire on each end of the coil.

Assembling the Circuit

When you've prepared your coil (as we describe in the preceding section), the next step is to assemble the various parts of the radio on your wooden base measuring around 15 by 23 cm. To make your radio look good, consider painting or staining the wood before you assemble the circuit.

Here's a list of the parts you need to assemble the circuit:

- Your coil
- A four-position terminal block
- A germanium diode (1N34A or similar)
- A tuning capacitor
- One length of hook-up wire, approximately 4 cm long
- Two lengths of hook-up wire, approximately 8 cm long

You need the following tools to build the crystal radio circuit:

- Phillips-head screwdriver
- Soldering iron and some solder
- Strong glue, such as a hot glue gun and some glue sticks
- Wire cutters
- Wire strippers

Figure 2-4 shows the layout for the assembled crystal radio circuit.

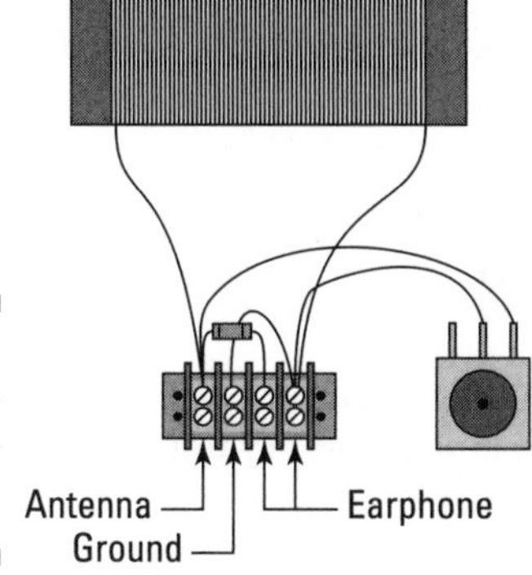

Figure 2-4: Layout for the crystal radio circuit.

We refer to the individual terminals of the terminal block by numbering them from left to right: 1 to 4. Terminal connectors in the top row are given the letter A and those in the bottom row are given the letter B. Thus, the terminal at the top left of the terminal block is terminal 1A, and the terminal at the bottom right is 4B.

To assemble the crystal radio circuit, follow these steps:

1. **Glue the terminal block, tuning capacitor and coil to the board.**

   Use Figure 2-4 to judge the placement of each of these parts. Be sure to give the glue enough time to cool and harden before you continue.

2. **Connect the diode between terminals 1A and 3A.**

   Interestingly enough, the direction in which you connect the diode doesn't matter in a crystal radio circuit.

3. **Strip about 1 cm of insulation from both ends of all three lengths of hook-up wire.**
4. **Connect one end of the 4 cm length of hook-up wire to terminal 2A on the terminal block, and then connect the other end to terminal 4A.**
5. **Use some sandpaper to scrape gently the enamel insulation off the ends of the wire.**
6. **Connect the two wires from the coil to terminals 1A and 4A.**
7. **Solder one end of one of the 8 cm hook-up wires to the centre lead of the capacitor and one end of the other 4 cm wire to either one of the other leads.**

   Which of the two outside leads you use doesn't matter.

8. **Connect the free ends of the wires you soldered in Step 7 to terminals 1A and 1D of the terminal block.**

   You're done!

When the radio circuit is assembled, look it over to make sure that you've connected all the pieces as shown in Figure 2-4.

Stringing up an Antenna

A long antenna is vital to the successful operation of a crystal radio. In general, the longer the antenna, the better. If possible, try to make your antenna at least 15 m.

You can make your antenna from just about any type of wire, insulated or not. A large roll of 18-AWG, solid hook-up wire is perfect.

The best configuration for a crystal radio antenna is to run the wire horizontally between two supports as high off the ground as you can get them, as shown in Figure 2-5. You probably won't find convenient poles as we show in the figure, but if you look around you should be able to locate two suitable points to which you can connect the ends of your antenna. Fence posts, trees, washing lines, a swing or almost any other tall structure does the trick.

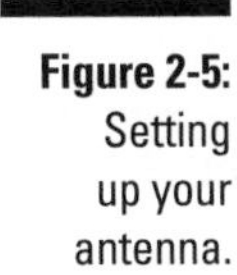

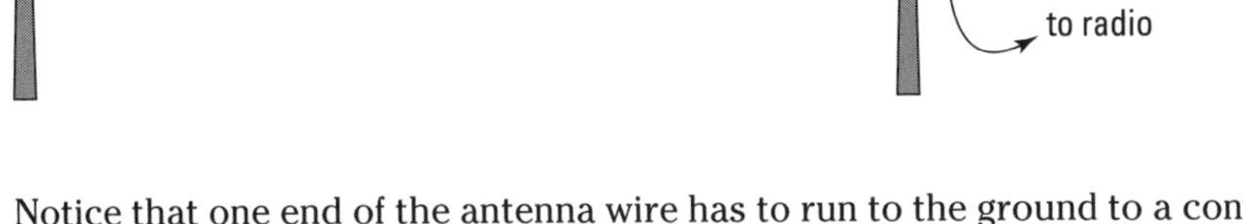

Figure 2-5: Setting up your antenna.

Notice that one end of the antenna wire has to run to the ground to a convenient place for connecting it to your crystal radio. You need to run this wire to the location where you intend to operate your radio.

Ensuring that your antenna is well insulated from the ground is vitally important. Remember that wood isn't a great insulator and most metals, of course, are excellent conductors. Therefore, be very careful about how you support the ends of the antenna wire to make sure that you don't inadvertently ground the antenna.

If you use insulated wire for the antenna, you can secure the ends to wood by using small eye screws available from any hardware store. Screw the eye screw into the wood and then simply tie the end of the antenna wire to it.

If the wire is uninsulated, you need to support it with something that doesn't conduct electricity. We suggest browsing the sprinkler parts department of your local hardware store to find a PVC pipe fitting. You can screw this fitting into wood or use duct tape or zip ties to secure it to metal, and then loop your antenna wire through the fitting and tie it.

A crystal radio is a relatively safe electronics project, but a few dangers are associated with the antenna. Here are some things to be careful of:

- Don't string up your antenna in a thunderstorm! Lightning loves wires, and you don't want to tempt Mother Nature by providing her with a convenient path to discharge her fury through.
- Don't operate your crystal radio in a thunderstorm!
- Don't under any circumstances run your antenna wire anywhere near a power cable or other utility line. That's a sure way to end up in hospital.

Connecting to Ground

An effective ground connection is every bit as important as a good antenna (see the preceding section). The best way to create a good ground connection is to use a metal cold-water pipe. Assuming that you've placed your antenna outdoors, you may be lucky enough to find an outdoor water tap near the end of the antenna. Then, you can connect one end of a length of hook-up wire to the water pipe and the other end to your crystal radio. (Note that this doesn't work with plastic pipe, only metal pipe.)

If you can't find a water pipe, hammer a length of metal bar into the ground. Fifteen centimetres will probably do it but the deeper you go, the better the ground connection.

The easiest way to connect a wire to a water pipe (or whatever you're connecting it to) is to use a pipe clamp, which you can find in the plumbing section of any hardware store. Use some coarse sandpaper to sand the pipe where you attach the clamp to improve the electrical connection, especially if the pipe has been painted or varnished. Strip 3–5 cm of insulation from the end of your ground wire and wrap it around the clamp. Then slip the clamp around the water pipe and tighten it up as shown in Figure 2-6.

Figure 2-6: A good ground connection.

Using Your Crystal Radio

When your crystal radio circuit is built, your antenna is up and your ground wire is connected, you're ready to put your crystal radio to the test. Follow these steps:

1. **Connect the two leads of the piezoelectric earphone to terminals 3B and 4B of the barrier strip.**
2. **Connect the antenna lead to terminal 1B of the terminal block.**
3. **Connect the ground lead to terminal 2B of the terminal block.**
4. **Put the earphone in your ear – you probably hear a radio station immediately.**
5. **Turn the knob on the tuning capacitor to hear other stations.**

The tuner on this crystal radio circuit isn't very sensitive, and so you can probably distinguish only two or three different stations.

If you haven't added a tuning capacitor, you can't tune to a specific station. Instead, you probably hear several stations at once. Even with a tuning capacitor, you may still hear several stations at the same time, because the tuning circuit for a simple crystal radio like this one isn't very discriminating.

Chapter 3

Working with Infrared

In This Chapter

- Transmitting information with infrared light
- Spotting infrared light
- Making infrared light
- Building proximity detector projects

In this chapter, we describe working with circuits that produce and detect the invisible light that's commonly called infrared.

Infrared light has all sorts of applications for wireless communication, the most common of which is the remote control for your television. Other uses for infrared include night-vision goggles and cameras as well as temperature detection.

Have fun!

Introducing Infrared Light

Infrared light is light whose frequency is just below the range of visible red light. Specifically, infrared is light whose frequency falls between 1 and 400 THz (1 THz is 1 trillion cycles per second). The infrared spectrum falls right between microwaves and visible light.

As we discuss in Chapter 1 of this minibook, an inverse relationship exists between *frequency* and *wavelength:* the lower the frequency, the longer the wavelength. If you describe infrared in terms of its wavelength rather than its frequency, infrared waves are longer than the waves of visible light, but shorter than microwaves. The wavelength of infrared light is between 0.75 and 300 microns (a millionth of a metre). Thus, at the very bottom edge of the infrared spectrum, the infrared waves are about one-third of a millimetre long. At the upper end, the waves are about one thousandth of a millimetre long. If the waves get any shorter than that, they become visible light.

Figure 3-1 shows the entire spectrum of electromagnetic radiation, so that you can see where infrared falls within the grand scheme of things radiation-wise.

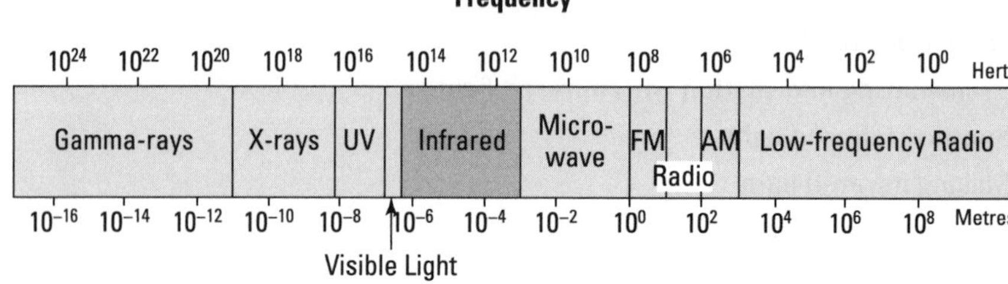

Figure 3-1: Infrared light falls between visible light and microwaves.

Infrared light isn't visible to human eyes. Perhaps Mother Nature thinks that we don't need to look at things in infrared, which is too bad because more than half of all the light energy emitted from the sun is in the form of infrared light. If human eyes could see infrared light as well as the visible light, sunny days would seem twice as bright.

Sometimes infrared light is confused with heat, because humans can't see infrared light waves but can feel them in the form of heat. In other words, infrared light waves heat surfaces that absorb them. Visible light has this effect too, which is why you feel cooler in the shade than in the sun. Because heat is an effect of infrared light, infrared light can be used as a heat source. But infrared light and heat aren't the same thing.

Infrared light is often used to detect objects that humans can't see in visible light. One common application of this ability is night vision. According to a principal of physics called *Planck's law,* all matter emits electromagnetic radiation if its temperature is above absolute zero. Some of that radiation is in the infrared spectrum, and so devices that can detect infrared light can literally see in the dark.

To enhance the effect, some night-vision devices illuminate an area with infrared light. Because the human eye can't see the infrared light, the area illuminated still appears dark, but to a detector sensitive to infrared light, the area is lit up and fully visible.

A remote history

The first wireless remote control was developed by Zenith in 1955. It used ordinary visible light to turn the TV on or off and to change channels, but it had one nasty defect: you had to position your television in the room so that light from an outside source (such as the setting sun shining through a window) didn't hit the light sensor. Otherwise, the TV may shut itself off right in the middle of the evening news when the sun reached just the right angle and hit the sensor.

Today remotes use complicated encoding schemes to avoid such random misfirings. You're probably familiar with the procedure you have to go through when programming a remote control to work with a particular television. This programming is necessary because no widely accepted standard exists for how the codes on a remote control should work, and so each manufacturer uses its own encoding scheme.

Another common application of infrared light is for wireless communications across short distances. The best known infrared devices are television remote controls. These units contain a bright infrared light source and the television itself includes an infrared detector. When you point the remote control at the television and push a button, the remote control turns on the infrared light source and encodes a message on it. The receiver picks up this signal, decodes the message and does whatever the message directs it to do – turns up the volume, changes the channel and so on.

Like visible light, solid objects can block infrared light and it can bounce off reflective objects. That's why the remote doesn't work when your cat is standing between you and the television. But it's also why you can get around your cat by pointing the remote at a window. The infrared waves bounce off the glass and, if the angle is right, arrive at the television.

Detecting Infrared Light

You can detect infrared light in an electronic circuit in several different ways, but the most common is with a device called a *phototransistor.* You can buy a phototransistor for less than a pound at any shop that stocks electronic components.

To understand how a phototransistor works, think about the workings of a transistor. As we describe in Book II, Chapter 6, a transistor has three terminals: the *base, collector* and *emitter.* A path exists within the transistor between the collector and emitter. How well this path conducts depends on whether voltage is applied across the base and the emitter. If voltage is applied, the collector-emitter path conducts well. If no voltage is on the base, the collector-emitter path doesn't conduct.

In a phototransistor, the base isn't a separate terminal connected to a voltage source in your circuit. Instead, the base is exposed to light. When infrared light hits the base, the energy in the light is converted to voltage and the emitter-collector path conducts.

Thus, infrared light hitting the base has the same effect as voltage on the base of a traditional transistor: the infrared light turns the transistor on. The brighter the infrared light, the better the emitter-collector path conducts.

Figure 3-2 shows a simple circuit that uses an infrared phototransistor to detect infrared light. When infrared light is present, the collector-emitter circuit conducts and the LED lights up. Thus, the LED lights when the phototransistor is exposed to infrared light.

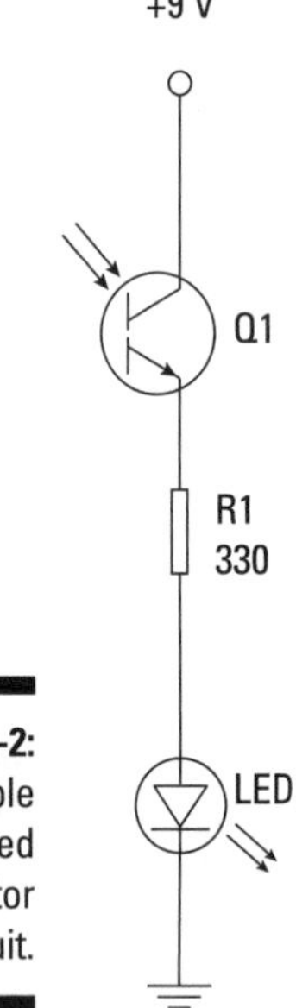

Figure 3-2: A simple infrared detector circuit.

Project 3-1 describes how to build this circuit on a solderless breadboard and Figure 3-3 shows the assembled circuit.

When you've assembled this circuit, try exposing the phototransistor to different light sources to see whether they emit infrared light. One sure source of infrared is your TV remote control. Point the remote at the phototransistor and press any button on it. You see the LED flash on and off quickly as it responds to the infrared signals being sent by the remote.

Another interesting source of infrared is an open flame. Be very careful, of course; we don't want you burning down your house just to see if the flames produce infrared light. If you have a small gas lighter, light it up and hold it near the phototransistor.

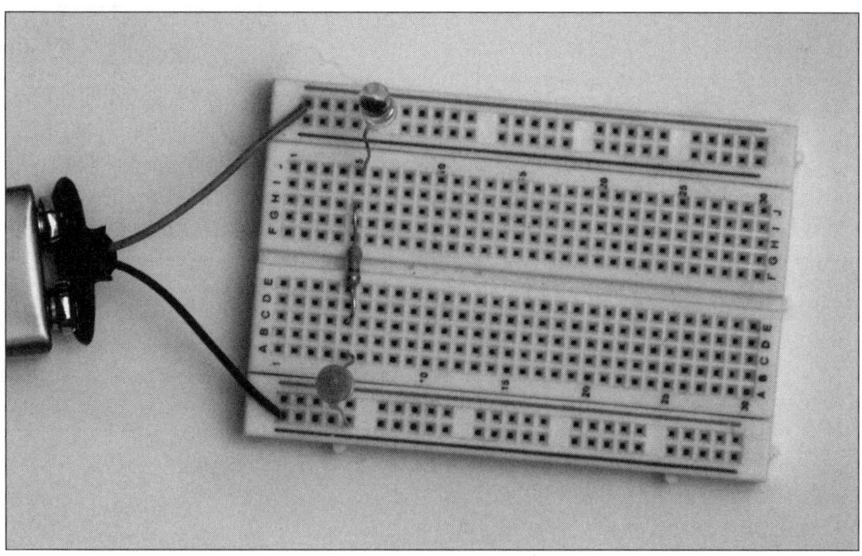

Figure 3-3: The assembled infrared detector circuit (Project 3-1).

Project 3-1: A Simple IR Detector

In this project, you build a simple infrared light detector using a phototransistor. When the phototransistor is exposed to infrared light, the LED lights up.

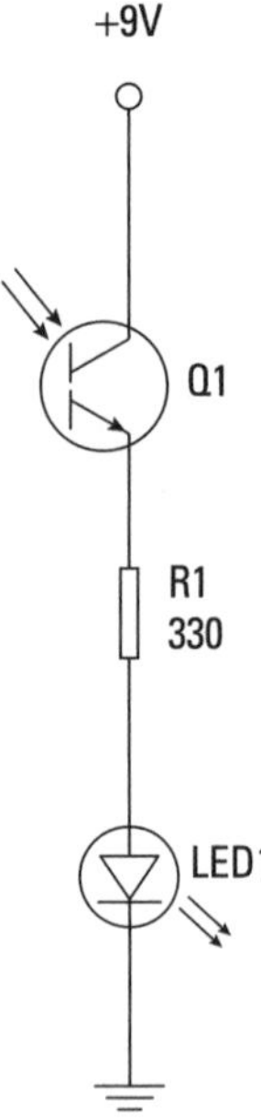

Parts List

1 9 V battery connector
1 9 V battery
1 IR phototransistor
1 Red LED
1 330 Ω ¼ W resistor (orange-orange-brown)

Steps

1. **Insert the photodiode.**

   Collector (short lead): Positive bus
   Emitter (long lead): J5

2. **Insert the resistor.**

   C5 to H5

3. **Insert the LED.**

   Cathode (short lead): Ground bus
   Anode (long lead): A5

4. **Connect the battery.**

   Red lead: Positive bus
   Black lead: Negative bus

5. **Expose the phototransistor to an infrared light source.**

   Try a variety of sources, including a TV remote control, a flame (be careful!) and sunlight. Try other sources that don't emit infrared, such as an LED tourch.

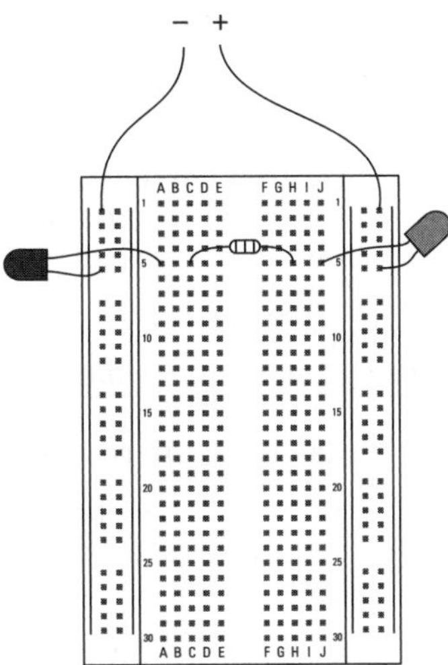

Creating Infrared Light

The easiest way to create infrared light is by using a special light-emitting diode (LED) that operates in the infrared spectrum. Infrared LEDs (often called IR LEDs) are readily available at electronics parts shops.

IR LEDs are similar to regular LEDs, except that you can't see the light they emit. The LED itself is usually a dark purple or blue colour. Like other LEDs, the cathode lead is shorter than the anode lead.

As with any LED, you have to use a resistor in series with an IR LED to prevent excess current from burning out the LED. To calculate the size of the resistor, you need to know three things:

- **Supply voltage:** For example, 9 V.
- **LED forward-voltage drop:** For most IR LEDs, the forward-voltage drop is 1.3 V.
- **Desired current through the LED:** Usually, the current flowing through the IR LED should be kept under 50 mA. However, IR LEDs are typically rated for more current than regular LEDs.

With these three facts to hand, you can calculate the correct resistor size by using Ohm's law (which we describe in Book II, Chapter 2):

1. **Calculate the resistor voltage drop.**

   Subtract the voltage drop of the IR LED (typically 1.3 V) from the total supply voltage. For example, if the total supply voltage is 9 V and the LED drops 1.3 V, the voltage drop for the resistor is 7.7 V.

2. **Convert the desired current to amperes.**

   In Ohm's law, you have to express the current in amperes. You can convert milliamperes to amperes by dividing the milliamperes by 1,000. Thus, if your desired current through the IR LED is 50 mA, you need to use 0.05 A in your Ohm's law calculation.

3. **Divide the resistor voltage drop by the current in amperes.**

   The result is the desired resistance in ohms. For example, if the resistor voltage drop is 7.6 V and the desired current is 50 mA, you need a 152 Ω resistor.

4. **Choose a standard resistor size that's close to the calculated resistance.**

   A 150 Ω resistor is close enough for a 9 V circuit. If you don't have a 150 Ω resistor, a 220 Ω does the job just fine.

When you've chosen the correct resistor size, simply wire it in series with the IR LED, as shown in the schematic in Figure 3-4.

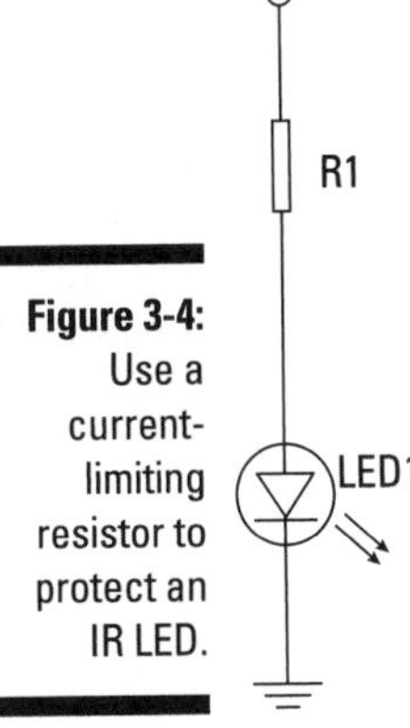

Figure 3-4: Use a current-limiting resistor to protect an IR LED.

Constructing Proximity Detectors

The combination of an IR LED and a photodiode is often used as a *proximity detector,* a gadget that detects when an object is nearby. You can build a proximity detector in two ways:

- **Common-emitter proximity detector:** Mount the IR LED and the phototransistor so that they face each other. The phototransistor detects the infrared light from the IR LED. If an object comes between the IR LED and the phototransistor, the light is blocked and the phototransistor turns off.
- **Common-collector proximity detector:** Mount the IR LED and the IR photodiode next to each other facing the same direction. When an object comes near the IR LED, some infrared light bounces off the object and the phototransistor detects it and turns on.

We show you how to build the two types of proximity detector circuits in the following two sections.

Building a common-emitter proximity detector

Figure 3-5 shows a schematic for a simple proximity circuit. Although it isn't shown in the schematic, this circuit assumes that IR LED and Q1 are oriented so that Q1 can detect the infrared light emitted by IR LED, indirectly (for a proximity detector) or directly (for an interrupter).

This circuit is called a *common-emitter* circuit, because the phototransistor's emitter is usually placed between the phototransistor side of the circuit and the output side of the circuit that's connected to the IR LED. In a common-emitter circuit, the output voltage is on when the phototransistor detects infrared light. Thus, the red LED lights up when the path between the IR LED and phototransistor isn't obstructed. If you block the path between the IR LED and phototransistor, the red LED goes dark.

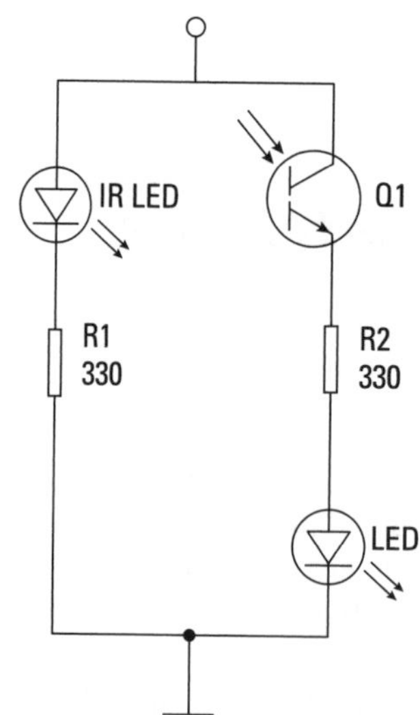

Figure 3-5: A common-emitter proximity detector circuit.

Project 3-2 shows you how to build this circuit configured as an interrupter and Figure 3-6 displays the finished project. When you connect this circuit to power, the red LED comes on. If you pass an object such as a piece of paper between the IR LED and the phototransistor, the red LED goes off.

The output in this circuit is simply a red LED, but you can just as easily connect the output to other circuit components. For example, the output can drive a mechanical relay if you want to use the proximity detector to turn on low voltage lights. Or you can connect the output to a digital logic circuit as we describe in Book VI, Chapter 4.

Figure 3-6: Using an IR LED and a phototransistor as a proximity detector (Project 3-2).

Project 3-2: A Common-Emitter Proximity Detector

In this project, you build a common-emitter proximity detector that lights a red LED whenever the path between an IR LED and a phototransistor is clear. If anything blocks the path, the red LED goes dark.

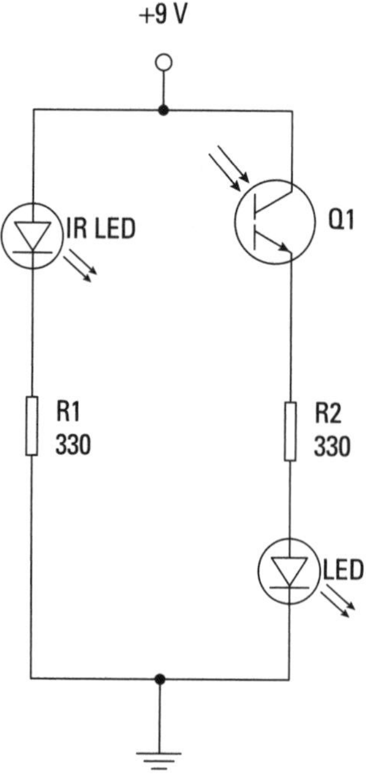

Parts List	
1	9 V battery connector
1	9 V battery
1	IR phototransistor
1	IR LED
1	Red LED
2	330 Ω ¼ W resistor (orange-orange-brown)

Steps

1. **Insert the photodiode.**
   Collector (short lead): Positive bus
   Emitter (long lead): A3
2. **Insert the red LED.**
   Cathode (short lead): Ground bus
   Anode (long lead): J3
3. **Insert the IR LED.**
   Cathode (short lead): A15
   Anode (long lead): Positive bus
4. **Insert the resistors.**
   R1: C3 to H3
   R2: E15 to ground bus
5. **Connect the battery.**
   Red lead: Positive bus
   Black lead: Negative bus
   The red LED illuminates when the power is connected.
6. **Use a piece of paper or other flat object to block the light path between the IR LED and the phototransistor.**
   The red LED goes dark when you interrupt the light path between the IR LED and the phototransistor.

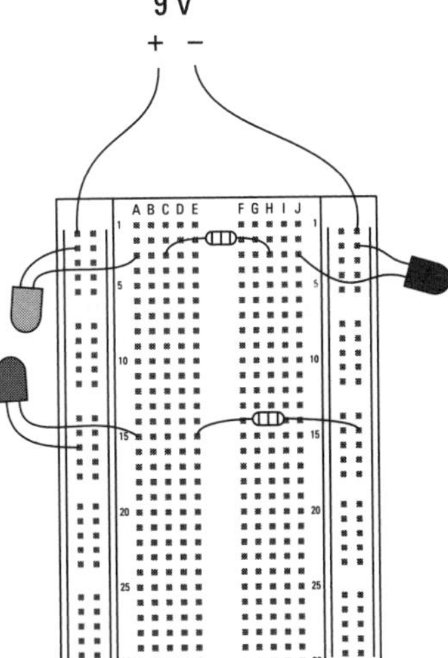

Building a common-collector proximity detector

Figure 3-7 shows a circuit that uses a *common-collector* circuit, in which the collector is the common point between the phototransistor circuit and the LED output circuit. When wired in this way, the LED is dark whenever the path between the IR LED and the phototransistor is clear. When something blocks the path and the phototransistor stops detecting infrared light, the red LED comes on.

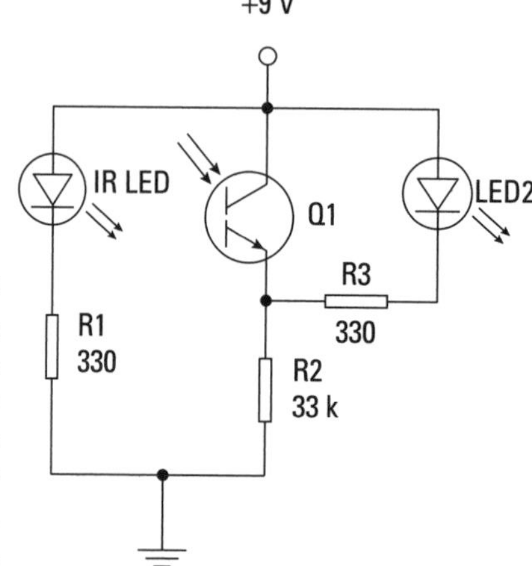

Figure 3-7: A common-collector proximity detector circuit.

Project 3-3 shows how to build this circuit configured as an interrupter and Figure 3-8 shows the finished project. When you connect this circuit to power, the red LED stays dark. But if you pass an object between the IR LED and the phototransistor, the red LED turns on.

Practical applications for this type of circuit include triggers for intruder alarms or automatic opening catflaps, or any circuit that makes something happen in response to an object coming between two points.

Figure 3-8: This circuit lights a red LED when the path between an IR LED and a phototransistor is blocked (Project 3-3).

Project 3-3: A Common-Collector Proximity Detector

In this project, you build a common-collector proximity detector that lights a red LED whenever something comes between an infrared LED and a phototransistor.

Parts List
1 9 V battery connector
1 9 V battery
1 IR phototransistor
1 IR LED
1 Red LED
2 330 Ω ¼ W resistor (orange-orange-brown)
1 33 k Ω ¼ W resistor (orange-orange-orange)
2 Short lengths of jumper wire

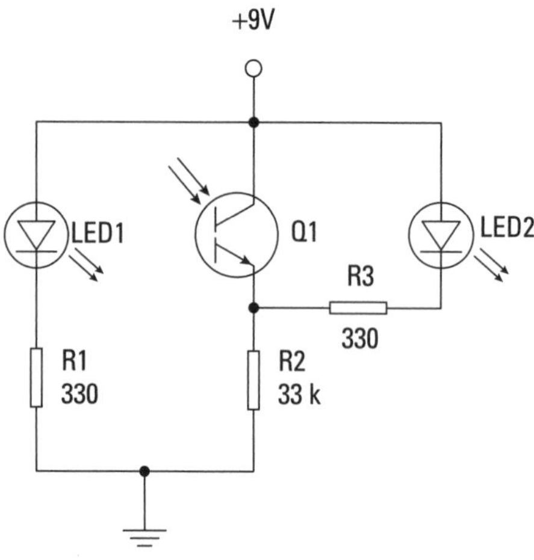

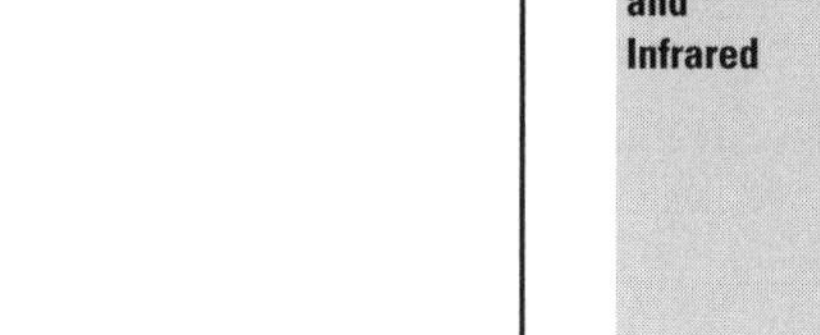

Steps

1. **Insert the red LED.**

   Anode (long lead): D5
   Cathode (short lead): F5

2. **Insert the photodiode.**

   Collector (short lead): E10
   Emitter (long lead): F10

3. **Insert the IR LED.**

   Anode (long lead): E22
   Cathode (short lead): F22

4. **Insert the resistors.**

   R1 (330 Ω): J22 to ground bus
   R2 (33 k Ω): J10 to ground bus
   R3 (330 Ω): H10 to H5

5. **Insert the jumper wires.**

   A5 to positive bus
   A10 to positive bus
   A22 to positive bus

6. **Connect the battery.**

   Red lead: Positive bus
   Black lead: Negative bus

7. **Use a piece of paper or other flat object to block the light path between the IR LED and the phototransistor.**

   The red LED lights up when you interrupt the light path between the IR LED and the phototransistor. When you remove the obstacle, the red LED goes dark.

8. **If the LED does not light, try changing R2 to 330 Ω to allow a larger current to pass through the LED.**

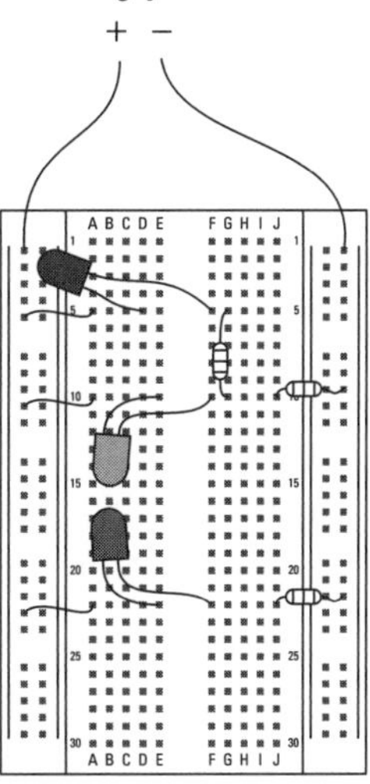

Book VI

Doing Digital Electronics

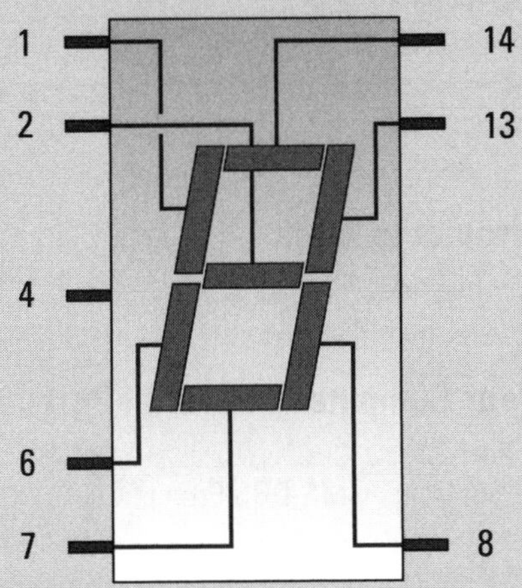

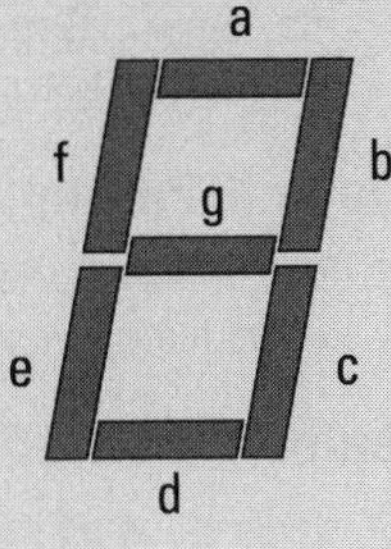

To save time typing and checking them, you can download two of the scripts from Chapter 4 of this book from `www.dummies.com/go/electronicsaiouk` for free.

Contents at a Glance

Chapter 1

Understanding Digital Electronics

In This Chapter

- Comparing analogue and digital circuits
- Counting in binary code
- Building simple logic circuits with switches and lamps

Welcome to the world of digital electronics. In this chapter, you discover the fundamental building-block circuits that you can use ultimately to create computers and other advanced electronic devices. This material sets you up nicely for the other chapters in this minibook, where you explore the amazing world of digital electronics.

This area of electronics is a complex, almost-overwhelming topic, but fortunately the essentials are fairly simple. We introduce you to the basic principles, such as the differences between analogue and digital circuits and how the binary system and basic logic operations work.

Hang on in there! Digital electronics really isn't as hard as you may expect.

Distinguishing Analogue and Digital Electronics

Electronics can be divided into two broad categories:

- **Analogue electronics:** Refers to circuits in which quantities such as voltage or current vary at a continuous rate. When you turn the dial of a potentiometer, for instance, you change the resistance by a continuously varying rate. The resistance of the potentiometer can be any value between the minimum and maximum allowed by the pot. We talk more about potentiometers in Book II, Chapter 6.

If you create a voltage divider by placing a fixed resistor in series with a potentiometer, the voltage at the point between the fixed resistor and the potentiometer increases or decreases smoothly as you turn the knob on the potentiometer.

- **Digital electronics:** Quantities are counted rather than measured and they vary in discrete steps. The term digital is a reference to your digits. The most natural way for humans to count is with their fingers.

A good everyday example of this difference is in clock faces. Figure 1-1 shows two clocks: one analogue and the other digital. On the analogue clock, the time is shown by hands that spin around a dial. The hands move continuously and therefore show an infinite number of positions. On a digital clock, a numeric display indicates the time in numbers. It counts the time in seconds and minutes, for example.

Figure 1-1: An analogue clock (left) and a digital clock (right).

Another example is the thermometer. A traditional glass–mercury thermometer has a small amount of liquid mercury inside a glass column. Mercury expands when it gets warm, and so the warmer the mercury, the higher it climbs in the glass column. Little tick marks are printed on the column to help you read the temperature. On a digital thermometer, the temperature is indicated by a numeric display.

Understanding Binary

Most digital electronic circuits work with the binary number system. Therefore, to comprehend the details of how digital circuits work you need to understand the workings of the binary numbering system.

Saying that a system is digital isn't the same as saying that it's binary. *Binary* is a particular type of digital system in which the counting is all done with the binary number system, using only the numerals 0 and 1. Nearly all digital systems are also binary systems, but the two terms aren't interchangeable.

Many systems are a combination of binary and analogue systems. In a system that combines binary and analogue values, special circuitry is required to convert from analogue to digital or vice versa. An input voltage (analogue) may be converted to a sequence of pulses, one for each volt. The pulses can then be counted to determine the voltage.

Knowing your number systems

A *number system* is simply a way of representing numeric values. Number systems use symbols called *numerals* to represent numeric quantities. The numerals 1, 2 and 3 represent the numeric quantities commonly known as one, two and three.

In most number systems, numerals can be strung together to create larger numeric values, and the position of each numeral in the string determines its relative value. For example, in the number 12, numeral 1 represents the quantity ten and numeral 2 represents the quantity two. In the number 238, numeral 2 represents the quantity two hundred, numeral 3 represents the quantity thirty and numeral 8 represents the quantity eight.

This system is so familiar that you can easily miss its brilliance and overlook the fact that it's completely arbitrary.

People have ten numerals in the everyday counting system – called the *decimal system* or *base 10* – probably as a result of the fact that humans have ten fingers. If humans had evolved with 12 fingers, you may well have learned how to count in base 12 and be using two additional invented numerals.

Different number bases may seem strange, but you encounter them every day without thinking about it. One common example is the system for keeping time, which works in several different number bases. An hour contains 60 minutes, for example, and you recognise straight away that 1:30 is halfway between 1:00 and 2:00. You may not realise it, but you're thinking in base 60 when you tell the time.

Counting by ones

Binary is one of the simplest of number systems because it has only two numerals: 0 and 1. In the decimal system, you use ten numerals: 0 through 9. In an ordinary decimal number, such as 3,482, the rightmost digit represents ones, the next digit to the left tens, the next hundreds, the next thousands and so on. These digits represent powers of ten: first 10^0 (which is 1); next 10^1 (10); then 10^2 (100); then 10^3 (1,000) and so on.

In binary, you have only two numerals rather than ten, which is why binary numbers look somewhat monotonous, as in 110011, 101111 and 100001.

The positions in a binary number (called *bits* rather than *digits*) represent powers of two rather than powers of ten: 1, 2, 4, 8, 16, 32 and so on. To figure out the decimal value of a binary number, you multiply each bit by its corresponding power of two and then add the results. You calculate the decimal value of binary 10111, for example, as follows:

$$\begin{aligned} 1\times 2^0 &= 1\times 1 = 1 \\ +1\times 2^1 &= 1\times 2 = 2 \\ +1\times 2^2 &= 1\times 4 = 4 \\ +0\times 2^3 &= 0\times 8 = 0 \\ +1\times 2^4 &= 1\times 16 = \underline{16} \\ & \qquad\qquad\quad 23 \end{aligned}$$

Fortunately, converting a number between binary and decimal is something that a computer is good at – so good, in fact, that you're unlikely ever to need to do any conversions yourself. The point of understanding binary isn't to be able to look at a number such as 1110110110110 and say instantly, 'Ah! Decimal 7,606!' (If you can do that, Hollywood may make a film about you with Dustin Hoffman in the lead role.)

Instead, the point is to have a basic understanding of how computers store information and – most importantly – of how the binary counting system works.

If you do find that you need to convert binary numbers to decimal, or vice versa, and you have access to a computer, you can use the Calculator program that comes free with Windows to do the conversion for you. For more information, see the later sidebar 'Using Windows Calculator for binary conversions'.

Here are some interesting characteristics of binary, which explain how the system is similar to and different from the decimal system:

- **In decimal, the number of decimal places allotted for a number determines how large the number can be.** If you allot six digits, for example, the largest number possible is 999,999. Because 0 is itself a number, however, a 6-digit number can have any of 1 million different values.

  Similarly, the number of bits allotted for a binary number determines how large that number can be. If you allot 8 bits, the largest value that number can store is 11111111, which happens to be 255 in decimal. Thus, a binary number that's 8 bits long can have any of 256 different values (including 0).

- **To work out quickly how many different values you can store in a binary number of a given length, use the number of bits as an exponent of two.** An 8-bit binary number, for example, can hold 2^8 values. Because 2^8 is 256, an 8-bit number can have any of 256 different values. This is why a *byte* – 8 bits – can have 256 different values.
- **The 'powers of two' thing is why digital systems don't use nice, even, round numbers for measuring such values as memory capacity**. A value of 1 k, for example, isn't an even 1,000 bytes: it's 1,024 bytes because 1,024 is 2^{10}. Similarly, 1 MB isn't an even 1,000,000 bytes, but 1,048,576 bytes, which happens to be 2^{20}.

All good geeks know their powers of two because they play such an important role in binary numbers. Just for the fun of it, but not because you need to know, Table 1-1 lists the powers of two up to 32. The table also shows the common shorthand notation for various powers of two. The abbreviation *k* represents 2^{10} (1,024). The *M* in *MB* stands for 2^{20}, or 1,024 k, and the *G* in *GB* represents 2^{30}, which is 1,024 MB.

Table 1-1 Powers of Two

Power	*Bytes*	*Kilobytes*	*Power*	*Bytes*	*k, MB or GB*
2^1	2		2^{17}	131,072	128 k
2^2	4		2^{18}	262,144	256 k
2^3	8		2^{19}	524,288	512 k
2^4	16		2^{20}	1,048,576	1 MB
2^5	32		2^{21}	2,097,152	2 MB
2^6	64		2^{22}	4,194,304	4 MB
2^7	128		2^{23}	8,388,608	8 MB
2^8	256		2^{24}	16,777,216	16 MB
2^9	512		2^{25}	33,554,432	32 MB
2^{10}	1,024	1 k	2^{26}	67,108,864	64 MB
2^{11}	2,048	2 k	2^{27}	134,217,728	128 MB
2^{12}	4,096	4 k	2^{28}	268,435,456	256 MB
2^{13}	8,192	8 k	2^{29}	536,870,912	512 MB
2^{14}	16,384	16 k	2^{30}	1,073,741,824	1 GB
2^{15}	32,768	32 k	2^{31}	2,147,483,648	2 GB
2^{16}	65,536	64 k	2^{32}	4,294,967,296	4 GB

A saying among computer programmers goes as follows: there are ten types of people in this world – those who know binary and those who don't. If you understand that, you're already thinking in binary!

Doing the logic thing

One of the great things about binary is that it's very efficient at handling special operations called *logical operations,* which compare two binary bits and render a third binary bit as a result. Logic circuits can use combinations of these logical operations to do very complex tasks.

In total, 16 possible logical operations exist and you can discover all 16 in Book VI, Chapter 2. Here, however, we introduce you to three basic logical operations:

- **AND:** An AND operation compares two binary values. If both values are 1, the result of the AND operation is 1. If one value is 0 or both of the values are 0, the result is 0.
- **OR:** An OR operation compares two binary values. If at least one of the values is 1, the result of the OR operation is 1. If both values are 0, the result is 0.
- **XOR:** An XOR operation compares two binary values. If exactly one of them is 1, the result is 1. If both values are 0 or if both values are 1, the result is 0.

Table 1-2 summarises how AND, OR and XOR work.

Table 1-2 Logical Operations for Binary Values

First Value	*Second Value*	*AND*	*OR*	*XOR*
0	0	0	0	0
0	1	0	1	1
1	0	0	1	1
1	1	1	1	0

You can apply logical operations to binary numbers that have more than one binary digit by applying the operation one bit at a time. The easiest way to do this manually is to line the two binary numbers on top of one another and then write the result of the operation below each binary digit. The following example shows how to calculate 10010100 AND 11011101:

```
    10010100
AND 11011101
    10010100
```

As you can see, the result is 10010100.

Using Windows Calculator for binary conversions

If you have a computer, you can use the free Calculator program that comes with all versions of Windows to work with binary numbers. The Calculator program has a special Programmer mode that many users don't know about. When you flip the Calculator into this mode, you can do instant binary and decimal conversions, which occasionally come in handy when you're working with IP addresses.

To use the Windows Calculator in Programmer mode, launch the Calculator by choosing Start⇨All Programs⇨Accessories⇨Calculator. Then choose View⇨Programmer from the Calculator menu. The Calculator changes to a fancy programmer model – the kind that was very expensive indeed 30 years ago. All kinds of buttons appear:

You can select the Hex, Dec, Oct and Bin radio buttons to switch from decimal to the different numbering systems commonly used in digital electronics: hexadecimal, octal and binary. For example, to find the binary equivalent of decimal 155, type **155** and then select the Bin radio button. The value in the display changes to 10011011.

Here are a few other interesting things about the Programmer mode:

- The Programmer Calculator has several features that are designed specifically for binary calculations, such as AND, XOR, NOT and NOR.
- The Programmer Calculator can also handle hexadecimal conversions. Hexadecimal doesn't come into play when you're dealing with IP addresses, but it's used for other types of binary numbers, and so this feature sometimes proves useful.
- The calculator in earlier versions of Windows (prior to Windows 7) doesn't have Programmer mode. However, the older versions do have a Scientific mode, which includes features for working with binary numbers.

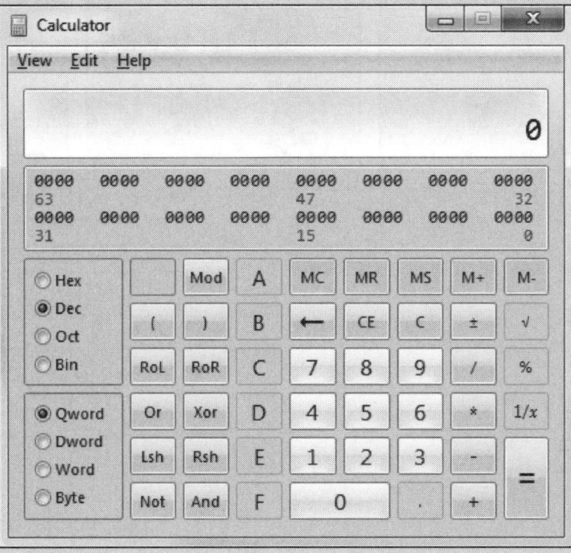

Using Switches to Build Gates

In Book II, Chapter 6, you read about gates in relation to transistors. But to give you a good idea of how basic gates work in digital circuits, we provide Projects 1-1, 1-2 and 1-3, which show you how to assemble an AND gate, an OR gate and an XOR gate, respectively (see the preceding section for descriptions of these three logical operations). Each gate uses simple DPDT (double pole, double throw) knife switches (for descriptions of DPDT and knife switches, turn to Book II, Chapter 1). Figure 1-2 shows these three projects fully assembled:

- **AND gate circuit (Project 1-1):** On the left of the figure. As you can see, the AND gate is implemented by connecting two switches together in series. Both switches need to be closed for the lamp to light.
- **OR gate circuit (Project 1-2):** In the middle of the figure. To implement an OR gate with switches, you simply wire the switches in parallel. The lamp lights if either of the switches is closed.
- **XOR gate circuit (Project 1-3):** On the right of the figure. This wiring is a bit trickier. The DPDT switches are cross-connected in such a way that the circuit to the lamp is completed if one of the switches is in the A position and the other is in the B position. If both switches are in the same position, the circuit isn't complete and so the lamp doesn't light.

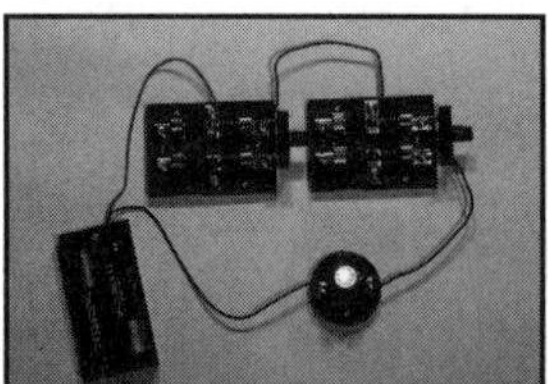

AND gate

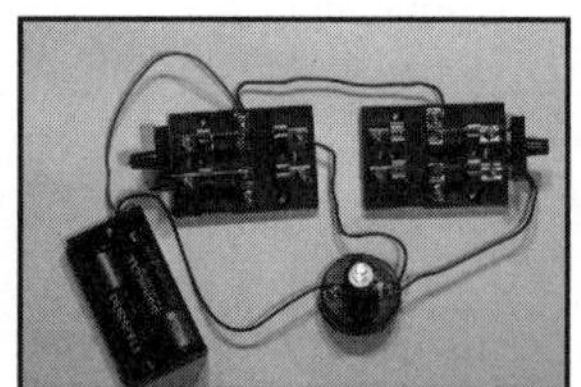

OR gate

XOR gate

Figure 1-2: Implementing AND, OR and XOR gates with knife switches.

Project 1-1: Building a Simple AND Circuit

In this project you use two switches and a lamp to create a simple circuit that performs a logical AND operation.

The two switches represent the two binary values that are input to the AND operation. A closed switch represents the binary value 1 and an open switch represents the binary value 0.

The lamp represents the binary output of the AND operation. When lit, the lamp represents the binary value 1. When it's not lit, the lamp represents the binary value 0.

Parts List

- 2 AA batteries
- 1 Battery holder
- 1 Lamp holder
- 1 3 V lamp
- 2 DPDT knife switches
- 2 12-13 cm lengths of 22-AWG stranded wire, stripped 1 cm on each end
- 1 Small Phillips-head screwdriver
- 1 Wire cutter
- 1 Wire stripper

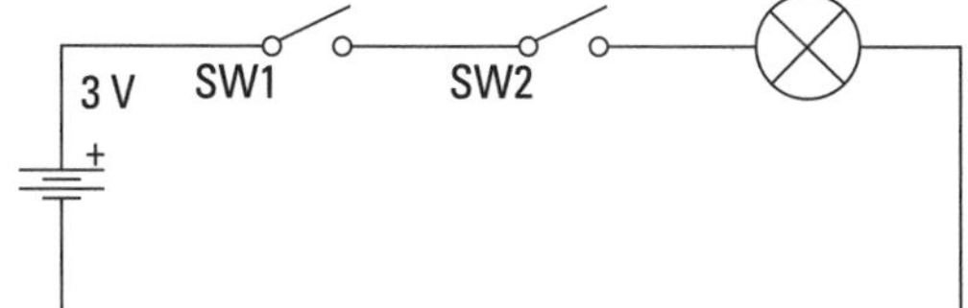

Steps

1. **Open both switches.**

   Move the handles to the upright position so that none of the contacts is connected.
2. **Attach the red lead from the battery holder to terminal 1X of one of the switches.**
3. **Attach the black lead from the battery holder to one of the terminals on the lamp holder.**
4. **Connect the two 12–13 cm wires as follows:**

Connect From	*Connect To*
Terminal 1A of the first switch	Terminal 1X of the second switch
Terminal 1A of the second switch	The unused terminal of the lamp holder
5. **Insert the batteries into the holder.**
6. **Manipulate the switches to verify the correct operation of the AND circuit.**

   The switches should operate as follows:

Switch 1	Switch 2	Lamp
Open	Open	Off
Open	Closed	Off
Closed	Open	Off
Closed	Closed	On

   Notice that the lamp comes on only when both switches are closed, which is exactly how an AND circuit should work.
7. **You're done!**

   Give yourself a pat on the back, because you've just built your first logic circuit.

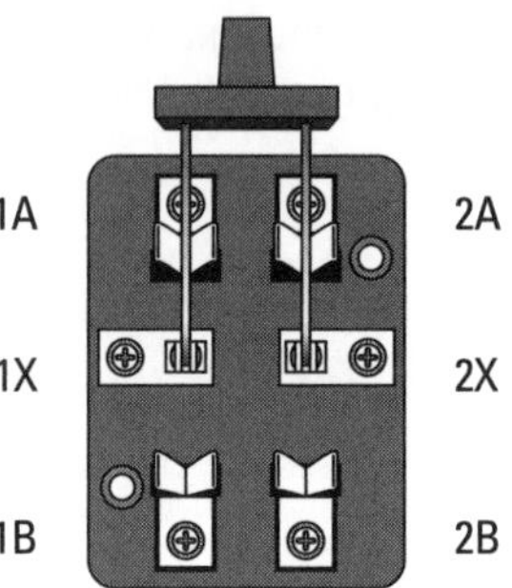

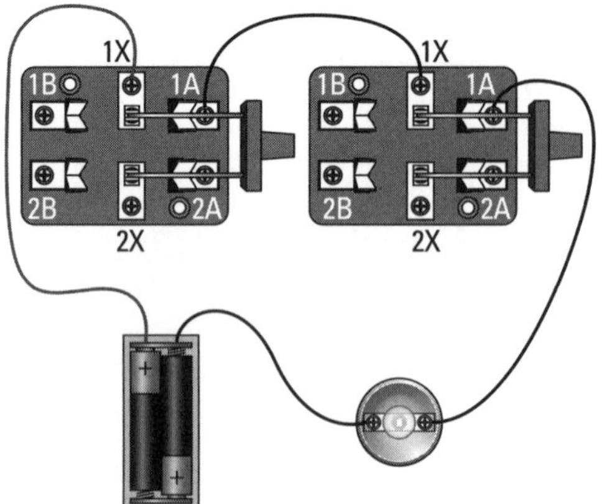

Project 1-2: Building a Simple OR Circuit

In this project, you build a simple OR circuit by wiring two switches in parallel to control a lamp. In this circuit, a lamp is lit when either of the two switches is closed. The switches represent the two binary inputs to the OR operation and the lamp represents the binary output.

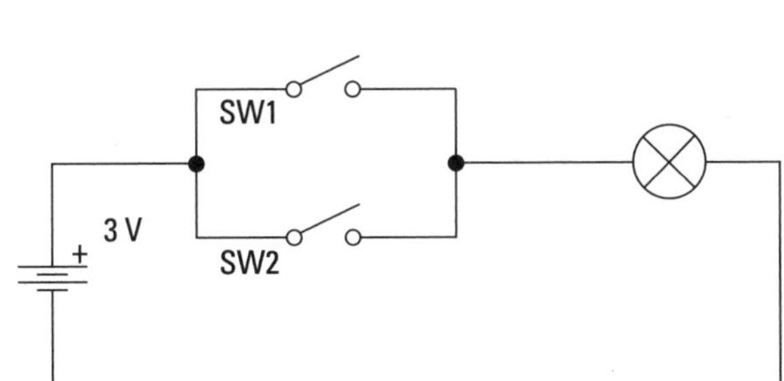

Parts List

- 2 AA batteries
- 1 Battery holder
- 1 Lamp holder
- 1 3 V lamp
- 2 DPDT knife switches
- 3 12–13 cm lengths of 22-AWG stranded wire
- 1 Small Phillips-head screwdriver
- 1 Wire cutter
- 1 Wire stripper

Steps

1. **Open both switches.**

   Move the handles to the upright position so that none of the contacts is connected.

2. **Attach the red lead from the battery holder to terminal 1X of the first switch.**
3. **Attach the black lead from the battery holder to the first terminal on the lamp holder.**
4. **Connect the first jumper wire from terminal 1A of the first switch to terminal 1X of the second switch.**
5. **Connect the second jumper wire from terminal 1A of the first switch to the second terminal on the lamp holder.**
6. **Connect the third jumper wire from terminal 1A of the second switch to the second terminal on the lamp holder.**
7. **Insert the batteries into the holder.**
8. **Manipulate the switches to verify the correct operation of the AND circuit.**

   The switches should operate as follows:

Switch 1	*Switch 2*	*Lamp*
Open	Open	Off
Closed	Closed	On
Closed	Open	On
Closed	Closed	On

   Notice that the lamp is on whenever at least one of the two switches is closed. The lamp is off only when both switches are open. That's the way an OR circuit should operate.

9. **You're done!**

   Congratulate yourself on a job well done. You've completed your second logic circuit.

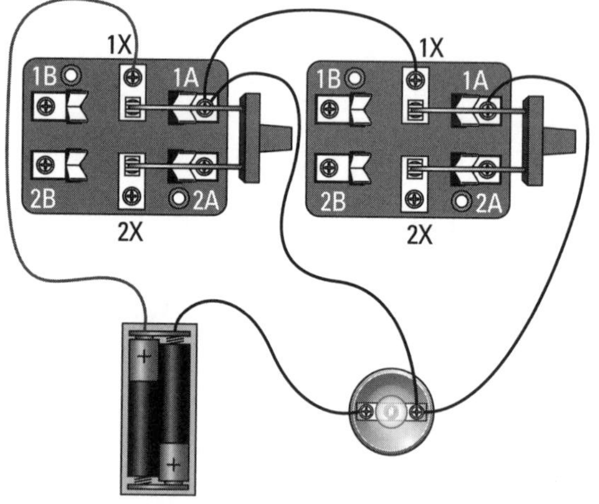

Project 1-3: Building a Simple XOR Circuit

In this project, you build a simple XOR circuit by wiring two switches to control a lamp. In this circuit, a lamp lights when one or the other switch is closed. If both switches are open or both are closed, the lamp doesn't light. The switches represent the two binary inputs to the XOR operation, and the lamp represents the binary output.

Unlike the other two projects in this chapter, the switches in this project use the A and B positions. Thus, a switch represents a binary 1 when it's in the A position. In the B position, the switch represents a binary 0.

Notice how in the schematic diagram the A and B terminals are cross-connected between the two switches. That way, the circuit to the lamp is closed only if the switches are in opposite positions. In other words, the lamp lights if SW1 is in the A position and SW2 is in the B position, or if SW1 is in the B position and SW2 is in the A position. If both switches are in the A position or if both switches are in the B position, the lamp doesn't light.

Parts List

- 2 AA batteries
- 1 Battery holder
- 1 Lamp holder
- 1 3 V lamp
- 2 DPDT knife switches
- 3 12–13 cm lengths of 22-AWG stranded wire
- 1 Small Phillips-head screwdriver
- 1 Wire cutter
- 1 Wire stripper

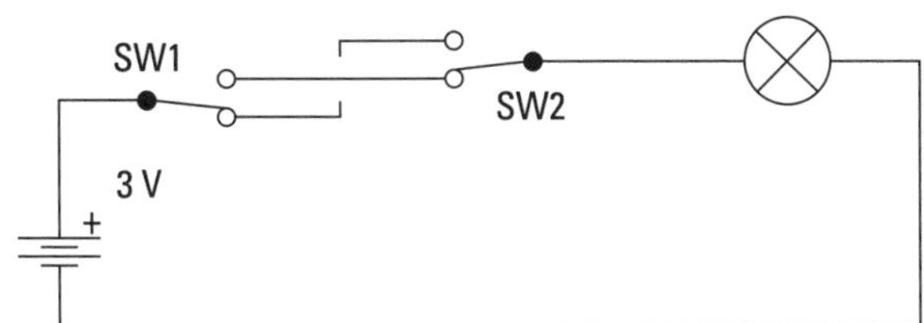

Steps

1. **Open both switches.**

   Move the handles to the upright position so that none of the contacts is connected.
2. **Attach the black lead from the battery holder to terminal 2X of the first switch.**
3. **Attach the red lead from the battery holder to the first terminal on the lamp holder.**
4. **Connect the first jumper wire from the second terminal on the lamp holder to terminal 1X on the second switch.**
5. **Connect the second jumper wire from terminal 2A on the first switch to terminal 1B on the second switch.**
6. **Connect the third jumper wire from terminal 2B on the first switch to terminal 1A on the second switch.**
7. **Insert the batteries into the holder.**
8. **Manipulate the switches to verify the correct operation of the XOR circuit.**

   The switches should operate as follows:

Switch 1	*Switch 2*	*Lamp*
Open	Open	Off
Open	Closed	On
Closed	Open	On
Closed	Closed	Off
9. **You're done!**

   Congratulate yourself. You've completed your third logic circuit!

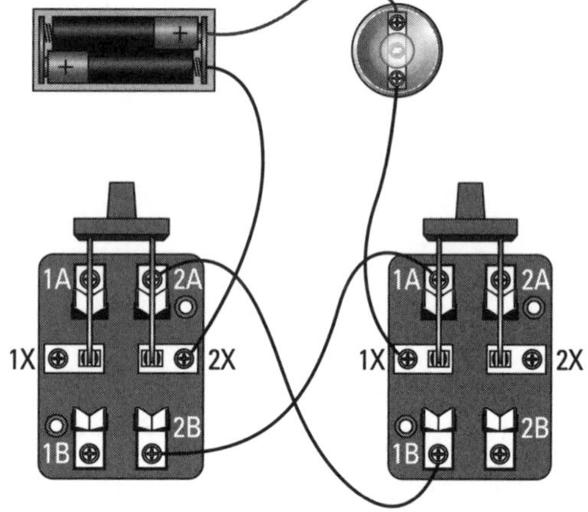

Chapter 2

Getting Logical

In This Chapter

- Discovering Boolean logic
- Examining the different types of logic gates
- Building logic circuits with only NAND or NOR gates
- Stimulating logic gates with software

Mr Spock doesn't just stand out in *Star Trek* because of his pointy ears. His uniqueness is also due to what's between them: his brain. His part Vulcan ancestry makes him colder and less emotional than the rest of the Starship Enterprise crew. Spock thinks like a computer, because they both employ a lot of logic.

In this chapter, we look at the basic principles of logic, which are the underpinnings of digital electronics. In particular, we explore *logic gates,* which (rather than being Bill's daughter!) are handy devices that perform a logical operation on one or more binary inputs and produce a single binary output result (we describe binary in this minibook's Chapter 1).

The overwhelming complexity of modern computer systems is built on the simple concept of logic gates. A modern computer processor consists of millions of individual logic gates connected in a way that enables the processor to perform complicated operations at amazing speed.

In this chapter we ask you to think a little like Mr Spock, as we investigate Boolean logic and how you can employ logic gates to great effect. Are you ready to boldly go (split infinitive and all)?

Introducing Boolean Logic and Logic Gates

In digital electronics, *Boolean logic* refers to the manipulation of binary values in which a 1 represents the concept of *true* and a 0 represents the concept of *false.* In electronic circuits that implement logic, binary values are represented by voltage levels. In the most common convention, a binary value of 1 is represented by +5 V (also called *high*) and a binary 0 is represented by 0 V (also called *low*).

Logical operations (also called *logical functions*) are functions that can be applied to one or more logic inputs and produce a single logic output. One of the most common types of logic operations is NOT, which simply inverts the state of its input. In other words, with the NOT operation, if the input is true, the output is false; if the input is false, the output is true.

A *gate* is a circuit or device that implements a logical function. Thus, a NOT gate is a circuit or device that implements the logical NOT operation. NOT gates are very common in digital circuits.

You can create gate switches in a variety of ways. The most common method uses transistors as switches, arranged in such a way that the correct output is generated based on the logical inputs and the type of gate being implemented. In Chapter 3 of this minibook, you see how gates are implemented, but in this chapter we discuss what the various types of gates do and how you can combine them in real-world circuits.

Regardless of the method used to create gate circuits, all logic circuits depend on different voltage ranges to represent 1 (by approximately +5 V, or high) and 0 (by approximately 0 V, or low).

George Boole becomes a father

This type of logic is called *Boolean* because it was invented in the 19th century by George Boole, an English mathematician and philosopher. In 1854, he published a book entitled *An Investigation of the Laws of Thought,* which laid out the initial concepts that eventually came to be known as Boolean algebra, also called Boolean logic.

Boolean logic is among the most important principles of modern computers. Thus, most people consider Boole to be the father of computer science.

Entering Through the Different Types of Logic Gates

In this section, we look at seven of the most common types of logic gates – NOT, AND, OR, NAND, NOR, XOR and XNOR – plus a theorem that combines two gates. All these gates except NOT use at least two inputs; the NOT gate has just one input. To help you get your bearings, Table 2-1 provides a brief overview of these gate types.

Table 2-1 Most Common Types of Logic Gates

Gate	*Description*
NOT	Inverts the input (high becomes low, low becomes high)
AND	Outputs high if all the inputs are high; otherwise, outputs low
OR	Outputs high if at least one of the inputs is high; otherwise, outputs low
NAND	Outputs high if all the inputs are low; otherwise, outputs low
NOR	Outputs high if at least one of the inputs is low; otherwise, outputs low
XOR	Outputs high if one, and only one, of the inputs is high; otherwise, outputs low
XNOR	Outputs high if one, and only one, of the inputs is low; otherwise, outputs low

Noting NOT gates

The simplest of all gates is the *NOT gate,* which is also called an *inverter.* A NOT gate has just one input, and its output is the opposite of the input. If the input is low, the output is high. If the input is high, the output is low.

Table 2-2 shows the truth table of an inverter. A *truth table* is simply a table that lists every possible combination of input values and shows the resulting output for each combination. For an inverter, the truth table is simple. Only one input exists and so only two possibilities exist: the input is low or it's high. As you can see in the table, the output is simply the opposite of the input.

In truth tables, a common convention is to use 0s and 1s to represent logic values rather than the terms *high* and *low.*

Table 2-2 Truth Table for an Inverter

Input	*Output*
0	1
1	0

Figure 2-1 shows the standard logic symbol for a NOT gate. Symbols such as this one are often used in schematic diagrams for circuits that use gates. The NOT symbol is simply a triangle with the input at one end and the output at the other. The small circle on the output is called a *negation bubble,* which indicates that the output is inverted.

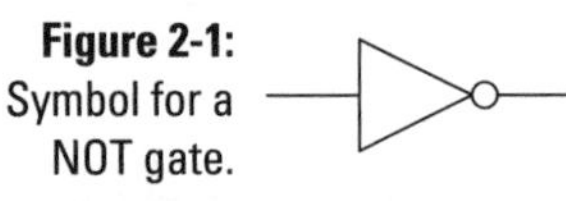

Figure 2-1: Symbol for a NOT gate.

Appraising AND gates

A *two-input AND gate* is a gate with two inputs and one output. The output is high only if both the inputs are high. Any other combination of inputs results in the outputs being low. Table 2-3 shows the truth table for an AND gate with two inputs.

Table 2-3 Truth Table for a Two-Input AND Gate

Input A	*Input B*	*Output*
0	0	0
0	1	0
1	0	0
1	1	1

Figure 2-2 shows the standard symbol for an AND gate. The inputs are on the left and the output is on the right.

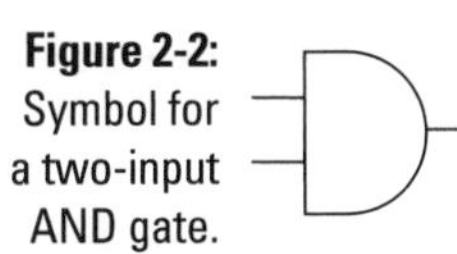

Figure 2-2: Symbol for a two-input AND gate.

You can create AND gates with more than two inputs. For each additional input that you add to a gate, the number of possible input combinations doubles: a two-input gate has 4 possible input combinations; a three-input gate has 8 possible combinations; a four-input gate has 16 possible input combinations and so on.

Table 2-4 shows the truth table for a three-input AND gate. As you can see, the output is high only if all the inputs are high. Any other combination of inputs produces low output.

Table 2-4 Truth Table for a Three-Input AND Gate

Input A	*Input B*	*Input C*	*Output*
0	0	0	0
0	0	1	0
0	1	0	0
0	1	1	0
1	0	0	0
1	0	1	0
1	1	0	0
1	1	1	1

You can combine gates in a circuit to create logic networks that are more complicated than a single gate can produce. For example, you create a three-input AND gate by using two two-input AND gates as shown in Figure 2-3. In the figure, the first AND gate produces a high output only if the inputs A and B are true. Then the output of the first AND gate is used as one of the inputs to the second AND gate; the other input is the input C.

Thinking outside the gate

You can think of an AND gate as being a form of multiplication. Consider that when you multiply any quantity of single-bit binary numbers, only two possible results exist: 0 or 1. Look at the truth tables in Tables 2-3 and 2-4, and try multiplying the binary inputs in each row. In each case, the answer is 0 for any combination that contains a 0 in any of the inputs. The answer is 1 only if all the inputs are 1.

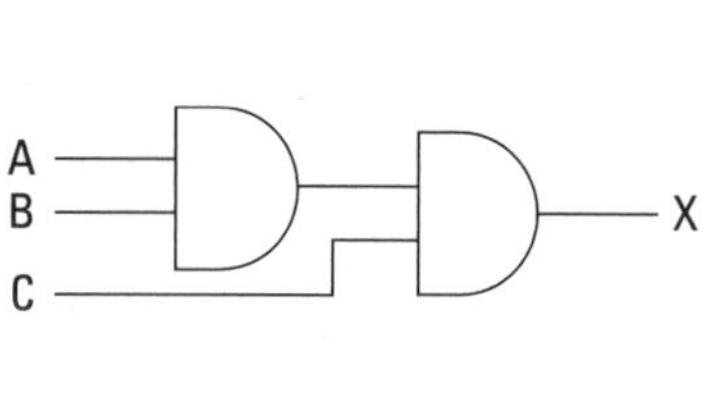

Figure 2-3: You can use a pair of two-input AND gates to create a logic network that operates like a three-input AND gate.

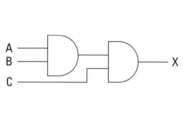

Because the output of the second gate is high only if both its inputs are high, and because the first input to the second gate is the output from the first gate (high only if both its inputs are high), the output of the entire circuit (designated as X) is high only if all three inputs (A, B and C) are high.

Figure 2-4 shows how AND gates can be used in a home-alarm system. Here, the inputs for the various sensors placed on the home's doors and windows are processed by the sensor circuit, which sends a high signal to one of the inputs of the AND gate if any of the sensors indicates an intrusion anywhere in the house. Then the arming circuit sends a 1 to the other input of the AND gate if the system is armed. Finally, the alarm circuit sounds an audible alarm if the AND gate's output is 1. As a result, the alarm sounds if an intrusion is detected and the system is armed.

When used in this way, an AND gate is often called an *enable input,* because one of the inputs to the AND gate enables the other input to be processed. When the enable input is high, the controlled input is allowed to pass through the AND gate. When the enable is low, the controlled input is inhibited.

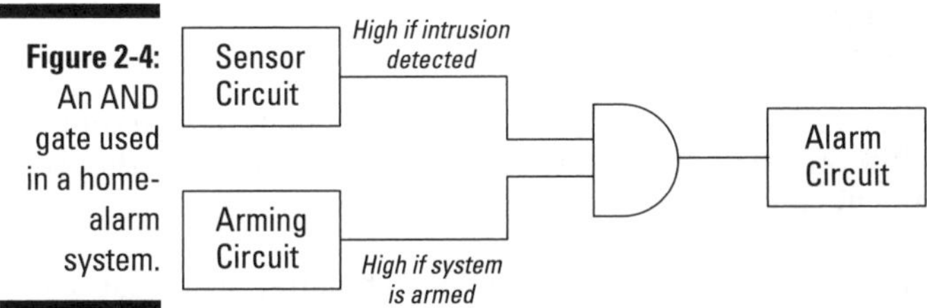

Figure 2-4: An AND gate used in a home-alarm system.

Figure 2-5 shows a more developed version of the home-alarm system, which uses an enable input. Here, the alarm doesn't sound immediately when an intrusion is detected. Instead, the alarm sounds 30 seconds after it's triggered, giving you time to disable the alarm before waking up the neighbours. When the sensor circuit detects an intrusion, it sends a trigger pulse to a 555 timer circuit, which then generates a 30-second high pulse. The high signal from the 555 timer output is routed through a NOT gate, which inverts the signal to low. The low signal is sent to one of the inputs of the second AND gate. The other input to the second AND gate is the output of the first AND gate, which indicates that an intrusion has been detected, and the system is armed.

For the first 30 seconds after the intrusion is detected, the first input to the second AND gate is low and the second input is high, so that the output from the second AND gate is low. Thus, the alarm doesn't sound. In effect, the timer output inhibits the alarm from sounding. But when the output from the 555 timer circuit goes low after the 30-second pulse ends, the NOT gate inverts the signal, sending a high output to the second AND gate. This situation causes the second AND gate's output to go high, and so the alarm sounds. Thus, the inverted pulse from the timer circuit enables the alarm circuit.

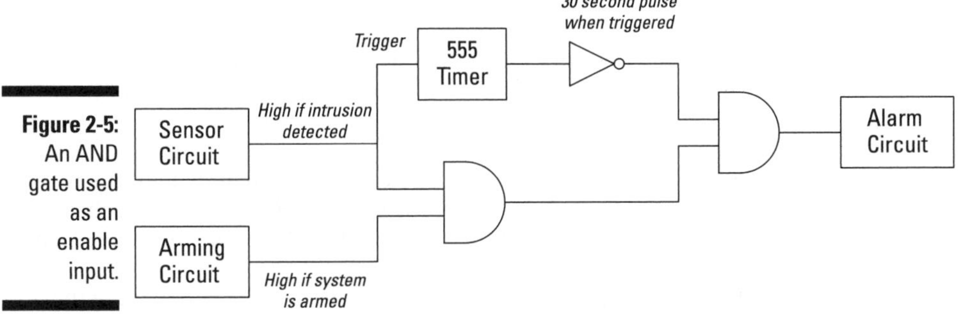

Figure 2-5: An AND gate used as an enable input.

Observing OR gates

An *OR gate* produces a high output if any of the inputs is high. The output from an OR gate is low only if all the inputs are low. Table 2-5 shows the truth table for a two-input OR gate.

In an OR gate, how many of the inputs are high doesn't matter. If at least one input is high, the output is high. Thus, in a two-input OR gate, the output is high if either input is high or both inputs are high. In a three-input OR gate, the output is high if any one, any two or all three of the inputs are high.

Table 2-5 Truth Table for a Two-Input OR Gate

Input A	*Input B*	*Output*
0	0	0
0	1	1
1	0	1
1	1	1

Figure 2-6 shows the standard symbol for an OR gate, with inputs on the left and output on the right.

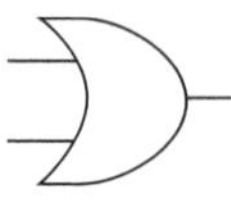

Figure 2-6: Symbol for a two-input OR gate.

Like AND gates, OR gates with more than two inputs are easy to create. No matter how many inputs the OR gate has, the output is high if any of the inputs is high.

Multiple-input OR gates are simple to create by combining two input OR gates. Figure 2-7 shows a logic network with three OR gates, effectively acting like a four-input OR gate: if any of the four inputs is high, the output is high.

You can combine two-input OR gates in a circuit to create OR networks that have more than two inputs. For example, Figure 2-8 shows how OR gates can be used in the sensor circuit of a home-alarm system with more than two inputs. In this circuit, eight separate alarm sensors are fed into a network of OR gates. If any one of the inputs is high, the output from the sensor circuit is high.

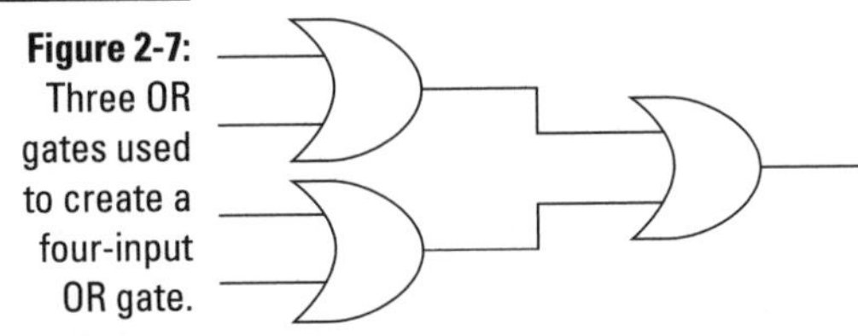

Figure 2-7: Three OR gates used to create a four-input OR gate.

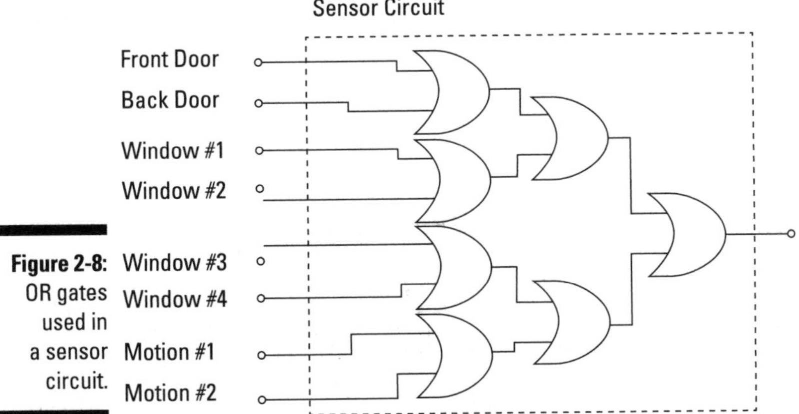

Figure 2-8: OR gates used in a sensor circuit.

The sensor circuit uses seven OR gates to create an eight-input OR gate. Take a close look at the way the OR gates are arranged in this network to make sure that you understand how it works. Each of the eight inputs is routed to one of the four OR gates in the first tier of gates. These four OR gates reduce the eight inputs to four outputs, which are then sent to the two OR gates in the second tier. These two OR gates reduce their four inputs to two outputs, which are sent to the final OR gate. Then the output of the last OR gate becomes the output of the entire sensor circuit.

Looking at NAND gates

A *NAND* gate is a combination of an AND gate and a NOT gate. In fact, the name NAND is a contraction of NOT and AND. As you can see in Table 2-6, the output of a NAND gate is low when both the inputs are high. Otherwise, the output of the NAND gate is high.

Table 2-6 Truth Table for a Two-Input NAND Gate

Input A	*Input B*	*Output*
0	0	1
0	1	1
1	0	1
1	1	0

Figure 2-9 shows the standard symbol for a NAND gate. This symbol is the same as the symbol for an AND gate (see the earlier Figure 2-2), with the addition of a circle at the output. As in the symbol for a NOT gate (check out the earlier Figure 2-1), the circle indicates that the output is inverted. In other words, a NAND gate is an AND gate whose output is inverted.

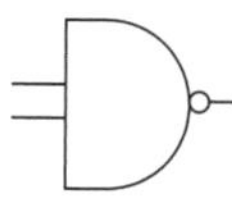

Figure 2-9: Symbol for a two-input NAND gate.

The NAND gate is special because you can use various combinations of NAND gates to create AND, OR or NOT gates. Thus, a logic network that consists of a combination of NOT, AND and OR gates can be created with an equivalent combination of just NAND gates. For this reason, the NAND gate is called a *universal gate.* To discover more about this characteristic of NAND gates, check out the 'Using universal NAND gates' section later in this chapter.

Another interesting thing about the NAND gate is that it can be used as a kind of OR gate (see the preceding section) that tests for low inputs instead of high inputs. In other words, the output of a NAND gate is high when either of the inputs is low.

This characteristic of NAND gates is useful in many situations. For example, consider the alarm sensor circuit in the preceding section and in Figure 2-8. Suppose that all the alarm sensors produce a high signal without an intrusion, and then go low following an intrusion. In the real world, many alarm sensors work exactly in this way. For instance, a sensor that detects whether a door is open is essentially a simple switch that's closed when the door is closed and open when the door is open. When the door is closed, current flows through the switch and the signal from the sensor is high. When the door is opened, current stops flowing and the signal from the sensor goes low.

Figure 2-10 shows how you can use NAND gates to test for a low input on any of eight sensors. This circuit is identical to the circuit in the earlier Figure 2-8, except that all the OR gates have been replaced by NAND gates.

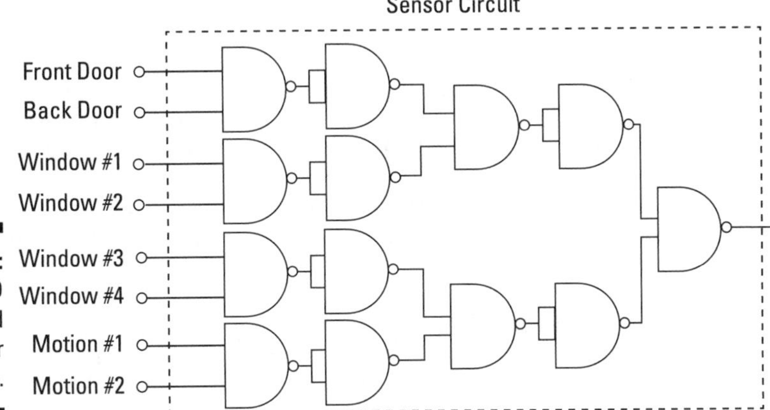

Figure 2-10: NAND gates used in a sensor circuit.

Checking out NOR gates

A *NOR* gate is a combination of an OR gate and a NOT gate (see the earlier sections 'Observing OR gates' and 'Noting NOT gates'). As with NAND in the preceding section, the name NOR is a contraction of NOT and OR.

Table 2-7 shows the truth table for a NOR gate. As you can see, the output of a NOR gate is low when any of its inputs are high. Otherwise, the output of the NAND gate is high.

Table 2-7 Truth Table for a Two-Input NOR Gate

Input A	*Input B*	*Output*
0	0	1
0	1	0
1	0	0
1	1	0

As Figure 2-11 shows, the standard symbol for a NOR gate is the same as the symbol for the OR gate in Figure 2-6, with a negation circle added to the output. The circle simply indicates that the output is inverted.

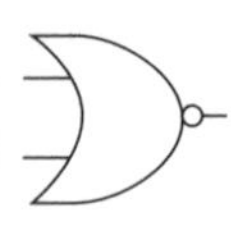

Figure 2-11: Symbol for a two-input NOR gate.

Like the NAND gate, the NOR gate is a universal gate. You can recreate any logic network that consists of NOT, AND and OR gates with just NOR gates. For more information, see the later section 'Combining universal NOR gates'.

Just as a NAND gate is like an OR gate for low inputs, a NOR gate is like an AND gate for low inputs. In other words, the output of the NOR gate is high when both the inputs are low.

Figure 2-5, earlier in this chapter, shows an alarm circuit that sounds an alarm 30 seconds after an input sensor detects an intrusion. In that circuit, the output pulse from a 555 timer circuit is inverted by a NOT gate and then sent to an AND gate, which sends a high output to an audible alarm circuit when the output from a sensor circuit is high and when the 30-second timer pulse ends.

Figure 2-12 shows a version of the same circuit, which uses a NOR gate instead of an AND gate to feed the audible alarm circuit. In this circuit, the alarm sounds when the output from the sensor circuit is low and the output from the timer circuit is also low. You don't need a NOT gate for the 555 timer output, but you do require a NOT gate on the output from the AND gate to invert its signal, so that it emits low when the system is armed and the alarm is tripped.

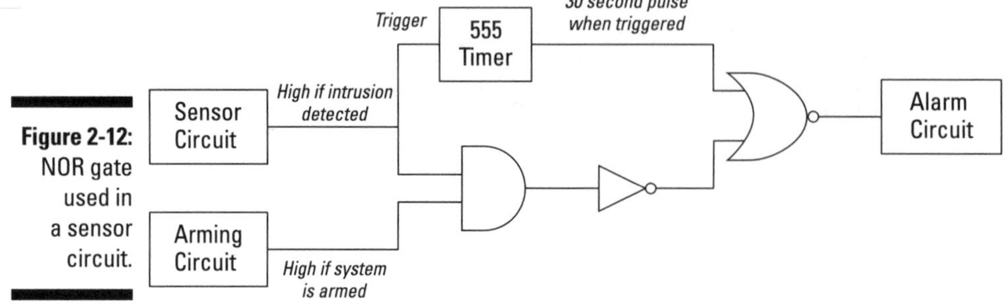

Figure 2-12: NOR gate used in a sensor circuit.

Going over XOR and XNOR gates

We cover two gate types in this section: *XOR,* which stands for *Exclusive OR,* and *XNOR,* which stands for *Exclusive NOR.*

In an XOR gate, the output is high if one, and only one, of the inputs is high. If both inputs are low or both are low, the output is low. Table 2-8 contains the truth table for an XOR gate.

Table 2-8 Truth Table for a Two-Input XOR Gate

Input A	*Input B*	*Output*
0	0	0
0	1	1
1	0	1
1	1	0

Here's another way to explain an XOR gate: the output is high if the inputs are different; if the inputs are the same, the output is low.

The XOR gate has a lesser-known cousin called the XNOR gate. An *XNOR gate* is an XOR gate whose output is inverted. Table 2-9 lists the truth table for an XNOR gate.

Table 2-9 Truth Table for a Two-Input XNOR Gate

Input A	*Input B*	*Output*
0	0	1
0	1	0
1	0	0
1	1	1

Figure 2-13 shows the symbols used for XOR and XNOR gates. As you can see, the only difference between these two symbols is that the XNOR has a circle on its output to indicate that the output is inverted.

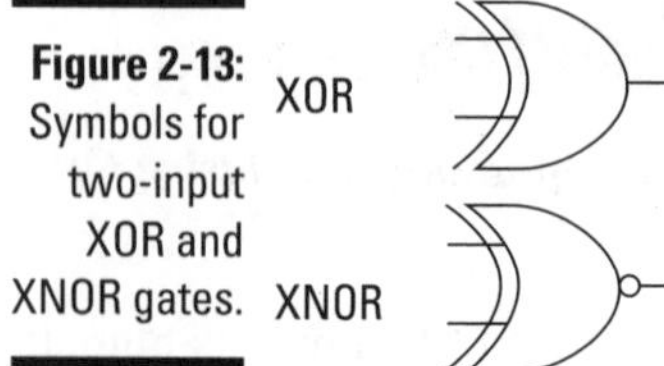

Figure 2-13: Symbols for two-input XOR and XNOR gates.

One of the most common uses for XOR gates is to add two binary numbers. For this operation to work, the XOR gate has to be used in combination with an AND gate, as shown in Figure 2-14.

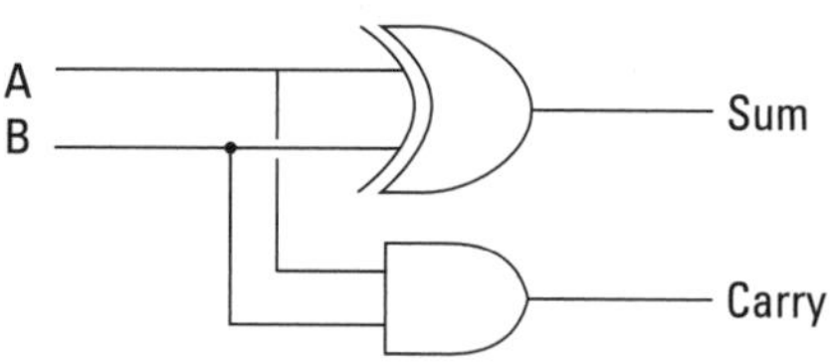

Figure 2-14: You can use an XOR gate and an AND gate to add two binary numbers.

To help understand how the circuit shown in Figure 2-14 works, we review briefly how binary addition works:

$0 + 0 = 0$

$0 + 1 = 1$

$1 + 0 = 1$

$1 + 1 = 10$

If you want, you can write the results of each of the preceding addition statements by using two binary digits:

$0 + 0 = 00$

$0 + 1 = 01$

$1 + 0 = 01$

$1 + 1 = 10$

When results are written with two binary digits, as in this example, you can easily see how to use an XOR and an AND circuit in combination to perform binary addition. If you consider just the first binary digit of each result, you notice that it looks just like the truth table for an AND circuit (see Table 2-3) and that the second digit of each result looks just like the truth table for an XOR gate (in Table 2-8).

The adder circuit shown in Figure 2-14 has two outputs. The first is called the *Sum* and the second is called the *Carry*. The Carry output is important when several adders are used together to add binary numbers that are longer than 1 bit.

Diving into De Marvellous De Morgan's Theorem

The logic concept *De Morgan's Theorem* was created by Augustus De Morgan, a 19th-century British mathematician who developed many of the concepts that make Boolean logic work. In fact De Morgan's theorem comprises two related theorems that have to do with how NOT gates are used in conjunction with AND and OR gates:

- An AND gate with inverted output behaves the same as an OR gate with inverted inputs.
- An OR gate with inverted output behaves the same as an AND gate with inverted inputs.

An AND gate with inverted output is also called a NAND gate, of course, and an OR gate with inverted output is also called a NOR gate. Thus, De Morgan's laws can also be stated as follows:

- A NAND gate behaves the same as an OR gate with inverted inputs.
- A NOR gate behaves the same as an AND gate with inverted inputs.

An OR gate with inverted inputs is called a *negative OR gate* and an AND gate with inverted inputs is called a *negative AND gate.*

In case you're not persuaded, review for a moment the truth table for a NAND gate:

A	*B*	*X*
0	0	1
0	1	1
1	0	1
1	1	0

Now look at the truth table for an OR gate, with an extra set of columns added to show the inverted inputs:

A	*B*	*NOT A*	*NOT B*	*X*
0	0	1	1	1
0	1	1	0	1
1	0	0	1	1
1	1	0	0	0

Here, the A and B columns represent the inputs. The NOT A and NOT B columns are the inputs after they've been inverted. Finally, the X column represents an OR operation applied to the NOT A and NOT B values.

As you can see, the final output column of these truth tables is the same. Thus, a NAND gate is equivalent to a negative OR gate.

Any time you see a NAND gate in a circuit diagram, you can substitute a negative OR gate.

Now take a look at the other side of De Morgan's Theorem. Here's a truth table for a NOR gate:

A	*B*	*X*
0	0	1
1	0	0
0	1	0
1	1	0

And here's the output of a negative AND gate:

A	*B*	*NOT A*	*NOT B*	*X*
0	0	1	1	1
0	1	1	0	0
1	0	0	1	0
1	1	0	0	0

Again, you can see that these two truth tables give the same output.

Just as a circle is used on the output of a NAND or NOR gate to indicate that the output is inverted, you can use a circle on the inputs to an OR or AND gate to indicate that the inputs are inverted. Figure 2-15 shows these symbols. The figure also shows that the negative OR and AND gates are interchangeable with the NAND and NOR gates.

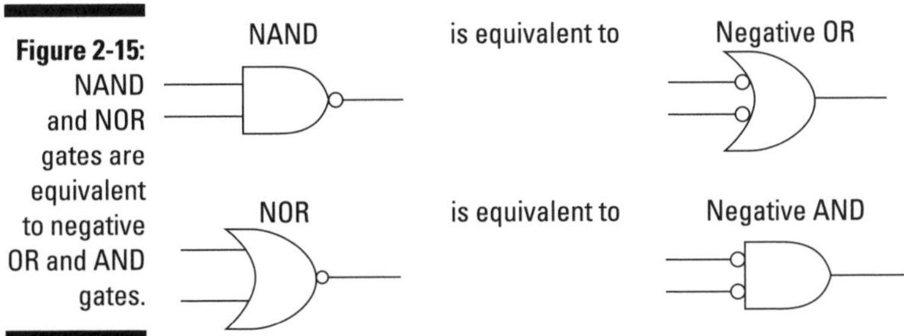

Figure 2-15: NAND and NOR gates are equivalent to negative OR and AND gates.

Understanding that All You Need is NAND (or NOR)

In 'Looking at NAND gates' earlier in this chapter, we mention that the NAND gate is a universal gate because any other type of gate (such as AND, OR, XOR and XNOR) can be constructed solely from combinations of NAND gates.

This fact is incredibly useful, because it enables you to build any logic circuit, simple or complex, by using just NAND gates. When you begin to construct your own digital circuits, you can stock up on integrated circuits that contain just NAND gates safe in the knowledge that you can build even the most complex circuits with your stock of NAND gates.

Also, in the earlier section 'Checking out NOR Gates', we reveal that the NOR gate is a universal gate, too. Thus, you can also build any logic circuit by using nothing but NOR gates.

In the following sections, we explain how you can use NAND and NOR gates to build other types of gates.

Using universal NAND gates

Figure 2-16 shows how you can use NAND gates in various combinations to create NOT, AND, OR and NOR gates (which we explain in the earlier section 'Entering Through the Different Types of Logic Gates').

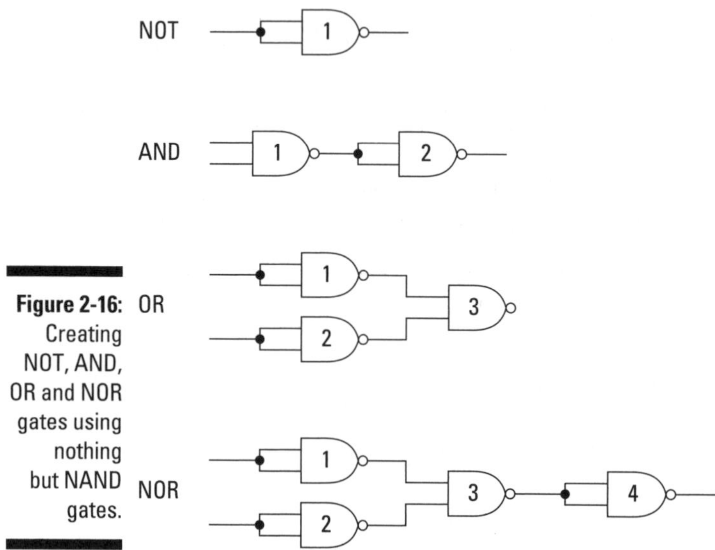

Figure 2-16: Creating NOT, AND, OR and NOR gates using nothing but NAND gates.

The following list describes how the circuits work:

- **NOT:** You can create a NOT gate from a NAND gate simply by tying the two inputs of the NAND gate together (we describe a project using this technique in Chapter 5 of this minibook). Because the two inputs of the NAND gate are tied together, only two input combinations are possible: both high or both low. If both inputs are high, the NAND gate outputs a low. If both inputs are low, the NAND gate outputs high. Thus, the circuit behaves exactly as a NOT gate.
- **AND:** You can create an AND gate by using two NAND gates. The first NAND gate does what NAND gates do: returns low if both inputs are high and returns high if both inputs are anything else. The second NAND gate is configured as a NOT gate to invert the output from the first NAND gate.

  One of the basic rules of NOT gates is that if you invert a signal twice, you end up with the same signal. If the original input is high, and you invert it, the signal becomes low. Invert the input again, and it returns to high. With this rule in mind, you can see how the two NAND gates work together to create an AND gate.
- **OR:** You need three NAND gates to create an OR gate. You use a pair of NAND gates configured as NOT gates to invert the two inputs. Then the third NAND gate produces a low output if both the original inputs are low. If one of the original inputs is high, or if both the original inputs are high, the output of the third gate is high.

- **NOR:** You need four NAND gates to create a NOR gate. As you can see in Figure 2-16, this circuit is the same as the OR gate circuit, with the addition of another NOT gate to invert the output from the third NAND gate. Inverting the output changes the overall function of the circuit from OR to NOR.

Combining universal NOR gates

Like NAND, NOR is a universal gate. Figure 2-17 shows how you can combine NOR gates in various ways to create NOT, AND, OR and NAND gates.

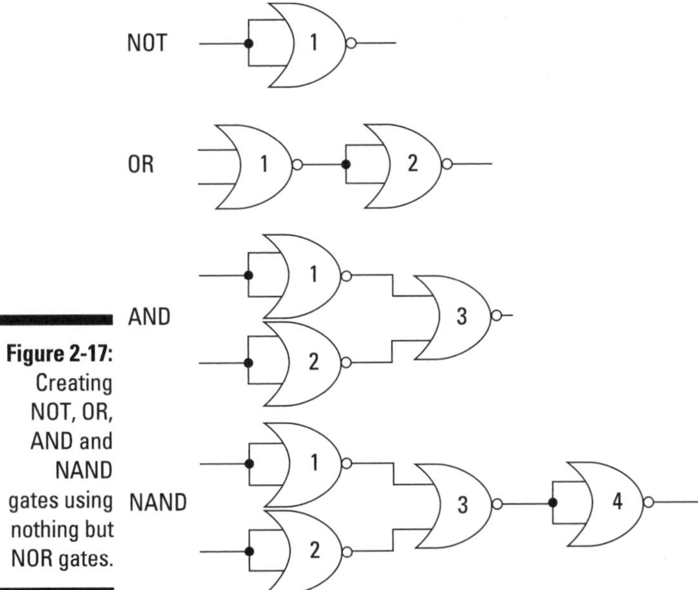

Figure 2-17: Creating NOT, OR, AND and NAND gates using nothing but NOR gates.

Here's how the circuits work:

- **NOT:** Creating a NOT gate from a NOR gate is the same as creating a NOT gate from a NAND gate: you simply tie the two inputs of the NOR gate together (see the preceding section). If both inputs are low, the NOR gate outputs a high. Otherwise, the output is low.
- **OR:** You need two NOR gates to create an OR gate. The first NOR gate returns low if either input is high or both inputs are high. The second NOR gate is configured as a NOT gate to invert the output of the first NOR gate.

- **AND:** You need three NOR gates to create an AND gate. The first two are configured as NOT gates, and so they invert the inputs. The third NOR gate produces a high output if both the original inputs are high.
- **NAND:** You require four NOR gates to create a NAND gate. The first three NOR gates are configured just as they are for an AND gate. Then a fourth NOR gate configured as a NOT gate inverts output from the third NOR gate.

Playing with Gates in Software

One of the best ways of finding out how logic gates work is to download one of the many software logic gate simulators that are available free on the Internet. You can find these programs by firing up your search engine and searching for keywords such as 'logic gate simulator'.

One of our favourite programs is Logic Circuit Designer, written by Ivan Andrei. You can download it from `http://download.cnet.com/Logic-Circuit-Designer/3000-2054_4-10840569.html`. Figure 2-18 shows this useful program in action.

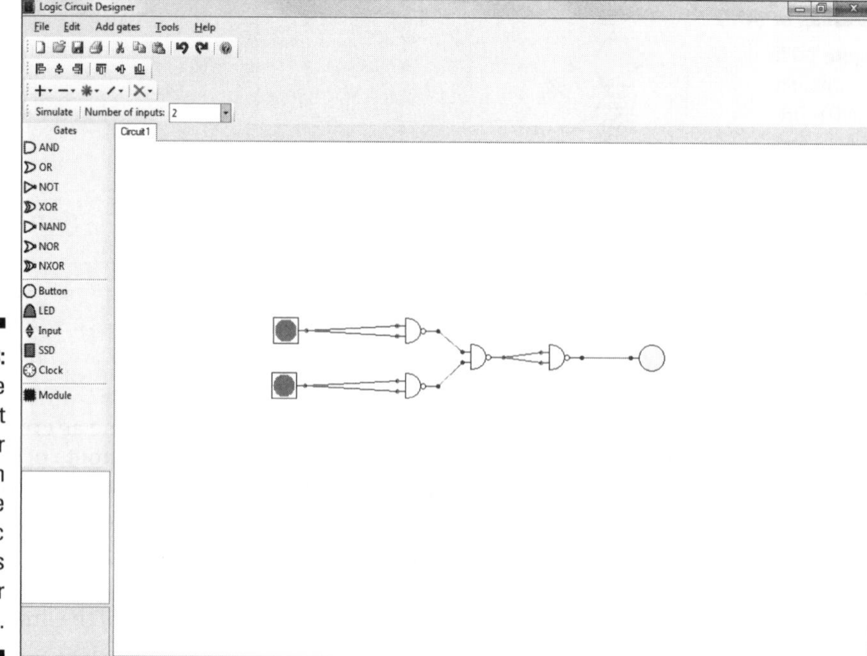

Figure 2-18: Using the Logic Circuit Designer program to simulate logic circuits on your computer.

Here are just a few of the features of the Logic Circuit Designer program:

- You can add AND, OR, NOT, XOR, NAND, NOR and XNOR gates to your circuit simply by dragging them from a toolbar.
- You can connect the gate inputs and outputs by clicking an input or output and then dragging it to another gate input or output.
- You can add simple switches to the circuit for external inputs.
- You can add LEDs at any point in the circuit to indicate the status of an output or an input.

Noting Notations

You can show input and output variables in the following different ways:

- If an input, we'll call it A, is changed by a NOT gate, the inverted output is represented by A with a short line or bar over the top, like this: $\bar{A}$.
- The AND operation is sometimes shown with a dot between two variables. So, if the input variables are A and B, the output is A.B, but the dot is sometimes left out so you could also see it shown as just AB.
- The OR operation is sometimes shown as a plus sign, like this: A+B = Q.

Many other symbols are used for logic notation which you can read about online at `http://en.wikibooks.org/wiki/Practical_Electronics/Logic`.

Chapter 3

Putting Logic Circuits to Work

In This Chapter

- Implementing logic gates with transistors
- Looking at TTL and CMOS integrated circuits
- Building simple logic circuits

In this chapter you get the chance to work with circuits using the popular logic gates that we introduce in this minibook's Chapter 2. We show you the basics of creating logic gates from simple transistor circuits and look at two popular integrated circuit (IC) families that provide prebuilt logic gates: transistor-transistor logic and complementary metal-oxide semiconductor logic, which sound less intimidating when abbreviated to TTL and CMOS.

Unless you're already well up-to-speed on logic gates, we suggest that you read Chapter 2 of this minibook before tackling the projects in this chapter. It isn't going to make much sense if you aren't familiar with the different types of gates we describe in that chapter.

Creating Logic Gates with Transistors

In Book II, Chapter 6, we discuss how you can use transistors as switches (if you want to know how transistors work, check out that chapter). In a nutshell, a voltage applied to the base of a transistor allows current to flow from the collector to the emitter. Thus, by applying an input signal to a transistor's base, you can control an output signal taken from the collector-emitter path.

You can build any logic gate you want by cobbling together a few transistors and resistors in just the right way. In this section, we look at simple transistor circuits for five gate types: NOT, AND, OR, NAND and NOR.

All the circuits in this chapter assume that a high signal (logical 1) is represented by a DC voltage of +5 V or more. Low (logical 0) is represented by near-zero voltage. This minibook's Chapter 1 has more details on these terms.

Note that you don't often build your own logic gates with transistors and resistors. Instead, you use ICs that contain prefabricated logic gates. But before you use logic ICs in your circuits, you need to have a basic understanding of how the gates inside them work. So we recommend that you take a look at simple transistor circuits for basic logic gates, and then examine the logic ICs in the later 'Introducing Integrated Circuit Logic Gates' section.

Discovering a transistor NOT gate circuit

A NOT gate simply inverts its input. If the input is high, the output is low, and if the input is low, the output is high. Such a circuit is easy to build using a single transistor and a pair of resistors. Figure 3-1 shows the schematic. The bar over the output is a common notation to indicate 'NOT'.

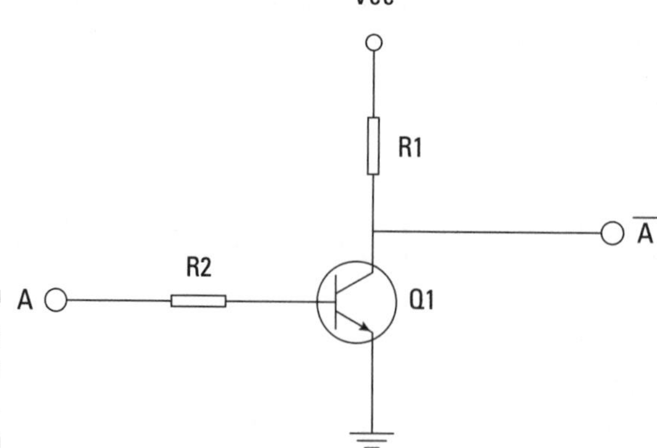

Figure 3-1: A transistor NOT gate.

The operation of this circuit is simple. The input is connected through resistor R2 to the transistor's base. When no voltage is present on the input, the transistor turns off. When the transistor is off, no current flows through the collector-emitter path. Thus, current from the supply voltage (Vcc in the schematic, typically between +5 V and +9 V) flows through resistor R1 to the output. In this way, the circuit's output is high when its input is low.

When voltage is present at the input, the transistor turns on, allowing current to flow through the collector-emitter circuit directly to ground, which causes the output to go low.

In this way, the output is high when the input is low and low when the input is high.

Project 3-1 demonstrates how to assemble a simple transistor NOT gate on a solderless breadboard. For this project, a normally open pushbutton is used as the input. When the button isn't pressed, the input is low and the output is high, which causes the light-emitting diode (LED) to light. When you press the button, the input goes high, the output goes low and the LED goes out. Figure 3-2 shows the assembled project.

Figure 3-2: A transistor NOT gate assembled on a breadboard (Project 3-1).

Project 3-1: A Transistor NOT Gate

In this project, you build a simple NOT gate by using a bipolar transistor. A NOT, also known as an inverter, reverses the logic level of its input. Thus, if the input is high, the output of the NOT gate is low. If the input is low, the output is high.

The input to this gate is controlled by the pushbutton switch SW1. When the switch is open, the input is low. When the switch is closed, the input is high.

The output from this gate is sent through an LED, and so the LED is on when the output is high and off when the output is low.

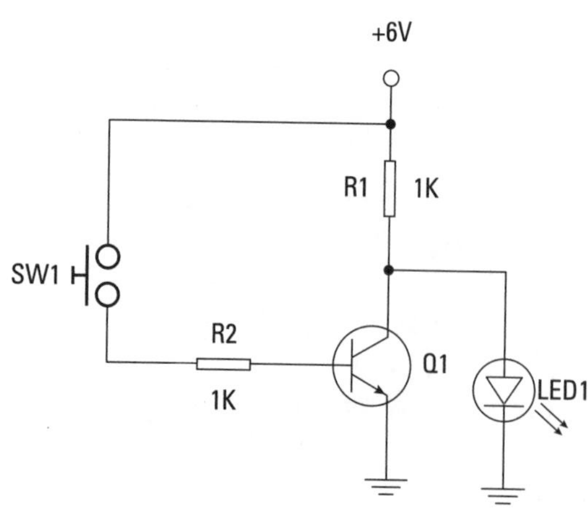

Parts List
1 Four-cell AAA battery holder
4 AAA batteries
1 NPN transistor (2N2222 or equivalent)
1 Red LED
2 1 kΩ ¼ W resistors (brown-black-red)
1 Normally open pushbutton
Miscellaneous jumper wires

NOT Gate Truth Table

Input	Output
0	1
1	0

Steps

1. **Insert transistor Q1:**
   Collector: G9
   Base: G10
   Emitter: G11
2. **Insert resistors R1 and R2:**
   R1 – 1 kΩ:F9 to positive bus
   R2 – 1 kΩ:F10 to D12
3. **Insert LED1:**
   Cathode (short lead): Ground bus
   Anode (long lead): J9
4. **Insert the jumper wire:**
   From: J11
   To: Ground bus
5. **Insert pushbutton SW1:**
   From: A12
   To: Positive bus
6. **Connect the batteries:**
   Red lead: Positive bus
   Black lead: Negative bus

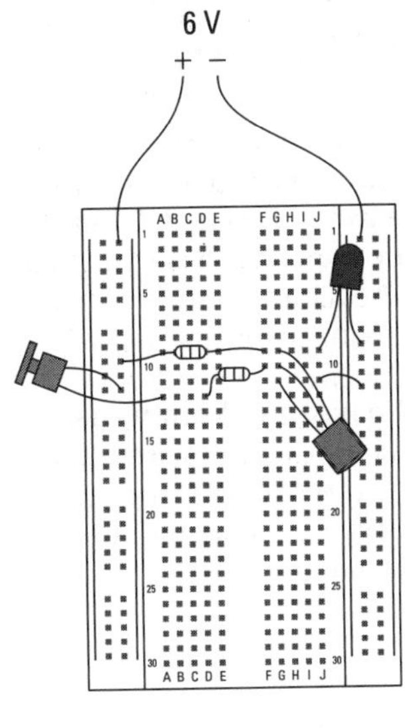

Transistor
Collector
Base
Emitter

LED
Cathode
Anode

Going high with a transistor AND gate circuit

A two-input AND gate produces a high output if both inputs are high. You can create a two-input AND gate by using two transistors and three resistors, as shown in Figure 3-3. In this circuit, the output current has to flow from the Vcc supply voltage through the collector-emitter circuits of both transistors to reach the output. Thus, current flows to the output only if both transistors are on.

The bases of both transistors are fed through R2 and R3 from the two inputs. So if both inputs are high, current flows through the base-emitter path of both transistors, turning them on and allowing current to flow through to the output. If either input is low, the corresponding transistor turns off and the output goes low.

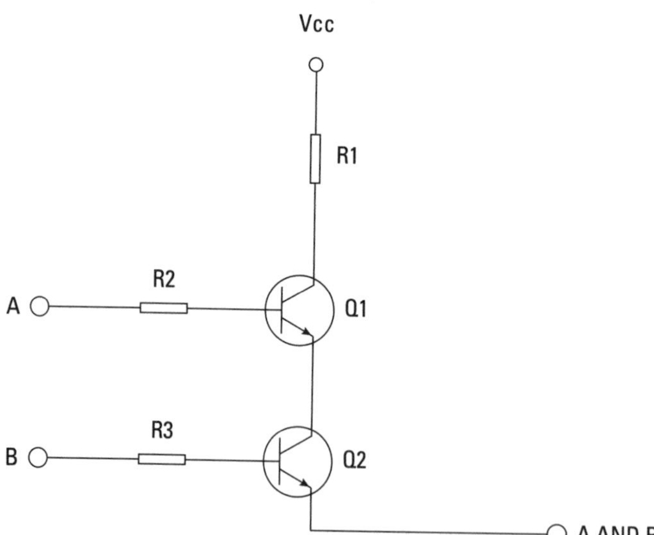

Figure 3-3: A transistor AND gate.

Staying low with a transistor NAND gate circuit

A two-input NAND gate produces a low output if both its inputs are high. Although you can create a NAND gate by combining the circuits shown in Figure 3-1 and Figure 3-3, so that the output from the AND gate is used as input to the NOT gate, that combination requires three transistors. In fact, creating a NAND gate using just two transistors is pretty easy, as shown in Figure 3-4.

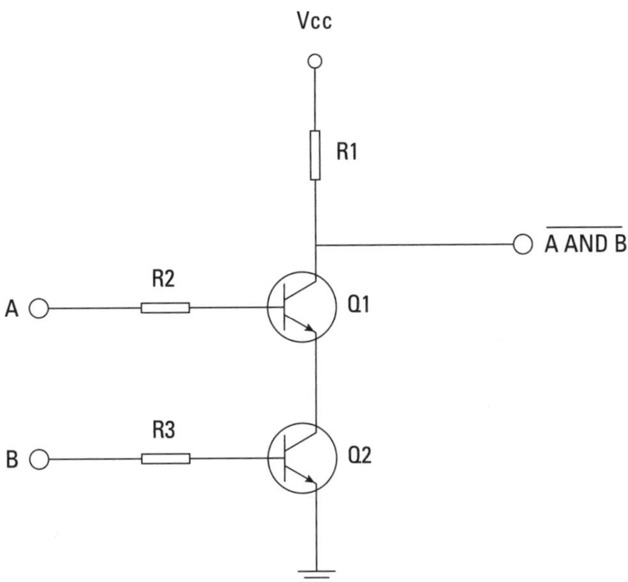

Figure 3-4: A transistor NAND gate.

This NAND gate circuit is almost identical to the AND gate circuit shown in the earlier Figure 3-3. The only difference is that instead of connecting the output to the emitter of the second transistor, the output is obtained before the collector of the first transistor. If both inputs are high, both transistors conduct through their collector-emitter paths, which creates a short circuit to ground and in turn causes the output to go low.

If either transistor turns off, however, the supply current can't flow through the transistors to ground, and so it flows through the output circuit instead. Thus, the output is high if either one of the inputs is low. If both inputs are high, the output is low.

Project 3-2 shows you how to assemble a simple transistor NAND gate on a solderless breadboard. Normally open pushbuttons are used for the two inputs. The LED is on until you press both the pushbuttons. This action causes both inputs to go high, which causes the output to go low and the LED to go dark. The completed project is shown in Figure 3-5.

Figure 3-5: A two-transistor NAND gate on a breadboard (Project 3-2).

Project 3-2: A Transistor NAND Gate

In this project, you build a simple NAND gate by using two bipolar transistors. The output of a NAND gate is low if both inputs are high; otherwise, the output is high.

The inputs to this gate are controlled by the pushbutton switches SW1 and SW2. When a switch is open, the corresponding input is low. When the switch is depressed, the corresponding input is high.

The output from this gate is sent through an LED, and so the LED is on when the output is high and off when the output is low.

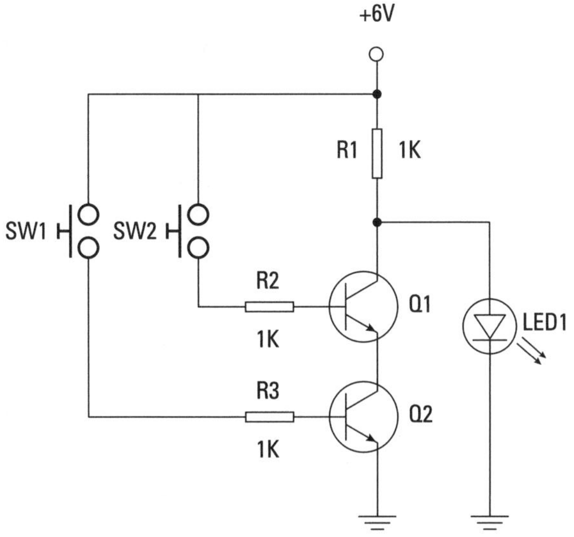

Parts List

- 1 Four-cell AAA battery holder
- 4 AAA batteries
- 2 NPN transistors (2N2222 or equivalent)
- 1 Red LED
- 3 1 kΩ 1/4 W resistors (brown-black-red)
- 2 Normally open pushbuttons
- Miscellaneous jumper wires

NAND Gate Truth Table

Input A	Input B	Output
0	0	1
0	1	1
1	0	1
1	1	0

Steps

1. **Insert transistors Q1 and Q2:**

Lead	**Q1**	**Q2**
Collector:	G9	G13
Base:	G10	G14
Emitter:	G11	G15

2. **Insert resistors R1, R2 and R3:**
   R1 – 1 kΩ:F9 to positive bus
   R2 – 1 kΩ:F10 to C10
   R3 – 1 kΩ:F14 to C14
3. **Insert LED1:**
   Cathode (short lead): Ground bus
   Anode (long lead): J9
4. **Insert the jumper wires:**
   Jumper 1: From H11 to H13
   Jumper 2: From J15 to ground bus
5. **Insert switches SW1 and SW2:**
   SW1: From A10 to positive bus
   SW2: From A14 to positive bus
6. **Connect the batteries:**
   Red lead: Positive bus
   Black lead: Negative bus

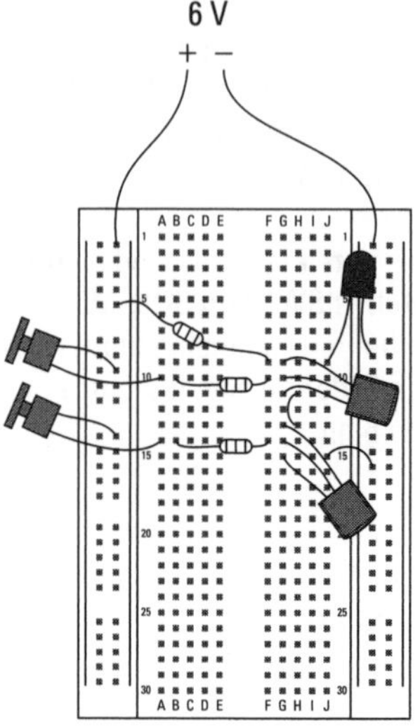

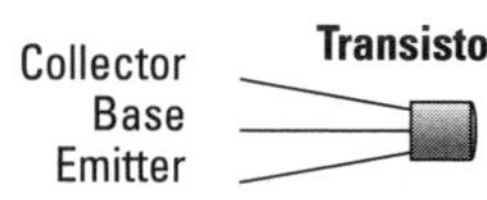

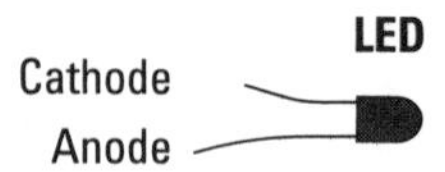

Looking at a transistor OR gate circuit

A two-input OR gate produces a high output if either input is high or both its inputs are high. Figure 3-6 shows a schematic for an OR gate created with two transistors and three resistors.

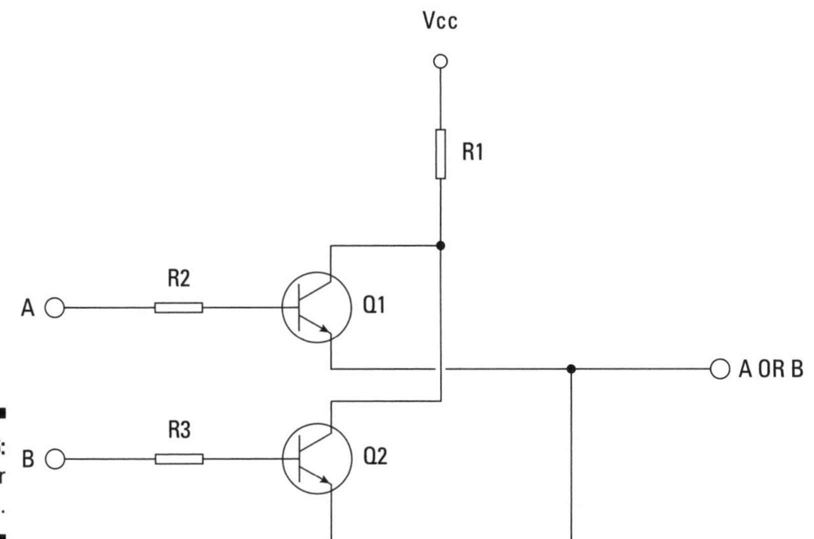

Figure 3-6: A transistor OR gate.

In the OR gate circuit, the supply voltage is connected separately to the collector of each transistor. Then the emitters of both transistors are connected to the output. That way, if voltage is applied to the base of either one of the transistors, that transistor turns on and passes current through to the output.

Thus, the output is high if one input is high or both inputs are high. The output is low only if both inputs are low.

Knowing about transistor NOR gate circuits

A NOR gate is an inverted OR gate. If at least one of the inputs is high, the output is low. If both inputs are low, the output is high.

Figure 3-7 shows a schematic for a NOR gate. This circuit is similar to the circuit shown in Figure 3-6 in the preceding section, except that the output is connected to the collector of both transistors and the emitter of each transistor is connected to ground. If either one of the transistors is on, current from Vcc is short-circuited to ground, bypassing the output. As a result, the output is high only when both inputs are low. If either input is high or both inputs are high, the output is low.

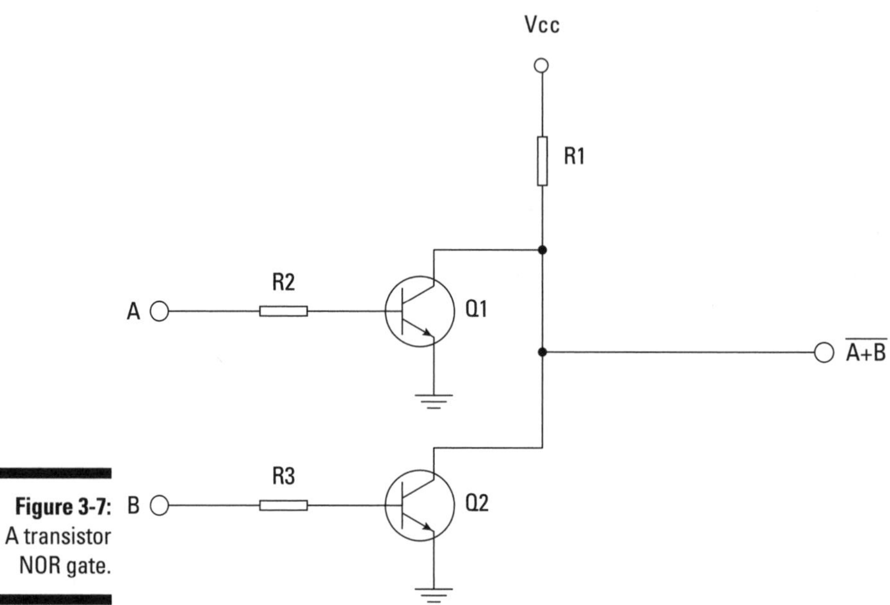

Figure 3-7: A transistor NOR gate.

You can build a two-transistor NOR gate by following the steps outlined in Project 3-3. As with the other projects in this chapter, Project 3-3 uses normally open pushbuttons to control the input circuits. When power is applied to this circuit, both inputs are initially low and the output is high. Pressing either one of the switches causes that switch's input to go high, which in turn causes the output to go low. The assembled project is shown in Figure 3-8.

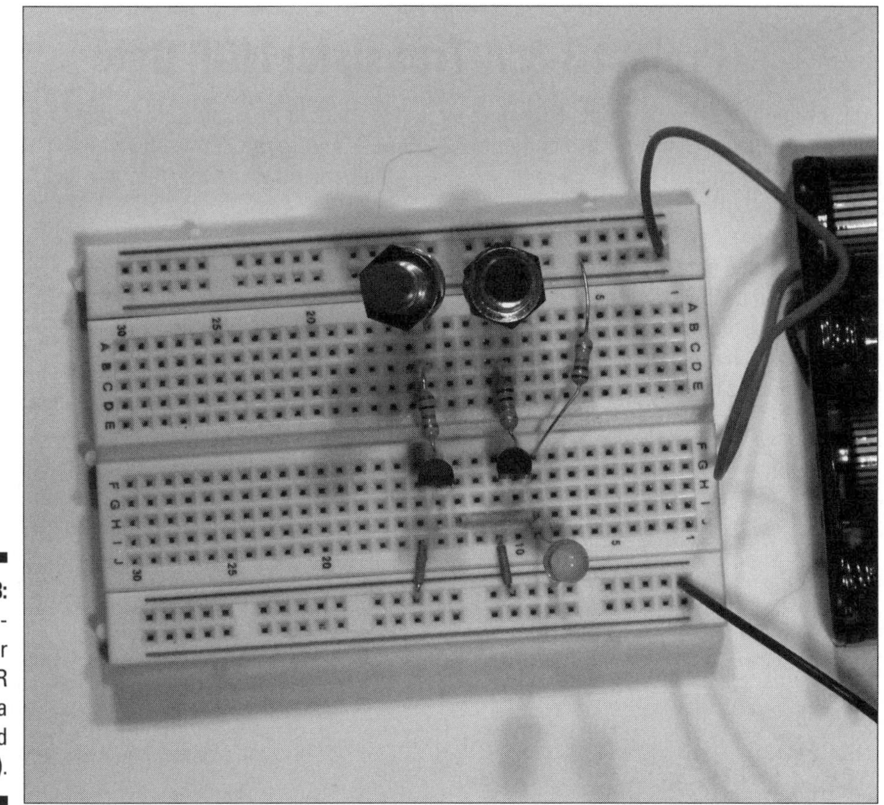

Figure 3-8: A two-transistor NOR gate on a breadboard (Project 3-3).

Project 3-3: A Transistor NOR Gate

In this project, you build a simple NOR gate by using two bipolar transistors. The output of a NOR gate is high if both inputs are low. Otherwise the output is low.

The inputs to this gate are controlled by normally open pushbutton switches. When a switch is open, the corresponding input is low. When the switch is depressed, the corresponding input is high.

The output from this gate is sent through an LED, and so the LED is on when the output is high and off when the output is low.

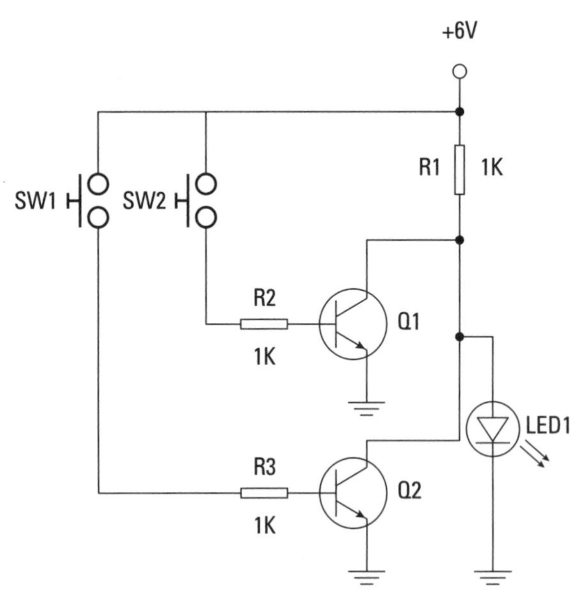

Parts List

- 1 Four-cell AAA battery holder
- 4 AAA batteries
- 2 NPN transistors (2N2222 or equivalent)
- 1 Red LED
- 3 1 kΩ ¼ W resistors (brown-black-red)
- 2 Normally open pushbuttons
- Miscellaneous jumper wires

NOR Gate Truth Table

Input A	Input B	Output
0	0	1
0	1	0
1	0	0
1	1	0

Steps

1. **Insert transistors Q1 and Q2:**

Lead	Q1	Q2
Collector:	G9	G13
Base:	G10	G14
Emitter:	G11	G15

2. **Insert resistors R1, R2 and R3:**
   R1 – 1 kΩ: F9 to positive bus
   R2 – 1 kΩ: F10 to C10
   R3 – 1 kΩ: F14 to C14
3. **Insert LED1:**
   Cathode (short lead): Ground bus
   Anode (long lead): J9
4. **Insert the jumper wires:**
   Jumper 1: From J10 to ground bus
   Jumper 2: From J15 to ground bus
   Jumper 3: From I9 to I13
5. **Insert switches SW1 and SW2:**
   SW1: From A10 to positive bus
   SW2: From A14 to positive bus
6. **Connect the batteries:**
   Red lead: Positive bus
   Black lead: Negative bus

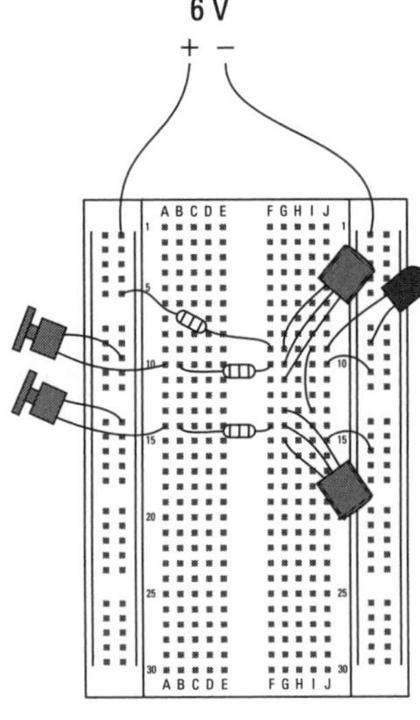

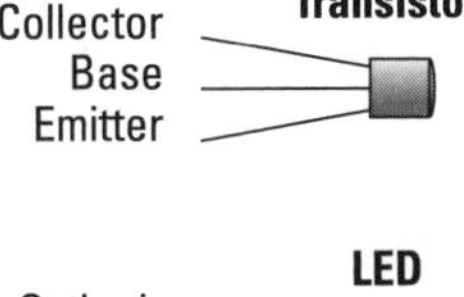

LED

Cathode

Anode

Introducing Integrated Circuit Logic Gates

Although you can build your own logic gates using transistors and resistors as we describe in the preceding sections, buying prepackaged ICs that implement logic gates is far easier. The main advantage of using IC logic gates is that you don't have to design the individual gates yourself or waste time assembling them.

The logic gate circuits in the preceding sections are among the simplest circuits for creating logic gates, but they're by no means the only ways to create logic gates – and not necessarily the best. Over the 50 years or so that circuit designers have been working on semiconductor-based logic circuits, many designs have been developed for creating logic gates.

Each approach to designing logic circuits results in an entire family of logic circuits for the various types of gates (NOT, AND, OR, NAND, NOR, XOR and XNOR), and for this reason the different designs are often referred to as *design families.* Each family has an easy-to-remember three- or four-letter acronym. Here are the most popular:

- **RTL:** *Resistor-transistor logic,* which uses resistors and bipolar transistors. The circuits we present in the preceding sections are examples of RTL circuits.
- **DTL:** *Diode-transistor logic,* which is similar to RTL but adds a diode to each input circuit.
- **TTL:** *Transistor-transistor logic* uses two transistors, one configured to work as a switch and the other configured to work as an amplifier. The switching transistor is used in the input circuits and the amplifier transistor is used in the output circuits. The amplifier allows the gate's output to be connected to a larger number of inputs than RTL or DTL circuits.

  In a TTL circuit, the switching transistors are special transistors that have two or more emitters. Each input is connected to one of these emitters so that the separate inputs all control the same collector-emitter circuit. The switching transistor's base is connected to the Vcc supply voltage and the collector is connected to the base of the amplifying transistor. Figure 3-9 shows a typical TTL circuit.

  Although you can build TTL circuits using individual transistors, ICs with TTL circuits are readily available. The most popular types of TTL ICs are designated by four-digit numbers starting with '74'. Several hundred types of 7400-series ICs are available, though many provide advanced logic circuits that you aren't likely to use for home electronics projects.

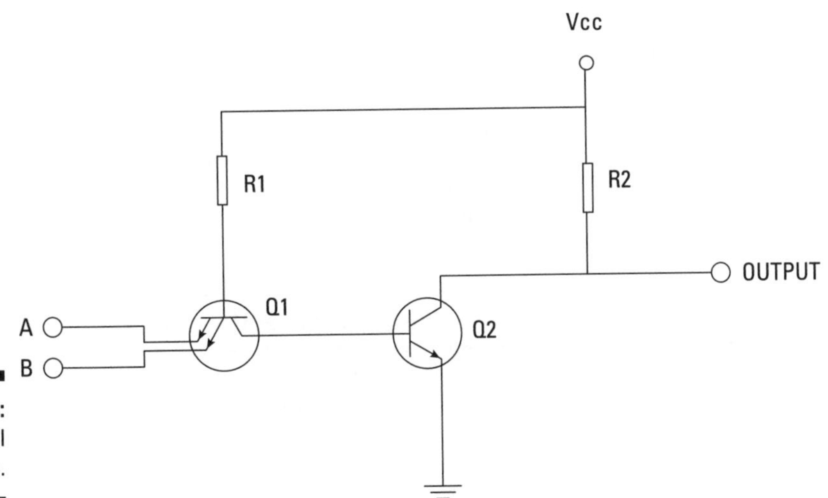

Figure 3-9: A typical TTL gate.

The ICs listed in Table 3-1 provide several basic logic gates in a single package. *Quad* means 'four' in the first part listed; *hex* means 'six' in the third part and, as we explained in the preceding chapter, a NOT gate is also known as an *inverter.*

- **CMOS:** *Complementary metal-oxide semiconductor logic* refers to logic circuits built with a special type of transistor called a *metal oxide semiconductor field effect transistor* (fortunately known as MOSFET for short). The physics of how a MOSFET differs from a standard bipolar transistor aren't all that important unless you want to become an IC designer. What is important is that MOSFETs use much less power, can switch states much faster and are significantly smaller than bipolar transistors. These differences make MOSFETs ideal for modern IC, which often require millions of transistors on a single chip.

Table 3-1 7400-Series TTL Logic Gates

Number	*Description*
7400	Quad two-input NAND gate (four NAND gates)
7402	Quad two-input NOR gate (four NOR gates)
7404	Hex inverter (six NOT gates)
7408	Quad two-input AND gate (four AND gates)
7432	Quad two-input OR gate (four OR gates)
7486	Quad two-input XOR gate (four XOR gates)

Apart from drawing less power and operating more quickly than TTL circuits, CMOS circuits work much like TTL circuits. In fact, CMOS chips are designed to be interchangeable with comparable TTL chips.

CMOS logic chips have a four-digit part number that begins with the number '4' and so are often called 4000-series chips. As with the 7400 series of TTL logic chips, several hundred types of 4000-series chips are available. Table 3-2 lists the ones that provide basic logic gates.

Table 3-2 4000-Series CMOS Logic Gates

Number	*Description*
4001	Quad two-input NOR gate (four NOR gates)
4009	Hex inverter (six NOT gates)
4011	Quad two-input NAND gate (four NAND gates)
4030	Quad two-input XOR gate (four XOR gates)
4071	Quad two-input OR gate (four OR gates)
4077	Quad two-input XNOR gate (four XNOR gates)
4081	Quad two-input AND gate (four AND gates)

CMOS logic circuits are very sensitive to static electricity. As a result, take special precautions when you handle them. Make sure that you discharge yourself properly by touching a grounded metal surface before you touch a CMOS chip. For maximum protection, wear an antistatic wrist band. For more information about static precautions, refer to Book I, Chapter 4.

The following sections describe several popular 4000-series ICs and present a few projects that use them.

Making use of the versatile 4000-series logic gates

The 4000-series CMOS logic circuits include ICs that provide several logic gates in a single package. Figure 3-10 shows the *pinout connections* (a schematic of what each input and output pin does) for six popular 4000-series chips. Each of these six chips contains four two-input logic gates in a 14-pin *DIP* (that is, a dual in-line package). Power, which can range from +3 V to +15 V, is connected to pin 14, and ground is connected to pin 7.

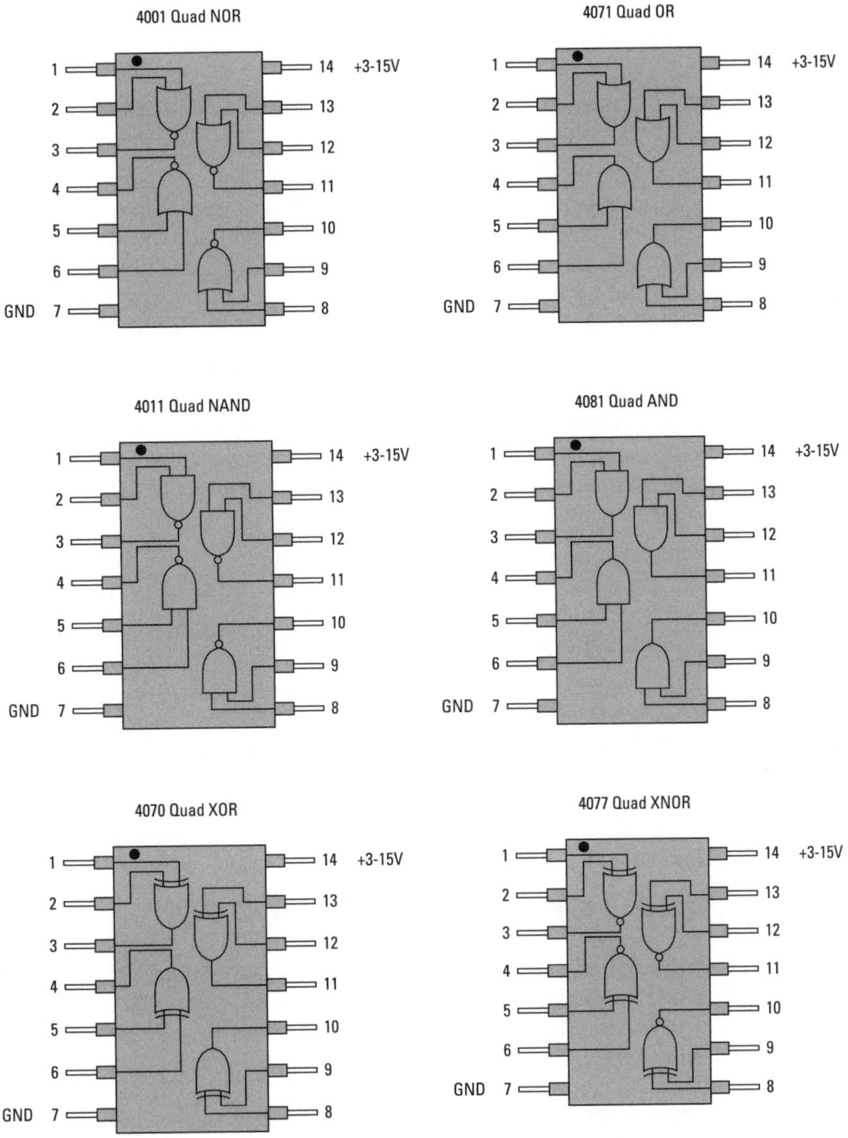

4001 Quad NOR

1 2 3 4 5 6 GND 7 — 14 +3-15V 13 12 11 10 9 8

4071 Quad OR

1 2 3 4 5 6 GND 7 — 14 +3-15V 13 12 11 10 9 8

4011 Quad NAND

1 2 3 4 5 6 GND 7 — 14 +3-15V 13 12 11 10 9 8

4081 Quad AND

1 2 3 4 5 6 GND 7 — 14 +3-15V 13 12 11 10 9 8

4070 Quad XOR

1 2 3 4 5 6 GND 7 — 14 +3-15V 13 12 11 10 9 8

4077 Quad XNOR

1 2 3 4 5 6 GND 7 — 14 +3-15V 13 12 11 10 9 8

Figure 3-10: Pinout chart for 4000-series quad two-input logic gate chips.

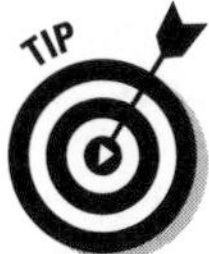

The 4001 and 4011 chips contain four NOR and NAND gates respectively, and because they're universal gates you can use them in combination to create other types of gates. So make sure that you stock up on 4001 or 4011 chips so that you can create any type of logic circuit you need.

Before you start building circuits with CMOS logic chips, here are a few tips for working with 4000-series chips:

- The outputs can source as much as 10 milliamperes (mA) with a 9 V power supply. At 6 V, the maximum is about 5 mA, which is just enough to light an LED.

  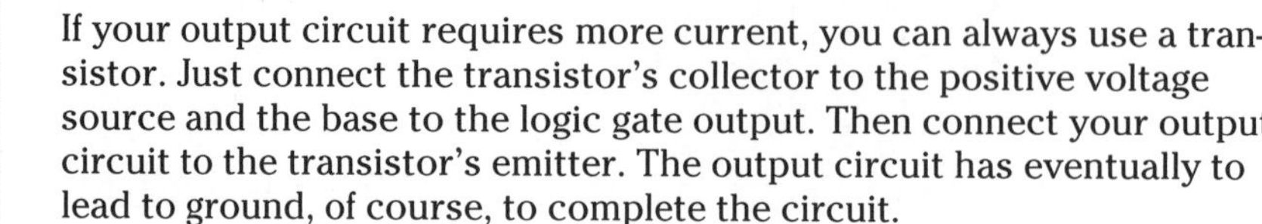

  If your output circuit requires more current, you can always use a transistor. Just connect the transistor's collector to the positive voltage source and the base to the logic gate output. Then connect your output circuit to the transistor's emitter. The output circuit has eventually to lead to ground, of course, to complete the circuit.

- The input pins of CMOS logic chips are notorious for picking up stray signals in the form of electrical noise. Although you don't need to do so in experimental breadboard circuits, in a real-world circuit you have to connect all unused input pins to the positive supply voltage or to ground. (You don't need to connect the unused outputs to anything – just the inputs.)
- In addition to connecting unused inputs to positive voltage or ground, a good idea is to place a small capacitor (47 nF is typical) across the power leads (pins 7 and 14). This capacitor helps to ensure that the input voltage stays constant.

- Don't forget that CMOS chips are very susceptible to damage from small amounts of static electricity. Be sure to ground yourself by touching a metal object before you touch CMOS chips.

Building projects with the 4011 Quad Two-Input NAND Gate

The 4011 Quad Two-Input NAND Gate is a popular CMOS logic gate IC. As its name suggests, this IC contains four two-input NAND gates. The earlier Figure 3-10 illustrates its pinouts, along with those for several other quad two-input gate chips.

You can purchase a 4011 IC at Maplin or online from any of the large electronic component distributors such as Farnell (`http://uk.farnell.com/`), RS Components (`http://uk.rs-online.com`) or Digi-Key (`www.digikey.co.uk`). Make sure you order one in a DIP package for breadboard insertion rather than a surface-mount package for production printed circuit boards.

As we describe in this minibook's Chapter 2, NAND gates (along with NOR gates) are *universal gates,* which means that you can construct any other type of gate using nothing but NAND gates combined in various ways. Projects 3-4 to 3-7 walk you through the step-by-step process of building various types of gate circuits using only NAND gates:

- **Project 3-4** uses just one NAND gate in a 4011. The two inputs of the NAND gate are connected to pushbuttons and the output is connected to an LED. When you build this project, you can visualise the operation of a NAND gate: the LED is on unless you press both buttons.

  Figure 3-11 shows Project 3-4 assembled on a solderless breadboard. This figure gives you a good idea of how to connect the components to the breadboard. The project description itself provides detailed instructions. The circuits for Projects 3-5, 3-6 and 3-7 are similar enough in appearance that you can use Figure 3-11 as a guide for the overall appearance of your finished projects.

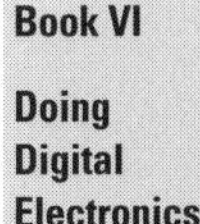

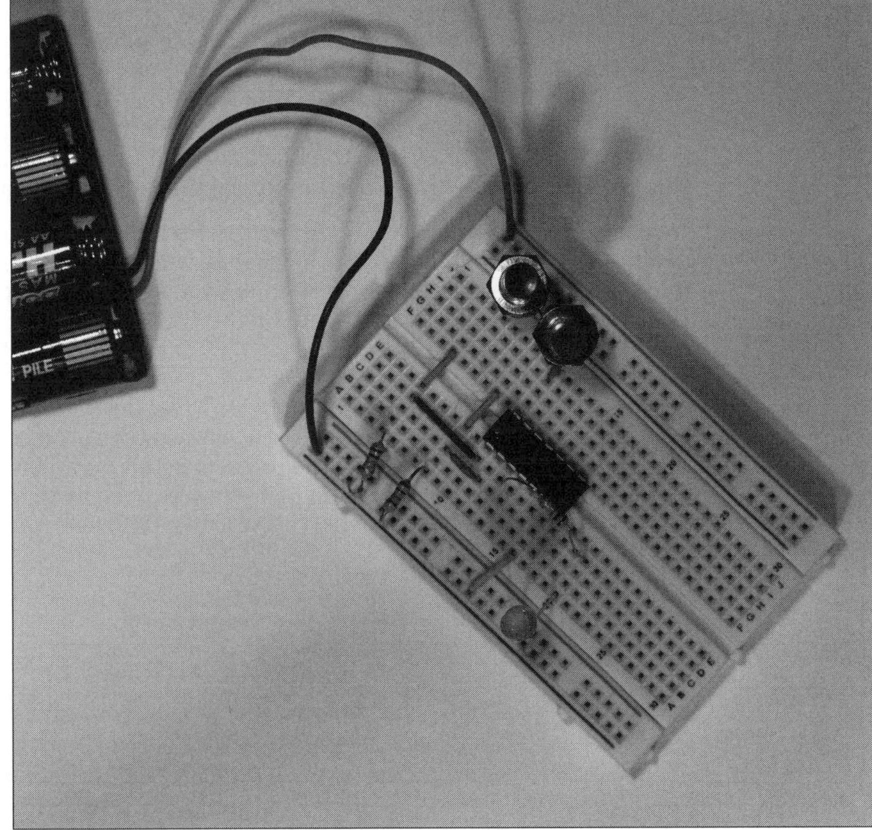

Figure 3-11: A NAND gate circuit that uses a CMOS logic chip (Project 3-4).

- **Project 3-5** uses two NAND gates on the 4011 to create an AND gate. Because a NAND gate is nothing more than an AND gate whose output is inverted, you can create an AND gate from a NAND gate by inverting the NAND gate's output.

  This inversion works because of a fundamental rule of logic: if you invert a value twice, you get the original value. Thus, if you invert an AND gate once, you get a NAND gate; if you invert it again, you're back to an AND gate.

  Luckily you can easily turn a NAND gate into a single-input inverter (that is, a NOT gate) by connecting the single input to both inputs of the NAND gate. This connection causes the two inputs to always be the same: either both are high or both are low. In a NAND gate, if both inputs are high, the output is low, and if both inputs are low, the output is high. Thus, wiring the inputs of a NAND gate together has the effect of inverting the input.

- **Project 3-6** shows you how to create an OR gate using three NAND gates. As Chapter 2 of this minibook explains, a NAND gate is the same as an OR gate whose inputs are inverted. Therefore, to create an OR gate using NAND gates, you invert the two inputs with the NAND gates configured as inverters (that is, with their inputs wired together). The output from these inverters is sent to the inputs of the third NAND gate.
- **Project 3-7** uses all four NAND gates on the 4011 chip to create a NOR gate, which is nothing more than an OR gate whose output is inverted. Thus, you create an OR gate by using the technique in Project 3-6 and then configure the fourth NAND gate on the 4011 chip as an inverter to invert the output of the OR gate.

These projects build on one another, and so don't tear down your breadboard after completing each project. Instead, use each assembled project as a starting point for the next one. If you build the projects in this way, you can simply scan the steps to see what resistors and jumper wires need to be moved for each project. (The 4011 IC itself, the LED and the two switches are in the same locations for all four projects.)

Project 3-4: A CMOS NAND Gate

In this project, you use a 4011 CMOS chip to build a NAND gate. The output of a NAND gate is high unless both inputs are high. When both inputs are high, the output goes low.

The 4011 IC has four independent gate circuits. In this project, you use just one of the four.

The inputs to this gate are fed through two normally open pushbutton switches. When a switch isn't pressed, the corresponding input is low. When the switch is pressed, the corresponding input goes high.

The output from this gate is sent through an LED, and so the LED is on when the output is high and off when the output is low.

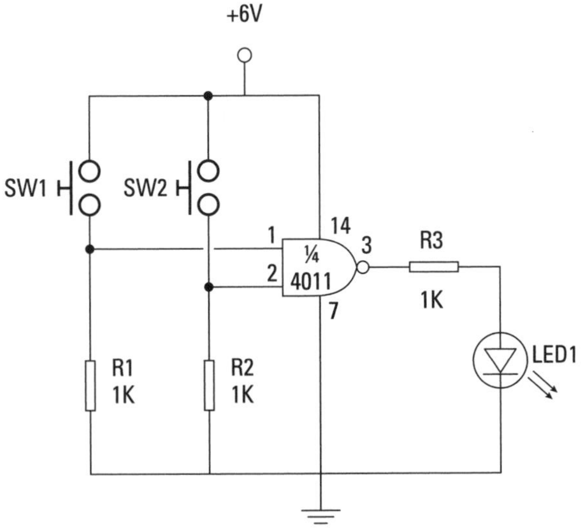

Parts List

1 Four-cell AAA battery holder

4 AAA batteries

1 4011 CMOS Quad Two-Input NAND Gate

1 Red LED

3 1 kΩ ¼ W resistors (brown-black-red)

2 Normally open pushbuttons

Miscellaneous jumper wires

NAND Gate Truth Table

Input A	Input B	Output
0	0	1
0	1	1
1	0	1
1	1	0

Steps

1. **Insert the 4011 IC:**

   Pin 1 goes in hole E10.

2. **Insert resistors R1, R2 and R3:**

   R1 – 1 kΩ: B4 to ground bus
   R2 – 1 kΩ: B8 to ground bus
   R3 – 1 kΩ: D12 to D20

3. **Insert LED1:**

   Cathode (short lead): Ground bus
   Anode (long lead): A20

4. **Insert the jumper wires:**

   Jumper 1: E4 to F4
   Jumper 2: E8 to F8
   Jumper 3: D4 to D10
   Jumper 4: C8 to C11
   Jumper 5: A16 to ground bus
   Jumper 6: J10 to positive bus

5. **Insert switches SW1 and SW2:**

   SW1: J4 to positive bus
   SW2: J8 to positive bus

6. **Connect the batteries:**

   Red lead: Positive bus
   Black lead: Negative bus

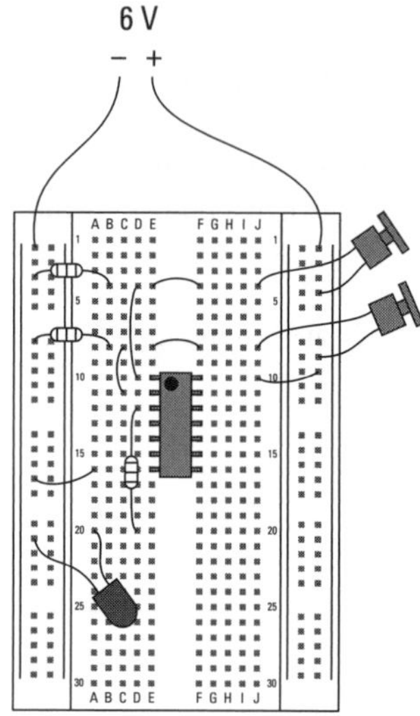

Project 3-5: A CMOS AND Gate

In this project, you use two of the NAND gates in a 4011 Quad NAND Gate IC to build an AND gate. The output of an AND gate is high if both inputs are high. If either input is low or both inputs are low, the output is low.

To create the AND gate, you send the output from the first NAND gate to the second NAND gate, which you configure as an inverter by tying its inputs together. Inverting the output of a NAND gate creates an AND gate.

The inputs to this gate are fed through two normally open pushbutton switches. When a switch isn't pressed, the corresponding input is low. When the switch is pressed, the corresponding input goes high.

The output from this gate is sent through an LED, and so the LED is on when the output is high and off when the output is low.

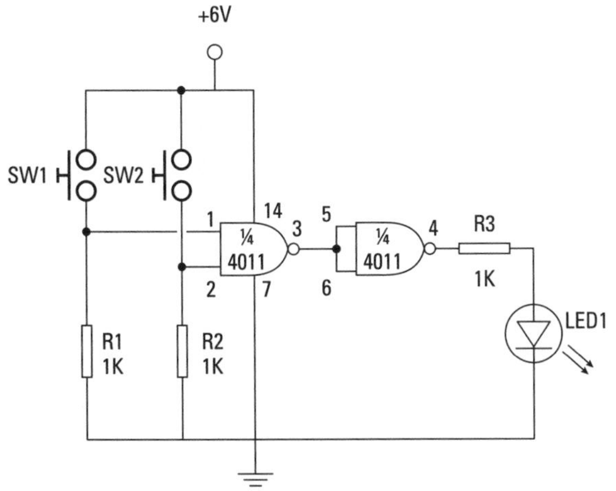

Parts List

- 1 Four-cell AAA battery holder
- 4 AAA batteries
- 1 4011 CMOS Quad Two-Input NAND Gate
- 1 Red LED
- 3 1 kΩ ¼ W resistors (brown-black-red)
- 2 Normally open pushbuttons
- Miscellaneous jumper wires

AND Gate Truth Table

Input A	Input B	Output
0	0	0
0	1	0
1	0	0
1	1	1

Steps

1. **Insert the 4011 IC:**
   Pin 1 goes in hole E10.
2. **Insert resistors R1, R2 and R3:**
   R1 – 1 kΩ: B4 to ground bus
   R2 – 1 kΩ: B8 to ground bus
   R3 – 1 kΩ: B13 to B20
3. **Insert LED1:**
   Cathode (short lead): Ground bus
   Anode (long lead): A20
4. **Insert the jumper wires:**
   Jumper 1: E4 to F4
   Jumper 2: E8 to F8
   Jumper 3: D4 to D10
   Jumper 4: C8 to C11
   Jumper 5: D12 to D14
   Jumper 6: C14 to C15
   Jumper 7: A16 to ground bus
   Jumper 8: J10 to positive bus
5. **Insert switches SW1 and SW2:**
   SW1: J4 to positive bus
   SW2: J8 to positive bus
6. **Connect the batteries:**
   Red lead: Positive bus
   Black lead: Negative bus

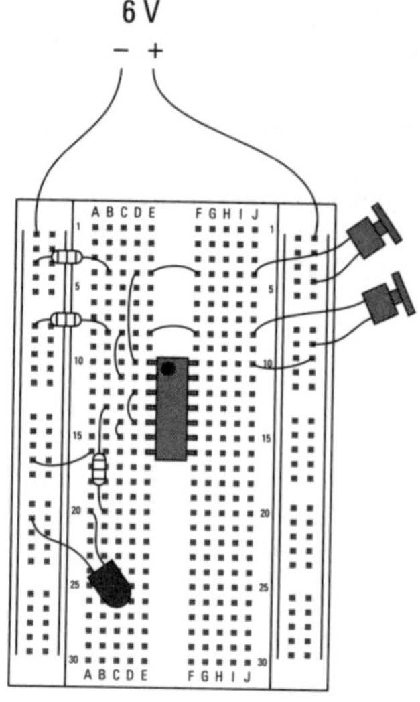

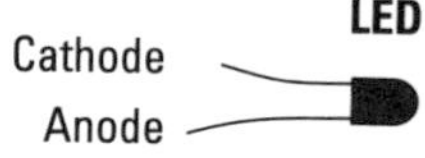

Project 3-6: A CMOS OR Gate

In this project, you use three of the NAND gates in a 4011 Quad NAND Gate IC to build an OR gate. The output of an OR gate is high if either input is high or both of the inputs are high. If both inputs are low, the output is low.

An OR gate can be built from NAND gates because an OR gate is the same thing as a NAND gate whose inputs are inverted. Thus, each of the inputs for this circuit is connected to a NAND gate configured as an inverter. Then the outputs from these inverters are sent to a third NAND gate.

The inputs to the OR gate are fed through two normally open pushbutton switches. When a switch isn't pressed, the corresponding input is low. When the switch is pressed, the corresponding input goes high.

The output from the OR gate is sent through an LED, and so the LED is on when the output is high and off when the output is low.

Parts List

1 Four-cell AAA battery holder
4 AAA batteries
1 4011 CMOS Quad Two-Input NAND Gate
1 Red LED
3 1 kΩ ¼ W resistors (brown-black-red)
2 Normally open pushbuttons
Miscellaneous jumper wires

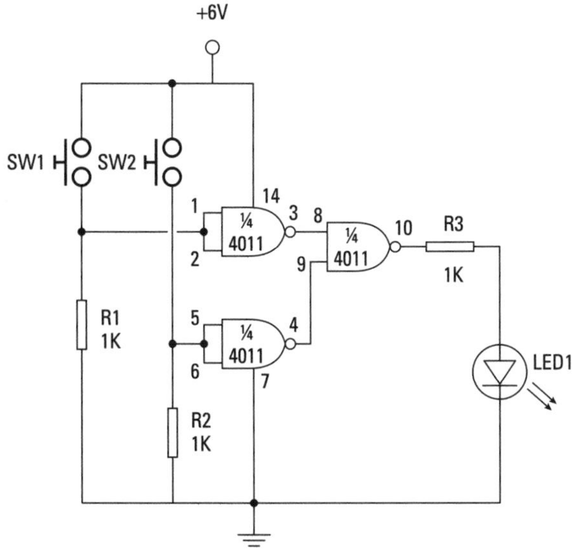

OR Gate Truth Table

Input A	Input B	Output
0	0	0
0	1	1
1	0	1
1	1	1

Steps

1. **Insert the 4011 IC:**

   Pin 1 goes in hole E10.

2. **Insert resistors R1, R2 and R3:**

   R1 – 1 kΩ: B4 to ground bus
   R2 – 1 kΩ: B8 to ground bus
   R3 – 1 kΩ: B20 to G20

3. **Insert LED1:**

   Cathode (short lead): Ground bus
   Anode (long lead): A20

4. **Insert the jumper wires:**

   Jumper 1: E4 to F4
   Jumper 2: E8 to F8
   Jumper 3: D4 to D10
   Jumper 4: C8 to C14
   Jumper 5: B10 to B11
   Jumper 6: B14 to B15
   Jumper 7: D12 to G16
   Jumper 8: D13 to G15
   Jumper 9: H14 to H20
   Jumper 10: A16 to ground bus
   Jumper 11: J10 to positive bus

5. **Insert switches SW1 and SW2:**

   SW1: J4 to positive bus
   SW2: J8 to positive bus

6. **Connect the batteries:**

   Red lead: Positive bus
   Black lead: Negative bus

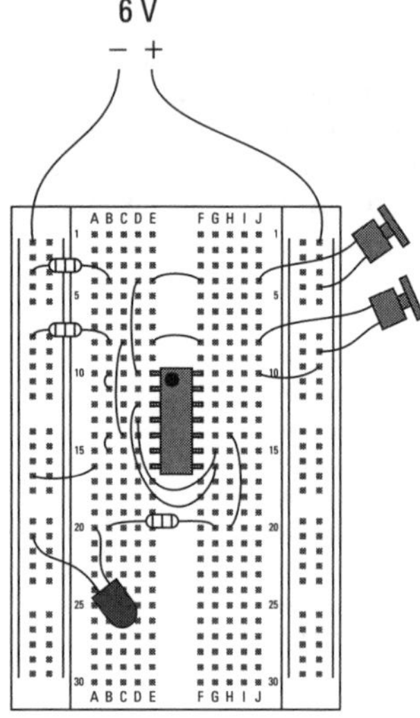

Project 3-7: A CMOS NOR Gate

In this project, you use all four of the NAND gates in a 4011 Quad NAND Gate IC to build a NOR gate. The output of a NOR gate is low if either input is HIGH or both the inputs are high. If both inputs are low, the output is high.

A NOR gate is simply an OR gate whose output is inverted by a NOT gate. Thus, you use the first three NAND gates of the 4011 to create an OR gate, following the steps in Project 3-6. Then you use the fourth NAND gate to invert the OR gate's output.

The inputs to the NOR gate are fed through two normally open pushbutton switches. When a switch isn't pressed, the corresponding input is low. When the switch is pressed, the corresponding input goes high.

The output from the NOR gate is sent through an LED and so the LED is on when the output is high and off when the output is low.

Parts List

1 Four-cell AAA battery holder
4 AAA batteries
1 4011 CMOS Quad Two-Input NAND Gate
1 Red LED
3 1 kΩ 1/4 W resistors (brown-black-red)
2 Normally open pushbuttons
Miscellaneous jumper wires

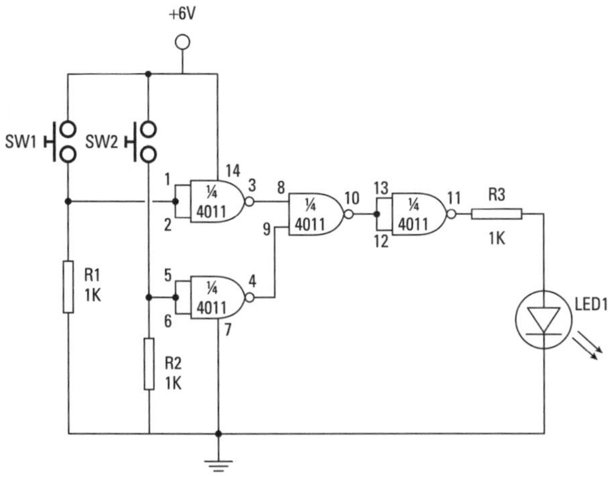

NOR Gate Truth Table

Input A	Input B	Output
0	0	1
0	1	0
1	0	0
1	1	0

Steps

1. **Insert the 4011 IC:**

   Pin 1 goes in hole E10.

2. **Insert resistors R1, R2 and R3:**

   R1 – 1 kΩ: B4 to ground bus
   R2 – 1 kΩ: B8 to ground bus
   R3 – 1 kΩ: B20 to G20

3. **Insert LED1:**

   Cathode (short lead): Ground bus
   Anode (long lead): A20

4. **Insert the jumper wires:**

   Jumper 1: E4 to F4
   Jumper 2: E8 to F8
   Jumper 3: D4 to D10
   Jumper 4: C8 to C14
   Jumper 5: B10 to B11
   Jumper 6: B14 to B15
   Jumper 7: D12 to G16
   Jumper 8: D13 to G15
   Jumper 9: G11 to G12
   Jumper 10: H12 to H14
   Jumper 11: I13 to I20
   Jumper 12: A16 to ground bus
   Jumper 13: J10 to positive bus

5. **Insert switches SW1 and SW2:**

   SW1: J4 to positive bus
   SW2: J8 to positive bus

6. **Connect the batteries:**

   Red lead: Positive bus
   Black lead: Negative bus

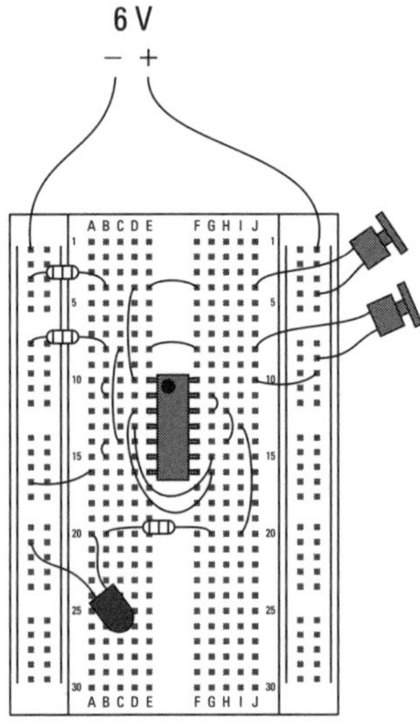

Chapter 4

Interfacing to Your Computer's Parallel Port

In This Chapter

- Getting to know the parallel port
- Programming the parallel port
- Displaying numbers
- Increasing the current

Electronics by itself can do some pretty amazing things, but when you combine it with the power of your computer you can start to dream big: complicated toys such as remote-controlled automatons, useful equipment such as watering routines for the garden and special lighting schemes for your living room.

Just imagine: you can entertain the children by giving them a fright on Halloween; or rig up some simple home automation to let potential burglars believe that someone's at home when you're out; or create something more often found in a Bond villain's lair. Almost anything is possible – the only limit is your imagination (though devising over-complicated ways of disposing of spies is way beyond the scope of this book!)

Building such systems may take you some time, depending on how complex you want them to be, but it needn't cost you loads of money. To start with all you need is an old computer – the sort you can find thrown away in a skip – and a circuit board that activates mechanical relays in response to output from the computer. You then run a simple software program on the computer to send signals to the relays via the computer's parallel-printer port, to make electronic events happen. Alternatively, you could add parallel ports to a more up-to-date computer using an expansion card or USB adaptor (available online and from stores like Maplin – just search for 'parallel card' or 'USB parallel'). To install a card you'd have to follow the manufacturer's instructions in opening up the back of your computer and inserting the card into the right slot. The adaptor just plugs into a USB port but may require you to install driver software for it to work. Overall, it's a much better idea - and much less work! - to get hold of an old computer from somewhere.

In this chapter, you discover how to create your own circuits that you can control via a parallel-printer port, so that you can build your own impressive contraptions. We introduce you to the parallel port and how to program it, and suggest kit for increasing the current for more powerful applications to amaze your friends and family. They may even come to think of you as a genius – even if it's an evil one bent on world domination (white cat on your lap is optional)!

Understanding the Parallel Port

Until a few years ago, all computers came equipped with a *parallel port,* which was used mostly to connect to a printer. Today, most printers connect to computers via USB (Universal Serial Bus) ports. USB has many advantages over parallel port, the most significant being faster data transfer and smaller cables.

The lowly parallel port has one advantage over USB ports, however: it makes creating your own circuits that interface directly with the port easy. These circuits can control low-current devices such as light-emitting diodes (LEDs), or activate transistors or mechanical relays that in turn activate high-current devices such as motors, incandescent lamps or sound systems.

After you build the circuits to connect to a parallel port, creating a software program on the computer that sends data to the parallel port is a doddle. When you run this program, your circuit detects the data sent to the parallel port to control LEDs or other low-power circuits.

Peering into the makeup of a parallel port

A standard parallel port has eight data pins, which are essentially transistor-transistor logic (TTL) outputs, with +5 V high representing 1 and 0 V low representing 0. In fact, the first parallel-printer ports designed for the IBM PC back in 1980 used 7400-series logic chips. Check out Chapter 3 of this minibook for all about TTL and logic chips.

The TTL logic levels used by the parallel port allow you to create logic circuits that interface with the output from a parallel port. You build a circuit that connects to a parallel port, use software on the computer to send data to the parallel port and your circuit can respond to the data you send.

Unfortunately, few computers today come with a built-in parallel port. Before you waste your time building circuits to interface with a parallel port, you need to find a computer that has one. Your best bet is to scavenge for an old machine. Look for one from before the year 2000, because most computers of that age have a parallel port, and one running Windows 95 or Windows 98 operating systems (O/S). If you can't find such an old Windows O/S, consider installing one of the free distributions of the Linux O/S available (just fire up a search engine and search for 'Linux' to get started).

A computer that old costs you next to nothing. Ask friends and family first and then look in car boot sales or local junk or computer repair shops. The computer isn't going to be much good for anything else, but dedicating a computer solely to your electronics work is a good idea.

If you wire the circuit that interfaces with the parallel port incorrectly, you risk damaging the computer's internal circuitry. Far better to fry an old computer than your latest laptop with all your precious files on it!

If you prefer, you can choose controllable circuit boards that use the more modern USB interface instead of the old parallel port. This chapter explains parallel-port programming to give you a basic understanding of how to control circuit boards from a PC, but USB boards like the Arduino are very capable and more widely available. Go to `www.arduino.cc` to find out more about Arduino and where to buy the boards. Or look out for the Wiley book on the subject, *Arduino Projects For Dummies* by Brock Craft.

Connecting with the DB25 connector and its pins

A parallel-printer port uses a standard type of connector called a *DB25 connector,* which has 25 pins. Each of these pins serves a different purpose in the port's communication with a printer.

Like most data connectors, DB25 connectors come in male and female variants. The male DB25 connectors consist of 25 pins and the female connectors have 25 holes. Figure 4-1 shows how the pins and holes of a DB25 female and male connector are numbered. As you can see, the connector consists of two rows of pins. The top row has 13 pins and the bottom row has 12. The pin in the top-right corner of the female connector is designated as pin 1; the pin at top left is pin 13; pin 14 is at bottom right; and pin 25 is at bottom left.

The pin numbers for a male connector are the mirror image of the numbers of the pins in the female connector. This arrangement of pins – or *pinout* – is necessary so that when the male connector is plugged into the female connector the pins connect properly (for example, pin 1 connects to pin 1, pin 2 connects to pin 2 and so on).

DB25 Parallel Port Pins

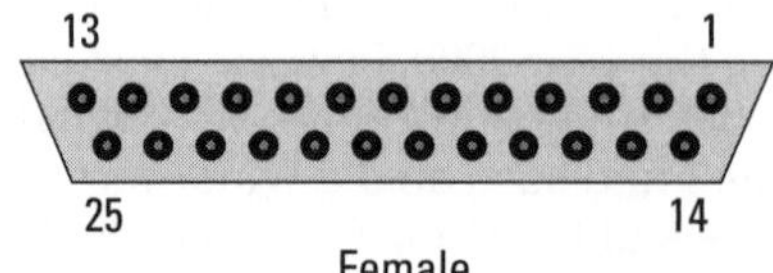

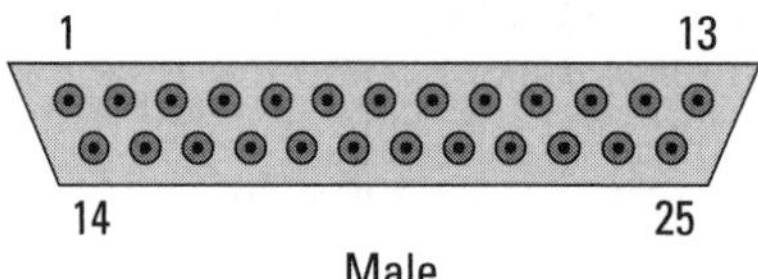

Figure 4-1: Pinouts for a DB25 connector.

Pinning your hopes on the pinout assignments

Table 4-1 lists the pinout assignments for a standard parallel port.

Table 4-1 Pinouts for a Standard Parallel-Printer Port

Pin	Name	Input or Output	Description
1	STROBE	Output or Input	Low when data is present on the data pins
2	D0	Output	Data bit 0
3	D1	Output	Data bit 1
4	D2	Output	Data bit 2
5	D3	Output	Data bit 3
6	D4	Output	Data bit 4
7	D5	Output	Data bit 5
8	D6	Output	Data bit 6
9	D7	Output	Data bit 7
10	ACK	Input	Low when data has been read
11	BUSY	Input	High when the printer is busy
12	PE	Input	High when the printer is out of paper
13	SEL	Input	High when the printer is ready

Pin	*Name*	*Input or Output*	*Description*
14	LINEFEED	Output or Input	Advances the printer
15	ERROR	Input	High when an error condition exists
16	RESET	Output or Input	High when the printer is reset
17	SELECT	Output or Input	High when the printer is offline
18	GND0	Neither	Ground connection
19	GND1	Neither	Ground connection
20	GND2	Neither	Ground connection
21	GND3	Neither	Ground connection
22	GND4	Neither	Ground connection
23	GND5	Neither	Ground connection
24	GND6	Neither	Ground connection
25	GND7	Neither	Ground connection

For the purposes of this chapter, the pins you're most interested in are numbered 2–9, which are collectively called the *data port.* When you connect the data port to a printer, its eight pins are capable of sending 1 byte of data at a time to the printer. When you connect it to a circuit of your own design, its pins operate as eight separate logic outputs, which you can use as inputs to your own logic circuits.

A parallel port also features four additional output pins (numbered 1, 14, 16 and 17) called the *control port,* which you can also use for output. When you connect the control port to a printer, these pins are used to control the operation of the printer. One of them, called the *strobe,* indicates that a new byte of data is available on the data pins; when the strobe pin goes high, the printer reads a byte of data from the data pins. Another control-port pin resets the printer.

The five pins numbered 10–13 and 15 that make up the *status port* allow the printer to send information back to the computer. One of the status-port pins lets the printer tell the computer that it's ready to receive data via the data port. Another pin lets the printer know that it has finished reading data from the data port. A third pin informs the computer that the printer is out of paper. The other status pins have similar functions.

When you connect the status port to a circuit of your own design, its pins can be used to send information to the computer. You can use the status pins to tell the computer that a switch has been closed, that a light sensor detects light or that a water-level sensor has reached a certain level.

Unfortunately, the programming required to detect the presence of input on one of the status-port pins is a bit too complex for this chapter. If you're an experienced C or C++ programmer, you'll have no trouble writing programs to read the status-port pins. But C and C++ programming are beyond the scope of this book, and so we focus on sending output to the parallel-port data pins.

The output pins of a parallel port use a +5 V high signal to represent 1 and 0 V to represent 0. The amount of current that each pin can source is relatively small – typiacally around 10–12 milliamperes (mA). That's enough current to drive an LED, but for anything more demanding you need a way to isolate the output load from the parallel port itself. To do that, you can use individual transistors or an integrated circuit (IC) designed specifically for this purpose. For details, read 'Using Darlington Arrays to Drive High-Current Outputs', later in this chapter.

Designing a parallel-port circuit

Figure 4-2 shows the two methods for connecting output devices to a parallel-port data pin. The top circuit drives an LED directly. A current-limiting resistor is required to prevent the LED from pulling too much current and damaging the LED, the parallel port itself or both.

The bottom circuit in Figure 4-2 shows how to use a transistor to switch a circuit by using the output from a parallel-port data pin. As you can see, the data pin is connected through a resistor to the transistor's base. When the data pin goes high, the transistor is turned on, allowing current to flow through the collector-emitter circuit.

Using a transistor driver this way provides two benefits:

- The transistor can switch more than the 10–12 mA that the parallel port's data pin can source directly.
- The collector-emitter circuit can operate at a different voltage from the one that the data pin's +5 V high signal provides.

When you build Project 4-1 later in this chapter, you see an example of a circuit that uses the first method to connect to parallel-port output. This project connects eight LEDs to the eight data pins. With the cable to connect this circuit to a computer's parallel port and the right software in place on the computer, you can write simple scripts that flash the LEDs on and off in just about any sequence you desire.

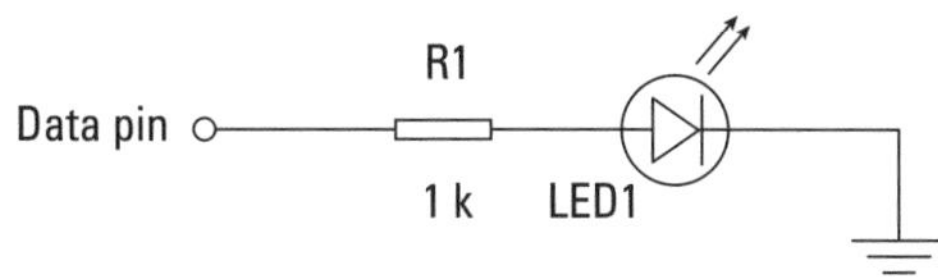

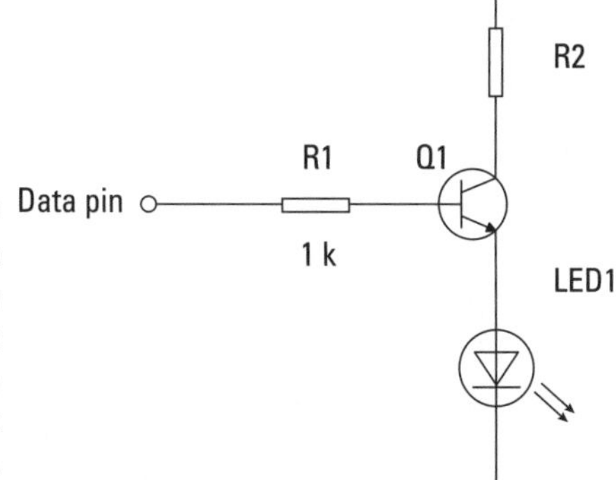

Figure 4-2: Two ways to connect an LED to a parallel-port output pin.

Working with DB25 connectors

To build your own circuits that interface with a parallel port, you need to use a DB25 connector that attaches to your circuit. The standard parallel port on a computer uses a DB25 female connector, which means that you need to provide a DB25 male connector for your circuits.

You have several ways to fashion a DB25 connector for your circuits. The simplest method is to find an old parallel printer cable, cut it six or seven centimetres past the male end, strip back the insulation and separate the 25 wires that run inside the cable. Strip the ends of each of these wires and then use your ohmmeter (see Book I, Chapter 8) to match up wire with the individual pins on the connector.

Keep good notes so that you know which wires to use for your circuits.

Alternatively, you can purchase a male DB25 connector and solder wires to each of the pins. Figure 4-3 shows a connector that we made to use along with Projects 4-1 and 4-2, which appear later in this chapter. We solder eight red wires to pins 2–9 (the data pins) and one black wire to pin 25 (one of the ground pins).

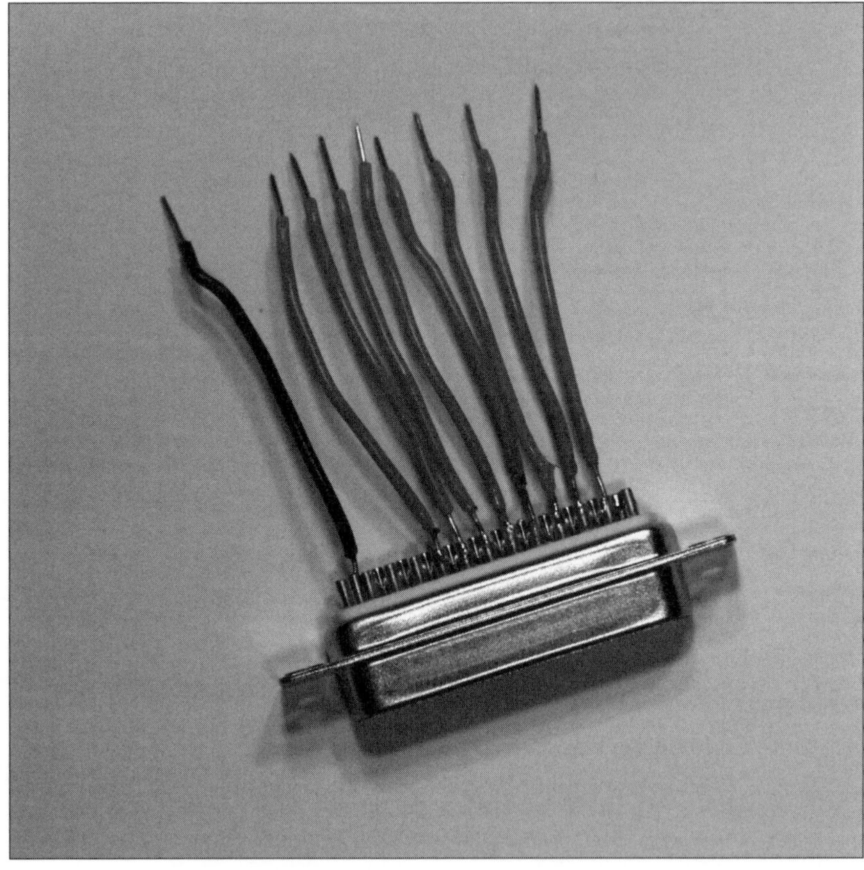

Figure 4-3: A male DB25 connector with leads soldered to the data pins and one of the ground pins.

If you plan to complete Project 4-1 or Project 4-2, you need to assemble a similar cable yourself. All you need are a male DB25 connector, some 18-AWG or 22-AWG solid wire, some solder, a soldering iron and a vice to hold the connector while you solder (check out Book I, Chapter 7, for all about soldering). You may also want to get some magnifying goggles; although they make you look like the mad professor in *Back to the Future,* they do make the job much easier by helping you see the project more closely.

Controlling Parallel-Port Output from an MS-DOS Prompt: Programming from a PC

Project 4-1 (see 'Building a Parallel-Port LED Flasher', later in this chapter) covers the details of building a simple breadboard project that lets you control eight LEDs from a computer's parallel port.

You can plug the project directly into the parallel port on the back of your computer, or you can use a short male-to-female DB25 printer cable to connect the project to the computer (see the earlier section 'Connecting with the DB25 connector and its pins'). Either way, you need some special software on the computer to send data to the parallel port to turn on the LEDs. Fortunately, you don't have to be a programmer to create this software, because software that does this job is readily available on the Internet.

The simplest software is designed to work with a popular kit that lets you connect relays to the parallel port. We describe this kit, called *Kit 74,* in 'Using a Kit 74 Relay Controller', later in this chapter. To find the software online, just search for 'Kit 74 software'. The software is contained in an archive file named `k74_dos.zip`. Download it to your computer and extract its contents to a folder on your hard drive named `C:\k74_dos`.

To use the commands in an O/S prior to Windows 7, click the Windows Start button, click Run, type `cmd` and press the Enter key. An MS-DOS command window opens (see Figure 4-4).

Figure 4-4: You can run an MS-DOS command to send data to the parallel port.

Type the following command, and press Enter:

```
cd \k74_dos
```

This command transfers you to the folder in which you saved the Kit 74 commands. Then you can type and run the commands to test your parallel-port circuits.

The Kit 74 DOS software consists of three commands – `RELAY`, `DELAY` and `WAITFOR` – that you can run from a command prompt. We explain these three commands in the following sections.

Using the RELAY command

The `RELAY` command sends a single byte of data to the parallel port. Each of the eight output pins is set high or low, depending on the byte you send. This command sets all eight pins to high:

```
RELAY FF
```

The following command sets all eight outputs to low:

```
RELAY 00
```

Unfortunately, most versions of the `RELAY` command available on the Internet have a bug that requires you to issue the command twice to get it to work. So you have to enter the command `RELAY FF` twice in sequence to turn on all the output pins.

You need to specify the output data as a single *hexadecimal* (or *hex*) number (which is a special number made of the numerals 0–9 and the letters A–F). Table 4-2 shows the hexadecimal numbers to use for each of the eight output pins.

Table 4-2 Hex Values for Data Output Pins

Data Pin	*Hex Value*
1	01
2	02
3	04
4	08
5	10
6	20
7	40
8	80

To turn all the pins on, use the value FF. To turn them all off, use the value 00.

To turn more than one pin on or off, you need first to calculate the eight-bit binary number equivalent of the pins you want to set. To turn on pins 1, 2, 3 and 8, for example, you use the binary value 10000111. (Notice that pin 1 is represented by the rightmost bit of the binary number and that pin 8 is the leftmost bit.) To bone up on binary, flip to Chapter 1 of this minibook.

After you work out the binary number for the pins you want to set, split it in half so that you have two four-bit numbers. In the example that sets pins 1, 2, 3 and 8, the first binary number is 1000 and the second is 0111.

Finally, look up each four-bit number in Table 4-3 to determine the single hexadecimal digit to use. For this example, the first four-bit number converts to 8 and the second four-bit number converts to 7. Combining these two numbers gives you the hexadecimal number 87. Thus, the command to turn on pins 1, 2, 3 and 8 is:

```
RELAY 87
```

You have to enter this command twice to get it to work.

Table 4-3 Converting Binary Values to Hexadecimal Digits

Binary Value	*Hexadecimal Digit*	*Binary Value*	*Hexadecimal Digit*
0000	0	1000	8
0001	1	1001	9
0010	2	1010	A
0011	3	1011	B
0100	4	1100	C
0101	5	1101	D
0110	6	1110	E
0111	7	1111	F

Creating a command script

You can easily create command scripts that can execute a series of RELAY commands in sequence. Just use Notepad (the simple text editor that comes free with all versions of Windows) to enter a series of RELAY commands, one on each line. Save the file to the C:\k74_dos folder, using the file extension .bat. Then you can run your script by typing the name of your script file (without the .bat extension) in the MS-DOS command window.

Figure 4-5 shows a simple script created with Notepad. Notice that this script contains two identical `RELAY` commands because of the bug in the `RELAY` command that requires you to run the command twice to get it to work.

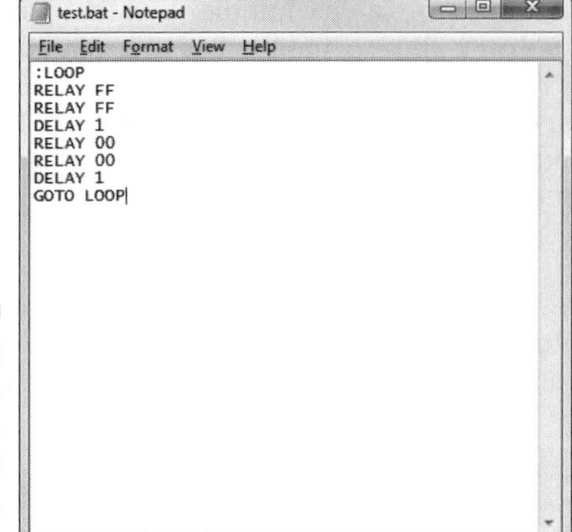

Figure 4-5: A simple script in a Notepad window.

On an older Windows computer (the kind that's likely to have a parallel port to play with), you find Notepad in the Start menu under Accessories.

Besides the `RELAY` command (and the `DELAY` and `WAITFOR` commands, which we explain in the following section), you often use two special MS-DOS commands in command scripts. The first command is called a *label,* which lets you give a name to a line in your script. Labels are indicated by a colon followed by a short word. `:LOOP` is a typical label.

The second command, called `GOTO`, creates a program loop by telling the script to jump to a label. Labels and `GOTO` commands are always used together, as follows:

```
:LOOP
RELAY FF
RELAY FF
GOTO LOOP
```

This sequence of commands causes the two `RELAY` commands to be executed. Then the `GOTO` command sends the script back to the `:LOOP` label command, which executes the `RELAY` commands again. The commands between the `GOTO` command and the label are executed again and again until you stop the script by pressing Ctrl+C or closing the command window.

Listing 4-1 shows a simple script that quickly flashes the LEDs in sequence, beginning with LED1. When the script gets to LED8, it reverses the direction and flashes the LEDs back to LED1. Then a `GOTO` command sends the script back to the `:LOOP` label to repeat the flashing. The resulting effect is that the LEDs sweep back and forth indefinitely.

This script is named `CYLON.BAT` because it resembles the flashing eyes of the evil Cylons from the classic science-fiction TV series *Battlestar Galactica.*

Listing 4-1: The `CYLON.BAT` Script

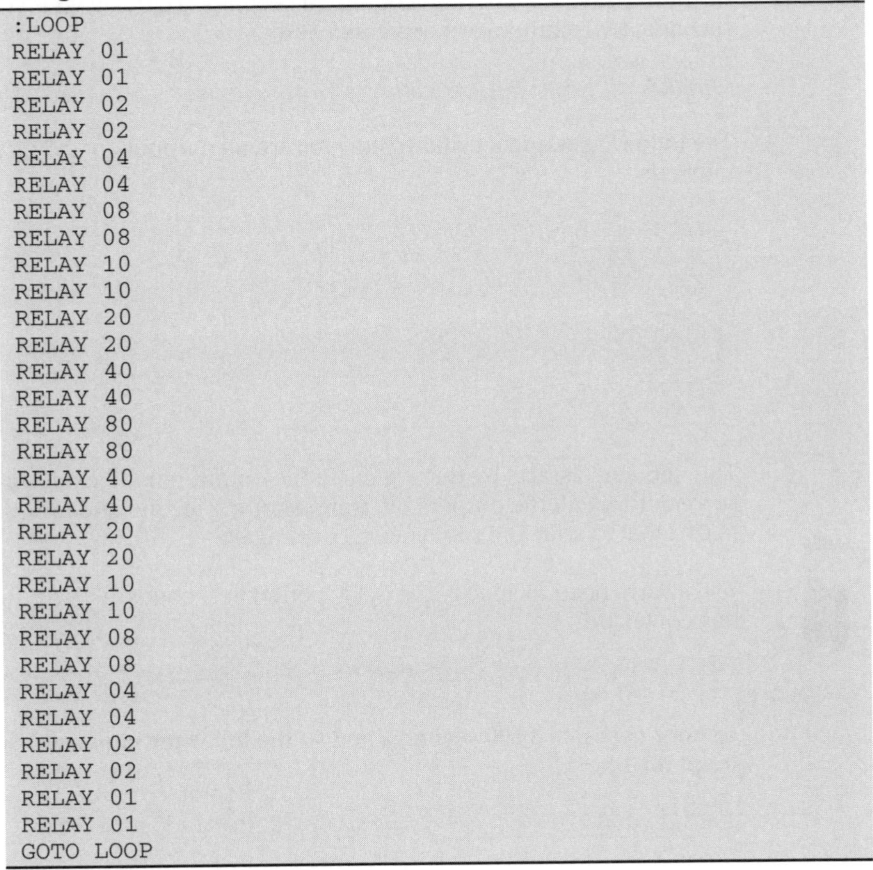

```
:LOOP
RELAY 01
RELAY 01
RELAY 02
RELAY 02
RELAY 04
RELAY 04
RELAY 08
RELAY 08
RELAY 10
RELAY 10
RELAY 20
RELAY 20
RELAY 40
RELAY 40
RELAY 80
RELAY 80
RELAY 40
RELAY 40
RELAY 20
RELAY 20
RELAY 10
RELAY 10
RELAY 08
RELAY 08
RELAY 04
RELAY 04
RELAY 02
RELAY 02
RELAY 01
RELAY 01
GOTO LOOP
```

Seeing why timing is everything: DELAY and WAITFOR

The Kit 74 software includes two timing commands that let you add delays to your scripts. By incorporating delays, you can control the timing of the devices controlled by your parallel-port circuit. For example, you can turn pin 1 on, wait 5 minutes and then turn it off again.

The most useful of the timing commands is `DELAY`, which simply causes your script to pause for a certain number of seconds. To delay your script for 10 seconds, for example, use this command:

```
DELAY 10
```

The following sequence shows how to turn all outputs on and off at 1-second intervals:

```
:LOOP
RELAY FF
RELAY FF
DELAY 1
RELAY 00
RELAY 00
DELAY 1
GOTO LOOP
```

This sequence starts by turning on all the output pins. Then it waits 1 second, turns all the outputs off, waits another second and jumps to the `LOOP` label to start the sequence all over again.

You always need to specify the delay period in seconds. To wait 1 minute, use this command:

```
DELAY 60
```

An hour contains 3,600 seconds, and so the following command delays the script for 1 hour:

```
DELAY 3600
```

The second timing command is `WAITFOR`, which waits to execute until a certain time of day arrives. To stop your script until 10:30 a.m., for example, use this command:

```
WAITFOR 10:30
```

Here's a sequence that turns all outputs on at 10:30 a.m. every day, leaves them on for an hour and turns them off:

```
:LOOP
WAITFOR 10:30
RELAY FF
RELAY FF
DELAY 3600
RELAY 00
RELAY 00
GOTO LOOP
```


Building a parallel-port LED flasher

Project 4-1 presents a simple breadboard circuit that connects eight LEDs to the eight output pins of a parallel port. To complete this project, you need to have a computer with a parallel port and install the Kit 74 parallel-port software (refer to 'Controlling Parallel-Port Output from an MS-DOS Prompt: Programming the Port', earlier in this chapter).

You also need to build a parallel-port connector that you can attach to your circuit. Refer to the earlier Figure 4-3 for information on building the required connector.

After you assemble the circuit, you can test it by connecting it to your computer's parallel port, opening a command prompt and using the `RELAY` command to send data to the port. To turn on all eight LEDs, for example, run this command:

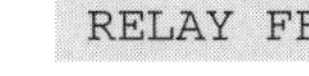

```
RELAY FF
```

The flaw in the `RELAY` command means that you need to run the command twice.

After the circuit works properly when you enter the `RELAY` command from the command prompt, try running the `CYLON.BAT` script from Listing 4-1 earlier in this chapter. This script flashes the LEDs sequentially from left to right and back again. The LEDs run repeatedly until you terminate the batch file by pressing Ctrl+C or closing the command window.

Figure 4-6 shows the finished parallel-port LED flasher circuit. You can see in the photo that we connect the male DB25 connector to one end of a parallel printer cable. The other end of this cable is connected to the computer.

Figure 4-6: The finished parallel-port LED flasher circuit (Project 4-1).

Project 4-1: A Parallel-Port LED Circuit

In this project, you build a circuit that connects an LED to each of the eight data pins on a parallel port. Then you write a program on the computer that flashes the LEDs to make sure that the circuit is working properly.

To complete this project, you need a male DB25 connector with solid 18-gauge wire soldered to pins 2–9 and to pin 25. You can build a connector like this yourself (refer to Figure 4-3, earlier in this chapter) or you can cut an old parallel printer cable apart, separate the wires and use an ohmmeter to single out the wires connected to required pins. For more information, refer to `Working with DB 25 Connectors´ earlier in this chapter.

You also need to locate and download the Kit 74 software. You can get it on several websites; just search for `Kit 74 software´. For more information, refer to `Controlling Parallel-Port Output from an MS-DOS Prompt: Programming the Port´ earlier in this chapter.

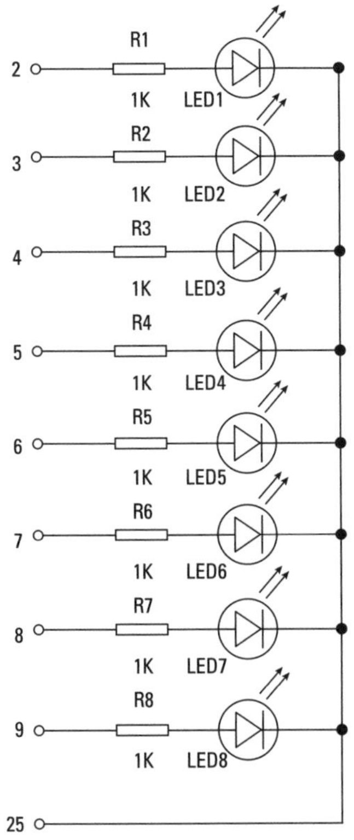

Parts List
1 Male DB25 connector with 18-AWG or 22-AWG wire on pins 2–9 and to 25
8 1 kΩ resistors (brown-black-red)
8 Red LEDs
1 5 cm jumper wire
1 Computer with a parallel-printer port

Steps

1. **Insert resistors R1 to R8:**

R1 – 1 kΩ:	D3 to F3
R2 – 1 kΩ:	D5 to F4
R3 – 1 kΩ:	D7 to F5
R4 – 1 kΩ:	D9 to F6
R5 – 1 kΩ:	D11 to F7
R6 – 1 kΩ:	D13 to F8
R7 – 1 kΩ:	D15 to F9
R8 – 1 kΩ:	D17 to F10

2. **Insert LEDs LED1 to LED8:**

   Cathode (short lead): Ground bus

   Anode (long lead): A3, A5, A7, A9, A11, A13, A15, A17
3. **Insert the DB25 wires:**

   Pin 2 to J3

   Pin 3 to J4

   Pin 4 to J5

   Pin 5 to J6

   Pin 6 to J7

   Pin 7 to J8

   Pin 8 to J9

   Pin 9 to J10

   Pin 25 to J11
4. **Insert the jumper wire:**

   I11 to ground bus
5. **Plug the male DB25 connector into the parallel port on the computer.**
6. **Turn the computer on.**
7. **Click the Windows Start button, choose Run, type CMD and press the Enter key.**

   An MS-DOS command window opens.
8. **Type the command CD \K74_DOS.**

   This command switches you to the folder that contains the Kit 74 software.
9. **Enter the command RELAY FF and press the Enter key; then do it again.**

   All eight LEDs come on, verifying that your circuit works.

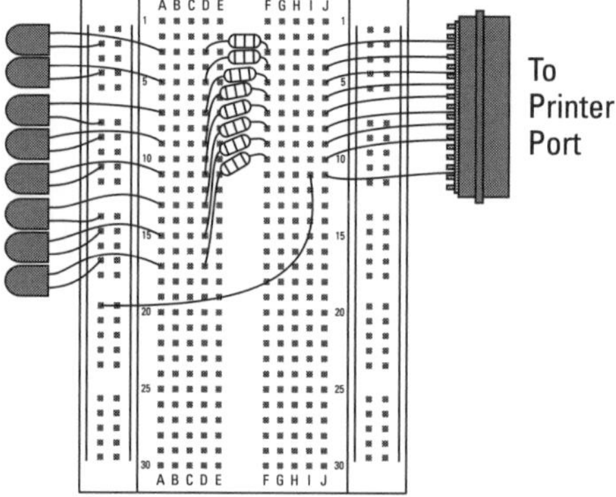

Sussing out Seven-Segment Displays

A *seven-segment display* is an array of seven LEDs arranged in a way that can display numerals as well as some alphabetic characters. You see this type of display most often on calculators, which have several seven-segment displays arranged side by side to display numbers. You can purchase an inexpensive (less than $2) seven-segment display at your local electronics-parts shop.

Introducing the seven-segment display

You may be wondering why we're talking about seven-segment displays in a chapter on parallel-port interfacing. The reason is that to use a seven-segment display for any practical purpose you have to connect the display to a digital circuit that's capable of controlling the individual segments to display meaningful information such as numerals or alphabetic letters. You *can* do that with logic gates (which we describe in this minibook's Chapter 3), but the resulting circuits are very complicated. A much easier option is to use the power of a computer to control the individual segments via a parallel-port connection.

Figure 4-7 shows how a single-digit seven-segment display module is usually wired up. As you can see, the segments themselves are referred to by the letters *a* to *g*. This particular display module is contained in a 14-pin *DIP* (a dual inline package), but only 8 of the pins are used. The anode of each LED segment is connected to one of the pins. The cathodes for all the segments are connected at pin 4. (This arrangement is called *common-cathode* wiring. You can also get seven-segment displays in which the anodes are connected to a common pin, called *common-anode* wiring.)

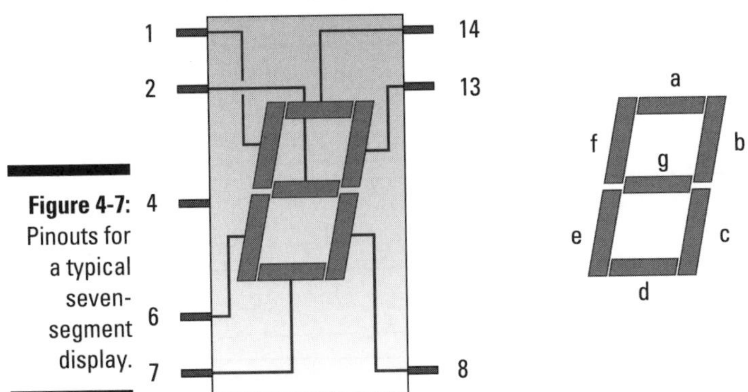

Figure 4-7: Pinouts for a typical seven-segment display.

To control a seven-segment display, you must first connect a positive voltage source to the anode of each of the seven segments. The cathode has to be connected to ground.

Be sure to use a current-limiting resistor in series with each anode to limit the current that flows through the LEDs.

To drive a seven-segment display from your computer's parallel port, connect the anode of each segment through a current-limiting resistor (1 kΩ is typical) to one of the data output pins. The most straightforward way to do that is to connect DATA1 (pin 2) to the *a* segment, DATA2 (pin 3) to the *b* segment and so on until DATA7 is connected to the *g* segment.

After you've connected a seven-segment display to the parallel port, you can form numerals or some alphabetic characters by sending the right data to the parallel port. Table 4-4 shows the data byte you need to send to display numerals 0–9.

Table 4-4 Data Values to Display Numerals on a Seven-Segment Display

Numeral	*Data Value*
0	3F
1	06
2	5B
3	4F
4	66
5	6D
6	7D

Numeral	Data Value
7	07
8	7F
9	6F

Thus, to display the numeral 5, use this `RELAY` command:

```
RELAY 6D
```

To understand why the data values in Table 4-4 are required to display numerals on a seven-segment display, remember that each of the segments in the display is connected to one of the data output pins of the parallel port. Thus, to light a particular combination of segments to form a numeral, you need to set the parallel port's output so that the data pins corresponding to the segments you want lit are high and the remaining pins are low.

To form the numeral 3, for example, segments *a, b, c, d* and *g* need to be turned on. Those segments are connected to data output pins 1, 2, 3, 4 and 7. Thus, you need to send a byte of data to the parallel port with the bit positions corresponding to pins 1, 2, 3, 4 and 7 set to the binary value `1` and the other bit positions set to binary `0`.

In a binary number, the bit positions are numbered right to left, and so the binary pattern you need to send to the parallel port to form the numeral 3 is:

```
01001111
```

The hexadecimal equivalent for this binary number is `4F`. Thus, the following command displays the numeral 3:

```
RELAY 4F
```

Listing 4-2 shows a script called `COUNTDOWN.BAT` that displays a NASA-style countdown from 9 to 0 at 1-second intervals. When the script reaches 0 that numeral flashes repeatedly until you cancel the batch file by pressing Ctrl+C or closing the command window. (See the 'Using the RELAY command' section earlier in the chapter to find out why each command is repeated.)

Listing 4-2: The COUNTDOWN.BAT Script

```
RELAY 6F
RELAY 6F
DELAY 1
RELAY 7F
RELAY 7F
DELAY 1
RELAY 07
RELAY 07
DELAY 1
RELAY 7D
RELAY 7D
DELAY 1
RELAY 6D
RELAY 6D
DELAY 1
RELAY 66
RELAY 66
DELAY 1
RELAY 4F
RELAY 4F
DELAY 1
RELAY 5B
RELAY 5B
DELAY 1
RELAY 06
RELAY 06
DELAY 1
:LOOP
RELAY 3F
RELAY 3F
RELAY 00
RELAY 00
GOTO LOOP
```

In this script, the `:LOOP` label appears near the end of the listing, not at the beginning. You can place labels anywhere you want in a script. The `GOTO LOOP` command at the end of the script causes the script to repeat the last four commands over and over until you interrupt the script by pressing Ctrl+C or closing the command window.

Building a seven-segment display countdown timer

Project 4-2 presents a breadboard circuit that connects a seven-segment display to seven of the eight output pins of a parallel port. As with the earlier Project 4-1, you need a computer with a parallel port and the Kit 74 software installed. You also need a parallel-port connector with wires soldered to the data pins and one of the common data ground pins. You can find information about the required software and the required cable earlier in the section 'Controlling Parallel-Port Output from an MS-DOS Prompt: Programming the Port'.

To test your assembled circuit, connect it to your computer's parallel port, open a command prompt and use the following `RELAY` command:

```
RELAY FF
```

All seven of the display's segments should light up.

If one doesn't light, carefully double-check your wiring. Also verify that your seven-segment display uses the same pinouts as the one listed in the project; some display modules have different packaging with different pin connections. Also, remember that you have to run the `RELAY` command twice to get it to work.

When the circuit checks out, run the `COUNTDOWN.BAT` script (in Listing 4-2, earlier in this chapter). The seven-segment display counts down the digits 9 to 0 at 1-second intervals. When it gets to zero, the display flashes until you terminate the batch file by pressing Ctrl+C or closing the command window.

Figure 4-8 shows the finished seven-segment-display circuit.

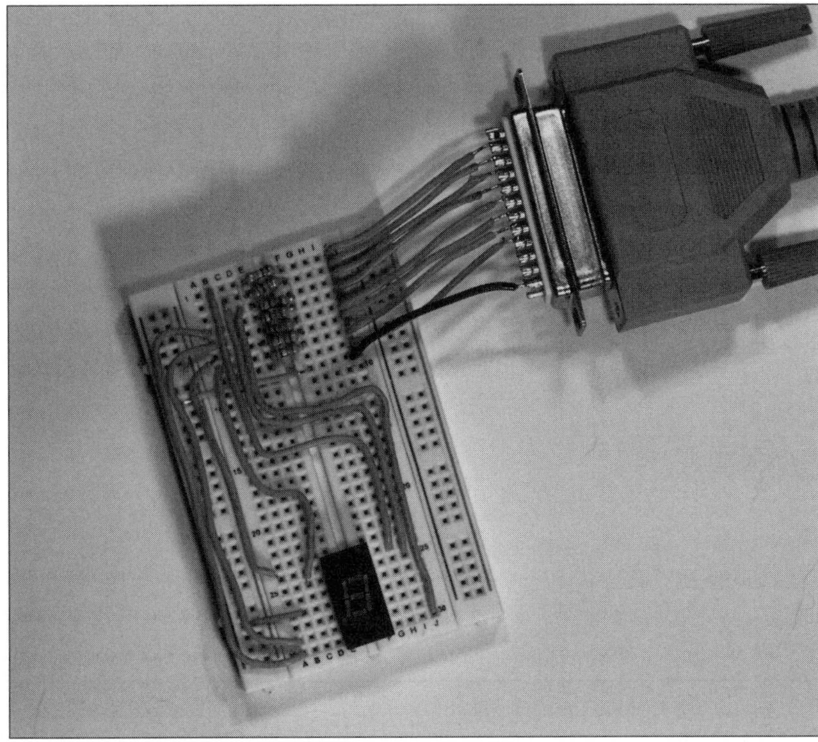

Figure 4-8: The finished parallel-port seven-segment display (Project 4-2).

Project 4-2: A Seven-Segment Display Circuit

In this project, you build a circuit that connects a seven-segment LED display to seven of the eight data pins on a parallel port. Then you write a program on the computer that flashes the segments in the correct sequence to count down the digits 9 to 0.

To complete this project, you need a male DB25 connector with solid 18-gauge wire soldered to pins 2–9 and to pin 25. You can build a connector like this yourself (refer to Figure 4-3, earlier in this chapter) or you can cut an old parallel printer cable apart, separate the wires and use an ohmmeter to single out the wires connected to required pins. For more information, check out 'Working with DB 25 Connectors' earlier in this chapter.

You also need to locate and download the Kit 74 software. You can get it on several websites; just search for 'Kit 74 software'. For more information, refer to 'Controlling Parallel-Port Output from an MS-DOS Prompt: Programming the Port' earlier in this chapter.

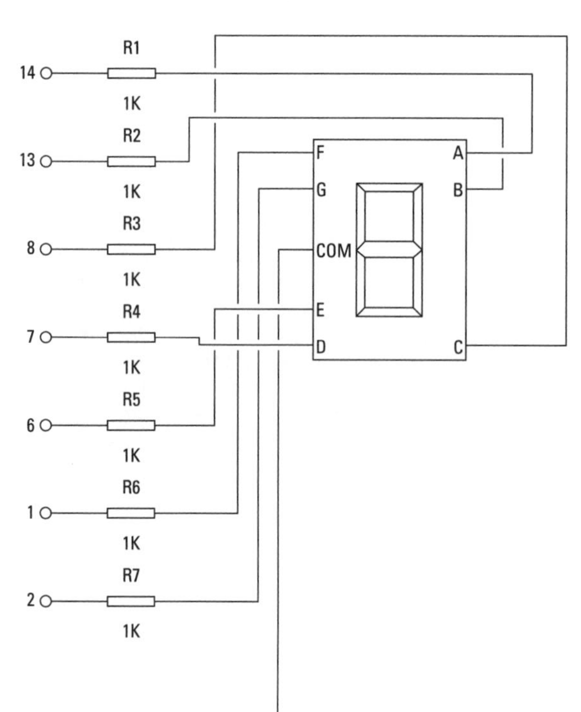

Parts List	
1	Male DB25 connector with 18-AWG or 22-AWG wire on pins 2–9 and 25
7	1 kΩ resistors (brown-black-red)
1	Seven-segment display module
9	Jumper wires (various lengths)
1	Computer with a parallel-printer port

Steps

1. **Insert resistors R1 to R8:**

   R1 – 1 kΩ: E1 to F1

   R2 – 1 kΩ: E2 to F2

   R3 – 1 kΩ: E3 to F3

   R4 – 1 kΩ: E4 to F4

   R5 – 1 kΩ: E5 to F5

   R6 – 1 kΩ: E6 to F6

   R7 – 1 kΩ: E7 to F7

2. **Insert the seven-segment display.**

   Pin 1 goes in E24 and pin 14 goes in F24.

3. **Insert the wires from the DB25 connector:**

   Pin 2 to J1

   Pin 3 to J2

   Pin 4 to J3

   Pin 5 to J4

   Pin 6 to J5

   Pin 7 to J6

   Pin 8 to J7

   Pin 9 to J8

   Pin 25 to J9

4. **Insert the jumper wires:**

   F9 to the ground bus

   B27 to the ground bus

   B1 to H24

   C2 to I25

   D3 to J30

   A4 to A30

   A5 to A29

   A6 to A24

   A7 to C25

5. **Plug the male DB25 connector into the parallel port on the computer.**

6. **Turn the computer on.**

7. **Click the Windows Start button, choose Run, type CMD and press the Enter key.**

   An MS-DOS command window opens.

8. **Type the command CD \K74_DOS.**

   This command switches you to the folder that contains the Kit 74 software.

9. **Enter the command RELAY FF and press the Enter key; then do it again.**

   All eight LEDs come on, verifying that your circuit works.

10. **Use Notepad or any other text editor to create a batch file with the program shown in Listing 4-2. Then, run it to verify that the program and your circuit work properly.**

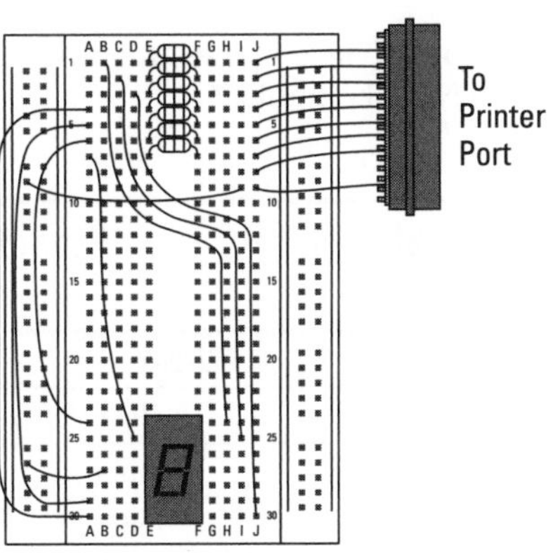

Driving up the Current

Two transistors can be connected one into the other so that the current amplified by the first transistor is amplified further by the second one. This arrangement, known as a *Darlington transistor* (sometimes called a *Darlington pair*), can switch much more current than the collector-emitter circuit of a standard transistor. You can use Darlington transistors to switch up to 500 mA from the output of a parallel-port data pin, which is enough current to drive a mechanical relay or a small electric motor.

But as the next section explains, you don't have to use individual Darlington transistors.

Using Darlington arrays for high-current outputs

You can use a specially designed IC for driving high-current loads from TTL-level inputs. The most common ICs of this type are the ULN2003, which has seven Darlington drivers in a 16-pin DIP package, and the ULN2803, which has eight drivers in an 18-pin DIP package. You can find these useful ICs on the Internet by searching for 'ULN2003' or 'ULN2803'.

Table 4-5 lists the pinouts for the ULN2003. As you can see, pins 1–7 are the input pins, which you can connect directly to the output pins from the parallel port. Pins 10–16 are the output pins, which you can connect to the circuit you want to control. Pin 8 connects to ground and pin 9 connects to a voltage source.

Table 4-5 also shows the pinouts for the ULN2803, which are similar to the ULN2003 pinouts.

Table 4-5 Pinouts for a ULN2003 and ULN2803 Darlington Array IC

ULN2003		*ULN2803*	
Pin	***Description***	***Pin***	***Description***
1	Input 1	1	Input 1
2	Input 2	2	Input 2
3	Input 3	3	Input 3

ULN2003		*ULN2803*	
Pin	***Description***	***Pin***	***Description***
4	Input 4	4	Input 4
5	Input 5	5	Input 5
6	Input 6	6	Input 6
7	Input 7	7	Input 7
8	Common ground	8	Input 8
9	Vss	9	Common ground
10	Output 7	10	Vss
11	Output 6	11	Output 8
12	Output 5	12	Output 7
13	Output 4	13	Output 6
14	Output 3	14	Output 5
15	Output 2	15	Output 4
16	Output 1	16	Output 3
		17	Output 2
		18	Output 18

The output circuit for a ULN2003/2803 is a little different from what you may expect. Instead of sourcing current for the load, the Darlington array sinks the current. Thus, the output pin is on the ground side of the load circuit, as shown in Figure 4-9. The voltage source (Vss) feeds the load circuit (in this case, a small motor) and the ULN2003.

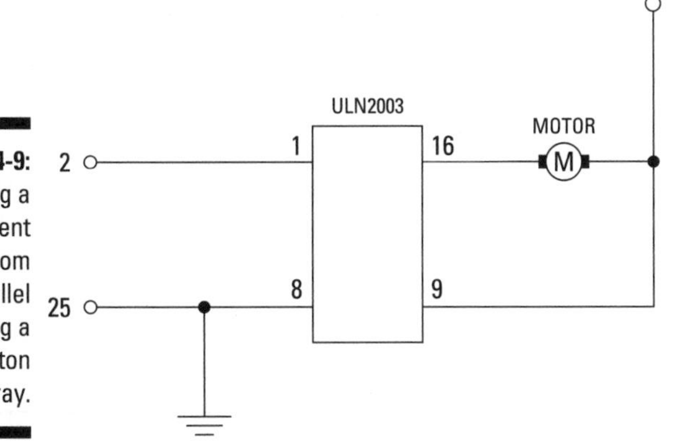

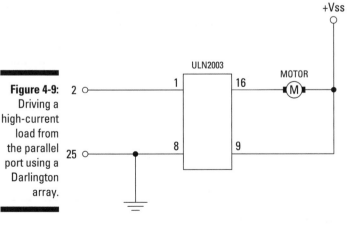

Figure 4-9: Driving a high-current load from the parallel port using a Darlington array.

If you want to use a ULN2003 or ULN2803 to drive an inductive load such as a relay or motor, use a Zener diode on the Vss pin, by connecting the cathode to the Vss pin and the anode to the GND pin. This diode prevents current from flowing in the wrong direction into the ULN2803 in case the relay or motor coil creates a large backward voltage spike, as coils are apt to do. Use a Zener diode with a voltage value just above the supply voltage and rated for 500 mW or above.

Building a motor driver

Project 4-3 presents a breadboard circuit that drives a small 3 V DC motor from a parallel port. This motor uses much more current than a parallel port can handle, and so you use a ULN2003 Darlington array IC to drive the motor.

As with the Projects 4-1 and 4-2 earlier in this chapter, you need a computer with a parallel port and the Kit 74 software installed, plus a parallel-port connector with wires soldered to the data pins and one of the common data ground pins. Please refer to those earlier projects for more information.

When the circuit is assembled and connected to your computer, you can test it by running the following command twice in a command window:

```
RELAY 01
```

The motor starts running. (If it doesn't start, check all your connections and check your commands.) To stop the motor, use this command twice:

```
RELAY 00
```

You can easily write a script to run the motor for a certain interval. Listing 4-3 shows a simple example. Here, the `MOTOR.BAT` file simply runs the motor for 30 seconds, turns it off for 30 seconds and then jumps to the `:LOOP` label to repeat the cycle.

Listing 4-3: The `MOTOR.BAT` Script

```
:LOOP
RELAY 01
RELAY 01
DELAY 30
RELAY 00
RELAY 00
DELAY 30
GOTO LOOP
```

Figure 4-10 shows the finished motor drive circuit.

Figure 4-10: The finished parallel-port motor driver (Project 4-3).

Project 4-3: A Parallel-Port Motor Driver

In this project, you build a circuit that connects a small 3 V motor to one of the eight data pins on a parallel port. Then you write a program on the computer that turns the motor on and off at 30-second intervals.

To complete this project, you need a male DB25 connector with solid 18-AWG or 22-AWG wire soldered to pins 2–9 and to pin 25. You can build a connector like this yourself (refer to Figure 4-3, earlier in this chapter), or you can cut an old parallel printer cable apart, separate the wires, and use an ohmmeter to single out the wires connected to required pins. For more information, refer to `Working with DB 25 Connectors´ earlier in this chapter.

You also need to locate and download the Kit 74 software. You can get it on several websites; just search for `Kit 74 software´ to find it. For more information, refer to `Controlling Parallel-Port Output from an MS-DOS Prompt: Programming the Port´ earlier in this chapter.

Parts List	
1	Male DB25 connector with 18-AWG or 22-AWG wire on pins 2 to 9 and 25
1	ULN2003 Darlington array
1	3 V DC motor
5	Jumper wires (various lengths)
1	Two-AA-battery holder
2	AA batteries
1	Computer with a parallel-printer port

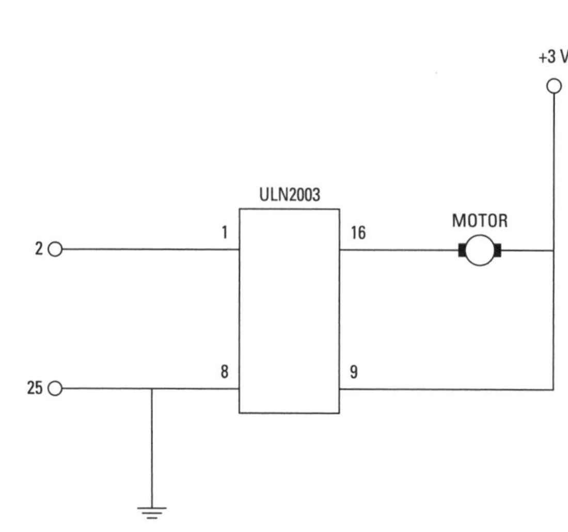

Steps

1. **Insert the ULN2003:**
   Pin 1 goes in hole E15 and pin 16 goes in hole F15.
2. **Insert the wires from the DB25 connector:**

Pin 2 to J1	Pin 7 to J6
Pin 3 to J2	Pin 8 to J7
Pin 4 to J3	Pin 9 to J8
Pin 5 to J4	Pin 25 to J9
Pin 6 to J5	

3. **Insert the jumper wires:**
   D1 to D15
   E1 to F1
   F9 to the ground bus
   A27 to the ground bus
   J27 to the positive bus
4. **Connect the motor.**
   One lead goes in the positive bus; the other lead goes in J15.
5. **Connect the batteries.**
   Positive goes anywhere in the positive bus; negative goes anywhere in the ground bus.
6. **Plug the male DB25 connector into the parallel port on the computer.**
7. **Turn the computer on.**
8. **Click the Windows Start button, choose Run, type CMD and press the Enter key.**
   An MS-DOS command window opens.
9. **Type the command CD \K74_DOS.**
   This command switches you to the folder that contains the Kit 74 software.
10. **Enter the command RELAY 01 and press the Enter key; then do it again.**
    The motor starts running.
11. **Enter the command RELAY 00 to stop the motor.**
12. **Use Notepad or any other text editor to create a batch file with the program shown in Listing 4-3. Then, run it to verify that the program and your circuit work properly.**

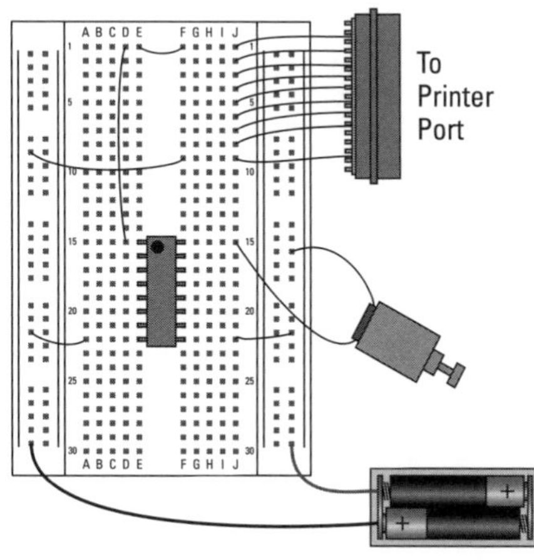

Chapter 5

Getting the Hang of Flip-Flops

In This Chapter

- Unlocking the key to latches
- Finding out about flip-flops
- Building basic flip-flop circuits

If the title of this chapter brings up images of strolling on the beach in high summer without tripping over, we apologise. The flip-flops that we're concerned with aren't the sort you wear on your feet, but the electronic kind. A *flip-flop* is a circuit that stores data, and as such flip-flops are the basis of modern computers.

In this chapter, you discover how to work latches which allow you to make logic circuits, and then go on to build simple flip-flops which allow you to make much faster logic circuits.

Looking at Latches

A *latch* is a logic circuit that has two inputs and one output. One of the inputs is called the *SET input* and the other is called the *RESET input.*

Latch circuits can be active-high or active-low. The difference is determined by whether the operation of the latch circuit is triggered by high or low signals on the inputs (refer to Chapter 2 of this minibook for an explanation of high and low in the context of logic circuits):

- **Active-high circuit:** Both inputs are normally tied to ground (low) and the latch is triggered by a momentary high signal on either of the inputs.
- **Active-low circuit:** Both inputs are normally high and the latch is triggered by a momentary low signal on either input.

In an active-high latch, both the SET and RESET inputs are connected to ground. When the SET input goes high, the output also goes high. When the SET input returns to low, however, the output remains high. The output of the active-high latch stays high until the RESET input goes high. Then, the output returns to low and goes high again only when the SET input is triggered again.

In other words, the latch remembers that the SET input has been activated. If the SET input goes high for even a moment, the output goes high and stays high, even after the SET input returns to low. The output returns to low only when the RESET input goes high.

On the other hand, in an active-low latch the inputs are normally held at high. When the SET input momentarily goes low, the output goes high. The output then stays high until the RESET input momentarily goes low.

In fact most latch circuits have a second output that's simply the first output inverted. So whenever the first output is high, the second output is low, and vice versa. These outputs are usually referred to as *Q* and $\overline{Q}$.

The notation $\overline{Sl}$ is usually pronounced 'bar Q' or 'Q bar', though just to confuse you some people pronounce it 'not Q'. The horizontal bar symbol over a label is a common logical shorthand for inversion. That is, $\overline{Q}$ is the inverse of Q. If Q is high, $\overline{Q}$ is low, and if Q is low, $\overline{Q}$ is high.

You can create an active-high latch fairly easily from a pair of NOR gates, as shown in Figure 5-1. (As we explain in Chapter 2 of this minibook, the output of a NOR gate is high if both inputs are low; otherwise, the output is low.) In this circuit, the SET input is connected to one of the inputs of the first NOR gate and the RESET input is connected to one of the inputs of the second NOR gate. The trick of the latch circuit is that the output of the NOR gates are cross-connected to the remaining NOR gate inputs. In other words, the output from the first NOR gate is connected to one of the inputs of the second NOR gate, and the output from the second NOR gate is connected to one of the inputs of the first NOR gate.

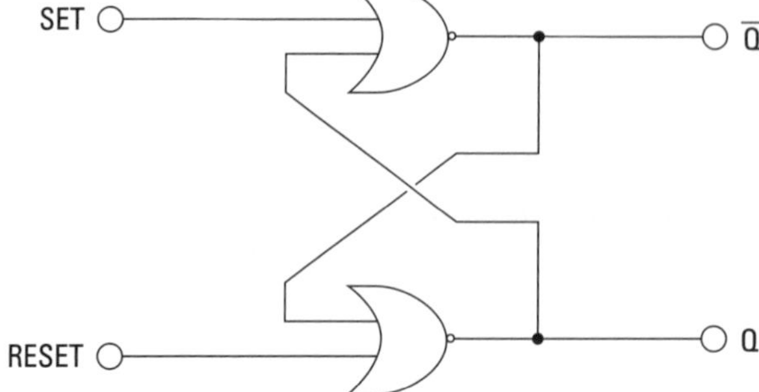

Figure 5-1: Schematic diagram for an active-high latch.

Figure 5-2 shows the schematic for an active-low latch. As you can see, the only difference between this schematic and the one shown in Figure 5-1 is that the active-low latch uses NAND gates instead of NOR gates. Notice also in this diagram that the inputs are referred to as $\overline{\text{SET}}$ and $\overline{\text{RESET}}$ rather than SET and RESET, which indicates that the inputs are active-low.

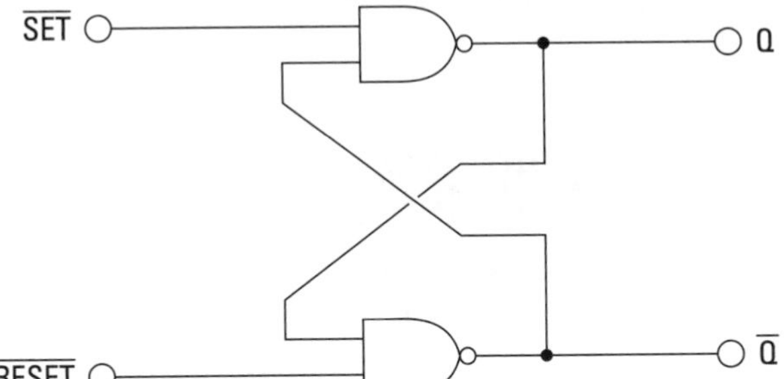

Figure 5-2: Schematic diagram for an active-low latch.

Projects 5-1 and 5-2, later in this section, show you how to build simple active-high and active-low latch circuits using a 4001 Quad 2-Input NOR Gate IC (integrated circuit) and a 4011 Quad 2-Input NAND Gate IC. The Q and $\overline{\text{Q}}$ outputs are used to drive light-emitting diodes (LEDs) so that you can see the state of the latch, and both inputs are controlled by normally open pushbuttons so that you can trigger the latch by momentarily pressing the buttons. Figures 5-3 and 5-4 depict the assembled active-high latch and the assembled active-low latch, respectively.

When you compare the schematic diagrams between these two projects, you see only two differences between them:

- **Gates:** The active-high circuit uses a 4001 IC, which contains NOR gates, whereas the active-low uses a 4011 IC, which contains NAND gates.
- **Resistor and switch positions:** The positions of R1 and R2 and SW1 and SW2 are reversed. In the active-high circuit, the resistors connect the two gate inputs to ground and the switches short the gate inputs to +6 V. In the active-low circuit, the resistors connect the gate inputs to +6 V and the switches short the gate inputs to ground.

Both these circuits use simple pushbutton switches to provide the trigger inputs, but we're sure that you can easily imagine other sources for the trigger pulse. For example, in a home alarm system, the $\overline{\text{SET}}$ input in an active-low latch may come from a window switch that breaks contact when the window is open, and the $\overline{\text{RESET}}$ input may come from a key lock on the alarm system's control panel.

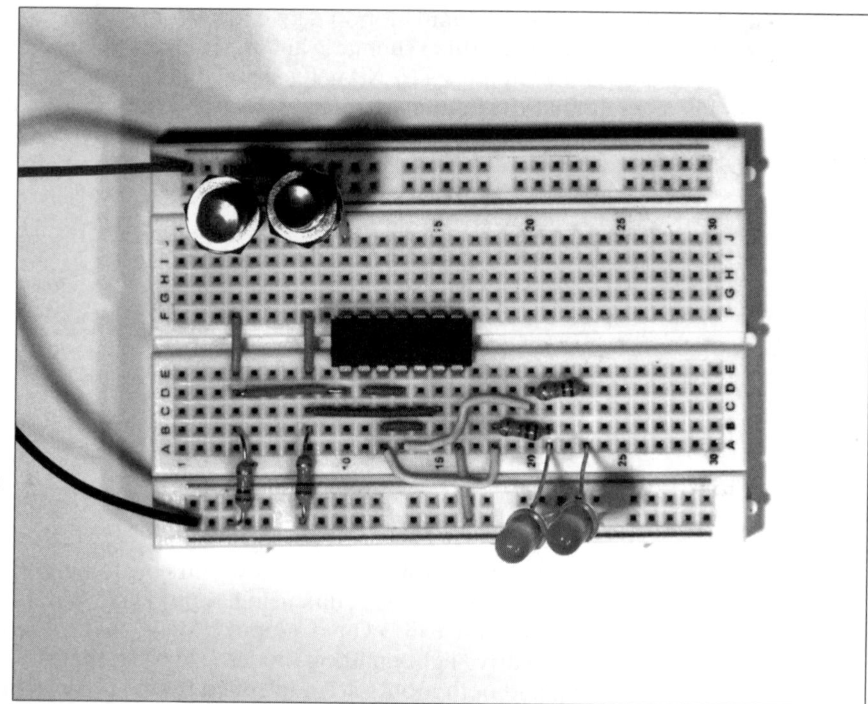

Figure 5-3: The assembled active-high latch (Project 5-1).

Figure 5-4: The assembled active-low latch (Project 5-2).

A latch with a SET and RESET input is often called an *SR latch*. The term *RS latch* is also used.

In some cases, you may need a latch in which one of the inputs is active-high and the other is active-low. For example, in the alarm system the key lock may send a high signal when the alarm is to be reset. Thus, the $\overline{Q}$ input for the alarm latch is active-low, but the RESET input is active high.

You can easily meet this requirement by adding an inverter to one of the inputs, as shown in Figure 5-5. Here, we use NAND gates to create an active-low latch but add a NOT gate to invert the RESET input. Thus, the $\overline{\text{SET}}$ input of this inverter is active-low and the RESET input is active-high.

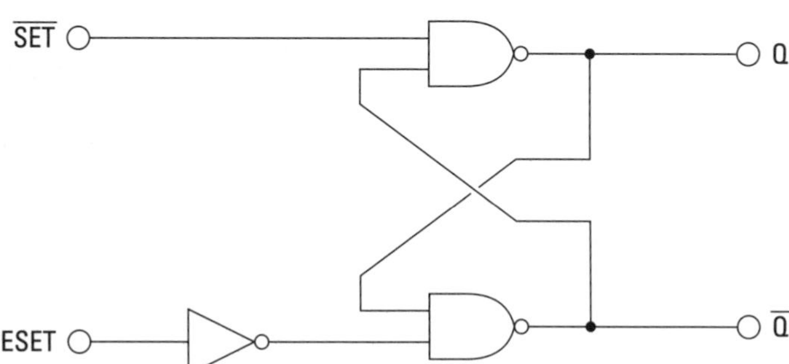

Figure 5-5: A latch in which $\overline{\text{SET}}$ is active-low and RESET is active-high.

Project 5-1: An Active-High Latch

In this project, you build a simple active-high latch circuit using a pair of NOR gates from a 4001 Quad 2-Input NOR Gate IC. You connect switches to the SET and RESET inputs and LEDs to the Q and $\overline{Q}$ outputs so that you can see the operation of the circuit. SW1 is RESET, SW2 is SET; LED1 is Q; and LED2 is $\overline{Q}$.

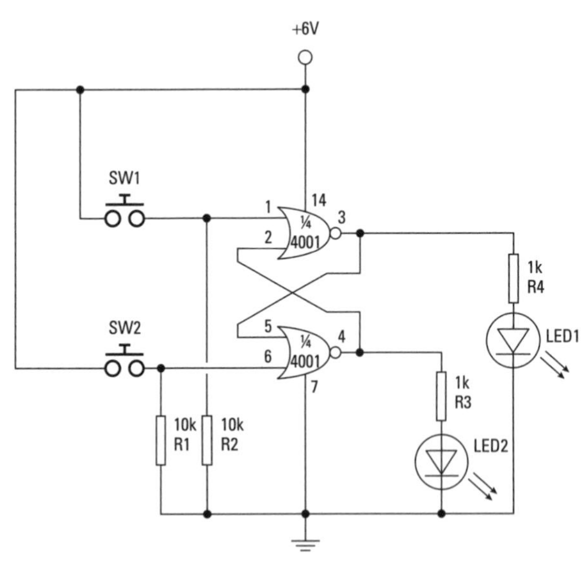

Parts List	
1	4001 Quad 2-Input NOR Gate
2	1 kΩ resistors (brown-black-red)
2	10 kΩ resistors (brown-black-orange)
2	Red LEDs
2	Normally open pushbuttons
4	AAA batteries
1	Four AAA-battery holder
10	Jumper wires (various lengths)

Steps

1. **Insert the 4001 IC:**
   Pin 1 goes in E10; pin 14 in F10
2. **Insert R1 to R4:**
   R1 (10 kΩ): B4 to ground bus
   R2 (10 kΩ): B8 to ground bus
   R3 (1 kΩ): D20 to D23
   R4 (1 kΩ): B18 to B21
3. **Insert the LEDs:**
   LED1: A21 to ground bus
   LED2: A23 to ground bus
   Connect the cathode (short lead) to the ground bus.
4. **Insert the jumper wires:**
   D4 to D10
   E4 to F4
   C8 to C15
   E8 to F8
   J10 to positive bus
   D11 to D13
   B12 to B14
   A12 to A18
   A13 to C20
   A16 to ground bus
5. **Insert the switches:**
   SW1: J4 to positive bus
   SW2: J8 to positive bus
6. **Connect the battery holder:**
   Red lead to positive bus
   Black lead to negative bus
7. **Insert the batteries.**
8. **Press the pushbuttons and watch the operation of the LEDs.**

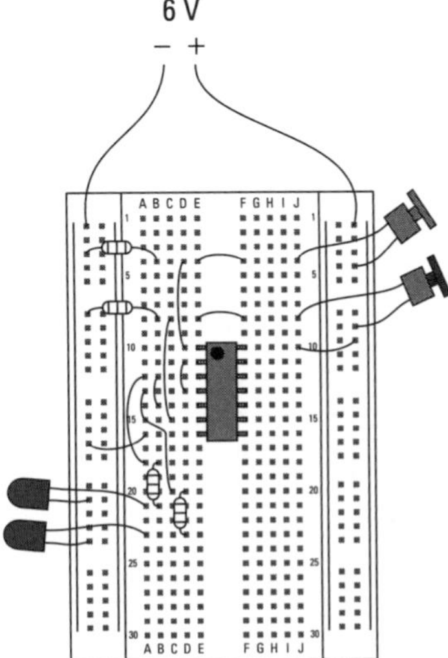

Project 5-2: An Active-Low Latch

In this project, you build a simple active-low latch circuit using a pair of NAND gates from a 4011 Quad 2-Input NAND Gate IC. Connect switches to the $\overline{\text{SET}}$ and $\overline{\text{RESET}}$ inputs and LEDs to the Q and $\overline{\text{Q}}$ outputs so that you can see the operation of the circuit. SW1 is $\overline{\text{SET}}$, SW2 is $\overline{\text{RESET}}$; LED1 is Q; and LED2 is $\overline{\text{Q}}$.

Parts List
1 4011 Quad 2-Input NAND Gate
2 1 kΩ resistors (brown-black-red)
2 10 kΩ resistors (brown-black-orange)
2 Red LEDs
2 Normally open pushbuttons
4 AAA batteries
1 Four AAA-battery holder
10 Jumper wires (various lengths)

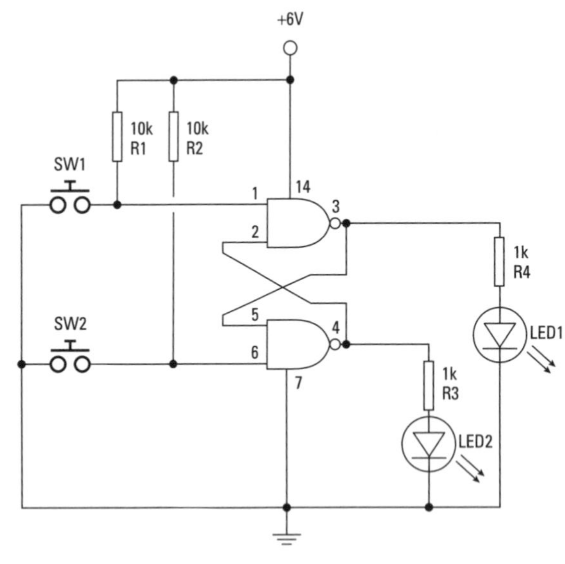

Steps

1. **Insert the 4011 IC:**
   Pin 1 goes in E10; pin 14 in F10
2. **Insert R1 to R4:**
   R1 (10 kΩ): I4 to positive bus
   R2 (10 kΩ): I8 to positive bus
   R3 (1 kΩ): D20 to D23
   R4 (1 kΩ): B18 to B21
3. **Insert the LEDs:**
   LED1: A21 to ground bus
   LED2: A23 to ground bus
   Connect the cathode (short lead) to the ground bus.
4. **Insert the jumper wires:**
   D4 to D10
   E4 to F4
   C8 to C15
   E8 to F8
   J10 to positive bus
   D11 to D13
   B12 to B14
   A12 to A18
   A13 to C20
   A16 to ground bus
5. **Insert the switches:**
   SW1: A4 to ground bus
   SW2: A8 to ground bus
6. **Connect the battery holder:**
   Red lead to positive bus
   Black lead to negative bus
7. **Insert the batteries.**
8. **Press the pushbuttons and watch the operation of the LEDs.**

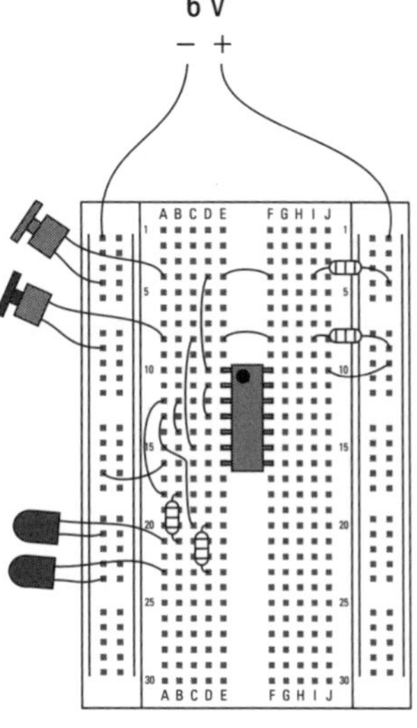

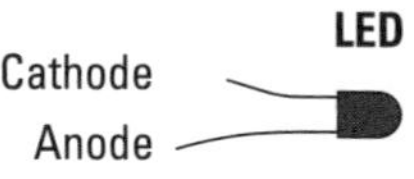

Going Over Gated Latches

A *gated latch* is a latch (check out the preceding section for details) with a third input that must be active in order for the SET and RESET inputs to take effect. This third input is sometimes called *ENABLE,* because it enables the operation of the SET and RESET inputs.

You can connect the ENABLE input to a simple switch. Then, when the switch is closed, the SET and RESET inputs are enabled; when the switch is open, any changes in the SET and RESET inputs are ignored.

Alternatively, you can connect the ENABLE input to a clock pulse. For example, you can connect the output of a 555 timer circuit to the ENABLE input, so that the latch inputs are operational only when the 555 timer's output is high. Note that the ENABLE input is often called the *CLOCK input.* (For more information about 555 timer circuits, refer to Book III, Chapter 2.)

You can add an ENABLE input to a latch easily enough, by adding a pair of NAND gates as shown in Figure 5-6. Here the SET and RESET inputs (the SR latch) are connected to one input of each of the two NAND gates. The ENABLE input is connected to the other input of each NAND gate. Then, the output from these gates is used as the inputs to the basic latch circuit.

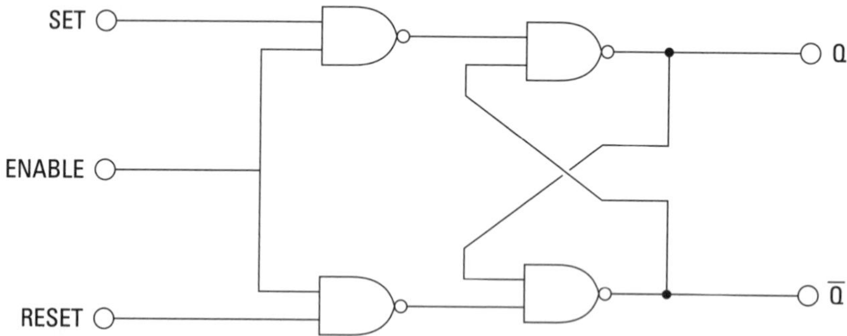

Figure 5-6: A gated SR latch.

Another common type of gated latch is called a *gated D latch,* which has just two inputs: DATA and ENABLE. When a high is received at the ENABLE input, the DATA input is copied to the output. Even if the ENABLE input then goes low, the output remains unchanged. The output can't be changed until the ENABLE input goes high.

To create a gated D latch from a gated SR latch, you simply connect the SET and RESET inputs together through an inverter, as shown in Figure 5-7. Thus, the SET and RESET inputs are always opposite of one another. When the DATA input is high, the SET input is high and the RESET input is low. When the DATA input is low, the SET input is low and the RESET input is high.

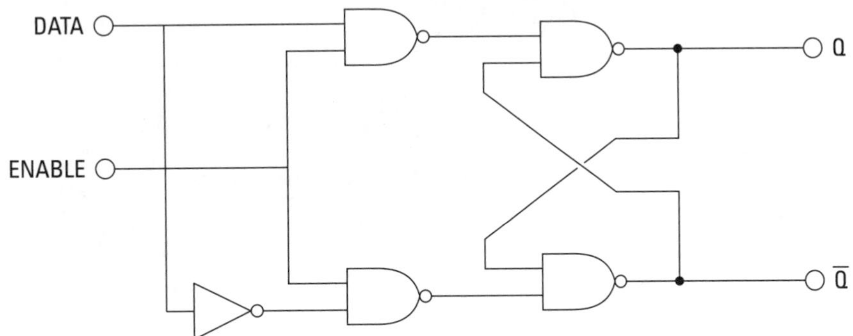

Figure 5-7: A gated D latch.

Project 5-3 describes how to build a gated D latch using two 4011 Quad 2-Input NAND Gates. Two 4011 chips are required because the NAND gate requires a total of five gates (four NAND gates and one NOT gate) and each 4011 provides just four gates. In Chapter 2 of this minibook, you discover that you can create a NOT gate from a NAND gate by tying the two inputs of the NAND gate together. In this project, you use that technique to create the NOT gate.

Figure 5-8 shows you the assembled gated D latch. To operate it, use the first button (the one in row 4) as the DATA input and the second button (in row 8) as the ENABLE input, as follows:

- **Set the Q output to high:** Press and hold the DATA input button, and then press and release the ENABLE input button to activate the latch. The first LED lights to indicate that the output is high.
- **Set the Q output to low (which sets the $\overline{Q}$ output to high):** Press and release the ENABLE button without pressing the DATA button. The first LED goes out to indicate that the Q output is low and the second LED lights to indicate that the $\overline{Q}$ output is high.

Figure 5-8: The assembled gated D latch (Project 5-3).

Project 5-3: A Gated D Latch

In this project, you build a gated D latch circuit using a pair of 4011 Quad 2-Input NAND gates. The circuit needs four NAND gates, plus one NOT gate which is made from a NAND gate by tying its two inputs together.

Parts List

- 2 4011 Quad 2-Input NAND Gate
- 2 1 kΩ resistors (brown-black-red)
- 2 10 kΩ resistors (brown-black-orange)
- 2 Red LEDs
- 2 Normally open pushbuttons
- 4 AAA batteries
- 1 Four AAA-battery holder
- 18 Jumper wires (various lengths)

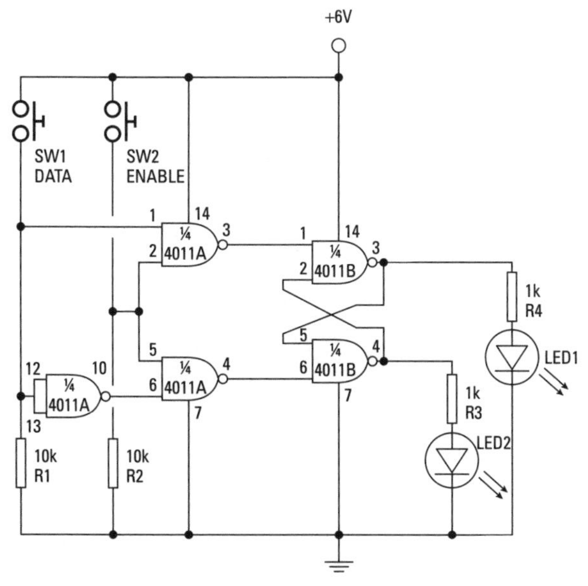

Steps

1. **Insert the two 4011 ICs:**

   4011A: Pin 1 in E10; pin 14 in F10
   4011B: Pin 1 in E18; pin 14 in F18

2. **Insert R1 to R4:**

   R1 (10 kΩ): B4 to ground bus
   R2 (10 kΩ): B8 to ground bus
   R3 (1 kΩ): D27 to D30
   R4 (1 kΩ): B25 to B28

3. **Insert the LEDs:**

   LED1: A28 to ground bus
   LED2: A30 to ground bus
   Connect the cathode (short lead) to the ground bus.

4. **Insert the jumper wires:**

   D4 to D10
   E4 to F4
   H4 to H12
   C8 to C11
   E8 to F8
   J10 to positive bus
   D11 to D14
   G11 to G12
   B12 to B18
   C13 to C23
   G13 to D15
   A16 to ground bus
   J18 to positive bus
   D19 to D21
   A20 to A25
   B20 to B22
   A21 to C27
   A24 to ground bus

5. **Insert the switches:**

   SW1: J4 to positive bus
   SW2: J8 to positive bus

6. **Connect the battery holder:**

   Red lead to positive bus
   Black lead to negative bus

7. **Insert the batteries.**

8. **Press the pushbuttons and watch the operation of the LEDs.**

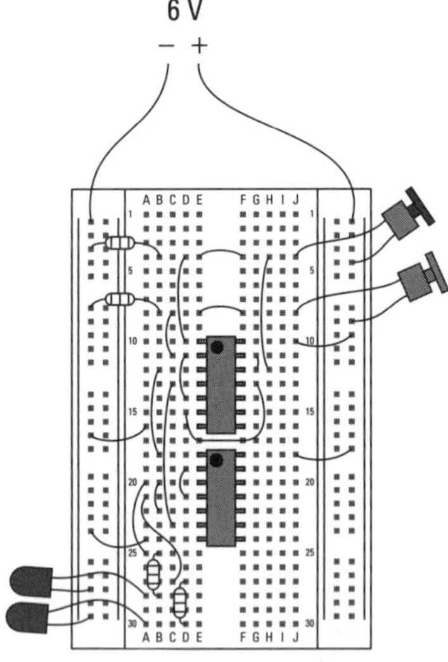

Introducing Flip-Flops

The fact that the name of one of the fundamental building blocks of modern computing machines suggests indecision is a little odd, but the reason for its moniker becomes clear in this section.

A *flip-flop* is a special type of gated latch (see the preceding section). The difference between a flip-flop and a gated latch is that in a flip-flop, the inputs aren't enabled merely by the presence of a high signal on the CLOCK input. Instead, they're enabled by the *transition* of the CLOCK input. Thus, at the moment that the CLOCK input transitions from low to high, the inputs are briefly enabled. When the clock stabilises at the high setting, the output state of the flip-flop is latched until the next clock pulse.

Flip-flops are often said to be *edge-triggered,* because the edge of the clock signal is what triggers the flip-flop. When used in clock-driven computer circuits, edge-triggering is an important characteristic because it helps circuit designers maintain better control over the timing in circuits that contain hundreds or perhaps thousands of flip-flops.

The circuitry that enables a flip-flop to respond to just the leading edge can be pretty complicated. One of the simplest methods is to feed the clock input into a NAND gate, passing one of the legs through an inverter (see Figure 5-9). This approach works because all logic gates have a very small delay between the time a signal arrives at the input and the correct signal arrives at the output.

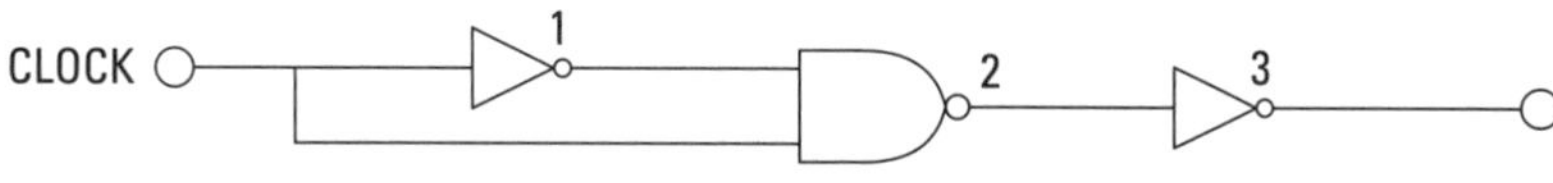

Figure 5-9: The clock transitions from low to high.

Here we guide you through what happens when the clock transitions from low to high in Figure 5-9:

1. Initially, the clock input is low. The inverter causes the first input to the NAND gate (marked '1' in the figure) to be high, while the second input is low. Because the inputs aren't both high, the output from the NAND gate at point 2 in the figure is high. The second inverter inverts the NAND gate output and so the final output from the circuit at point 3 is low, just like the clock input.

2. When the clock input goes high, the second input to the NAND gate goes high immediately. However, it takes a tiny fraction of a second for the inverter to respond, and so for that period the output from the inverter is still high. Thus, both inputs to the NAND gate are high for a few ms, which causes the output from the NAND gate at point 2 in the figure to go low. Then, the second NOT gate inverts the NAND gate output, causing the output at point 3 to go high for a brief moment.
3. When the first NOT gate catches up and its output goes low (at point 1 in the figure), the NAND gate responds to the low and high input by setting its output to high at point 2. The second NOT gate then inverts that output at point 3.

The net result of the circuit in Figure 5-9 is that long clock pulses are turned into short clock pulses. The duration between the pulses remains the same, but the high part of the pulse becomes much shorter.

Flip-flops are designed for use in circuits that use steady clock pulses. An easy way to provide clock pulses for a flip-flop circuit is to use a 555 timer IC, as we describe in Book III, Chapter 2. However, the input source for the CLOCK input of a flip-flop doesn't have to be an actual clock; it can also be a one-shot input triggered by a pushbutton.

As with latches, several different types of flip-flops exist. Here are the most common:

- **SR flip-flop:** Similar to an SR latch (check out the earlier Figure 5-6). Besides the CLOCK input, an SR flip-flop has two inputs, labelled SET and RESET. If the SET input is high when the clock is triggered, the Q output goes high. If the RESET input is high when the clock is triggered, the Q output goes low.

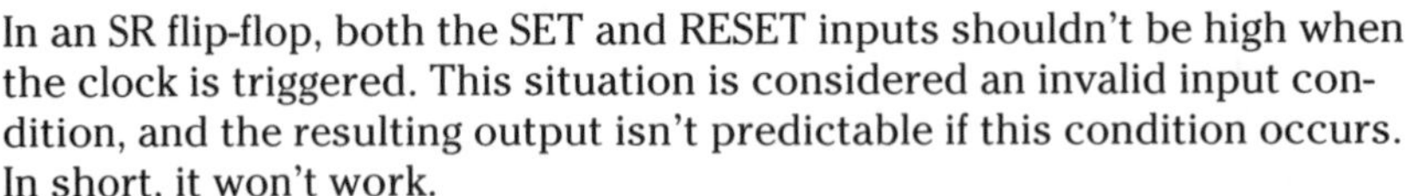

  In an SR flip-flop, both the SET and RESET inputs shouldn't be high when the clock is triggered. This situation is considered an invalid input condition, and the resulting output isn't predictable if this condition occurs. In short, it won't work.

- **D flip-flop:** Has just one input in addition to the CLOCK input. This input is called the *DATA input.* When the clock is triggered, the Q output is matched to the DATA input. Thus, if the DATA input is high, the Q output goes high, and if the DATA input is low, the Q output goes low.

  Most D-type flip-flops also include S and R inputs that let you set or reset the flip-flop. Note that the S and R inputs in a D flip-flop ignore the CLOCK input, and so if you apply a high to either S or R, the flip-flop is set or reset immediately, without waiting for a clock pulse.

- **JK flip-flop:** A common variation of the SR flip-flop. A JK flip-flop has two inputs, labelled *J* and *K.* The J input corresponds to the SET input in an SR flip-flop and the K input corresponds to the RESET input.

The difference between a JK flip-flop and an SR flip-flop is that both inputs can be high in the former. When both the J and K inputs are high, the Q output is *toggled,* which means that the output alternates between high and low. For example, if the Q output is high when the clock is triggered and J and K are both high, the Q output is set to low. If the clock is triggered again while J and K remain high, the Q output is set to high again, and so on, with the Q output alternating from high to low at every clock tick.

- **T flip-flop:** A JK flip-flop whose output alternates between high and low with each clock pulse. Toggles are widely used in logic circuits, because you can combine them to form counting circuits that count the number of clock pulses received.

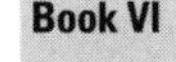

You can create a T flip-flop from a D flip-flop by connecting the $\overline{Q}$ output directly to the D input. As a result, whenever a clock pulse is received, the current state of the Q output is inverted (that's what the $\overline{Q}$ output is) and fed back into the D input, causing the output to alternate between high and low.

You can also create a T flip-flop from a JK flip-flop simply by hard-wiring both the J and K inputs to high. When J and K are high, the JK flip-flop acts as a toggle.

Although you can construct your own flip-flop circuits using NAND gates, an easier option is to use ICs that contain flip-flops. One common example is the 4013 Dual D Flip-Flop. This chip contains two D-type flip-flops in a 14-pin DIP (that is, a dual inline package). Table 5-1 lists the *pinouts* – what each pin does.

Table 5-1 Pinouts for the 4013 Dual D Flip-Flop IC

Pin	*Name*	*Explanation*	*Pin*	*Name*	*Explanation*
1	Q1	Flip-flop 1 Q output	8	SET2	Flip-flop 2 SET input
2	$\overline{Q1}$	Flip-flop 1 $\overline{Q}$ output	9	DATA2	Flip-flop 2 DATA input
3	CLOCK1	Flip-flop 1 CLOCK input	10	RESET2	Flip-flop 2 RESET input
4	RESET1	Flip-flop 1 RESET input	11	CLOCK2	Flip-flop 2 CLOCK input
5	DATA1	Flip-flop 1 DATA input	12	$\overline{Q2}$	Flip-flop 2 $\overline{Q}$ output
6	SET	Flip-flop 1 SET input	13	Q2	Flip-flop 2 Q output
7	GND	Ground	14	V_{DD}	+3 to 15 V

When you use one of the flip-flops in a 4013 IC, be sure to connect any unused inputs to ground. All unused inputs in CMOS logic chips need to be connected to ground, but for simple breadboard circuits the ground connections aren't usually required. However, the DATA and CLOCK inputs of a 4013 flip-flop don't work properly if you don't ground the SET and RESET inputs.

Building flip-flop projects

Project 5-4 shows you how to use a 4013 IC to create a basic D flip-flop. This circuit works much the same as the D-type latch you create in the earlier Project 5-3, except that it requires only one IC rather than two and the wiring is much simpler. That's because the engineers who designed the 4013 IC crammed all the wiring between the individual NAND gates in the IC, so that you don't have to wire the gates together on the breadboard. Instead, all you have to do is hook up the inputs and the outputs and watch the circuit work. Figure 5-10 shows the assembled circuit.

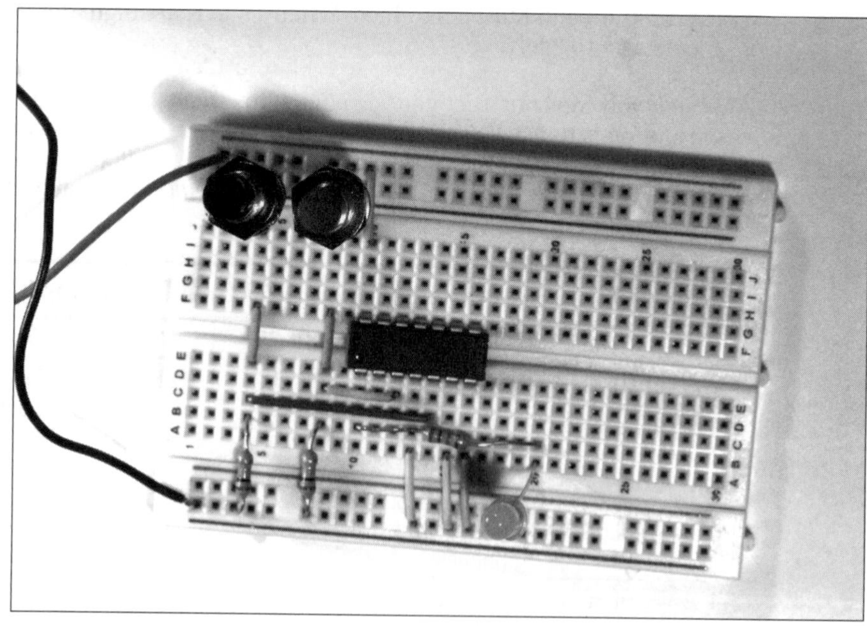

Figure 5-10: The assembled D flip-flop circuit (Project 5-4).

Project 5-5 shows you how to build a T flip-flop in which each press of a button causes an output LED to alternate between on and off. For this project, the $\overline{Q}$ output from the flip-flop is connected to the DATA input. Then, each time the CLOCK input goes high, the inverted output from the $\overline{Q}$ output is fed into the DATA input, causing the Q output to invert. Figure 5-11 shows the assembled circuit.

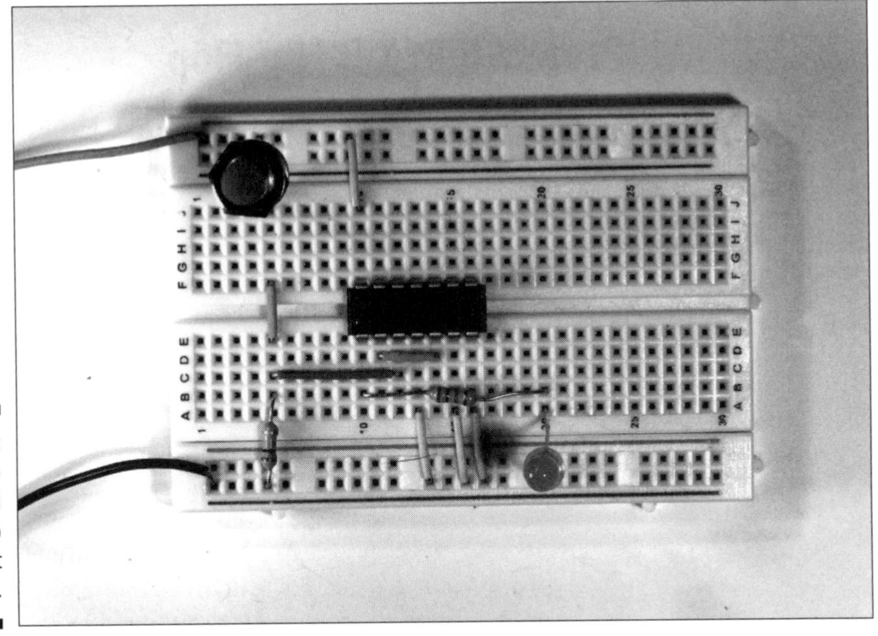

Figure 5-11: The assembled T flip-flop circuit (Project 5-5).

Debouncing a clock input

When you use a mechanical switch to trigger the clock input of a flip-flop, the switch is very likely to have some mechanical bounce. This bounce happens when the switch contacts don't close completely cleanly; instead, the contacts bounce a little bit when they first touch each other.

Even though these bounces are usually just a few ms apart, they can end up confusing the flip-flop, which thinks that each bounce of the switch contacts is a separate press of the button. So instead of just turning the LED attached to the Q output from off to on, a single press of the button may turn it from off to on, and then back off, and then on, and then off again and so on until the switch settles down into its fully-closed position.

You can *debounce* a mechanical switch – that is, eliminate the bounce effect – in a number of different ways. The easiest method is to connect the mechanical switch to a one-shot timer circuit that uses an RC network (with a resistor and a capacitor; see Book II, Chapter 3) to create a very short time interval such as 10 or 20 ms. Though short, this interval is enough to eliminate the negative bouncing effect.

For more information about how to build a one-shot circuit using a 555 timer IC, refer to Book III, Chapter 2.

Project 5-4: A D Flip-Flop

In this project, you build a circuit that demonstrates the operation of a type-D flip-flop. This circuit uses one of the two flip-flops on a 4013 Dual D Flip-Flop IC. The Data and Clock inputs are connected to pushbuttons and the $\overline{Q1}$ output is connected to an LED. The Q1 output isn't connected.

Parts List	
1	4013 Dual D Flip-Flop
1	1 kΩ resistor (brown-black-red)
2	10 kΩ resistors (brown-black-orange)
1	Red LED
2	Normally open pushbuttons
4	AAA batteries
1	Four AAA-battery holder
7	Jumper wires (various lengths)

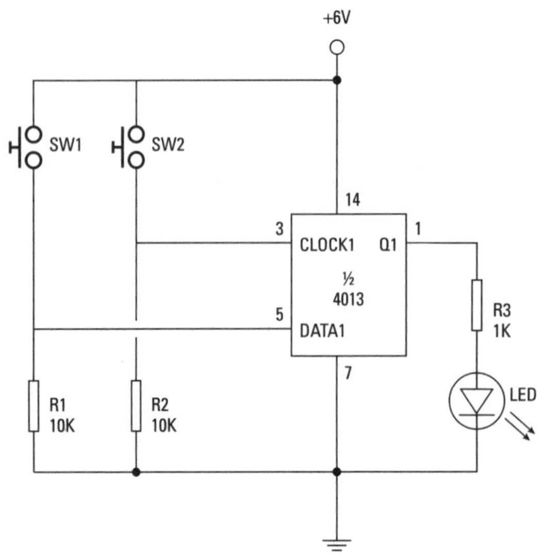

Steps

1. **Insert the 4013:**
   Pin 1 in E10; pin 14 in F10
2. **Insert R1 to R3:**
   R1 (10 KΩ): B4 to negative bus
   R2 (10 KΩ): B8 to negative bus
   R3 (1 KΩ): B10 to B20
3. **Insert the LED:**
   A20 to ground bus
   Connect the cathode (short lead) to the ground bus.
4. **Insert the jumper wires:**
   C4 to C14
   E4 to F4
   D8 to D12
   E8 to F8
   A13 to ground bus
   A15 to ground bus
   A16 to ground bus
   J10 to positive bus
5. **Insert the switches:**
   SW1: J4 to positive bus
   SW2: J8 to positive bus
6. **Connect the battery holder:**
   Red lead to positive bus
   Black lead to negative bus
7. **Insert the batteries.**
8. **Press the pushbuttons and watch the operation of the LEDs.**
   Press and hold the DATA button, and then momentarily press the CLOCK button. The LED comes on. Release the DATA button, and then momentarily press the Clock button. The LED goes out.

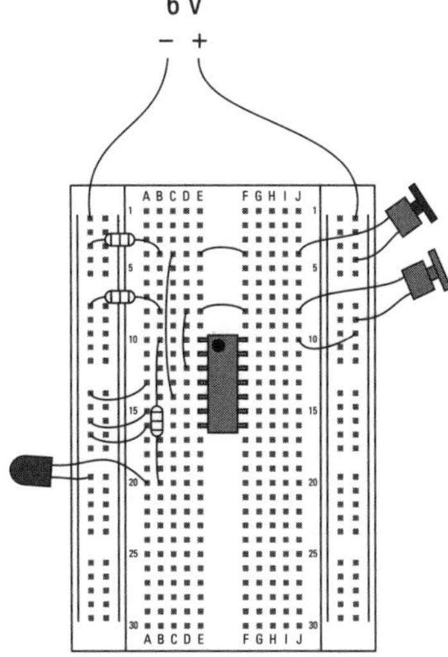

Project 5-5: A Toggle Flip-Flop

In this project, you build a toggle flip-flop using a 4013 Dual D Flip-Flop IC. The clock input is connected to a pushbutton and the $\overline{Q}$ output is connected to an LED. The circuit has Q1 connected to the LED – $\overline{Q1}$ is looped back to the DATA1 input. As a result, each time the switch is pressed the $\overline{Q1}$ output is read into the DATA input, which causes the Q1 output to invert.

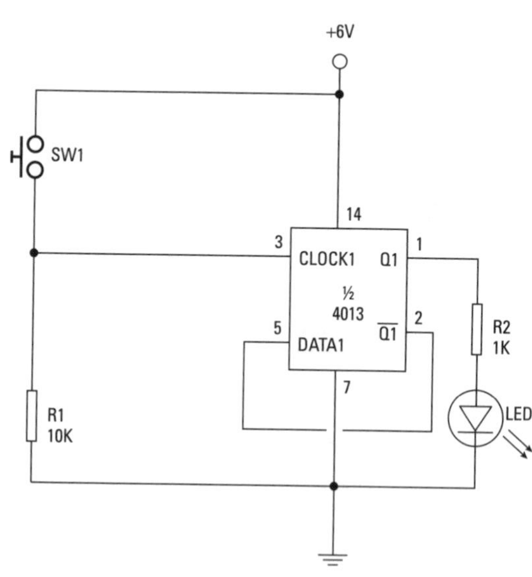

Parts List	
1	4013 Dual D Flip-Flop
1	1 kΩ resistor (brown-black-red)
1	10 kΩ resistors (brown-black-orange)
1	Red LED
1	Normally open pushbuttons
4	AAA batteries
1	Four AAA-battery holder
7	Jumper wires (various lengths)

Steps

1. **Insert the 4013:**
   Pin 1 in E10; pin 14 in F10
2. **Insert the resistors:**
   R1 (10 kΩ): B5 to ground bus
   R2 (1 kΩ): B10 to B20
3. **Insert the LED:**
   A20 to ground bus
   Connect the cathode (short lead) to the ground bus.
4. **Insert the jumper wires:**
   C5 to C12
   E5 to F5
   D11 to D14
   J10 to positive bus
   A13 to ground bus
   A15 to ground bus
   A16 to ground bus
5. **Insert the switch:**
   J5 to positive bus
6. **Connect the battery holder:**
   Red lead to positive bus
   Black lead to negative bus
7. **Insert the batteries.**
8. **Press the pushbutton and watch the operation of the LED:**
   Each time you press the pushbutton, the LED switches from on to off or from off to on. Note, however, that pressing the switch in a way that makes a single, clean contact can sometimes be difficult. Often the switch bounces, sending several very short pulses to the clock input. When that happens, the flip-flop may change state two or more times with a single push of the button. This is to be expected when a mechanical switch is used as the clock source.

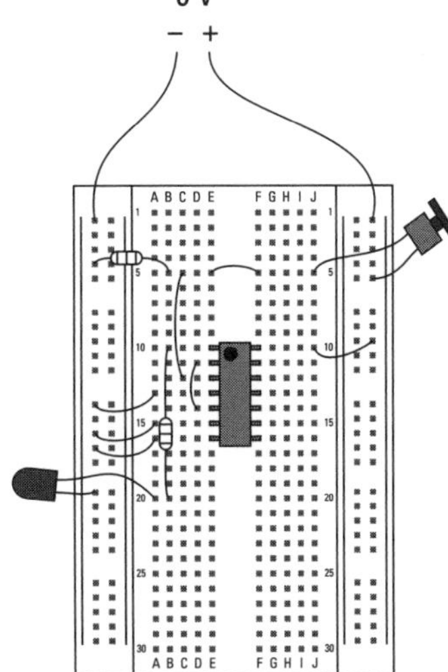

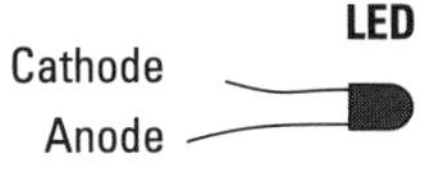

Book VII

Working with BASIC Stamp Processors

Check out www.dummies.com/go/electronicsaiouk for some great bonus code samples.

Contents at a Glance

Chapter 1

Introducing Microcontrollers and the BASIC Stamp

In This Chapter

- Looking at microcontrollers and how they work
- Examining the BASIC Stamp microcontroller
- Building simple BASIC Stamp projects

You can make a circuit board do almost anything you like using a computer with a parallel printer port and a bit of software, as you discover in Book VI, Chapter 4. But sometimes you want to dispense with that ugly, bulky PC and the wires running from it to your amazing electronic invention. For example, perhaps you'd prefer not to have to use a computer just to run some Christmas lights or you want to free your walking, talking robot from the reins of the PC. That's where microcontrollers such as the BASIC Stamp come into the picture.

In a nutshell, a *microcontroller* is a small computer on a single chip, which you can purchase for $25 or less. This chapter introduces you to microcontrollers in general, and describes the fundamentals of using the BASIC Stamp microcontroller in particular, one of the most popular models available today. In Chapters 2 to 4 of this minibook, we go into more details about using the BASIC Stamp microcontroller, including how to write the programs that control its operation.

You can buy many other different kinds of microcontrollers as well. We choose to work with the BASIC Stamp for this minibook primarily because its programming language is easier to pick up and use than the languages of other types of microcontrollers. As an added plus, you can purchase a BASIC Stamp Starter Kit for around $100, making the BASIC Stamp an easy and affordable way to discover microcontrollers.

Meeting the Mighty Microcontroller

In essence, a microcontroller is a complete computer on a single chip. Like all computer systems, microcontrollers consist of several basic subsystems, all interconnected together, but starting at its heart with:

- **Central processing unit (CPU):** The brains of the microprocessor. A CPU carries out the instructions provided to it by a program. The CPU can do basic arithmetic as well as other operations necessary to the proper functioning of the computer, such as moving data from one location of memory to another or receiving data as input from the outside world.

  The CPU of a microcontroller is usually much simpler than the CPU found in a desktop computer, but it's conceptually very similar. In fact, the CPUs found in many modern microcontrollers are as advanced as the CPUs used in desktop computers only a few years ago.

- **Clock:** Drives the CPU and other components of the microcontroller by providing timing pulses that control the pacing of program instructions as they're executed one at a time by the CPU. For most microcontrollers, the clock ticks along at a pace of a few million ticks per second. In contrast, the clock that drives a typical desktop computer ticks along at a few *billion* ticks per second.

- **Random access memory (RAM):** Provides a scratchpad area where the computer can store the data on which it's working. For example, if you want the computer to determine the result of a calculation (such as 2 + 2), you need to provide a location in the RAM where the computer can store the result.

  In a desktop computer, the amount of available RAM is measured in billions of bytes (GB for gigabytes). In a microcontroller, the RAM is often measured just in bytes. That's right: not billions (GB), millions (MB; megabytes) or even thousands (KB; kilobytes) of bytes, but plain old bytes. For example, the popular BASIC Stamp 2, which we use in this chapter and throughout this minibook, has a total of 32 bytes of RAM.

- **EEPROM:** A special type of memory that holds the program that runs on a microcontroller.

  EEPROM stands for *Electrically Erasable Programmable Read-Only Memory*, but that's not on the test!

  EEPROM is *read-only,* which means that when data has been stored in EEPROM, a program running on the microcontroller's CPU can't change it. However, you can write data to EEPROM memory by connecting the EEPROM to a computer via a USB port. Then, the computer can send data to the EEPROM.

In fact, this process is how you program microcontrollers. You use special software on a PC to create the program that you want to run on the microcontroller. Then, you connect the microcontroller to the PC and transfer the program from the PC to the microcontroller. The microcontroller then executes the instructions in the program.

Most microcontrollers have a few thousand bytes of EEPROM memory, which is enough to store relatively complicated programs downloaded from a PC.

One of the most important features of EEPROM memory is that it doesn't lose its data when you turn off the power. Thus, after you transfer a program from a PC to a microcontroller's EEPROM, the program remains in the microcontroller until you replace it with another program. You can turn the microcontroller off and put it away in a cupboard for years, and when you turn the microcontroller back on, the program recorded years ago runs again.

- **I/O pins:** A vitally important feature, which enables the microcontroller to communicate with the outside world. Although some microcontrollers have separate input pins and output pins, most have shared I/O pins that can be used for input and output.

  I/O pins normally use the basic Transistor-Transistor Logic (TTL) interface that we describe in Chapter 3 of Book VI: high (logic 1) is represented by +5 V (volts) and low (logic 0) is represented by 0 V.

Most microcontrollers can handle only a small amount of current directly through the I/O pins: 20–25 milliamperes (mA) is typical and is enough to light up a light-emitting diode (LED). But for circuits that require more current you need to isolate the higher current load from the microcontroller I/O pins, usually with a transistor driver.

Working with the BASIC Stamp

The *BASIC Stamp* is a microcontroller, made by Parallax, which is an essentially complete, self-contained computer system. Depending on the model, the Stamp (as it's usually called) can be a single chip or a small circuit board.

The key feature that sets BASIC Stamp microcontrollers apart from others is that they include a built-in programming language called *PBASIC* (Parallax's modified version of BASIC). This programming language makes creating programs that run on the Stamp easy. With most other microcontrollers, programming is much more difficult because you have to write the programs in a more complicated programming language.

Introducing the BASIC Stamp 2 Module

Figure 1-1 shows one of the more popular BASIC Stamps, called the BASIC Stamp 2 Module. This model is a complete computer system on a 24-pin *DIP* (dual inline package) that you can solder directly to a circuit board or (more likely) insert into a 24-pin DIP socket.

Figure 1-1: A BASIC Stamp 2 Module.

Here are the basic specifications for the BASIC Stamp 2 microcontroller (check out the earlier section 'Meeting the Mighty Microcontroller' for descriptions and roles of the various microprocessor parts):

- **Clock speed:** 20 MHz
- **Current draw:** 3 mA
- **EEPROM:** 2,048 bytes

- **Maximum I/O pin current:** 20 mA
- **Number of I/O pins:** 16
- **Power supply:** 5.5–15 V
- **RAM:** 32 bytes

On the fourth item in the checklist, note that a limit applies to the total amount of current the I/O pins combined can handle: 40 mA for the first group of 8 I/O pins and another 40 mA for the second group.

We show you how to work with the BASIC Stamp 2 microcontroller throughout this chapter and the others in this minibook.

Buying a BASIC Stamp

Although you can purchase a BASIC Stamp microcontroller by itself, the easiest way to get into BASIC Stamp programming is to buy a starter kit that includes a BASIC Stamp along with the software that runs on your PC for programming it and the USB cable that connects your PC to the Stamp. In addition, most starter kits come with a prototype board that makes designing and testing simple circuits that interface with the Stamp pretty easy.

One such starter package is the BASIC Stamp Activity Kit, which you can purchase for around £130 from one of the UK distributors listed by Parallax on its website (`www.parallax.com`). This kit comes with the following components, which are pictured in Figure 1-2, and which I list here in order of importance:

- **BASIC Stamp HomeWork board:** A prototyping board that includes a BASIC Stamp 2 microcontroller, a small solderless breadboard, a clip for a 9 V battery and a connector for the USB programming cable (see the following section for more details on the HomeWork board).
- **USB cable:** Connects the HomeWork board to a computer so that you can program the Stamp.
- **Useful electronic components:** Includes resistors, LEDs, capacitors, a seven-segment display, pushbuttons, a potentiometer and plenty of jumper wires, for building circuits that connect to the Stamp.
- **Servo:** A fancy motor that the Stamp can control.
- **Instruction book:** *What's a Microcontroller?* gives a detailed overview of programming the BASIC Stamp 2.

Figure 1-2: The BASIC Stamp Activity Kit.

Working with the BASIC Stamp HomeWork board

Figure 1-3 shows the BASIC Stamp HomeWork board. The chips near the centre of this board constitute the BASIC Stamp 2 module. On the left is a battery clip to which you can connect a 9 V battery for power. On the right is a small solderless breadboard, consisting of 17 rows of solderless connectors.

Immediately to the left of the breadboard is a row of 16 connectors that provide access to the 16 I/O pins of the BASIC Stamp 2, and immediately above the breadboard is a row of connectors that provide access to +5 V (identified as *Vdd* and *Vin* on the board) and ground (identified as *Vss*). This breadboard is designed to allow you to construct simple circuits that connect to the 16 I/O pins. You can connect pushbuttons and other devices that send input to the program running on the Stamp or LEDs or other output equipment that the Stamp program can control.

Figure 1-3: The BASIC Stamp HomeWork board.

Choosing the Board of Education

You can construct all the projects that we describe in this minibook on the BASIC Stamp HomeWork board. But if you want to create more elaborate projects, you may want to consider using an alternative called the *Board of Education,* as shown in Figure 1-4. This board costs about the same as the HomeWork board but provides a few additional handy features. In particular, the Board of Education includes the following:

- A prototype board similar to the one on the HomeWork board.
- A special 20-pin connector that allows you to connect external circuits permanently to the BASIC Stamp's 16 I/O pins – which is useful if you want to build circuits that are more permanent than the solderless breadboard allows.
- Special connectors for I/O pins 12, 13, 14 and 15 that are designed to connect the BASIC Stamp to servos (motors).

- A 9 V battery clip and an external power connector that allow you to power the BASIC Stamp with a 9 V battery or an external power supply.
- A removable 24-pin BASIC Stamp 2 Module, which allows you to use the Board of Education to program the Stamp. When you're certain that the program is working properly, you can remove the BASIC Stamp 2 Module from the Board of Education and plug it into your own circuit.

Figure 1-4: The Board of Education.

Connecting to BASIC Stamp I/O pins

Before you start writing programs to run on a BASIC Stamp, you have to build the circuit(s) that you're going to connect to the I/O pins on the Stamp. When the circuit is constructed, you can then write a program that controls the circuits you've connected to each pin.

The BASIC Stamp 2 has a total of 16 separate I/O pins, which means that you can connect as many as 16 separate circuits. These 16 I/O circuits are more than enough for most BASIC Stamp projects.

For now, start with a basic circuit that simply connects an LED to one of the Stamp's output pins. The program that you write and load onto a Stamp can turn each of the 16 I/O pins on (high) or off (low) by using simple programming commands. When an I/O pin is high, +5 V is present at the pin. When it's low, no voltage is present.

Each I/O pin can handle as much as 20 mA of current, and altogether they can handle up to 40 mA which is more than enough to light an LED. As with any LED circuit, you need to provide a current-limiting resistor in the circuit. If you forget to include this resistor, you destroy the LED and possibly fry the Stamp too – so always be sure to include the current-limiting resistor. A 470Ω resistor is usually appropriate for LED circuits connected to Stamp I/O pins.

Figure 1-5 shows a simple schematic diagram for a circuit that drives an LED from pin 15 of a BASIC Stamp. Notice in this schematic that the pin output is represented by a simple five-sided shape (to the left of the figure). Commonly, schematics do *not* draw the Stamp as a single rectangle as you would for other integrated circuits. Instead, each I/O connection in the circuit is shown using a connector shape.

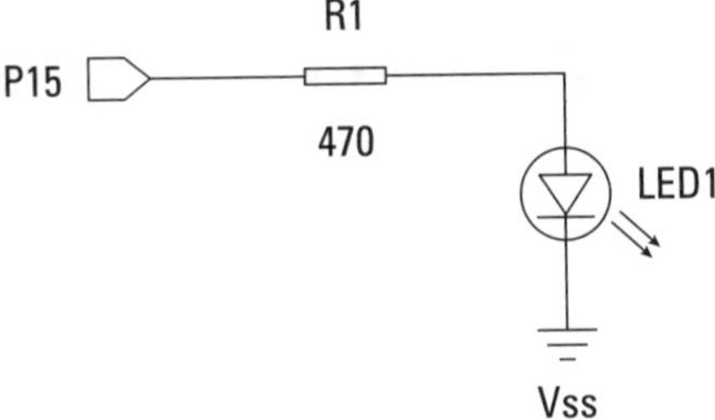

Figure 1-5: Schematic for an LED circuit connected to a BASIC Stamp I/O pin.

Book VII
Working with BASIC Stamp Processors

Figure 1-6 shows how you can build this circuit on a BASIC Stamp HomeWork board. Notice that the resistor is inserted into the P15 connection and the sixth hole in the second row. The LED's anode is inserted in the seventh hole in row 2, and the cathode is in one of the Vss connectors. Project 1-2, which appears in the later section 'Flashing an LED with a BASIC Stamp', walks you through creating this circuit.

Figure 1-6: Assembling an LED circuit on a BASIC Stamp HomeWork board.

Installing the BASIC Stamp Windows Editor

The BASIC Stamp Windows Editor is the software that you use on your computer to create programs that can be downloaded to a BASIC Stamp microcontroller. This software is available free from the Parallax website (`www.parallax.com`).

The easiest way to find the right page from which you can download the software is to use an Internet search engine to search for 'BASIC Stamp Windows Editor'.

Or you can download the software by opening a web browser such as Internet Explorer, going to `www.parallax.com` and following these instructions:

1. **Click the Resources tab.**
2. **Click Downloads in the menu on the left side of the page.**
3. **Click BASIC Stamp Software.**
4. **Select the version of the software for your computer and download and install the software.**

After you've installed the software, run it by choosing BASIC Stamp Editor from the Windows Start menu. Figure 1-7 shows how the editor appears when you first run it.

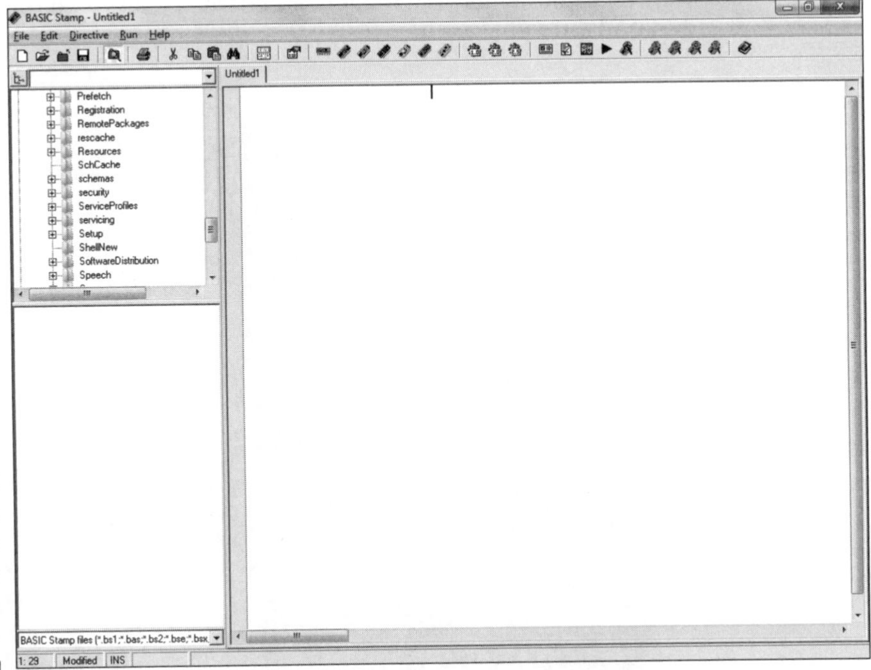

Figure 1-7: The BASIC Stamp Windows Editor.

Connecting to a BASIC Stamp

Before you can use the Stamp Editor to program a BASIC Stamp, you need to connect the Stamp to your computer and then configure the Stamp Editor so that it can communicate with the Stamp. The following steps describe the procedure for doing that with a BASIC Stamp HomeWork board. The procedure for connecting a Board of Education (which we describe in the earlier section 'Choosing the Board of Education') is very similar, and so you'll have no trouble following it if you have a Board of Education instead of a HomeWork board:

1. **Insert a 9 V battery in the HomeWork board.**

   You see a green LED light up on the board when the battery is inserted. If this light doesn't come on check the battery, because it may be dead.

2. **Plug the USB-to-serial adapter into the DB9 connector on the HomeWork board.**

This step is necessary because the HomeWork board uses an older-style serial connector to connect to your computer, but most computers don't have serial ports. The adapter converts the serial port connection on the HomeWork board to a USB port.

Figure 1-8 shows how the HomeWork board appears with the USB-to serial adapter plugged in.

Figure 1-8: The HomeWork board with the USB-to-serial adapter connected.

3. **Plug the smaller end of the USB cable into the USB-to-serial adapter and the larger end into an available USB port on your computer.**

   After a moment, your computer chirps happily to acknowledge the presence of the BASIC Stamp. A pop-up bubble may appear informing you that Windows is installing the driver needed to access the device. If so, just wait for the bubble to disappear.

In addition to the activity on your computer, you can also notice a green LED on the USB-to-serial adapter, indicating that the adapter is powered up and ready to adapt.

4. **In the BASIC Stamp Editor, choose Run⇨Identify.**

   Doing so brings up the Identification window, as shown in Figure 1-9.

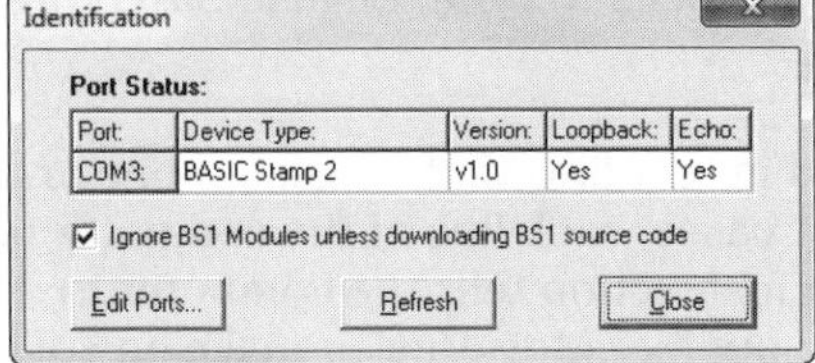

Figure 1-9: Identifying your BASIC Stamp.

   This window indicates that your BASIC Stamp 2 is connected. If it doesn't, consult the 'Connection Troubleshooting' section of the Stamp Editor's Help command (Help⇨BASIC Stamp Help).

5. **Click Close to dismiss the Identification window.**

6. **You're done and ready to write your first BASIC Stamp program!**

Here are a couple of points to know about connecting a BASIC Stamp to the Stamp Editor:

- If you're having trouble connecting, the most likely cause (other than a loose connection or forgetting to insert the battery) is an incorrect driver for the USB-to-serial adapter. To install the right driver software, take a look at `http://www.parallax.com/usbdrivers`. Follow the instructions that appear on that page to install the correct driver.
- Parallax makes a version of the Board of Education that has a USB port directly on the board. If you're using that board instead of the HomeWork board, you don't need a USB-to-serial adapter. Instead, you can plug the USB cable directly into the Board of Education.

Writing Your First PBASIC Programs

If you've never done any form of computer programming before, you're in for a fun and fascinating adventure, during which you can find out more about how computers work.

Discovering programming basics

In a nutshell, a computer program is a set of written instructions that a computer knows how to read, interpret and carry out. The instructions are written in a language that humans and computers can read. They aren't quite English, but they resemble English enough that English-speaking people can understand what they mean. (Of course, non-English programming languages are also available, but PBASIC happens to be an English-based programming language.)

Computer programs are stored in text files that consist of one or more lines of written instructions. In most cases, each line of the computer program contains one instruction. Each instruction tells the computer to do something specific, such as add two numbers together or make one of the output pins go high.

The trick of computer programming is to put the right instructions together in the right sequence to get the program to do exactly what you want it to do. Of course, to do that, you need to have a solid understanding of what you want the program to do and a knowledge of the variety of instructions that are available. The PBASIC programming language consists of about 70 different types of instructions. But don't be discouraged; you can write useful programs using only a handful of these commands.

Saying 'Hello World'

Pretty much the most basic computer program is called Hello World. This simple program displays the text string '`Hello, World!`' to demonstrate what a simple program looks like.

In PBASIC (the official name of the BASIC language that's used on BASIC Stamps), the Hello World program consists of three lines:

```
' {$STAMP BS2}
' {$PBASIC 2.5}
DEBUG "Hello, World!"
```

The first two lines are called *directives.* They don't tell the BASIC Stamp to do anything as such; instead, they provide information that the Stamp Editor needs to know to prepare your program so that it can be downloaded to the Stamp. The first line indicates that the microcontroller on which you're going to run the program is a BASIC Stamp 2 (BS2). The second line indicates that this program uses version 2.5 of PBASIC for this program (the current version).

Every program you write must include these two lines. Fortunately, you don't have to type them yourself. Instead, you can use menu commands or toolbar buttons to insert the directives automatically:

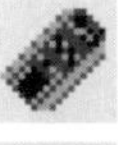

- **Directive⇨Stamp⇨BS2:** Inserts the `$STAMP BS2` directive to indicate that you're using BASIC Stamp 2.
- **Directive⇨PBASIC⇨Version 2.5:** Inserts the `$PBASIC 2.5` directive to indicate that you're using version 2.5 of PBASIC.

The third line of the Hello World program is the only line that tells the BASIC Stamp to do something. This command, called `DEBUG`, tells the BASIC Stamp to send a bit of text to the computer connected via the USB port. The `DEBUG` command always consists of two parts: the word 'DEBUG' followed by text that must be enclosed in quotation marks. For example:

```
DEBUG "Hello, World!"
```

This line sends the message 'Hello, World!' to your computer. The message is displayed in a window called the Debug Terminal window within the Stamp Editor.

Running the Hello World program

Project 1-1 shows you how to run the Hello World program on a HomeWork board. Figure 1-10 shows the resulting output as displayed in the Debug Terminal window.

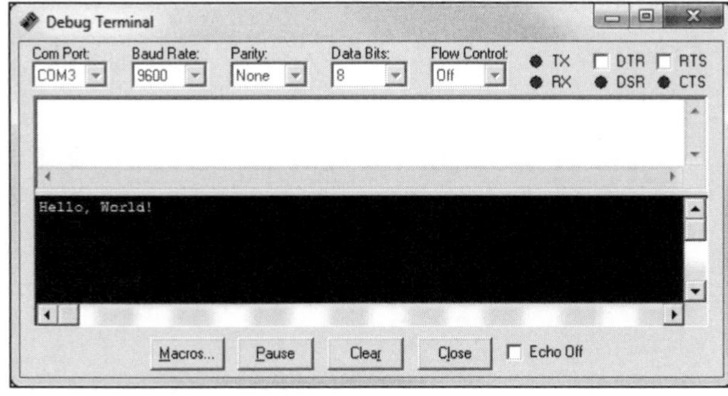

Figure 1-10: The Debug Terminal window displays the output from the Hello World program.

Project 1-1: Hello, World!

In this project, you run a simple Hello World program on a BASIC Stamp. The output is displayed in the Debug Terminal window with the Stamp Editor program.

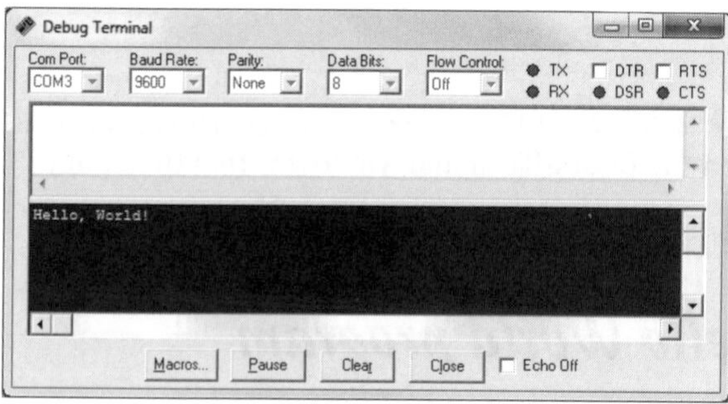

Parts List

- 1 Computer with BASIC Stamp Editor software installed.
- 1 BASIC Stamp HomeWork board
- 1 9 V battery
- 1 USB cable
- 1 USB to serial adapter

Steps

1. **Open the BASIC Stamp Editor.**
2. **Connect your BASIC Stamp to the computer and identify it in the Stamp Editor.**

   For more information about how to do this, refer to the earlier section `Connecting to a BASIC Stamp.´
3. **Type the Hello World program into the Editor program window.**

   The Hello World program is shown in the accompanying listing.
4. **Choose File➪Save.**

   This brings up a Save As dialog box.
5. **Navigate to the folder where you want to save the program, type a filename and click Save.**

   You can use any filename you like, but BASIC Stamp 2 program files must have the extension .bs2.
6. **Choose Run➪Run.**

   If you prefer, you can click the Run button on the toolbar or press F9. When you choose the Run command, the program is downloaded to the BASIC Stamp. When the program is downloaded, it starts to run automatically on the Stamp.
7. **View the Hello, World! message displayed in the Debug Terminal window.**

Hello World Program

```
'{$STAMP BS2}
'{$PBASIC 2.5}
DEBUG "Hello, World!"
```

Flashing an LED with a BASIC Stamp

A BASIC Stamp is serious overkill for a circuit that simply flashes an LED on and off: you can do that for a few quid with a 555 timer IC, a capacitor and a couple of resistors (as we describe in Book III, Chapter 2). But discovering how to flash an LED on and off with a BASIC Stamp is an important step towards completing more complex projects.

To flash an LED, you first have to connect an LED to an output pin in such a way that the Stamp can turn the LED on by taking the output pin high. You find out how to do that earlier in this chapter, in the section 'Connecting to BASIC Stamp I/O Pins'. So all that remains is discovering how to write a PBASIC program that flashes the LED.

To write the program, you need to know the following five PBASIC instructions:

- **`HIGH`:** Sets one of the Stamp's I/O pins to high. You use this instruction to turn the LED on.
- **`LOW`:** Sets one of the Stamp's I/O pins to low. You use this instruction to turn the LED off.
- **`PAUSE`:** Causes the Stamp to sit idle for a specified period of time. You use this instruction to delay the program a bit between `high` and `low` commands so that the LED stays on for a while before you turn it off, and then stays off for a while before you turn it back on.
- **`GOTO`:** Causes the program to loop back to a previously designated location. You use this instruction to cause the program to flash the LED on and off repeatedly, instead of flashing it on and off only once.
- **`Label`:** Marks the location that you want the `GOTO` statement to loop to.

Here's the complete program that flashes the LED:

```
' {$STAMP BS2}
' {$PBASIC 2.5}
Main:
 HIGH 15
 PAUSE 1000
 LOW 15
 PAUSE 1000
 GOTO Main
```

Take a look at how this program works, one line at a time:

Program Line	***What It Does***
`' {$STAMP BS2}`	Indicates that the program runs on a BASIC Stamp 2.
`' {$PBASIC 2.5}`	Indicates that the program uses version 2.5 of PBASIC.
`Main:`	Creates a label named Main that marks the location that the GOTO command loops back to.
`HIGH 15`	Makes I/O pin 15 high, which turns the LED on.
`PAUSE 1000`	Pauses the program for 1,000 milliseconds (ms), which is the same as one second, and allows the LED to stay on for one full second.
`LOW 15`	Makes I/O pin 15 low, which turns the LED off.
`PAUSE 1000`	Pauses the program for 1,000 ms, which allows the LED to stay off for one full second.
`GOTO Main`	Makes the program skip back to the Main label, which causes the program to loop through the HIGH, PAUSE, LOW and PAUSE instructions over and over again.

The net effect of this program is that the LED on pin 15 flashes on and off at one-second intervals.

Project 1-2 describes building a simple circuit that connects an LED to pin 15 and runs the LED Flasher program so that the LED flashes on and off. The completed circuit for this project is shown in the earlier Figure 1-6.

Project 1-2: An LED Flasher

In this project, you build a circuit that connects an LED to pin 15 of the BASIC Stamp processor. Then, you download and run a program that flashes the LED on and off at one-second intervals.

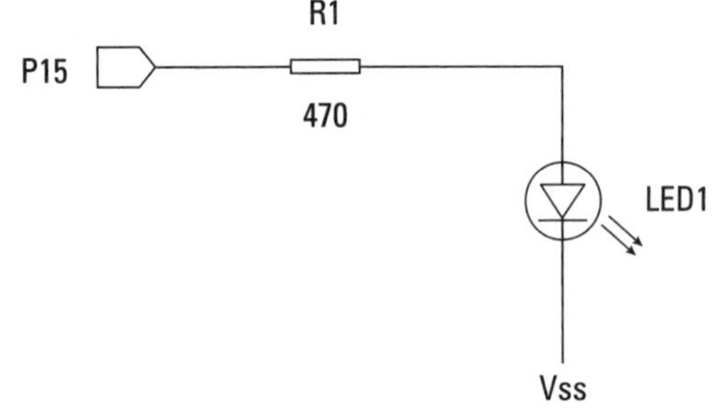

Parts List

1 Computer with BASIC Stamp Editor software installed.
1 BASIC Stamp HomeWork board
1 9 V battery
1 USB cable
1 USB to serial adapter
1 Red LED
1 460 Ω resistor (yellow-violet-brown)

LED Flasher

```
'{STAMP BS2}
'{$PBASIC 2.5}
Main:
    HIGH 15
    PAUSE 1000
    LOW 15
    PAUSE 1000
    GOTO Main
```

Steps

1. **Insert the resistor and LED in the HomeWork board's solderless breadboard as shown in the accompanying diagram.**

   The resistor needs to connect the P15 I/O pin to the anode (long lead) of the LED, and the cathode (short lead) of the LED needs to connect to one of the Vss connections.

2. **Open the BASIC Stamp Editor.**
3. **Connect your BASIC Stamp to the computer and identify it in the Stamp Editor.**

   For more information about how to do this, refer to the earlier section `Connecting to a BASIC Stamp.´

4. **Type the LED Flasher program into the Editor program window.**

   The LED Flasher program is shown in the accompanying listing.

5. **Choose File➪Save.**

   This brings up a Save As dialog box.

6. **Navigate to the folder where you want to save the program, type a filename, and click Save.**

   You can use any filename you want, but BASIC Stamp 2 program files must have the extension .bs2.

7. **Choose Run➪Run.**

   If you prefer, you can click the Run button on the toolbar or press F9. When you choose the Run command, the program is downloaded to the BASIC Stamp. When the program is downloaded, it starts to run automatically on the Stamp.

8. **Observe the LED flashing on and off.**

   If the LED doesn't flash on and off, recheck your wiring. The most likely error is that you inserted the LED backwards or connected the resistor to the wrong I/O pin.

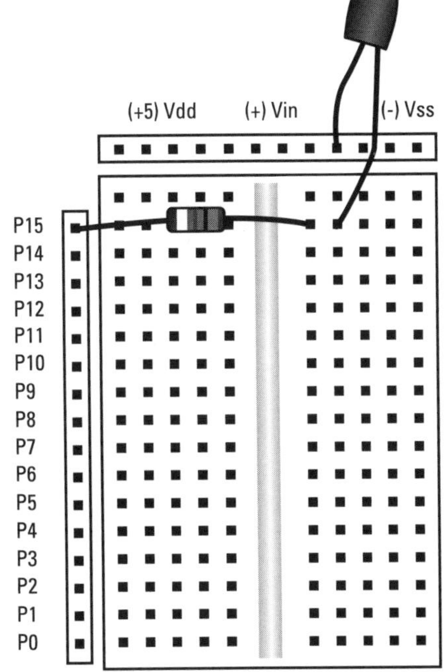

Chapter 2

Programming in PBASIC

In This Chapter

- Discovering the essentials of the PBASIC language
- Working with variables and constants
- Adding IF-THEN-ELSE logic to your programs
- Looping with `DO` and `FOR` loops

This chapter is about the exciting but somewhat intimidating topic of computer programming. Specifically, it covers programming the BASIC Stamp microprocessor (which we introduce in Chapter 1 of this minibook) using its built-in programming language called PBASIC.

This journey into programming is fascinating but may seem rocky at times, because you need to grasp certain important concepts, which can be difficult to get your head around. But trust us: programming a BASIC Stamp microprocessor isn't rocket science. You can and will get to grips with the tricky ideas, and when you do your imagination is the only limit to what you can get a BASIC Stamp microprocessor to do.

So grab your thinking cap, check in the mirror that it fits and you look suitably sharp, and get started.

PBASIC programming involves much more than this short chapter can cover. For complete information about the PBASIC programming language, download the 500+ page *BASIC Stamp Syntax and Reference Manual* from `www.parallax.com`.

Introducing and Using the PBASIC Language

As you can discover in Chapter 1 of this minibook, you use the BASIC Stamp Windows Editor (hereafter referred to simply as the Stamp Editor) to create your programs and download them to the BASIC Stamp.

If you're unclear on how to do that, please read that chapter before proceeding. You can't progress very far in PBASIC programming without writing some real programs, downloading them to a BASIC Stamp and observing how they work.

In this chapter, we focus on the PBASIC statements that you use to control BASIC Stamp output pins as well as the statements for controlling the execution of your program. With just the statements used in this chapter, you can set up complicated programs that can control output devices connected to the BASIC Stamp.

For information about how to use BASIC Stamp I/O ports with input devices, please see this minibook's Chapter 3.

The basics behind BASIC

PBASIC is a special version of the old programming language called BASIC, which was developed in the early 1960s by two professors at Dartmouth College (John Kemeny and Thomas Kurtz). It was originally designed to be a programming language for nonprogrammers.

Back in the 1960s, the names of programming languages were usually acronyms that indicated the intended purpose of the language. For example, FORTRAN, which was designed for the purpose of solving maths formulas, stands for *formula translator.* COBOL for business problems means *common business-oriented language.* BASIC was no exception: it stands for *beginners all-purpose symbolic instruction code,* and the *P* in PBASIC stands for *Parallax,* the company that invented the BASIC Stamp.

Building a test circuit for the programs

All the programs in this chapter assume that an LED is connected to output pins 0, 2, 4, 6, 8 and 10 of the BASIC Stamp HomeWork board. The programs work with just the even-numbered I/O pins simply because the holes on the breadboard in the BASIC Stamp HomeWork board are too close together to connect LEDs easily to adjacent pins.

Project 2-1 shows how you can build a test circuit that has six LEDs connected to pins 0, 2, 4, 6, 8 and 10 using components that come with the BASIC Stamp Activity Kit, which is available from online electronics stores. You can also easily assemble this circuit with a Board of Education (check out this minibook's Chapter 1) and your own LEDs and resistors.

Figure 2-1 shows the finished circuit, ready to be used for testing the programs you're going to write as you work your way through this chapter.

Figure 2-1: A BASIC Stamp HomeWork board with six LEDs.

Project 2-1: An LED Test Board

In this project, you connect six LEDs to a BASIC Stamp HomeWork or Board of Education board to test the programs presented in this chapter.

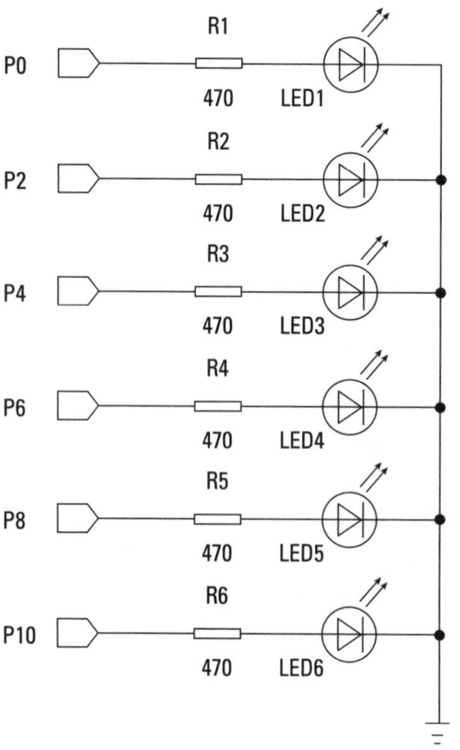

Parts List

- 1 Computer with BASIC Stamp Editor software installed.
- 1 BASIC Stamp HomeWork board or Board of Education
- 1 9 V battery
- 1 USB cable
- 1 USB-to-serial adapter
- 6 LEDs
- 6 470 Ω resistors (yellow-violet-brown)
- 6 Jumper wires

Test Flash Program

```
'{$PBASIC 2.5}
'{$STAMP BS2}

Main:
    HIGH 0
    HIGH 2
    HIGH 4
    HIGH 6
    HIGH 8
    HIGH 10
    PAUSE 500
    LOW 0
    LOW 2
    LOW 4
    LOW 6
    LOW 8
    LOW 10
    PAUSE 500
    GOTO Main
```

Steps

1. **Connect the six LEDs to the output pins as follows:**

Pin (Cathode)	Breadboard Hole
P0	A17
P2	A15
P4	A13
P6	A11
P8	A9
P10	A7

2. **Connect the resistors across the gap in the centre of the breadboard as follows:**

Resistor	From	To
R1	E17	F17
R2	E15	F15
R3	E13	F13
R4	E11	F11
R5	E9	F9
R6	E7	F7

3. **Use the jumper cables to connect the resistors to ground:**

   I17 to I15
   J15 to J13
   I13 to I11
   J11 to J9
   I9 to I7
   J7 to Vss strip

4. **Open the BASIC Stamp Editor.**
5. **Connect your BASIC Stamp to the computer and identify it in the Stamp Editor.**

   For more information about how to do this, refer to Chapter 1 of this minibook.
6. **Type the Test Flash program into the Editor program window, and then save the file.**

   The Test LED program is shown in the accompanying listing.
7. **Choose File⇨Save.**

   This brings up a Save As dialog box.
8. **Navigate to the folder where you want to save the program, type a filename and click Save.**

   You can use any filename you like, but BASIC Stamp 2 program files must have the extension .bs2.
9. **Choose Run⇨Run.**

   If you prefer, you can click the Run button in the toolbar or press F9.

   When you choose the Run statement, the program is downloaded to the BASIC Stamp. When the program is downloaded, it starts to run automatically on the Stamp.
10. **Watch the LEDs flash!**

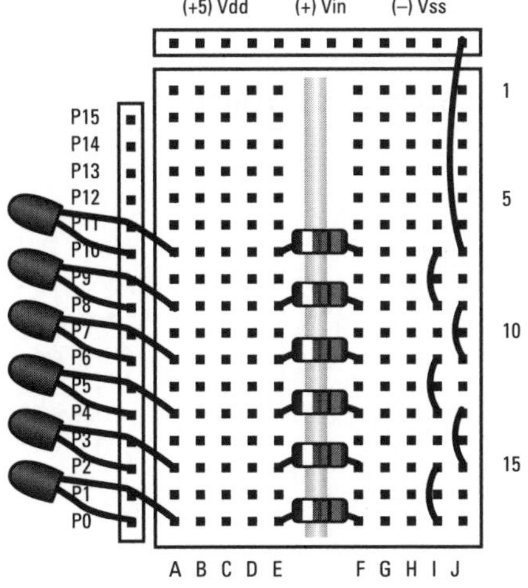

Flashing the LEDs

In this minibook's Chapter 1, you see a program that flashes a single LED on and off. In this chapter, we show you several variants of that program, which flash the six LEDs in the test project in various sequences. Along the way, you can add more PBASIC statements to your repertoire to provide more and more complex ways to control the flashing.

Keep in mind throughout this chapter that if you can turn an LED on or off with a PBASIC program, you can control *anything* that can be connected to a BASIC Stamp I/O port. The BASIC Stamp itself doesn't know or care what kind of circuit you connect to an I/O pin. All it knows is that when you tell it to, the BASIC Stamp makes the I/O port high or low. The external circuitry connected to the pin determines what happens when the pin goes high.

The only limitation is that the BASIC Stamp itself can source only about 20 milliamperes (mA) through its I/O pins. If the circuit connected to the pin requires more current than 20 mA, you have to isolate the higher-current portion of the circuit from the BASIC Stamp. The easiest way to do so is by using a transistor driver or a relay.

Listing 2-1 shows a simple program that flashes all six LEDs on and off at half-second intervals. This program uses nothing more than the `HIGH`, `LOW`, `PAUSE` and `GOTO` statements that we present in this minibook's Chapter 1. The program turns all six LEDs on, pauses for 500 milliseconds (ms; half a second), turns the LEDs off, waits another half-second and then jumps back to the `Main` label to start the whole process over. (The numbers and arrows to the right are not part of the program but are annotations which we explain after Listing 2-1.)

Listing 2-1: Flashing LEDs

```
' {$PBASIC 2.5}→ 1
' {$STAMP BS2} → 2

Main:          → 4
  HIGH 0       → 5
  HIGH 2
  HIGH 4
  HIGH 6
  HIGH 8
  HIGH 10
  PAUSE 500    → 11
  LOW 0        → 12
  LOW 2
  LOW 4
  LOW 6
  LOW 8
  LOW 10
  PAUSE 500    → 18
  GOTO Main    → 19
```

The following paragraphs summarise the operation of this program:

→ 1 This line indicates that the program is written in version 2.5 of PBASIC. Every program you write for the BASIC Stamp 2 has to include this line. We describe how to insert it automatically into a program in Chapter 1 of this minibook.

→ 2 This line indicates that the program is going to run on a BASIC Stamp 2, and it's required for every program you run on a BASIC Stamp 2 microcontroller. You can insert it automatically by choosing Directive⇨Stamp⇨BS2.

→ 4 The label `Main:` identifies the location to which the `GOTO` statement in line 19 jumps. `Main:` is known as a *label,* which is simply a named location in your program. To create a label, you just type a name followed by a colon. For more information about creating value names in PBASIC, see the section 'Creating Names' later in this chapter.

→ 5 This line sets the output of pin 0 to high, which in turn lights up the LED. The following lines (6 through 10) similarly turn on pins 2, 4, 6, 8 and 10.

→ 11 This line pauses the program for 500 ms (half a second).

→ 12 This line and the five that follow set the outputs of pins 0, 2, 4, 6, 8, and 10 to low, which in turn extinguishes the LEDs.

→ 18 This line pauses the program for an additional half-second.

→ 19 This line transfers control of the program back to the `Main` label in line 4 so that the program repeats.

Commenting to Clarify Your Code

A *comment* is a bit of text that provides an explanation of your code. PBASIC completely ignores comments, and so you can put any text you want in a comment. Using plenty of comments in your programs to explain what your program does and how it works is a good idea.

A comment begins with an apostrophe. When PBASIC sees an apostrophe on a line, it ignores the rest of the line. Thus, if you place an apostrophe at the beginning of a line, the entire line is considered to be a comment. If you place a comment in the middle of a line (for example, after a statement), everything after the apostrophe is ignored.

Common programming practice is to begin a program with a group of comments that indicates what the program does, who wrote it and when. This block of comments should also indicate what I/O devices are expected to be connected to the BASIC Stamp. Listing 2-2 shows a version of the LED Flasher program (from the preceding section) that includes both types of comments.

You may notice that the `$PBASIC` and `$STAMP` directives begin with an apostrophe. Technically, these lines are treated as comments.

Listing 2-2: LED Flasher with Comments

```
' LED Flasher Program
' Doug Lowe
' July 10, 2011
'
' This program flashes LEDs connected to pins 0, 2, 4, 6, 8, and 10
' at one-half second intervals.

' {$PBASIC 2.5}
' {$STAMP BS2}

Main:
  HIGH 0       'Turn the LEDs on
  HIGH 2
  HIGH 4
  HIGH 6
  HIGH 8
  HIGH 10
  PAUSE 500    'Wait one-half second
  LOW 0        'Turn the LEDs off
  LOW 2
  LOW 4
  LOW 6
  LOW 8
  LOW 10
  PAUSE 500    'Wait one-half second
  GOTO Main
```

Creating Names in your Programs

In PBASIC, you can create your own names to use as program labels and for constants and variables (check out the later 'Employing Constants as Substitute Values' and 'Creating Variables to use RAM Memory' sections). You can also give a name of your own to I/O pins, which makes remembering what kind of input or output is expected from each pin easier (see the section 'Assigning Names to I/O Pins' later in this chapter).

You need to follow a few simple rules when creating names in PBASIC:

- Names can consist of a combination of upper- and lowercase letters, numbers and underscore characters (_). Other special characters, such as dollar signs or exclamation marks, aren't allowed. Thus, `Timer_Routine` and `Relay7` are valid names, but `LED$` or `Bang!` aren't.
- Names have to begin with a letter or an underscore but can't begin with a number. Thus, `Timer1` and `_Timer1` are valid names, but not `1Timer`.
- Names may be as long as 32 characters.
- Names aren't case-sensitive, which is to say that PBASIC doesn't distinguish between upper- and lowercase letters. Thus, PBASIC considers all the following names to be identical: `TimerCheck`, `timercheck`, `TIMERCHECK` and `TiMeRcHeCk`.

Although nothing in PBASIC is case-sensitive, and you can write anything in upper- or lowercase, common PBASIC programming convention is that keywords such as `HIGH` and `GOTO` are written in all caps, while names are written with just the first letter capitalised.

Employing Constants as Substitute Values

A *constant* is a name that has been assigned a value, which allows you to use the constant name in your program rather than the value itself. Later, if you decide to change the value, you don't have to hunt through the program to find every occurrence of the constant. Instead, you simply change the line that defines the constant.

Here's a statement that creates a constant named `Delay` and assigns the value `500` to it:

```
Delay  CON 500
```

The `CON` keyword indicates that `Delay` is a constant whose assigned value is 500.

To use a constant, just substitute the name of the constant wherever you'd use the value. For example, the following line pauses the program for the value assigned to the `Delay` constant:

```
PAUSE Delay
```

Listing 2-3 shows a version of the LED Flasher program that uses a constant to determine how fast the LEDs are going to flash.

Listing 2-3: The LED Flasher Program with a Constant

```
' LED Flasher Program
' Doug Lowe
' July 10, 2011
'
' This program flashes LEDs connected to pins 0, 2, 4, 6, 8, and 10
' at one-half second intervals.
'
' This version of the program uses a constant
' for the time interval.

' {$PBASIC 2.5}
' {$STAMP BS2}

Delay CON 500

Main:
  HIGH 0
  HIGH 2
  HIGH 4
  HIGH 6
  HIGH 8
  HIGH 10
  PAUSE Delay
  LOW 0
  LOW 2
  LOW 4
  LOW 6
  LOW 8
  LOW 10
  PAUSE Delay
  GOTO Main
```

Assigning Names to I/O Pins

You can use the `HIGH` and `LOW` statements to set the output status of an I/O pin. For example, the following statement sets pin 6 to high:

```
HIGH 6
```

Here, the number 6 indicates that pin 6 is to be set to high.

The problem with using just the pin number to identify which pin you want to control is that you can't tell what kind of device is connected to pin 6 simply by looking at the statement. It may be an LED, but it may also be a motor or a servo or even a pneumatic valve that causes a Frankenstein creature to pop up.

To remedy this situation, PBASIC lets you assign a name to an I/O pin by placing a statement similar to this one near the beginning of your program:

```
Led1 PIN 0
```

Here, the name `Led1` is assigned to pin 0. Now, you can use the name `Led1` in a `HIGH` or `LOW` statement, as follows:

```
HIGH Led1
```

This statement sets the I/O pin referenced by the name `Led1` to high.

Listing 2-4 shows a version of the LED Flasher program that uses pin names instead of the pin numbers. The real advantage of creating pin names is that changing the pin configuration of your project later on becomes much easier.

For example, suppose that you decide that instead of connecting the six LEDs to pins 0, 2, 4, 6, 8 and 10, you want to connect them to pins 0, 1, 2, 3, 4 and 5. By using pin names, you need to change the pin assignments just once when you modify the program, in the `PIN` statements near the beginning of the program.

Listing 2-4: The LED Flasher Program with Pin Names

```
' LED Flasher Program
' Doug Lowe
' July 10, 2011
'
' This program flashes LEDs connected to pins 0, 2, 4, 6, 8, and 10
' at one-half second intervals.
'
' This version of the program uses pin names instead of numbers.

' {$PBASIC 2.5}
' {$STAMP BS2}

Led1  PIN 0
Led2  PIN 2
Led3  PIN 4
Led4  PIN 6
Led5  PIN 8
Led6  PIN 10

Main:
```

(continued)

Listing 2-4 *(continued)*

```
HIGH Led1
HIGH Led2
HIGH Led3
HIGH Led4
HIGH Led5
HIGH Led6
PAUSE 500
LOW Led1
LOW Led2
LOW Led3
LOW Led4
LOW Led5
LOW Led6
PAUSE 500
GOTO Main
```

Creating Variables to Use RAM Memory

The BASIC Stamp 2 microprocessor has 32 bytes of RAM memory available for your programs to use. That may not sound like a lot, but it's plenty of RAM for most of the programs you're likely to create for your BASIC Stamp projects.

To use RAM memory in PBASIC, you create variables. A *variable* is simply a name that refers to a location in RAM. When you've created a variable, you can use the variable name in your program code to set or retrieve the value of the variable, and you can use the variable name in *expressions,* which perform simple mathematical calculations on the value of variables.

To create a variable, you list the name you want to use for the variable, followed by the keyword `VAR`, followed by one of four keywords that indicates the *type* of the variable you're creating. For example, the following creates a variable named `Count`, using the variable type `BYTE`:

```
Count VAR BYTE
```

Here are the four options for the variable type:

- **`BIT`:** Uses just one binary bit. Thus, the BASIC Stamp can squeeze up to eight `BIT` variables in each of its 32 bytes of available RAM. `BIT` variables are mostly used to keep track of whether some event has occurred. For example, you can set up a `BIT` variable to remember whether a user has pressed an input button. The value 0 indicates that the button hasn't yet been pressed and the value 1 means that the button has been pressed.

- **BYTE:** Uses one of the 32 available bytes of RAM and can have a value ranging from 0 to 255. This type of variable is useful for simple counters that don't need to exceed the value 255. For example, if you're creating a timer to count down 60 seconds, a `BYTE` variable does the trick.
- **NIB:** If you have a very small counter whose value is never going to exceed 15, you can use a `NIB` variable, which requires only one-half of one byte of RAM. Thus, in theory you can create as many as 64 different `NIB` variables in a BASIC Stamp 2's 32 bytes of RAM.
- **WORD:** Uses two of the 32 available bytes and can have a value ranging from 0 to 65,535. You need to use a `WORD` variable whenever the value to be stored in the variable is greater than 255. For example, a `WORD` variable is ideal for holding the length of a delay used by the `PAUSE` statement.

When you've created a variable, you can use it in an *assignment statement* to assign it a value. An assignment statement consists of a variable name followed by an equals sign, followed by the value to be assigned. For example, this assignment statement assigns the value 500 to a variable named `Delay`:

```
Delay = 500
```

The value on the right side of the equals sign can be an arithmetic calculation. For example:

```
Delay = 500 + 10
```

In this example, the value 510 is assigned to the variable named `Delay`.

Little point exists in doing arithmetic using only numerals; after all, you can do the calculation yourself. Thus, the previous example can be written as follows:

```
Delay = 510
```

The real power of variable assignments happens when you use variables on the right side of the equals sign. For example, the following statement increases the value of the `Delay` variable by 10:

```
Delay = Delay + 10
```

In this example, the previous value of `Delay` is increased by 10. For instance, if the `Delay` variable's value was 150 before this statement executed, it becomes 160 after.

Listing 2-5 shows a program that uses a variable to change the speed at which the LEDs flash each time the `GOTO` statement causes the program to loop. As you can see, a variable named `Delay` is used to provide the number of milliseconds that the `PAUSE` statement is to pause. Each time through the loop, the value of the `Delay` variable is increased by 10. Thus, the LEDs flash very quickly when the program starts, but the flashing gets progressively slower as the program loops.

Listing 2-5: The LED Flasher Program with a Variable

```
' LED Flasher Program
' Doug Lowe
' July 10, 2011
'
' This program flashes LEDs connected to pins 0, 2, 4, 6, 8, and 10
' at one-half second intervals.
'
' This version of the program uses a variable delay.

' {$PBASIC 2.5}
' {$STAMP BS2}

Led1  PIN 0
Led2  PIN 2
Led3  PIN 4
Led4  PIN 6
Led5  PIN 8
Led6  PIN 10

Delay VAR Word
Delay = 10

Main:
  HIGH Led1
  HIGH Led2
  HIGH Led3
  HIGH Led4
  HIGH Led5
  HIGH Led6
  PAUSE Delay
  LOW Led1
  LOW Led2
  LOW Led3
  LOW Led4
  LOW Led5
  LOW Led6
  PAUSE Delay
  Delay = Delay + 10
  GOTO Main
```

PBASIC lets you use a variable in a `HIGH` or `LOW` statement to indicate which pin is to be controlled. For example:

```
Led VAR BYTE
Led = 0
HIGH Led
```

This sequence of statements creates a variable named `Led`, assigns the value 0 to it and then uses it in a `HIGH` statement. The result is that I/O pin 0 is set to high.

Carrying out Maths Functions

PBASIC lets you perform addition, subtraction, multiplication and division using the symbols (called *operators*) +, –, * and /. Here's an example of an assignment that uses all some of these symbols:

```
X VAR BYTE
X = 10 * 3 / 2 + 5
```

In this example, the value 20 is assigned to the variable `X` (10 × 3 = 30, 30 / 2 = 15, and 15 + 5 = 20).

Here are a few things you need to know about how PBASIC does maths:

- Unlike most programming languages, PBASIC performs mathematical operations strictly on a left-to-right basis. For example, consider the following assignment:

  ```
  X = 10 + 3 * 2
  ```

  Most programming languages first multiply the 3 by the 2, giving a result of 6, and then add the 6 to the 10, giving the final result 16. That's because multiplication is ordinarily done before addition in equations. But PBASIC calculates the expression left to right, and so it first adds 10 and 3, giving the result 13, and then multiplies the 13 by 2, giving the result 26.

- You can use parentheses to force PBASIC to calculate a certain part of the formula first. For example:

  ```
  X = 10 + (3 * 2)
  ```

  Here, PBASIC first does the calculation inside the parenthesis, giving a result of 6. It then adds the 6 to the 10 to give the final result, 16.

- When PBASIC does division, it discards the remainder and returns the result as a whole number. For example:

  ```
  X = 8 / 3
  ```

  This statement assigns the value 2 to `X`, because 8 divided by 3 is 2 with a remainder of 2. PBASIC discards the remainder and returns the result 2.

Making use of IF Statements

An IF statement lets you add *conditional testing* to your programs. In other words, it lets you execute certain statements only if a particular condition is met. This type of conditional processing is an important part of any but the most trivial of programs.

Every IF statement must include a *conditional expression* that lays out a logical test to determine whether the condition is true or false. For example:

```
X = 5
```

This condition is true if the value of the variable X is 5. If X has any other value, the condition is false.

You can use less-than or greater-than signs in a conditional expression, as follows:

```
Led < 10
Speed > 1000
```

The first expression is true if the value of Led is less than 10. The second expression is true if the value of Speed is greater than 1,000.

In its simplest form, the IF statement causes the program to jump to a label if a condition is true. For example:

```
IF Led < 11 THEN Main
```

Here, the program jumps to the Main label if the value of the Led variable is less than 11.

Listing 2-6 shows a program that flashes the LEDs in sequence. This program uses a variable named Led to represent the output pin to be used. On each pass through the loop, it adds 2 to the Led variable to determine the next LED to be fired. Then, an IF statement is used to loop back to the Main label if the Led variable is less than 11. This design sets up the basic loop that first flashes the LED on pin 0, and then the LED on pin 2, and then pins 4, 6 and 8, and finally 10.

After the program flashes the LED in pin 10, the program adds 2 to the Led variable, setting this variable to 12. Then, the conditional expression in the IF statement (X < 11) tests false instead of true, and so the IF statement doesn't skip to the Main label at this point. Instead, the statement after the IF statement is executed, which resets the Led variable to zero. Then, a GOTO statement sends the program back to the Main label, where the first LED is flashed again.

Listing 2-6: The LED Flasher Program with an `IF` statement

```
' LED Flasher Program
' Doug Lowe
' July 10, 2011
'
' This program flashes LEDs connected to pins 0, 2, 4, 6, 8, and 10
' in sequence.
'
' This version of the program uses a simple IF statement.

' {$PBASIC 2.5}
' {$STAMP BS2}

Speed VAR BYTE
Led VAR BYTE

Speed = 50
Led = 0

Main:
  HIGH Led
  PAUSE Speed
  LOW Led
  PAUSE Speed
  Led = Led + 2
  IF Led < 11 THEN Main
  Led = 0
  GOTO Main
```


A second and more useful form of the `IF` statement lets you list one or more statements to be executed if the condition is true. For example:

```
IF Led < 10 THEN
    Led = Led + 2
ENDIF
```

In this example, 2 is added to the `Led` variable if the value of the `Led` variable is less than 10.

You can place as many statements as you want between the `IF` and `ENDIF` statements. For example:

```
IF Led < 10 THEN
    Speed = Speed + 10
    Led = Led + 2
ENDIF
```

Here, the `Speed` variable is also increased if the condition expression is true.

The main difference between the `IF` statement with `ENDIF` and an `IF` statement without `ENDIF`, is that without the `ENDIF` the statement that's executed if the `IF` condition is true has to be on the same line as the `IF` and `THEN` keywords. If the `THEN` keyword is the last word on a line, PBASIC assumes that you're going to use an `ENDIF` to mark the end of the list of statements to be executed if the `IF` condition is true. If you forget to include the `ENDIF` statement, the program doesn't work properly.

One last trick that the `IF` statement lets you do is list statements that you want to execute if the condition *isn't* true. You do so by using an `ELSE` statement along with the `IF` statement. For example:

```
IF Led < 10 THEN
    Led = Led + 2
ELSE
    Led = 0
ENDIF
```

Here, `Led` is increased by 2 if its current value is less than 10. But if the current value of `Led` isn't less than 10, the `Led` variable is reset to 0.

Listing 2-7 shows a version of the LED Flasher program that uses an IF-THEN-ELSE statement to flash the LEDs in sequence.

Listing 2-7: LED Flasher with an IF-THEN-ELSE statement

```
' LED Flasher Program
' Doug Lowe
' July 10, 2011
'
' This program flashes LEDs connected to pins 0, 2, 4, 6, 8, and 10
' in sequence.
'
' This version of the program uses an IF-THEN-ELSE statement.

' {$PBASIC 2.5}
' {$STAMP BS2}

Speed VAR BYTE
Led VAR BYTE

Speed = 50
Led = 0
```

```
Main:
  HIGH Led
  PAUSE Speed
  LOW Led


 PAUSE Speed
  IF Led < 10 THEN
    Led = Led + 2
  ELSE
    Led = 0
  ENDIF
  GOTO Main
```

Pressing DO Loops into Service

The DO loop is a special PBASIC statement that performs essentially the same function as a label and a GOTO statement (check out the earlier section 'Flashing the LEDs'). For example, consider the following:

```
Main:
  HIGH 0
  PAUSE 500
  LOW 0
  PAUSE 500
  GOTO Main
```


You can accomplish the same function without the Main label or the GOTO statement by placing the lines that turn the LED on and off between DO and LOOP statements, as follows:

```
DO
  HIGH 0
  PAUSE 500
  LOW 0
  PAUSE 500
LOOP
```

The lines between the DO and LOOP statements are executed over and over again indefinitely.

Listing 2-8 shows the LED Flasher program implemented with a simple DO loop instead of a label and a GOTO statement.

Listing 2-8: LED Flasher with a DO loop

```
' LED Flasher Program
' Doug Lowe
' July 10, 2011
'
' This program flashes LEDs connected to pins 0, 2, 4, 6, 8, and 10
' in sequence.
'
' This version of the program uses a DO loop.

' {$PBASIC 2.5}
' {$STAMP BS2}

Speed VAR BYTE
Led VAR BYTE

Speed = 50
Led = 0

DO
  HIGH Led
  PAUSE Speed
  LOW Led
  PAUSE Speed
  IF Led < 10 THEN
    Led = Led + 2
  ELSE
    Led = 0
  ENDIF
LOOP
```

You can add a conditional test (for details, flip to the earlier section 'Making use of `IF` Statements') to the `LOOP` statement to make the loop conditional. For example:

```
Led = 0
DO
  HIGH Led
  PAUSE 500
  LOW Led
  PAUSE 500
  Led = Led + 2
LOOP UNTIL Led > 10
```

This code flashes the LEDs on pins 0, 2, 4, 6, 8 and 10. After the LED on pin 10 is flashed, the next-to-last line sets the `Led` variable to 12. Then, the `LOOP UNTIL` statement sees that `Led` is greater than 10, and so it stops looping.

Instead of the word `UNTIL`, you can use the word `WHILE` to mark the condition in a `DO` loop. A substantial difference exists between `UNTIL` and `WHILE`, just as the words suggest:

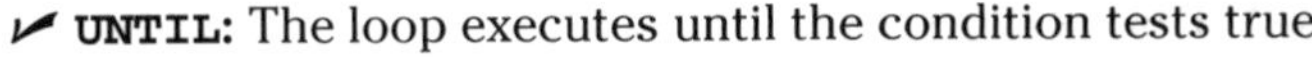

- **UNTIL:** The loop executes until the condition tests true.

- **WHILE:** The loop executes until the condition tests false.

You can also include the conditional test on the `DO` statement or on the `LOOP` statement:

- If you place the conditional test on the `DO` statement, the condition is tested *before* each execution of the loop.
- If you place it on the `LOOP` statement, the condition is tested *after* the completion of each loop.

A common programming technique is to place `WHILE` tests on the `DO` statement and `UNTIL` tests on the `LOOP` statement. For example:

```
Led = 0
DO WHILE Led < 11
  HIGH Led
  PAUSE 500
  LOW Led
  PAUSE 500
  Led = Led + 2
LOOP
```


Here, the value of `Led` is tested prior to each execution of the loop. The loop is executed as long as `Led` is less than 11.

You can *nest* `DO` loops, which simply means that one `DO` loop can contain another `DO` loop. Note that when you nest `DO` loops, the inner loop must have a conditional test (see the earlier 'Making use of `IF` Statements' section). Otherwise, it loops forever and the outer loop never has a chance to complete.

Listing 2-9 shows a program that uses two nested `DO` loops to flash the LEDs in sequence. The innermost `DO` loop flashes the six LEDs once. It uses an `UNTIL` condition to stop the loop after the last LED has flashed. The outermost `DO` loop continues endlessly, causing the flashing sequence to continue indefinitely.

Listing 2-9: The LED Flasher Program with Nested DO Loops

```
' LED Flasher Program
' Doug Lowe
' July 10, 2011
'
' This program flashes LEDs connected to pins 0, 2, 4, 6, 8, and 10
' in sequence.
'
' This version of the program uses nested DO loops.

' {$PBASIC 2.5}
```

(continued)

Listing 2-9 *(continued)*

```
' {$STAMP BS2}

Speed VAR BYTE
Led VAR BYTE

Speed = 50

DO
  Led = 0
  DO
    HIGH Led
    PAUSE Speed
    LOW Led
    PAUSE Speed
    Led = Led + 2
  LOOP UNTIL Led > 10
LOOP
```

Keeping Count with FOR Loops

A FOR loop is a special type of looping statement that automatically keeps a counter variable. FOR loops are ideal when you want to execute a loop a certain number of times or when you want to perform an action on multiple I/O pins. Therefore, a FOR loop is the ideal way to implement the LED Flasher program.

Here's the basic structure of a FOR loop:

```
FOR counter = start-value TO end-value
  Statements...
NEXT
```

Here's an example that flashes the LED on pin 0 ten times:

```
X VAR BYTE
FOR X = 1 TO 10
  HIGH 0
  PAUSE 500
  LOW 0
  PAUSE 500
NEXT
```

In this example, the loop is executed ten times. The value of the variable X is increased by 1 each time through the loop.

In the preceding example, the program didn't use the counter variable, which is common in FOR loops; sometimes the only purpose for the counter variable is to control how many times the loop is executed. But you can use the counter variable within the loop. For example, here's a loop that makes every I/O pin on the Stamp high for one-tenth of a second:

```
IO_Pin VAR BYTE
FOR IO_Pin = 0 TO 15
  HIGH IO_Pin
  PAUSE 100
  LOW IO_Pin
NEXT
```

Normally, the counter variable is increased by 1 on each pass through the loop. You can use the STEP keyword to specify a different step value if you want. When you use the STEP keyword, the basic structure of the FOR statement is as follows:

```
FOR counter = start-value TO end-value STEP step-value
  Statements...
NEXT
```

For example, you can flash LEDs on just the even-numbered pins as follows:


```
Led VAR Byte
FOR Led = 0 TO 10 STEP 2
  HIGH Led
  PAUSE 100
  LOW Led
NEXT
```

Another interesting feature of FOR loops is that they can count backward. All you have to do is specify a start value that's larger than the end value:

```
Led VAR Byte
FOR Led = 10 TO 0 STEP 2
  HIGH Led
  PAUSE 100
  LOW Led
NEXT
```

Listing 2-10 shows a version of the LED Flasher program that uses a pair of FOR loops to flash the LEDs first in one direction and then in the opposite direction. The created effect is similar to the spooky electronic eyes on the evil Cylons in the old TV series *Battlestar Galactica.* The first FOR loop flashes the LEDs on pins 0, 2, 4, 6 and 8. Then the second FOR loop flashes the LEDs on pins 10, 8, 6, 4 and 2. Both FOR loops are contained within a DO loop that keeps the LEDs bouncing back and forth indefinitely.

Listing 2-10: The LED Flasher Program with FOR Loops

```
' LED Flasher Program
' Doug Lowe
' July 10, 2011
'
' This program flashes LEDs connected to pins 0, 2, 4, 6, 8, and 10
' back and forth, like Cylon eyes.
'
' This version of the program uses FOR loops.

' {$STAMP BS2}
' {$PBASIC 2.5}

Led VAR Byte

Main:
  FOR Led = 0 TO 8 STEP 2
    HIGH Led
    PAUSE 100
    LOW Led
  NEXT
  FOR Led = 10 TO 2 STEP 2
    HIGH Led
    PAUSE 100
    LOW Led
  NEXT
  GOTO Main
```

Like `DO` loops (which we discuss in the preceding section), you can nest `FOR` loops. When you nest `FOR` loops, the innermost loop(s) complete their entire cycle each time through the outer loop. In other words, if a `FOR` loop that repeats 10 times is placed within an outer loop that repeats 10 times, the statements within the innermost loop execute a total of 100 times – 10 times for each of the 10 repetitions of the outer loop.

Listing 2-11 shows a variation on the Cylon eyes program in Listing 2-10. This one uses an outer `FOR` loop that varies the delay time for the `PAUSE` statements. The result is that the LEDs sweep very quickly at first, but slow by 10 ms on each repetition of the outer loop until the delay reaches 1 second per LED.

Listing 2-11: The LED Flasher Program with Nested FOR Loops

```
' LED Flasher Program
' Doug Lowe
' July 10, 2011
'
' This program flashes LEDs connected to pins 0, 2, 4, 6, 8, and 10
' back and forth, like Cylon eyes.
'
' This version of the program uses nested FOR-NEXT loops to slow the
' sweeping motion of the LEDs.

' {$STAMP BS2}
' {$PBASIC 2.5}

Led VAR Byte
Speed VAR Word

FOR Speed = 10 TO 1000 STEP 10
  FOR Led = 0 TO 8 STEP 2
    HIGH Led
    PAUSE Speed
    LOW Led
  NEXT
  FOR Led = 10 TO 2 STEP 2
    HIGH Led
    PAUSE Speed
    LOW Led
  NEXT
NEXT
```

Chapter 3

Discovering More PBASIC Programming Tricks

In This Chapter

- Connecting pushbuttons
- Generating random numbers
- Using a potentiometer as input
- Creating subroutines with the `GOSUB` command

In this chapter, we describe some more PBASIC programming techniques, to add to the fundamental ones we introduce in Chapter 2 of this minibook. These tips are sure to become invaluable in your BASIC Stamp projects. Specifically, you discover handling input data in the form of pushbuttons, generating random numbers (that make your programs more interesting by adding a degree of randomness), reading the value of a potentiometer and using the `GOSUB` command to organise your program into subroutines.

Pushing Buttons with a BASIC Stamp

In Chapter 2 of this minibook, we discuss connecting a light-emitting diode (LED) to a BASIC Stamp I/O pin and turning the LED on or off by using the `HIGH` and `LOW` commands in a PBASIC program. Those commands are designed to use BASIC Stamp I/O pins as output pins by setting the status of an I/O pin to high or low so that external circuitry (such as an LED) can react to the pin's status.

But what if you want to use an I/O pin as an input instead of an output? In other words, you want the BASIC Stamp to react to the status of an external circuit instead of the other way around. The easiest way to do that is to connect a pushbutton to an I/O pin. Then, you can add commands to your PBASIC program to detect whether the pushbutton is pressed.

You can connect a pushbutton to a BASIC Stamp I/O pin in two ways:

- **Active-high:** This connection places +5 V on the I/O pin when the pushbutton is pressed. When the button is released, the I/O pin sees 0 V.
- **Active-low:** This connection sees +5 V when the pushbutton isn't pressed. When you press the pushbutton, the +5 V is removed and the I/O pin sees no voltage.

Figure 3-1 shows examples of active-high and active-low pushbuttons. In the active-high circuit, the I/O pin is connected to ground through R1 and R2 when the pushbutton isn't pressed. Thus, the voltage at the I/O pin is 0. When the pushbutton is pressed, the I/O pin is connected to Vdd (+5 V) through R1, causing the I/O pin to see +5 V. As a result, the I/O pin is low when the button isn't pressed and high when the button is pressed.

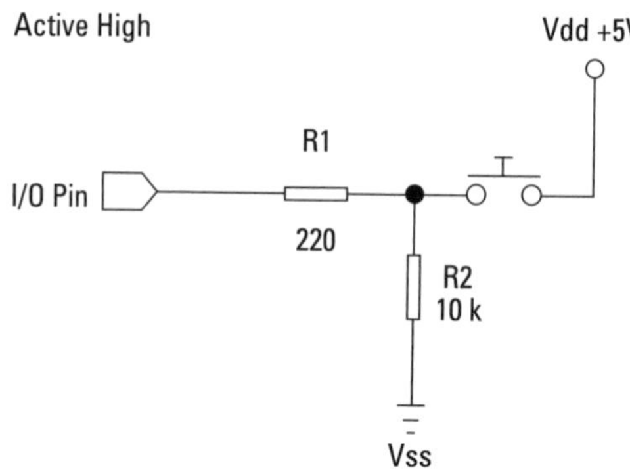

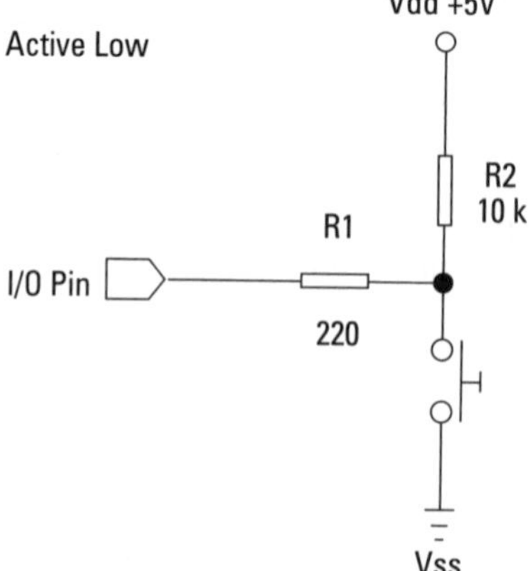

Figure 3-1: Active-high and active-low input circuits.

In the active-low circuit, the I/O pin is connected to Vdd (+5 V) through R1 and R2, causing the I/O pin to go high. But when the button is pressed, the current from Vdd is connected to ground through R2, causing the voltage at the I/O pin to drop to zero. Thus, the I/O pin is high when the button isn't pressed and low when the button is pressed.

Note that in both circuits, R1 is connected directly to the I/O pin to prevent excessive current flow when the switch is pressed. Without this resistor, the pin would be connected directly to Vdd (+5 V) or Vss (ground) when the button is pressed, which can damage your BASIC Stamp.

In an active-high circuit, R2 is called a *pull-down* resistor because it pulls the current from the I/O pin down to zero when the pushbutton isn't depressed. In an active-low circuit, R2 is called a *pull-up resistor* because it pulls the voltage at the I/O pin up to Vdd (+5 V) when the pushbutton isn't depressed.

Checking the Status of a Switch in PBASIC

When you've connected a switch to a BASIC Stamp I/O pin, you need to know how to determine whether the switch is open or closed from a PBASIC program. The easiest way to do so is to first assign a name to the pin you want to test using the PIN directive. For example, if an active-high input button (see the preceding section for an explanation) is connected to pin 14, you can assign it a name such as this:

```
Button1    PIN 14
```

Here, the name `Button1` is assigned to pin 14.

To determine whether the pushbutton is pressed, you can use an `IF` statement as follows:

```
IF Button1 = 1 THEN
  HIGH Led1
ENDIF
```

Here, the output pin designated as `Led1` is made high when the button is pressed.

If you want `Led1` to be high *only* when `Button1` is pressed, use this code:

```
IF Button1 = 1 THEN
  HIGH Led1
ELSE
  LOW Led1
ENDIF
```

Here, `Led1` is made high if the button is pressed and low if the button isn't pressed.

In addition, you can put the whole thing in a loop to test repeatedly the status of the button and turn the LED on and off accordingly:

```
DO
  IF Button1 = 1 THEN
    HIGH Led1
  ELSE
    LOW Led1
  ENDIF
LOOP
```

Listing 3-1 shows an interesting program that works with a BASIC Stamp that has a pushbutton switch connected to pin 14 and LEDs connected to pins 0 and 2. The program flashes the LED connected to pin 2 on and off at half-second intervals until the pushbutton switch is depressed. Then, it flashes the LED on pin 0.

Listing 3-1: The Pushbutton Program

```
' Pushbutton Program
' Doug Lowe
' July 13, 2011


' {$STAMP BS2}
' {$PBASIC 2.5}

Led1 PIN 0
Led2 PIN 2
BUTTON1 PIN 14

DO

  IF BUTTON1 = 1 THEN
```

```
    LOW Led2
    HIGH Led1
    PAUSE 100
    LOW Led1
    PAUSE 100
  ELSE
    LOW Led1
    HIGH Led2
    PAUSE 100
    LOW Led2
    PAUSE 100
  ENDIF

  PAUSE 100

LOOP
```

Project 3-1 shows how to build a simple circuit you can use to test the program in Listing 3-1.

Project 3-1: A Pushbutton-Controlled LED Flasher

In this project, you connect two LEDs to a BASIC Stamp HomeWork board or Board of Education to test the programs that we present in this chapter.

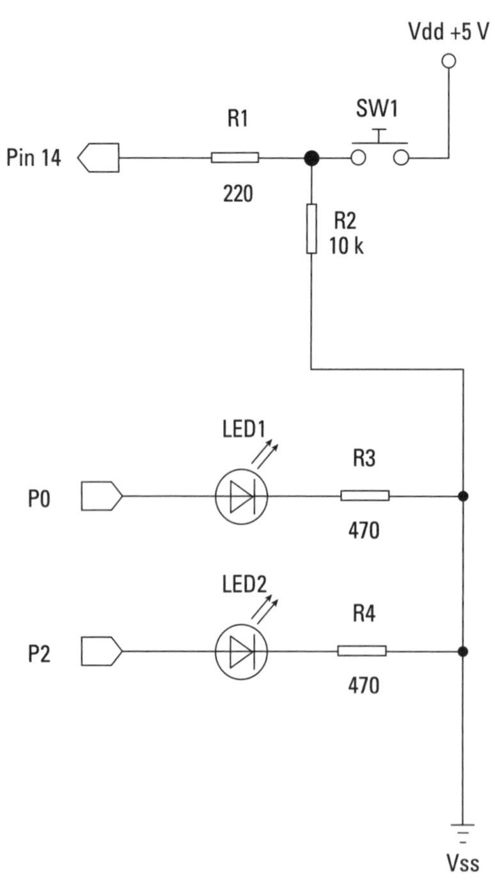

Parts List

1 Computer with BASIC Stamp Editor software installed
1 BASIC Stamp HomeWork board or Board of Education
1 9 V battery
1 USB cable
1 USB to serial adapter
2 LEDs
2 470 Ω resistors (yellow-violet-brown)
1 220 Ω resistors (red-red-brown)
1 10 kΩ resistor (brown-black-orange)
5 Jumper wires
1 Normally open push button switch

Switch Test Program

```
'{$STAMP BS2}
'{$PBASIC 2.5}

Led1 PIN 0
Led2 PIN 2
Button1 PIN 14


DO
  IF Button1 = 1 THEN
    HIGH Led1
    LOW Led2
  ELSE
    LOW Led1
    HIGH Led2
  ENDIF
LOOP
```

Steps

1. **Connect the two LEDs to the output pins as follows:**

Pin (Anode)	Breadboard Hole
P0	A17
P2	A15

2. **Connect the resistors across the gap in the centre of the breadboard as follows:**

Resistor	From	To
R1	P14	C3
R2	E5	F5
R3	E17	F17
R4	E15	F15

3. **Use the jumper wires to connect the resistors to ground.**

   I17 to I15

   J15 to Vss

   F5 to Vss

   D5 to D3

   D1 to Vdd

4. **Connect switch SW1 to E1, F1, E3 and F3.**
5. **Open the BASIC Stamp Editor.**
6. **Connect your BASIC Stamp to the computer and identify it in the Stamp Editor.**

   For more information about how to do this, refer to Chapter 1 of this minibook.

7. **Type the Switch Test program (see nearby listing) into the Editor program window, and then save the file.**
8. **Choose Run⇨Run.**

   If you prefer, you can click the Run button in the toolbar or press F9.

   When you choose the Run command, the program is downloaded to the BASIC Stamp. When the program is downloaded, it starts to run automatically on the Stamp.

9. **Press the button.**

   When you run the program, the LED connected to pin 2 lights up. When you push the button, the LED on pin 2 goes dark and the LED on pin 0 lights.

10. **Now try the programs in Listing 3-1.**

    This program is similar to the Switch Test program in this project but adds the complication that the LEDs are flashing on and off at half-second intervals.

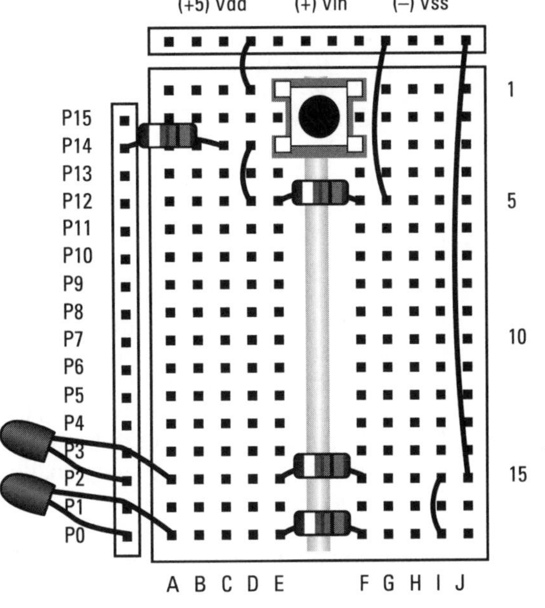

Randomising Your Programs

Many computer-controlled applications require a degree of randomness to their operation. A classic example is the *Indiana Jones* ride at Disneyland and Disney World, in which the adventure is slightly different every time. At the start, your vehicle can enter through one of three doors; the exact door chosen for your ride is determined randomly. Many other details of the ride are randomly varied in an effort to make the adventure slightly different every time you climb board.

You can add a bit of randomness to your own BASIC Stamp programs by using the `RANDOM` command. This command scrambles the bits of a variable (check out this minibook's Chapter 2 for a definition).

Although you can use any type of variable with a `RANDOM` command, you almost always want to use a `Byte` or `Word` variable:

- **`Byte`:** The `RANDOM` command generates a random number between 0 and 255.
- **`Word`:** The random value is between 0 and 65,535.

Here's an example of a simple way to use the `RANDOM` command to add a random pause:

```
Result VAR Word
RANDOM Result
PAUSE Result
```

This sequence of code creates a `Word` variable named `Result`, randomises the variable and then pauses for the number of milliseconds (ms) indicated by the `Result` variable.

In most cases, the value returned by the `RANDOM` command isn't really the random number you're looking for. Usually, you want to determine a random number that falls within a range of numbers. For example, if you're writing a dice program and want to simulate the roll of a single die, you need to generate a random number between 1 and 6. You can easily reduce the value of a full `Word` variable to a number between 1 and 6 by using a simple mathematical calculation that involves a special type of division operation called *modulus division.*

As we explain in Chapter 2 of this minibook, when PBASIC does division, it keeps the integer portion of the quotient and discards the remainder. For example, 10 / 3 = 3; the remainder (1) is simply discarded.

Modulus division, which is represented by two slashes (//) instead of one, throws away the quotient and keeps only the remainder. Thus, 10 // 3 = 1, because the remainder of 10 / 3 = 1.

You can put modulus division to good use when working with random numbers. For example:

```
Result VAR Word
Die    VAR Byte
RANDOM Result
Die = Result // 6
```

In fact, the preceding calculation isn't quite right. We include it to point out a common pitfall that happens if you forget that the remainder may happen to be 0. The above calculation actually returns a random number that falls between 0 and 6.

To find a random number that falls between 1 and 6, you need to calculate the modulus division by 5, not by 6, and then add 1 to the result. For example:

```
Result VAR Word
Die    VAR Byte
RANDOM Result
Die = Result // 5 + 1
```

The preceding calculation returns a random number between 1 and 6.

Listing 3-2 shows a sample program that lights `Led1` (pin 0) until the pushbutton on pin 14 is pressed. Then, it lights `Led2` (pin 2) for a random period of time between 1 and 10 seconds. It uses the following equation to calculate the number of seconds to pause:

```
Seconds = Result // 9 + 1
```

The program then multiplies the seconds by 1,000 to convert to milliseconds as required by the `PAUSE` command.

Listing 3-2: The Random Program

```
' Random Program
' Doug Lowe
' July 10, 2011
'
' This program turns on the LED at pin 2 for a random number
' of seconds between 1 and 10 when the pushbutton on pin 14
' is pressed.
```

(continued)

Listing 3-2 *(continued)*

```
' {$STAMP BS2}
' {$PBASIC 2.5}

Result      VAR Word
Seconds     VAR Byte
Button1     PIN 14
Led1        PIN 0
Led2        PIN 2

DO
  HIGH Led1
  IF Button1 = 1 THEN
    RANDOM Result
    Seconds = Result // 9 + 1
    LOW Led1
    HIGH Led2
    PAUSE Seconds * 1000
    HIGH Led1
    LOW Led2
  ENDIF
LOOP
```

Each time you press the pushbutton in this program, `Led2` lights up for a different length of time, between 1 and 10 seconds.

In fact, the `RANDOM` command isn't really all that random. The starting value of the variable used by the `RANDOM` command is called the *seed value.* The `RANDOM` command applies a complex mathematical calculation to the seed value to determine a result that appears to be random. However, the result isn't random: given a particular seed value, the `RANDOM` command always returns the same result.

For example, if you use a `Word` variable whose initial value is 0, the `RANDOM` command always changes the variable's value to 64,992. If you apply the `RANDOM` command to the same variable again, the result is always 9,072. The sequence of numbers generated from a given initial value is distributed randomly across the entire range of possible values (for example, 0 to 65,535), but the sequence is the same every time.

Thus, the program in Listing 3-2 always generates the same sequence of random delays. In particular, the sequence for the first ten button presses is always as follows:

- **First press:** 4 seconds
- **Second press:** 1 second
- **Third press:** 4 seconds

- **Fourth press:** 1 second
- **Fifth press:** 3 seconds
- **Sixth press:** 5 seconds
- **Seventh press:** 4 seconds
- **Eighth press:** 1 second
- **Ninth press:** 1 second
- **Tenth press:** 4 seconds

This sequence appears fairly random, but every time you reset the program and start over, the sequence is identical.

The easiest way around this problem is to make the initial value of the variable fed to the RANDOM command dependent on some external event, such as the press of a button. You can easily do that by creating a loop that counts while it waits for the user to press the button. Because your program has no way to determine exactly when the user is going to press the button, the number used to start the random number generator is truly random.

Listing 3-3 shows an improved version of the random program that uses this technique to create a truly random delay. Only one additional line of code is needed to randomise the delay properly:

```
Result = Result + 1
```

By adding 1 to the Result variable each pass through the loop, the seed value for the RANDOM command is unpredictable, and so a true random value is generated.

Listing 3-3: An Improved Version of the Random Program

```
' Improved Random Program
' Doug Lowe
' July 10, 2011
'
' This program turns on the LED at pin 2 for a random number
' of seconds between 1 and 10 when the pushbutton on pin 14
' is pressed.
'
' This version of the program uses a counter to create a truly random number.

' {$STAMP BS2}
' {$PBASIC 2.5}

Result      VAR Word
```

(continued)

Listing 3-3 *(continued)*

```
Seconds     VAR Byte
Button1     PIN 14
Led1        PIN 0
Led2        PIN 2

DO
  HIGH Led1
  Result = Result + 1
  IF Button1 = 1 THEN
    RANDOM Result
    Seconds = Result // 9 + 1
    LOW Led1
    HIGH Led2
    PAUSE Seconds * 1000
    HIGH Led1
    LOW Led2
  ENDIF
LOOP
```

Reading a Value from a Potentiometer

A *potentiometer* (often called a *pot*) is simply a variable resistor with a knob you can turn to vary the resistance (check out Book II, Chapter 2 for more about potentiometers). Pots of various types are often used as input devices for BASIC Stamp projects. For example, you can use a simple pot to control the speed of a pair of flashing LEDs: as you turn the pot's knob, the rate at which the LEDs flash changes.

Although the most common type of pot has a mechanical knob to vary its resistance, many pots use some other means to vary their resistance. For example, a joystick uses pots that are connected to a moveable stick. One of the pots measures the motion of the stick in the *x*-axis; the other measures the motion of the stick in the *y*-axis. You can also connect a pot to the hinge on a door, in which case the resistance of the pot indicates not only whether the door is opened or closed, but also the door's angle if it's partly opened.

The BASIC Stamp doesn't have the ability to read directly the value of a pot connected to one of its I/O pins. However, by cleverly combining the pot with a small capacitor, you can measure how long the capacitor takes to discharge. With this knowledge, you can calculate the resistance of the pot by using the resistor-capacitor (RC) time calculations that we present in Book II, Chapter 3 (flip to that chapter too to find out how RC circuits work and how to do the time calculations). For the purposes of this chapter, you don't need to perform the time calculations yourself, but brushing up on how RC circuits work isn't going to hurt.

Figure 3-2 shows a typical RC circuit connected to a pin on a BASIC Stamp. Here, a 10 kΩ pot is placed in parallel with a 0.1 μF capacitor. In addition, a 220 Ω resistor is placed in series with the pot to protect the BASIC Stamp from damage that may be caused by excess current if you turn the pot's knob so that the resistance of the pot drops to zero.

The capacitor in this circuit is small enough (0.1 μF) that the circuit charges and discharges very quickly – within about a millisecond or so, depending on where the pot knob is set. Thus, your program isn't delayed significantly while it waits for the capacitor to discharge so that it can determine the resistance of the pot.

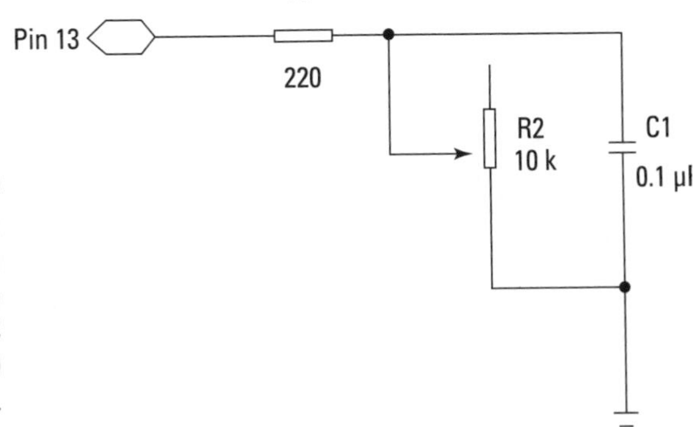

Figure 3-2: Connecting a pot to a BASIC Stamp I/O pin.

Given the circuit in Figure 3-2, how do you go about measuring the resistance of the pot? The answer requires a clever bit of programming. You set pin 13 to high, which charges the capacitor. Then, you set up a loop to monitor the input status of pin 13, which stops driving P13 high and allows the capacitor to discharge through the R2. Each time you check the status of pin 13, you add 1 to a counter. When the capacitor has discharged, pin 13 goes low. When pin 13 is low, the loop ends and the counter indicates how long discharging the capacitor took. Knowing the size of the capacitor and the length of time required to discharge the capacitor allows you to calculate the resistance of the pot.

Fortunately, PBASIC includes a command called `RCTIME` that does this process automatically. All you have to do is tell the `RCTIME` command what pin the RC circuit is on, whether you want to measure how long the RC circuit takes to charge or discharge, and the name of a variable in which to store the resulting time calculation.

Here's how to use the `RCTIME` command to determine how long an RC circuit on pin 13 takes to discharge, storing the answer in a variable named `Timer`:

```
RCTIME 13, 1, Timer
```

This RCTIME command sets the variable named Timer to a value that indicates how long it took the RC circuit to discharge. Immediately before this command, you need to set the I/O pin (in this case, pin 13) to high to charge the capacitor. You also need to pause for a short time (usually, 1 ms is enough) to allow the circuit to charge.

Although you can use this technique to calculate the actual resistance of a pot, you don't usually have to know the exact resistance. Instead, knowing that the counter increases when the resistance of the pot increases and decreases when the resistance of the pot decreases, is usually sufficient.

For the circuit shown in Figure 3-2, the RCTIME command calculates time values ranging from about 12 when the resistance of the pot is near 0 to about 54 when the resistance of the pot is at its maximum (10 kΩ).

Listing 3-4 shows a simple program that alternately flashes LEDs connected to pins 0 and 2. The rate at which the LEDs flash is set by a pot in an RC circuit on pin 13. As you can see, the program simply multiples the time value calculated by the RCTIME command by 10 to determine how long the program is to pause between flashes. As you turn the pot's knob to decrease the resistance of the pot, the LEDs flash at a faster rate.

Listing 3-4: An LED Flashing Program That Uses a Pot

```
' Potentiometer LED Flashing Program
' Doug Lowe
' July 10, 2011
'
' This program flashes LEDs connected to pins 0 and 2
' at a rate determined by an RC circuit on pin 13.

' {$STAMP BS2}
' {$PBASIC 2.5}

Time VAR Word
Led1 PIN 0
Led2 PIN 2
Pot  PIN 13
DO
  HIGH Pot
  RCTIME Pot, 1, Time
  HIGH Led1
  LOW  Led2
  PAUSE Time * 10
  LOW Led1
  HIGH Led2
  PAUSE Time * 10
LOOP
```

Project 3-2 shows you how to build a circuit that includes a 10 kΩ potentiometer and a capacitor so that you can test the code in Listing 3-4. Figure 3-3 shows the completed circuit.

Figure 3-3: A circuit that uses a potentiometer to control flashing LEDs (Project 3-2).

Project 3-2: Using a Potentiometer to Control Flashing LEDs

In this project, you create a BASIC Stamp circuit that uses a potentiometer to control the rate at which two LEDs alternately flash.

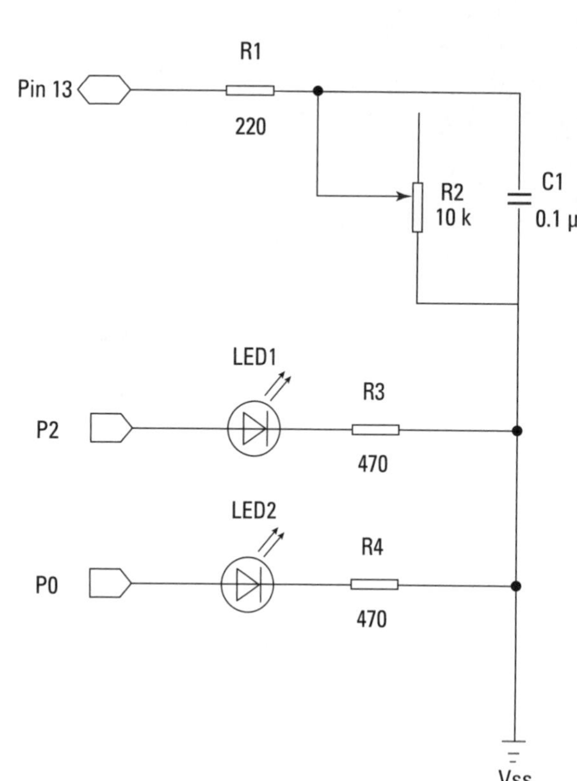

Parts List

- 1 Computer with BASIC Stamp Editor software installed
- 1 BASIC Stamp HomeWork board or Board of Education
- 1 9 V battery
- 1 USB cable
- 1 USB to serial adapter
- 2 LEDs
- 2 470 Ω resistors (yellow-violet-brown)
- 1 220 Ω resistor (red-red-brown)
- 1 10 kΩ potentiometer
- 1 0.1 µF capacitor
- 4 Jumper wires

Steps

1. **Connect the two LEDs to the output pins as follows:**

Pin (Anode)	Breadboard Hole
P0	A17
P2	A15

2. **Connect the resistors in the centre of the bread- board as follows:**

Resistor	From	To
R1	B4	PIN 13
R2	E15	F15
R3	E17	F17

3. **Insert the jumper wires.**

   F1 to C4
   H3 to Vss
   I17 to I15
   J15 to Vss
4. **Insert the capacitor.**

   One lead go in G1, the other in Vss.
5. **Insert the potentiometer.**

   The potentiometer has three leads, which go in D4, F3 and F5.
6. **Open the BASIC Stamp Editor.**
7. **Connect your BASIC Stamp to the computer and identify it in the Stamp Editor.**

   For more information about how to do this, refer to Chapter 1 of this minibook.
8. **Type the Potentiometer LED Flashing program into the Editor program window and then save the file.**

   The Potentiometer LED Flashing program is shown in Listing 3-4.
9. **Choose Run⇨Run.**

   If you prefer, you can click the Run button in the toolbar or press F9.

   When you choose the Run command, the program is downloaded to the BASIC Stamp. When the program is downloaded, it starts to run automatically on the Stamp.
10. **Turn the potentiometer knob.**

    The rate at which the LEDs flash varies as you increase or decrease the resistance of the potentiometer.

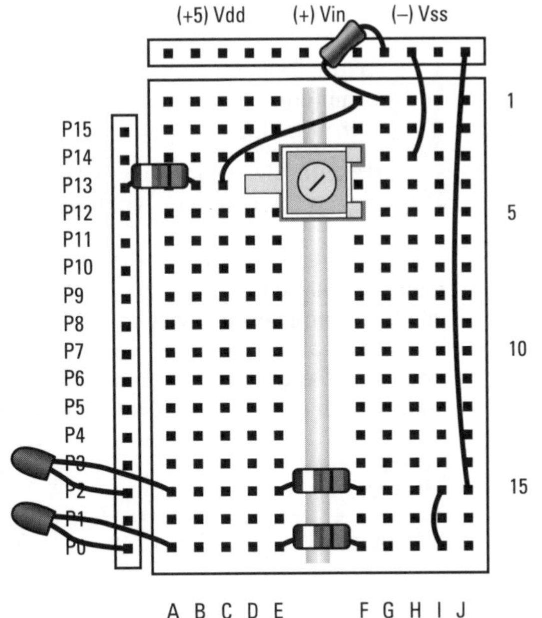

Using Subroutines and the GOSUB Command

A *subroutine* is a section of a program that can be called upon from any location in the program. When the subroutine finishes, control of the program jumps back to the location from which the subroutine was called. Subroutines are useful because they let you separate long portions of your program from the program's main loop, which simplifies the main program loop to make it easier to understand.

Another benefit of subroutines is that they can make your program shorter. Suppose that you're writing a program that needs to perform some complicated calculation several times. If you place the complicated calculation in a subroutine, you can call the subroutine from several places in the program. That way, you have to write the code that performs the complicated calculation only once. Without subroutines, you'd have to duplicate the complicated code each time you need to perform the calculation, which would be a real drag.

To create and use subroutines, you make use of two PBASIC commands:

- **GOSUB:** Calls the subroutine. You typically use the GOSUB command within your program's main loop whenever you want to call the subroutine.
- **RETURN:** Always the last command in the subroutine. RETURN jumps back to the command that immediately follows the GOSUB command.

To create a subroutine, you start with a label and end with a RETURN command. Between them, you write whatever commands you want to execute when the subroutine is called.

Here's an example of a subroutine that generates a random number between 1 and 999 and saves it in a variable named Rnd:

```
GetRandom:
  RANDOM Rnd
  Rnd = Rnd // 999 + 1
  RETURN
```

To call this subroutine, you simply use a GOSUB command as follows:

```
GOSUB GetRandom
```

This GOSUB command transfers control to the GetRandom label. Then, when the GetRandom subroutine reaches its RETURN command, control jumps back to the command immediately following the GOSUB command.

Listing 3-5 shows a complete program that uses a subroutine to get a random number between 1 and 1,000 and uses the random number to cause the LED on pin 0 to blink at random intervals. You can run this program on any BASIC Stamp circuit that has an LED on pin 0, including the circuits you build for Projects 3-1 and 3-2 earlier in this chapter.

Listing 3-5: Using a Subroutine to Blink an LED

```
' LED Blinker Program
' Doug Lowe
' July 10, 2011
'
' This program blinks the LED on pin 0 randomly.

' {$STAMP BS2}
' {$PBASIC 2.5}

Rnd   VAR Word
Led1  PIN 0

DO
  GOSUB GetRandom
  HIGH Led1
  PAUSE Rnd
  LOW Led1
  PAUSE 100
LOOP

GetRandom:
  RANDOM Rnd
  Rnd = Rnd // 999 + 1
  RETURN
```

When you use a subroutine, you have to prevent your program from accidentally 'falling into' your subroutine and executing it when you don't intend it to. For example, suppose that the program in Listing 3-5 uses a `FOR-NEXT` loop instead of a `DO` loop, because you want to blink the LED only 100 times. Here's an example of how *not* to write that program:

```
FOR Counter = 1 TO 100
  GOSUB GetRandom
  HIGH Led1
  PAUSE Rnd
  LOW Led1
  PAUSE 100
NEXT

GetRandom:
  RANDOM Rnd
  Rnd = Rnd // 999 + 1
  RETURN
```

We hope you can see the problem: after the FOR-NEXT loop blinks the LED 100 times, the program continues with the next command after the FOR-NEXT loop, which is the subroutine!

To prevent that from happening, you can use another PBASIC command, END, which simply tells the BASIC Stamp that you've reached the end of your program and it should stop executing commands. You place the END command after the NEXT command, as follows:

```
FOR Counter = 1 TO 100
  GOSUB GetRandom
  HIGH Led1
  PAUSE Rnd
  LOW Led1
  PAUSE 100
NEXT
END

GetRandom:
  RANDOM Rnd
  Rnd = Rnd // 999 + 1
  RETURN
```

The program stops after the FOR-NEXT loop finishes.

Most BASIC Stamp programs don't require an END command, because they're written so that they loop continuously as long as power is applied to the Stamp. Even in the case of programs that loop indefinitely, however, you have to be careful to make sure that your subroutines appear after the program's main loop. That way, your subroutines are executed only when you explicitly call them with a GOSUB command.

Chapter 4

Adding Sound and Motion to Your BASIC Stamp Projects

In This Chapter

- Making noise with a piezo speaker
- Creating movement with a servo

Everyone enjoys being able to create a din from time to time (except perhaps Trappist monks), and so in this chapter we lead you through working with devices that add sound (as well as motion) to your BASIC Stamp projects.

To make some noise, you can add a piezo speaker to create audible output tones, something that's useful when your BASIC Stamp program needs to get a person's attention or you want to create a sound effect. To produce movement, you can add a very useful device called a *servo,* which lets you control mechanical motion with a BASIC Stamp program.

Creating Sound with a Piezo Speaker and a BASIC Stamp

The BASIC Stamp Activity Kit (to which we introduce you in Chapter 1 of this minibook) comes with a small piezoelectric speaker, which you can connect directly to an I/O pin to create beautiful music. Well, perhaps the music is less than beautiful, but you can coax the BASIC Stamp into emitting a variety of squeaks, burps and squelches that resemble musical notes. Plus you can create interesting sound effects such as police sirens or chirping crickets. Figure 4-1 shows this handy little speaker.

Figure 4-1: The piezo-electric speaker that comes with the BASIC Stamp Activity Kit.

If you haven't purchased the BASIC Stamp Activity Kit, you can order a similar component (specification around 5 Vp-p rated voltage square wave, 1 mA, 4 kHz) from online component distributors for just a few pounds.

The piezo speaker is polarised, and so when you connect it to an I/O pin, be sure to connect the + terminal to the I/O pin and the other terminal to Vss (ground), as shown in the schematic in Figure 4-2.

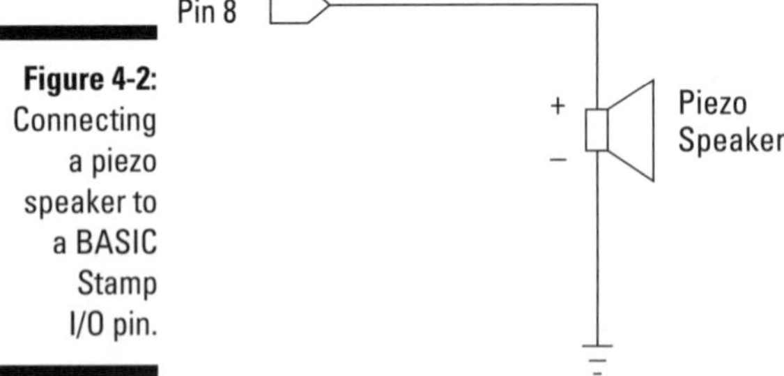

Figure 4-2: Connecting a piezo speaker to a BASIC Stamp I/O pin.

Freaking out with the FREQOUT command

Programming a piezo speaker is remarkably simple. PBASIC includes a command called FREQOUT that sends a frequency of your choice to an output pin. So you can create an audible tone on a piezo speaker by using the FREQOUT command and the following syntax:

```
FREQOUT pin, duration, frequency
```

Here's an explanation:

- *pin* is the pin number to which you want to send the frequency.
- *duration* is the length of time in milliseconds (ms) you want the frequency to play.
- *frequency* is the frequency in hertz that you want to generate.

For example, the following command generates a 2,000 hertz (Hz) frequency for 5 seconds on pin 8:

```
FREQOUT 8, 5000, 2000
```


You can easily create a beeping sound by alternately sending short bursts of a frequency to the speaker followed by a brief pause. For example:

```
DO
  FREQOUT 8, 250, 1500
  PAUSE 250
LOOP
```

This code repeatedly sends a 1,500 Hz signal for a quarter of a second, and then pauses for a quarter of a second. The result is a *beep-beep-beep* sound that isn't irritating at all!

Testing the piezo speaker

Project 4-1 shows how to build a simple circuit that connects a piezo speaker to a BASIC Stamp so that you can create audible output; two pushbuttons vary the sound output. Figure 4-3 illustrates the circuit.

The piezo speaker is very quiet. It draws just 1 milliampere (mA) and so can't make a lot of noise. The speaker is loudest with frequencies between 4,500 and 5,500 Hz.

Figure 4-3: A piezo speaker connected to a BASIC Stamp (Project 4-1).

Project 4-1: Creating Sound with a Piezo Speaker

In this project, you connect a piezo speaker to a BASIC Stamp HomeWork board or Board of Education to run programs that generate sound. The circuit also includes two pushbuttons (on pins 10 and 14) that you can use to vary the sound produced by the Stamp program.

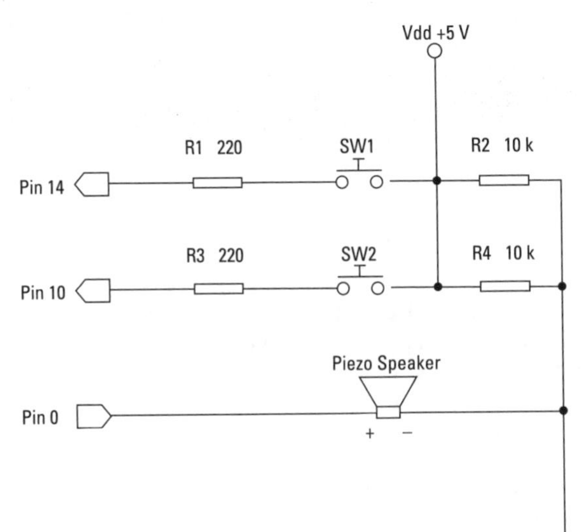

Parts List

- 1 Computer with BASIC Stamp Editor software installed
- 1 BASIC Stamp HomeWork board or Board of Education
- 1 9 V battery
- 1 USB cable
- 1 USB to serial adapter
- 1 Piezo speaker
- 2 Normally open pushbuttons
- 2 220 Ω resistors
- 2 10 kΩ resistors
- 8 Jumper wires

Beeper Program

```
' {$STAMP BS2.5}
' {$PBASIC 2.5}

DO
  FREQOUT 0,250,1500
  PAUSE 250
LOOP
```

Steps

1. **Insert the piezo speaker.**

   The positive lead goes in E17, the negative lead in F17.
2. **Insert the two pushbuttons:**

   SW1 goes in E1, F1, E3 and F3.

   SW2 goes in E5, F5, E7 and F7.
3. **Insert the resistors:**

   R1 (220 Ω) Pin 14 to B3.

   R2 (10 KΩ E4 to F4

   R3 (220 Ω) Pin 10 to B7.

   R4 (10 KΩ) E8 to F8
4. **Insert the jumper wires:**

   Vdd to D1

   Vss to G4

   C1 to C5

   D3 to D4

   H4 to H8

   D7 to D8

   J8 to J17

   Pin 0 to A17
5. **Open the BASIC Stamp Editor.**
6. **Connect your BASIC Stamp to the computer and identify it in the Stamp Editor.**

   For more information about how to do this, refer to Chapter 1 of this minibook.
7. **Type the Beeper program into the Editor program window, and then save the file.**

   The Beeper program is shown in the accompanying listing.
8. **Choose Run⇨Run.**

   If you prefer, you can click the Run button in the toolbar or press F9.

   When you choose the Run command, the program is downloaded to the BASIC Stamp. When the program is downloaded, it starts to run automatically on the Stamp.
9. **Now try the programs in Listings 4-1 and 4-2.**

   These programs vary the tone in different ways to demonstrate the versatility of the `FREQOUT` command.

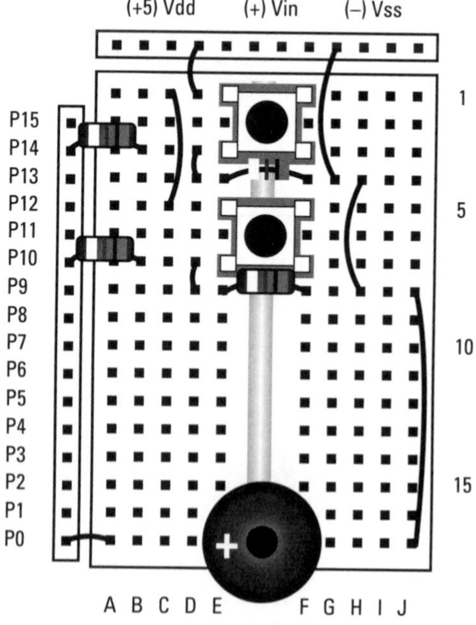

Playing with sound effects

With creative use of the FREQOUT command, PAUSE commands and FOR-NEXT loops, you can create some interesting and at times wonderfully annoying sound effects. The idea is to use short durations in the FREQOUT command and use FOR-NEXT loops or some other means to vary the frequency. You can also use PAUSE commands between tones to create beeping or clicking effects. We explain these commands in this minibook's Chapter 2.

The best way to find out what kinds of sound effects are possible with the FREQOUT command is to experiment. Listings 4-1, 4-2 and 4-3 give three sample programs you can run with the circuit you create in Project 4-1. Use these programs as starting points for your own experiments.

The program in Listing 4-1 plays two different beeping sounds when you press one of the pushbuttons. If you press Switch1 (on pin 14), a 5,000 Hz tone beeps twice a second. If you press Switch2 (on pin 10), a 5,000 Hz tone beeps five times a second.

Listing 4-1: Generating Two Different Types of Beeping Sounds

```
' Sound Program
' Doug Lowe
' July 15, 2011
'
' This program creates fast and slow beeping sounds.
' A piezo speaker must be connected to pin 0.
' The normally open pushbutton switches must be connected to pins 10 and 14.

' {$STAMP BS2}
' {$PBASIC 2.5}

Speaker   PIN 0
Switch1   PIN 10
Switch2   PIN 14
Frequency VAR Word
Time      VAR Word

DO
  IF Switch1 = 1 THEN
    FREQOUT Speaker,250, 5000
    PAUSE 250
  ELSEIF Switch2 = 1 THEN
    FREQOUT Speaker,100, 5000
    PAUSE 100
  ENDIF
LOOP
```

Listing 4-2 shows how you can use FREQOUT within a FOR-NEXT loop to create a continuously rising or falling tone, much like a police siren. The program varies the frequency from 3,000 to 5,000 Hz. When you press either of the pushbuttons, the rate at which the pitch rises and falls changes.

This rising or falling is governed by a variable named Time. Each time it runs through the FOR-NEXT loop, the program calls a subroutine named GetTime, which checks the status of the pushbutton switches and changes the Time variable if either of the switches is down. In this way the program alters the rate of the pitch change when the buttons are pressed.

Listing 4-2: Generating a Siren Effect

```
' Siren Effect Program
' Doug Lowe
' July 15, 2011
'
' This program generates a rising and falling pitch similar to a police siren.
' The rate at which the pitch rises and falls changes if you press either
' of the two pushbuttons.

' {$STAMP BS2}
' {$PBASIC 2.5}

Speaker   PIN 0
Switch1   PIN 10
Switch2   PIN 14
Frequency VAR Word
Time      VAR Word


DO

  FOR Frequency = 3000 TO 5000 STEP 15
    GOSUB SetTime
    FREQOUT 0, Time, Frequency
  NEXT
  FOR Frequency = 5000 TO 3000 STEP 15
    GOSUB SetTime
    FREQOUT 0, Time, Frequency
  NEXT

LOOP

SetTime:
  Time = 15
  IF Switch1 = 1 THEN
    Time = 5
  ENDIF
  IF Switch2 = 1 THEN
    Time = 2
  ENDIF
  RETURN
```

Listing 4-3 shows a program that plays two songs on the piezo speaker: 'Mary had a little lamb' and 'Good morning to all'. The former is played when you press `SW1` and the latter when you press `SW2`.

To simplify the code that generates the musical notes, the program defines several constants that represent the frequency for each of the notes required by the songs. For example, the constant `NoteC6` is 1,046, the frequency in Hz of C in the sixth octave of a piano keyboard. The constants span two full octaves, which is plenty of range for the songs to be played. Both songs are played in the key of C, and so no flats or sharps are required. (If this musical stuff makes no sense to you, don't worry about it – this is an electronics book after all, not one for budding songwriters.)

The program also sets up constants for the duration of a quarter note, half note and whole note. The constants make specifying a particular pitch for a particular duration easy in a `FREQOUT` command. Thus, playing a melody is simply a matter of writing a sequence of `FREQOUT` commands to play the correct notes for the correct durations in the correct order. That's precisely what the subroutines labelled `Mary` and `Morning` do.

Listing 4-3: Making Music with a BASIC Stamp

```
' Song Program
' Doug Lowe
' July 15, 2011
'
' This program plays one of two songs on the piezo speaker
' on pin 0.
' If SW1 on pin 14 is pressed, the program plays "Mary Had a Little Lamb."
' If SW2 on pin 10 is pressed, the program plays "Good Morning to All."

' {$STAMP BS2}
' {$PBASIC 2.5}

SW1       PIN 14
SW2       PIN 10
Speaker   PIN 0

NoteC6    CON 1046
NoteD6    CON 1175
NoteE6    CON 1318
NoteF6    CON 1370
NoteG6    CON 1568
NoteA6    CON 1760
NoteB6    CON 1975
NoteC7    CON 2093
NoteD7    CON 2349
NoteE7    CON 2637
NoteF7    CON 2794
NoteG7    CON 3136
```

(continued)

Listing 4-3 *(continued)*

```
NoteA7    CON 3520
NoteB7    CON 3951
NoteC8    CON 4186

Whole     CON 1000
Half      CON 500
Quarter   CON 250


DO
  IF SW1 = 1 THEN
    GOSUB Mary
  ENDIF
  IF SW2 = 1 THEN
    GOSUB Morning
  ENDIF
LOOP

Mary:
  FREQOUT Speaker, Quarter, NoteE7  ' Mar-
  FREQOUT Speaker, Quarter, NoteD7  ' y
  FREQOUT Speaker, Quarter, NoteC7  ' Had
  FREQOUT Speaker, Quarter, NoteD7  ' a
  FREQOUT Speaker, Quarter, NoteE7  ' Lit-
  FREQOUT Speaker, Quarter, NoteE7  ' tle
  FREQOUT Speaker, Quarter, NoteE7  ' Lamb
  PAUSE Quarter
  FREQOUT Speaker, Quarter, NoteD7  ' Lit-
  FREQOUT Speaker, Quarter, NoteD7  ' tle
  FREQOUT Speaker, Quarter, NoteD7  ' Lamb
  PAUSE Quarter
  FREQOUT Speaker, Quarter, NoteE7  ' Lit-
  FREQOUT Speaker, Quarter, NoteG7  ' tle
  FREQOUT Speaker, Quarter, NoteG7  ' Lamb
  PAUSE Quarter
  FREQOUT Speaker, Quarter, NoteE7  ' Mar-
  FREQOUT Speaker, Quarter, NoteD7  ' y
  FREQOUT Speaker, Quarter, NoteC7  ' Had
  FREQOUT Speaker, Quarter, NoteD7  ' a
  FREQOUT Speaker, Quarter, NoteE7  ' Lit-
  FREQOUT Speaker, Quarter, NoteE7  ' tle
  FREQOUT Speaker, Quarter, NoteE7  ' Lamb
  FREQOUT Speaker, Quarter, NoteE7  ' Its
  FREQOUT Speaker, Quarter, NoteD7  ' Fleece
  FREQOUT Speaker, Quarter, NoteD7  ' Was
```

```
  FREQOUT Speaker, Quarter, NoteE7  ' White
  FREQOUT Speaker, Quarter, NoteD7  ' As
  FREQOUT Speaker, Quarter, NoteC7  ' Snow
  PAUSE Half
  RETURN

Morning:
  FREQOUT Speaker, Half, NoteC7     ' Good
  FREQOUT Speaker, Half, NoteD7     ' Morn-
  FREQOUT Speaker, Half, NoteC7     ' ing
  FREQOUT Speaker, Half, NoteF7     ' To
  FREQOUT Speaker, Whole, NoteE7    ' You
  FREQOUT Speaker, Half, NoteC7     ' Good
  FREQOUT Speaker, Half, NoteD7     ' Morn-
  FREQOUT Speaker, Half, NoteC7     ' ing
  FREQOUT Speaker, Half, NoteG7     ' To
  FREQOUT Speaker, Whole, NoteF7    ' You
  FREQOUT Speaker, Half, NoteC7     ' Good
  FREQOUT Speaker, Half, NoteC8     ' Morn-
  FREQOUT Speaker, Half, NoteA7     ' ing
  FREQOUT Speaker, Half, NoteF7     ' Dear
  FREQOUT Speaker, Half, NoteE7     ' Child-
  FREQOUT Speaker, Whole, NoteD7    ' ren
  FREQOUT Speaker, Half, NoteB7     ' Good
  FREQOUT Speaker, Half, NoteA7     ' Morn-
  FREQOUT Speaker, Half, NoteF7     ' ing
  FREQOUT Speaker, Half, NoteG7     ' To
  FREQOUT Speaker, Whole, NoteF7    ' All
  RETURN
```


Moving by Degrees with a Servo and a BASIC Stamp

A *servo* is a special type of motor designed to rotate to a particular position and hold that position until told to rotate to a different one. Hobby servos are frequently used in radio-controlled vehicles such as airplanes, boats and cars, but many other uses for servos also exist.

The BASIC Stamp Activity Kit comes with a servo that you can use to practise writing programs. You can also purchase servos from the UK distributors listed by Parallax (`www.parallax.com`) or from electronics shops such as Maplin. Figure 4-4 shows a hobby servo.

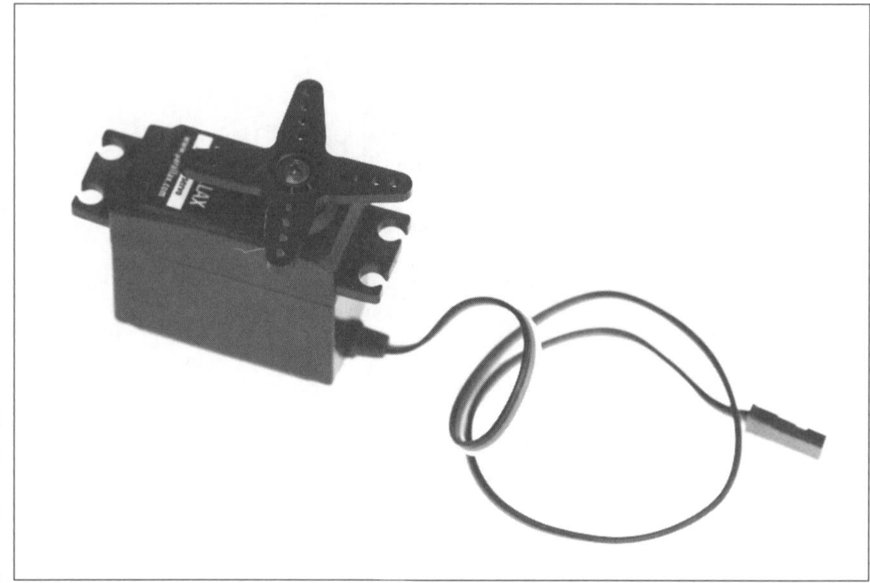

Figure 4-4: A typical hobby servo.

Connecting a servo to a BASIC Stamp

Servos use a special three-conductor cable to provide power and a *control signal* that tells the servo what position it should move to and hold. The cable's three wires are coloured red, black and white and have the following functions:

- **Red wire:** Supplies the voltage required to operate the servo. For most servos, this voltage can be anywhere between +4 V and +9 V. On a BASIC Stamp HomeWork board, you need to connect this cable to one of the Vdd pins.
- **Black wire:** The ground connection. On the BASIC Stamp HomeWork board, you connect it to a Vss pin.
- **White wire:** The control wire, which connects to one of the BASIC Stamp's I/O pins.

Figure 4-5 shows how to connect these wires in a BASIC Stamp circuit.

The control wire controls the position of the servo by sending it a series of pulses approximately 20 ms apart. The length of each one of these pulses determines the position that the servo rotates to and holds.

Most hobby servos have a range of motion of 180° – that is, half a complete revolution. The range of pulse durations is 0.5–2.5 ms, where 0.5 ms pulses move the servo to its minimum position (0°) and 2.5 ms pulses move the servo to its maximum position (180°). To hold the servo at the centre point of this range (90°), the pulses need to be 1.5 ms in duration.

To connect a servo to a BASIC Stamp HomeWork Board, you have to use a 3-pin header, which comes with the BASIC Stamp Activity Kit and is pictured in Figure 4-6. This header consists of three pins that you can plug in to the solderless breadboard. Then, you can plug the servo cable into the adapter.

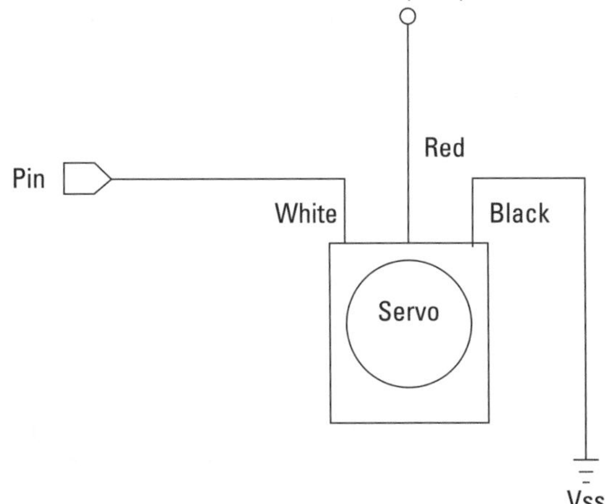

Figure 4-5: Connecting a servo to a BASIC Stamp.

Figure 4-6: 3-pin header for connecting the servo to a BASIC Stamp HomeWork Board.

Programming a servo in PBASIC

The easiest way to control a servo from a BASIC Stamp microcontroller is to use the `PULSOUT` command. This command sends a pulse of any duration you specify to an I/O pin of your choosing. The syntax of this command is as follows:

```
PULSOUT pin, duration
```

You specify the duration in units of 2 microseconds (μs). A microsecond is one-millionth of a second and so one thousand μs make up 1 ms: therefore 2 μs equal 500 μs. Thus, to send a 1.5 ms pulse with the `PULSOUT` command, you need to specify 750 (1.5 × 500) as the duration, as follows:

```
PULSOUT 0,750
```

Here, a 1.5 ms pulse is sent to pin 0.

To make your servo-programming life easier, Table 4-1 lists the duration values to use for a typical hobby servo for various angles.

Table 4-1: `PULSOUT` Duration Values for Servo Control

Angle	*Duration*	*Angle*	*Duration*
0	250	95	778
5	278	100	806
10	306	105	833
15	333	110	861
20	361	115	889
25	389	120	917
30	417	125	944
35	444	130	972
40	472	135	1000
45	500	140	1028
50	528	145	1056
55	556	150	1083
60	583	155	1111
65	611	160	1139
70	639	165	1167
75	667	170	1194

Angle	Duration	Angle	Duration
80	694	175	1222
85	722	180	1250
90	750		

For example, to move the servo on pin 0 to 75°, use this command:

```
PULSOUT 0,667
```

To hold its position, a servo needs a constant stream of pulses approximately 20 ms apart. Thus, `PULSOUT` commands are usually contained in `DO` loops or `FOR-NEXT` loops. For example, here's a bit of code that keeps the servo on pin 0 at 45° indefinitely:

```
DO
  PULSOUT 0,500
  PAUSE 20
LOOP
```

Listing 4-4 shows a complete program that moves the servo to 45° when `SW1` (a pushbutton on pin 14) is pressed and 135° when `SW2` (a pushbutton on pin 10) is pressed.

Listing 4-4: A Servo Control Program

```
' Servo Control Program
' Doug Lowe
' July 15, 2011
'
' This program moves a servo to one of two positions when
        SW1 is pressed
' and returns the servo to centre position when SW2 is
        pressed.

' {$STAMP BS2}
' {$PBASIC 2.5}


Servo PIN 0
SW1   PIN 14
SW2   PIN 10

Position VAR Word
Position = 500
DO
  IF SW1 = 1 THEN
    Position = 500
```

(continued)

Listing 4-4 *(continued)*

```
   ENDIF
   IF SW2 = 1 THEN
     Position = 1000
   ENDIF
   PULSOUT Servo, Position
   PAUSE 20
 LOOP
```

Building a servo project

Project 4-2 shows you how to build a complete circuit that uses a servo as well as two pushbuttons. This circuit is capable of running the program shown in Listing 4-4. Figure 4-7 depicts the assembled project.

Figure 4-7: A BASIC Stamp project that controls a servo (Project 4-2).

For projects that require multiple servos or a lot of other work besides managing the servo, consider using Parallax's Propeller chip instead of a BASIC Stamp. The Propeller processor is designed for programs that have to do several things at once, such as managing servos. The propeller costs a little more than a BASIC Stamp and its programming language is a little more complicated, but in the end controlling multiple servos is much easier with a Propeller than with a BASIC Stamp.

Project 4-2: Using a Servo with a BASIC Stamp

In this project, you connect a servo to a BASIC Stamp HomeWork Board or Board of Education. The circuit also includes two pushbuttons (on pins 10 and 14) that you can use to control the action of the servo.

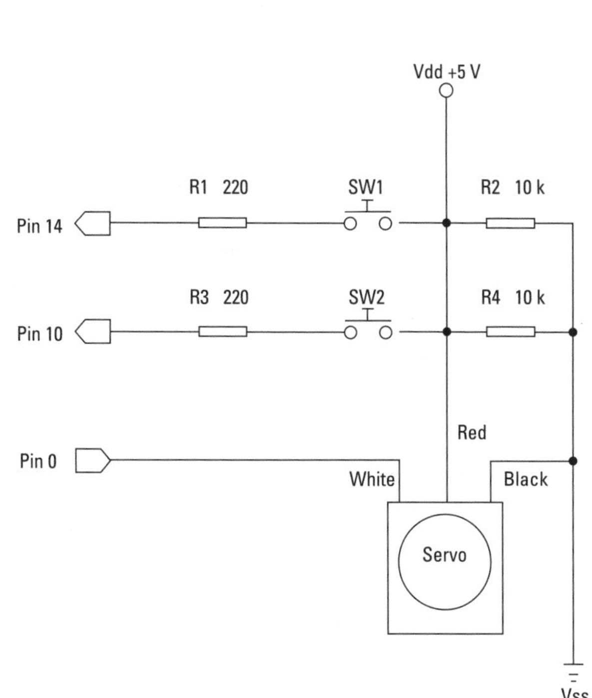

Parts List

1 Computer with BASIC Stamp Editor software installed
1 BASIC Stamp HomeWork board or Board of Education
1 9 V battery
1 USB cable
1 USB to serial adapter
1 Hobby servo
1 3-pin male-to-male header
2 Normally open pushbuttons
2 220 Ω resistors
2 10 kΩ resistors
10 Jumper wires

Servo Program

```
' {$STAMP BS2}
' {$PBASIC 2.5}

X VAR Byte
DO
  FOR X = 1 To 200
    PULSOUT 0, 350
    PAUSE 10
  NEXT
  FOR X = 1 To 200
    PULSOUT 0, 1150
  PAUSE 10
  NEXT
LOOP
```

Steps

1. **Insert the 3-pin header.**
   The three pins go in G15, G16 and G17.
2. **Insert the two pushbuttons:**
   SW1 goes in E1, F1, E3 and F3.
   SW2 goes in E5, F5, E7 and F7.
3. **Insert the resistors:**

R1 (220 Ω)	Pin 14 to B3
R2 (10 kΩ	E4 to F4
R3 (220 Ω)	Pin 10 to B7
R4 (10 kΩ	E8 to F8

4. **Insert the jumper wires:**
   Vdd to D1
   Vss to G4
   C1 to C5
   D3 to D4
   H4 to H8
   D7 to D8
   J8 to J15
   Pin 0 to A17
   E17 to F17
   B5 to I16
5. **Connect the servo.**
   Plug the 3-pin servo connector into the 3-pin male-to-male header. Connect the white wire to G17.
6. **Open the BASIC Stamp Editor.**
7. **Connect your BASIC Stamp to the computer and identify it in the Stamp Editor.**
   For more information about how to do this, refer to Chapter 1 of this minibook.
8. **Type the Servo program into the Editor program window and then save the file.**
   The Beeper program is shown in the accompanying listing.
9. **Choose Run⇨Run.**
   If you prefer, you can click the Run button in the toolbar or press F9. When you choose the Run command, the program is downloaded to the BASIC Stamp. When the program is downloaded, it starts to run automatically on the Stamp. The servo alternates from about 20° to about 165° once per second.
10. **Now try the program in Listing 4-4.**
    This program uses the switches to control the servo's action.

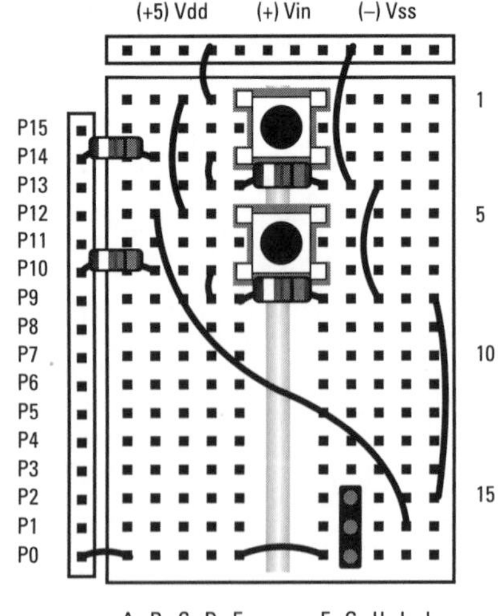

Book VIII

Having Fun with Special Effects

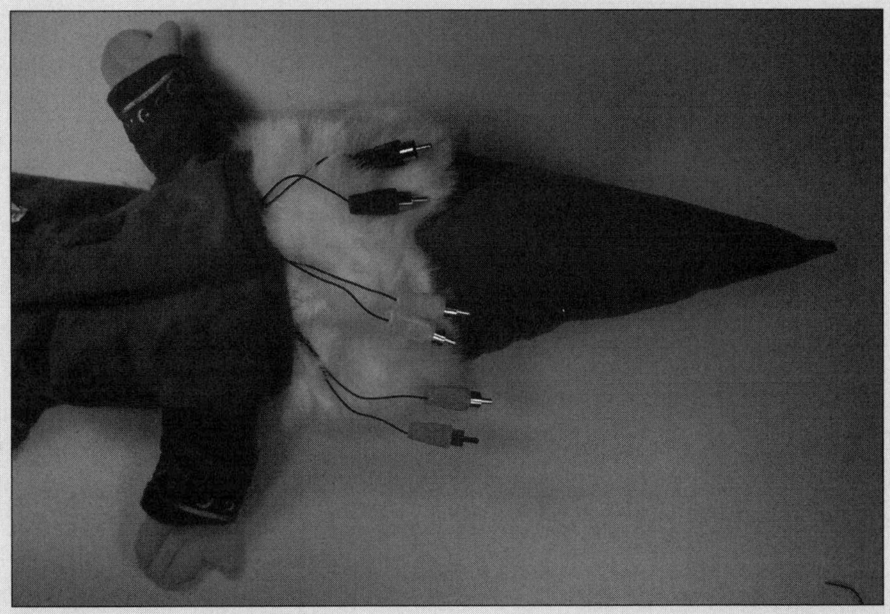

To help you get hold of the right parts for all your electronics projects, you can find an online list of sources from both the UK and further afield at `www.dummies.com/extras/electronicsaiouk`.

Contents at a Glance

Chapter 1

Synchronising Sight and Sound with a Colour-Organ Circuit

In This Chapter

- Preparing the project
- Constructing your colour organ
- Playing around with the circuit

One of the great things about theme parks these days is that sometimes the long wait to go on a particular ride is almost as good as the ride itself. One of the best examples is the famous *Indiana Jones Adventure: Temple of the Forbidden Eye* in Disneyland, California. Just outside an ancient temple, you pass by a rickety steam-powered generator that's barely running. The clickity-clickity sound of the generator alternately grows louder and softer as it sputters and threatens. When you're inside the temple, you pass through narrow tunnels and creepy caverns that are lit overhead by lights that appear to be powered by the ancient generator. The lights flicker and dim, grow brighter for a moment, and then flicker and dim again in sync with the generator.

In this chapter, you discover how to build an electronic circuit (called a *colour organ*) that can create this type of creepy lighting, as well as thunderstorm effects or a spooky red heartbeat for the heart of a plastic skeleton or other Halloween prop.

Considering the Colour-Organ Project

The operation of a colour-organ circuit is pretty simple: it converts the volume of an audio input into an output voltage that gets higher as the sound source gets louder. If you connect a light to the output, the light glows brighter when the audio input is louder and dimmer when the input is quieter.

The completed colour organ is shown in Figure 1-1; we house the project in a 130 x 68 x 44 mm box. It has terminals to which you can connect a 12 V power supply, terminals to which you can connect halogen lamps (maximum 12 V/50 watts (W)), an RCA-style audio input connector for connecting a sound source to (maximum 60 W), a knob for adjusting sensitivity and a power switch.

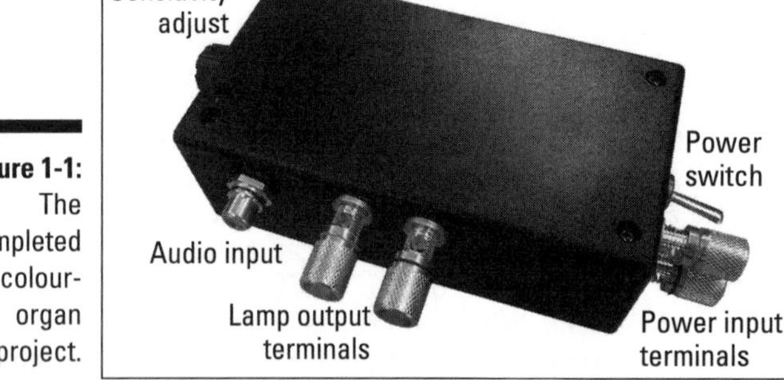

Figure 1-1: The completed colour-organ project.

To keep the project simple, you can do as we do in this chapter and build the electronics using an inexpensive kit – specifically, the snappily named Velleman MK114 Low Voltage Light Organ Minikit. This kit is available on the Internet for under £10; searching for 'Velleman MK114' brings up several sources.

Figure 1-2 shows how you can connect the colour-organ project to create light that varies in brightness with a sound source. Of course you need a source for the sound, such as a portable CD player or other sound system with external speaker outputs that you can tap into.

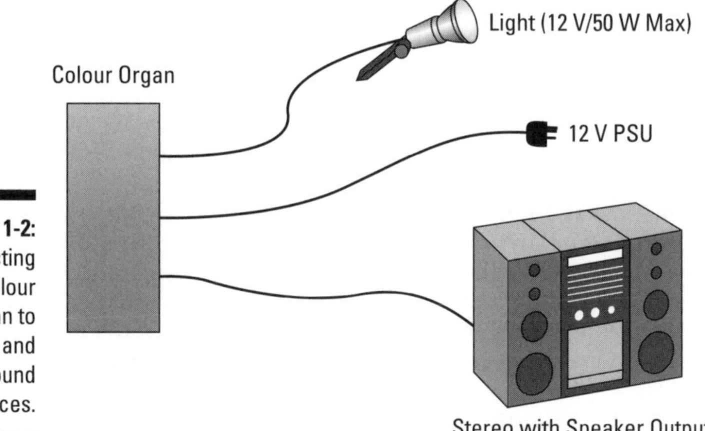

Figure 1-2: Connecting the colour organ to light and sound sources.

Note that the diagram in Figure 1-2 doesn't show the details of how to connect the sound system to the colour-organ circuit and the speakers. The easiest way to connect the circuit to the sound system is simply to replace one of the speakers with the colour organ. That way, the colour organ responds to one of the stereo channels while the speaker plays the sound coming from the other stereo channel (check out the later section 'Putting Your Colour Organ to Work' for more details). For this type of hookup, you need just a single cable that has an RCA connector on one end (to plug into the colour organ) and the proper connector on the other end to connect to the sound system's speaker output.

You also need a suitable recording for your sound source. For example, if you want to use the colour organ to create a thunderstorm effect, you need a recording of thunder. To make a red light flash in sync with a heartbeat, you need a recording of a heartbeat. Many such sound effects are available online, and so you can locate and download one to meet your needs without too much trouble. If you want to customise the sound effect, you can download a free audio editor called Audacity from `www.audacity.sourceforge.net`.

Understanding how the colour organ works

You can design a colour-organ circuit in several different ways. Most of the circuits that drive low-voltage DC lamps rely on a type of transistor, called a *MOSFET* (which stands for 'metal oxide semiconductor field effect transistor'), designed to handle large currents. It has three terminals: source, drain, and gate. A voltage at the gate above a given threshold allows the drain and source path (called the *channel*) to conduct. The larger the gate voltage, the more the channel will conduct. Conversely, if there is no or too little voltage on the gate, the source is disconnected from the drain. The load is connected between the positive supply and the drain, with the source connected to ground. The gate voltage is derived from the audio input.

The audio input isn't connected directly to the transistor gate, however. Instead, the colour organ uses an *optoisolator* to isolate the audio input from the supply-voltage side of the circuit. This device is a single component that consists of an infrared light-emitting diode (LED) and a photodiode or other light-sensitive semiconductor. Voltage on the LED causes it to emit light, which the photodiode detects and passes on to the output circuit.

The Velleman MK114 kit that we employ in this chapter uses an optoisolator transistor, in which the photosensitive semiconductor is in fact a transistor whose gate is stimulated by light rather than by current. The optoisolator is an integrated circuit (check out Book III, Chapter 1) in a 6-pin DIP package.

Figure 1-3 shows a simplified schematic diagram for the circuit that the Velleman MK114 kit uses. As you can see, the audio input is applied to the LED side of the optoisolator, controlled by a potentiometer (something we describe in Book II, Chapter 2), which lets you adjust the sensitivity of the circuit. The output from the optoisolator is applied to the gate of two MOSFETs (T4 and T5) via the amplifier/logic circuit formed by transistors T1-T3. The load is connected between the positive supply and the drains of the two MOSFETs. Thus, the volume of the audio input controls the current through the output circuit.

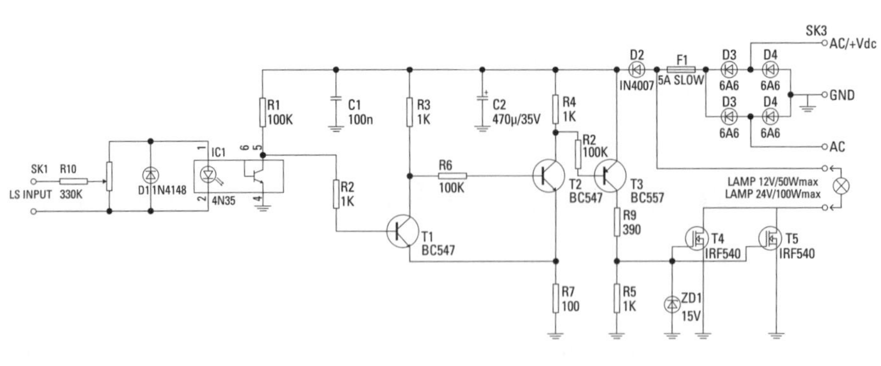

Figure 1-3: A simplified version of the colour-organ circuit. Make sure to follow the detailed schematic diagram that accompanies your circuit.

Getting your equipment together

Other than the Velleman kit itself (see the preceding section), you can purchase most of the materials you need to build the colour-organ project at your local Maplin shop or any other supplier of electronic components. The following table lists all the materials you need.

Quantity	*Description*
1	Velleman MK114 Low Voltage Light Organ kit, available from Quasar Electronics (`www.quasarelectronics.co.uk`) and other online suppliers
1	130 x 68 x 44 mm plastic enclosure (Quasar Electronics part WCAH2853)
4	M3 x 10 mm female-female Hex standoffs (to provide a little room between the board and the enclosure)
8	M3 x 6 mm screws
1	RCA-style chassis socket (Quasar Electronics part 755.280UK, or similar)
1	Control knob (Quasar Electronics part KB0295)
1	5 A rated single pole, single throw (SPST ON-OFF) toggle switch (Quasar Electronics part 785.660UK, or similar)
2	4 mm plug sockets/screw terminal, panel mount, Red (Quasar Electronics part 780.030UK, or similar)
2	4 mm plug sockets/screw terminal, panel mount, Black (Quasar Electronics part 780.030UK, or similar).
35-45 cm	0.8 mm diameter stranded wire
1	PSU (Power Supply Unit) 12 V DC 50 W or 12 V AC 50 VA. (This may be a wall wart type or bench power supply. Alternatively, use a 12 V car or motorcycle battery with suitable cabling. However, please be very careful to guard against short circuits if you do use a battery.)

To assemble the colour-organ project, you require the following tools:

- Drill with a range of bits
- Hobby vice
- Magnifying goggles
- Phillips screwdriver
- Pliers
- Small flat-edge jeweller's screwdriver
- Solder

- Soldering iron, preferably with both 20- and 40-W settings
- Wire cutters
- Wire strippers

Building the Colour Organ Project

When you have your tools and materials at the ready, as we describe in the preceding section, you can start to build your colour-organ circuit.

Assembling your colour organ

Here are the steps for constructing this project:

1. **Assemble the Velleman MK114 kit.**

   The kit comes with simple but accurate instructions. In essence, you just mount and solder all the components onto the circuit board. Pay special attention to the colour codes for the resistors and the orientation of the diodes and electrolytic capacitor. Do not fit the detachable rotor shaft to the potentiometer at this time (see step 4).

   Figure 1-4 shows the completed MK114 kit.

Figure 1-4: The assembled Velleman MK114 kit.

Mount the circuit board in a good hobby vice and use a crocodile clip or masking tape to hold the components in place while soldering.

2. **Drill all the mounting holes in the project box except the hole for the sensitivity control on the left side of the box.**

   Figure 1-5 shows the orientation of the approximate location of the mounting holes.

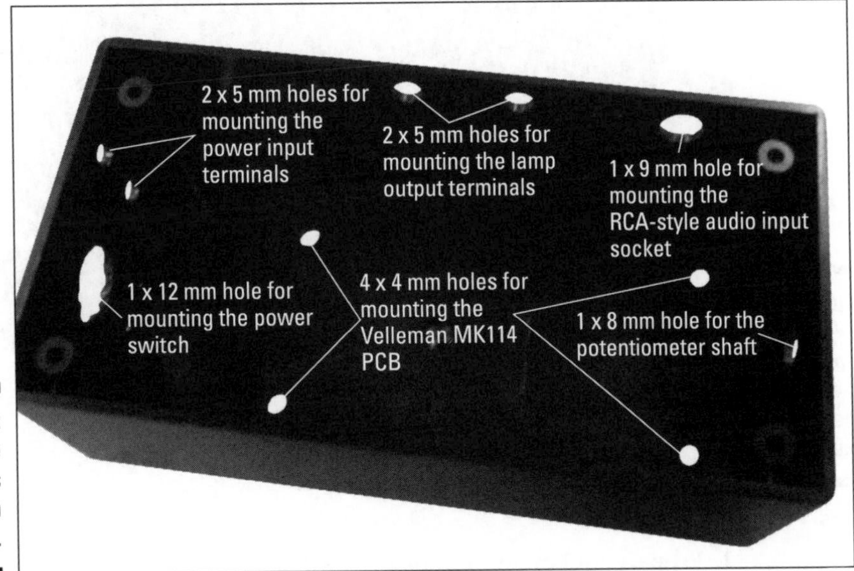

Figure 1-5: Drill the holes as indicated in this photo.

   Use the assembled circuit board to determine the exact drilling locations for the four holes that mount the circuit board. The position of the other holes isn't critical, with the exception of the hole for the potentiometer knob. Don't drill that hole until Step 4.

3. **Mount the four standoffs for the MK114 circuit board.**

   Use M3 screws through the four 4 mm holes in the bottom of the box to secure the standoffs in place.

4. **Drill the hole for the circuit board's potentiometer.**

   Position the circuit board, without the potentiometer rotor shaft fitted, on the four standoffs to determine the exact position for the 8 mm hole.

Thinking inside the (project) box

In the steps that follow, you assemble all the parts into the project box. Use Figure 1-6 as a guide for the proper placement of each part.

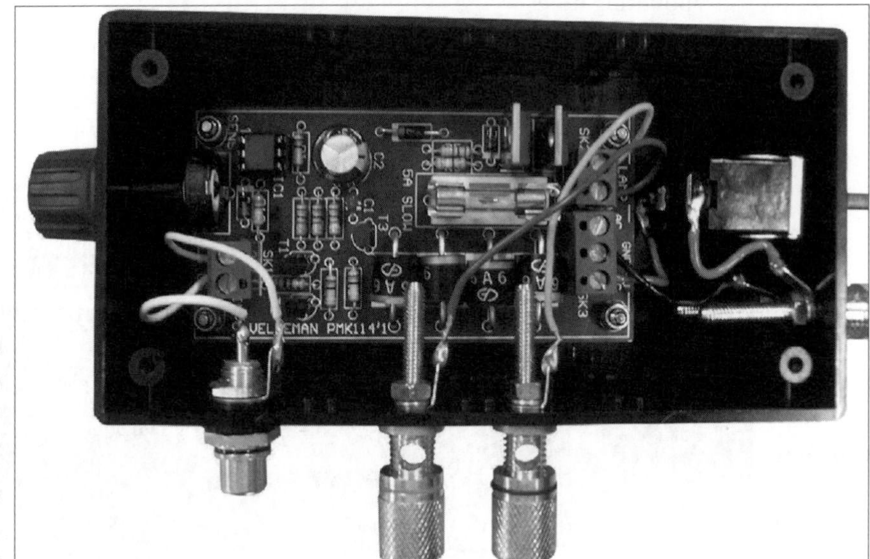

Figure 1-6: How the parts go together inside the project box.

1. **Prepare and mount the power switch.**

   Cut two 3.5 cm lengths of wire and strip 5 mm of insulation from both ends of each. Remove the screws from the two switch terminal posts and solder a wire to each terminal through the resulting holes.

   Remove the top mounting nut and ON-OFF indicator plate from the switch and push the switch lever/mounting thread through the 12 mm mounting hole from inside the box. Orientate the switch body so that the terminals are at the top and the bottom. (You may need to bend the terminals slightly to get the switch to fit in this orientation.) Reattach the indictor plate and mounting nut and secure the switch by tightening the nut with needle-nose pliers.

2. **Mount the black power socket.**

   Remove the mounting nut, washer and terminal tab from the treaded shaft of one of the black 4 mm plug sockets. Push the treaded shaft through the lower of the 5 mm holes adjacent to the toggle switch. Replace the terminal tab, washer and mounting nut. Secure in place by tightening the nut with needle-nose pliers.

Cut a 3.5 cm length of wire and strip 5 mm of insulation from each end. Solder one end of the wire to the terminal tab.

3. **Mount the red power socket.**

   Remove the mounting nut, washer and terminal tab from the treaded shaft of one of the red 4 mm plug sockets. Push the treaded shaft through the upper of the 5 mm holes adjacent to the toggle switch. Replace the terminal tab, washer and mounting nut. Secure in place by tightening the nut with needle-nose pliers.

4. **Connect the switch to the red power socket.**

   Solder the wire from the top terminal of the switch to the terminal tab of the red power socket already fitted.

5. **Mount the black lamp output socket.**

   Remove the mounting nut, washer and terminal tab from the treaded shaft of the remaining black 4 mm plug socket. Push the treaded shaft through the right-hand 5 mm hole on the long edge of the case. Replace the terminal tab, washer and mounting nut. Secure in place by tightening the nut with needle-nose pliers.

   Cut a 9 cm length of wire and strip 5 mm of insulation from each end. Solder one end of the wire to the terminal tab.

6. **Mount the red lamp output socket.**

   Remove the mounting nut, washer and terminal tab from the treaded shaft of the remaining red 4 mm plug socket. Push the treaded shaft through the left-hand 5 mm hole on the long edge of the case. Replace the terminal tab, washer and mounting nut. Secure in place by tightening the nut with needle-nose pliers.

   Cut a 9 cm length of wire and strip 5 mm of insulation from each end. Solder one end of the wire to the terminal tab.

7. **Prepare the RCA-style phono jack.**

   Cut two 6.5 cm lengths of wire and strip 5 mm of insulation from both ends of each. Solder one of the wires to the centre terminal of the RCA-style phono jack and the other to the ground terminal tab.

   At this point you have finished with the soldering iron, so you can turn it off.

8. **Mount the RCA-style phono jack in the 9 mm hole in the project box.**

   Remove the mounting nut, terminal tab and lock washer from the jack. Pass the wire connected to the centre terminal of the phono jack through the 9 mm hole and insert the threaded end of the phono jack into the hole. Slip the lock washer, the ground terminal and the nut over the wire connected to the centre terminal, and thread them onto the threaded part of the jack. Tighten with needle-nose pliers.

Wiring up your circuit

In this section, you attach wires to the MK114 circuit board.

Don't mount the circuit board to the standoffs quite yet. You have an easier time connecting the wires if the circuit board is loose. After the wires are all connected, you mount the board.

1. **Connect the wire from the bottom terminal of the toggle switch to the +V terminal of block SK3 on the Velleman MK114 circuit board.**

   Use a small, flat screwdriver to tighten the terminal. Make sure that you connect the wire securely.

2. **Connect the wire from the bottom (black) power supply input 4 mm plug socket to terminal block SK3 on the Velleman MK114 circuit board.**

   If using an AC power supply, connect the wire to the unoccupied terminal marked AC. If using a DC power source, connect the wire to the middle terminal, marked GND.

   Use a small, flat screwdriver to tighten the terminal. Make sure that you connect the wire securely.

3. **Connect the wire from the red lamp output 4 mm plug socket to terminal block SK2 on the Velleman MK114 circuit board.**

   The wire connects to the SK2 terminal adjacent to SK3.

   Use a small, flat screwdriver to tighten the terminal. Make sure that you connect the wire securely.

4. **Connect the wire from the black lamp output 4 mm plug socket to terminal block SK2 on the Velleman MK114 circuit board.**

   The wire connects to the SK2 terminal nearest the circuit board mounting hole.

   Use a small, flat screwdriver to tighten the terminal. Make sure that you connect the wire securely.

5. **Connect the two wires from the RCA-style phono jack to terminal block SK1 on the Velleman MK114 circuit board.**

   Connect the wire from the centre of the jack to the SK1 terminal nearest the potentiometer and the wire from the tab to the SK1 terminal nearest the PCB mounting hole.

   Use a small, flat screwdriver to tighten the terminals. Make sure that you connect the wires securely.

6. **Mount the MK114 circuit board on the standoffs.**

   Position the circuit board over the standoffs and use the remaining four M3 screws to secure it in place.

7. **Attach the rotor shaft to the potentiometer.**

   Slide the potentiometer's detachable rotor shaft through the 8 mm hole and engage the slotted end with the centre of the potentiometer until it clicks in place.

8. **Attach the knob to the potentiometer shaft now protruding from the box.**

   Use a small, flat screwdriver to tighten the screw on the side of the knob.

9. **Place the lid on the project box and secure it in place with the screws provided.**

   Now you really are done!

Putting Your Colour Organ to Work

When you've completed the colour organ, you can use it to make interesting sounds and lighting effects. To do that, you have to connect it to lights and a sound system:

1. **Connect a 12 V halogen lamp to the colour organ's lamp output terminals.**
2. **Connect a speaker-level audio input to the RCA-style phono connector.**
3. **Connect your power supply to the power input terminals.**

   If using a DC supply, the red terminal is positive and the black terminal is negative. If using an AC supply, the polarity is unimportant.

4. **Turn on the colour organ.**
5. **Play the sound.**
6. **Turn the knob on the colour organ to adjust the sensitivity.**

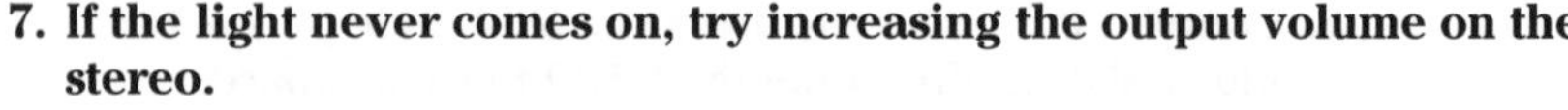

7. **If the light never comes on, try increasing the output volume on the stereo.**

The colour organ can handle only 50 W on the output circuit, and so be careful not to overload the circuit.

The easiest way to connect the colour organ to the sound system is to replace one of the speakers with the colour organ. The type of cable you need to do that depends on how the speakers connect to the sound system:

- If they connect with simple post connectors, you need a cable with bare wire on one end and an RCA plug on the other end.
- If they connect with RCA connectors, you need a cable with RCA plugs on both ends.

When you use the colour organ in this way, remember that the colour organ responds to one channel of the stereo recording while the speaker plays the other channel. In most cases, you don't notice much difference, but in some stereo recordings the sound on the left channel is very different from the sound on the right channel. This difference can affect the quality of the sound, and it can also prevent the light from flashing in perfect sync with the sound, because the light is responding to a different sound source from the one you're hearing through the speakers.

In some cases, you can use this situation to improve the effect you're trying to achieve with the colour organ. For example, in a real thunderstorm, the lightning flashes well before the thunder is heard. To reproduce this effect, all you need is a sound recording of a thunderstorm in which the thunder is heard on the left channel before it's heard on the right channel. Then, if you connect the colour organ to the left channel and the speaker to the right channel, the light flashes before the sound is heard.

Play around and have fun!

Chapter 2

Unearthing Hidden Treasure with a Handy Metal Detector

In This Chapter

- Exploring the detector circuit
- Assembling the project
- Finding metal!

Everyone likes discovering something for nothing, whether it's free sample music tracks to download from the Internet or treasure in your own home or garden. With a metal detector, you can scout out coins under the rug, nail heads covered by paint, keys in somebody's pocket or many other hidden bits of metal.

In this project, we show you how to build a small, handheld metal detector that can detect certain kinds of metal – especially ferrous (iron-containing) metals – even if that metal is hidden under a little paint, plaster, sand or soil (but never water!). So here's your chance to become known as your town's Iron Man or Woman.

Uncovering the Big Picture: Project Overview

In this project, you use an integrated circuit (IC; check out Book III, Chapter 1 for details) that generates an alternating current (AC) signal that goes through a coil (also called an *inductor,* which we discuss in Book II, Chapter 4). Because metal objects conduct electricity, you can induce a current in those objects. When the coil in the metal detector comes near to a metal object, the electromagnetic field in the coil induces currents in that object. This generated electromagnetic field changes the current in the coil. When the signal changes, the IC turns on a light-emitting diode (LED), alerting you to the presence of metal.

You can see the finished metal detector in Figure 2-1.

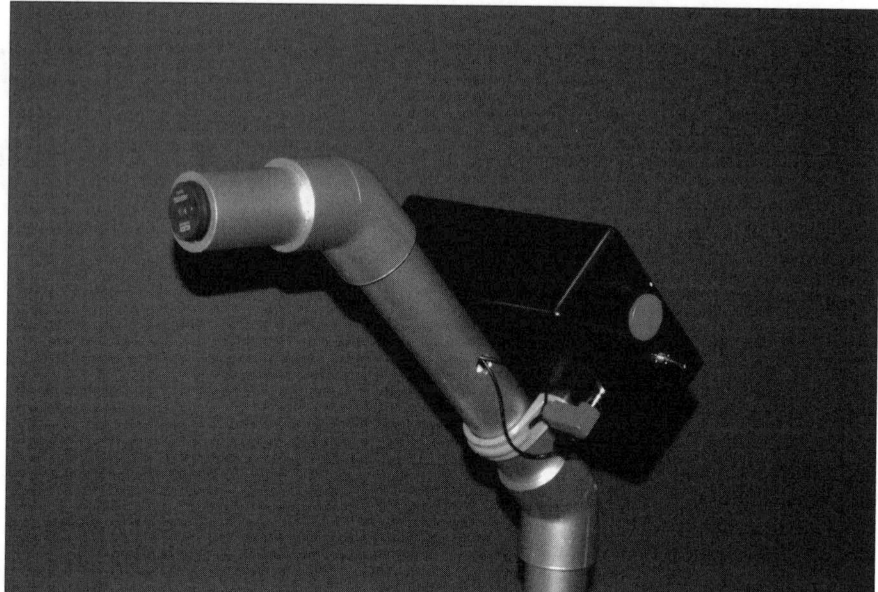

Figure 2-1: The handheld metal detector.

Here are the general tasks you need to do to build the metal detector:

1. **Construct a simple electronic circuit containing a coil, a proximity detector IC, a transistor (which we describe in Book II, Chapter 6), a couple of resistors (see Book II, Chapter 2) and an LED.**
2. **Install the circuit in a box with batteries and an on/off switch.**
3. **Mount the box on a handle made from plastic pipe.**

Scoping out the schematic

You need to put together only one breadboard for this project (we introduce breadboards in Book I, Chapter 6). Figure 2-2 shows the schematic for the board.

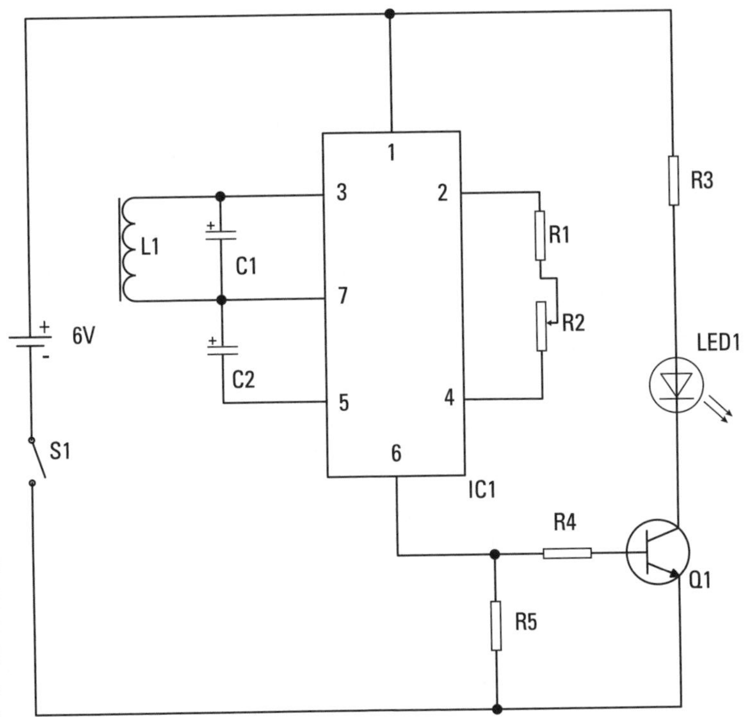

Figure 2-2: Schematic of the metal detector.

Here's a tour of the schematic elements:

- **L1** is a coil (or inductor) wired in parallel to capacitor C1; the combination of this dynamic duo is a parallel *LC* (inductor/capacitor) circuit (for more on capacitors, turn to Book II, Chapter 3). When a signal that oscillates at several kilohertz (kHz) passes through this circuit, the signal creates an electric field around the coil. When you bring the coil near to a metallic object, that electric field induces an oscillating signal in the object, which in turn creates an electric field that induces current in the coil. This current changes the oscillating signal running through the LC parallel circuit.
- **IC1** is a TDA0161 proximity detector. This IC is designed to supply the oscillating signal that's sent through the LC parallel circuit. The IC also responds to any changes in the signal: the IC has an output of 1 milliamp (mA) or less if the coil is far from a metallic object and an output of 10 mA or higher if the coil is near to a metallic object.
- **R1** and **R2** are resistors that are used to calibrate the circuit in IC1 to the LC circuit. You carry out this calibration by adjusting the value of R2 (which is a variable resistor) when the coil isn't near to any metal objects.

- **R5** is a resistor connected between the output of IC1 and ground. When the output of IC1 is on, current flows through this resistor and provides a positive voltage to the base of Q1.
- **Q1** is a 2N3904 transistor that you connect to the output of IC1. When the output of IC1 is high, Q1 turns on and allows current to flow through LED1.
- **LED1** is the indicator light used to indicate that the device has detected metal in the vicinity.
- **R3** is a resistor that limits the amount of current flowing through LED1, preventing it from burning out.
- **S1** is the on/off switch.

Sticking to adhesive precautions

You use PVC glue to connect the PVC pipe fittings that form the detector's handle. You can buy various types of such adhesive at building supply and DIY shops.

PVC adhesives are very powerful chemicals. Wear some form of work gloves to avoid getting the glue on your hands. Read the label for safety tips, such as using the glue in a well-ventilated area and what to do if some does come in contact with your skin.

Perusing the parts

Even though this project doesn't require all that many parts, you still have to go out and buy or assemble them. Several of the parts are shown in Figure 2-3. Here's what you need:

- Two 1 kΩ (kohm) resistors rated for 1/2 W, 5% tolerance (R1, R4)
- One 10 kΩ 1/2 W or 1 W knob-operated, panel mounting potentiometer (R2) such as the Vishay P16NM103KAB15 from Mouser (online, at `uk.mouser.com`) (We describe potentiometers in Book II, Chapter 2.)
- One 330 Ω resistor rated for 1/2 W, 5% tolerance (R3)
- One 120 Ω resistor rated for 1/2 W, 5% tolerance (R5)
- Two 0.0047 microfarad (μF) ceramic capacitors (C1, C3)
- One 2N3904 transistor (Q1)
- One T-1 ¾ LED (LED1)
- One LED panel-mount socket (T-1 ¾)

- One TDA0161 proximity detector (IC1)

- One battery pack for 4 AA batteries
- One 680 microhenrys (680μH) bobbin-type inductor (L1)

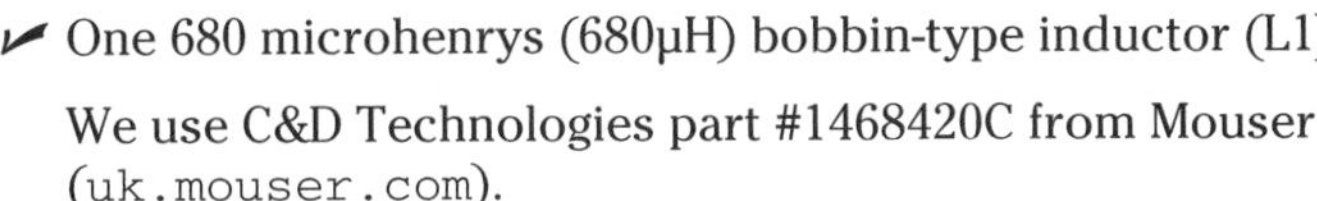

TIP: We use C&D Technologies part #1468420C from Mouser (`uk.mouser.com`).

- One single-pole, single-throw (SPST) toggle switch, used as the on/off switch
- One 400-pin breadboard
- Four 2-pin terminal blocks (5.08mm pitch)
- One knob (for the potentiometer)
- Two phono jacks
- Two right-angle phono plugs

TIP: We use right-angle plugs to avoid having a loop of wire coming out of the box. You can also use banana plugs and jacks.

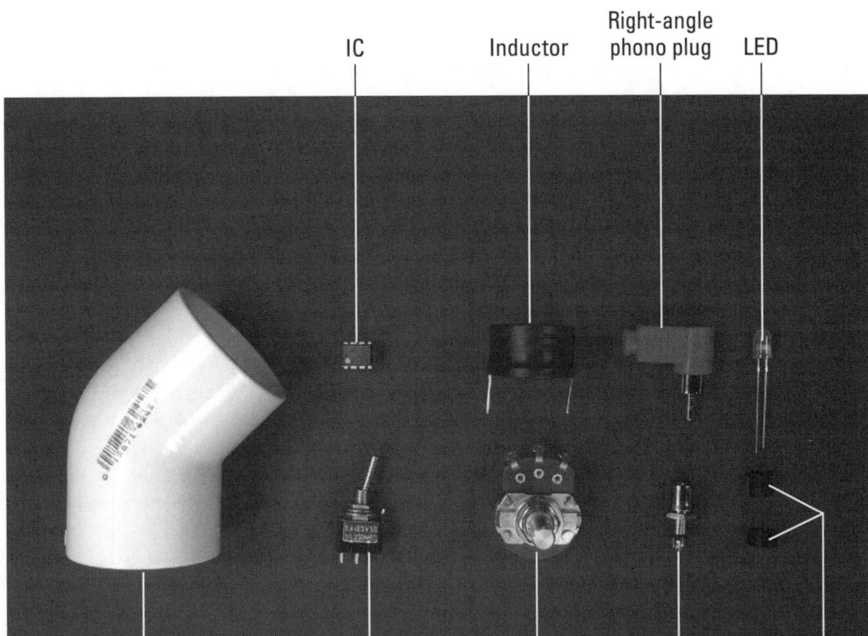

Figure 2-3: Key components of the metal detector.

- An enclosure (we suggest a plastic box, 15 x 10 x 5 cm)
- An assortment of different lengths of prestripped, short 0.7 mm diameter wire
- Two PVC 45° joints with 2.5 cm slip fitting on both ends (one male and one female)
- One PVC 2.5 cm end cap with 2.5 cm female slip fitting
- One 2.5 cm diameter PVC pipe, 2.5 cm in length
- One 2.5 cm clamp to attach the enclosure to the pipe
- Two 1.3 cm 8-32 (6.35 mm) panhead screws
- Two 8-32 (6.35 mm) nuts

Taking Construction Step by Step

Although the metal detector circuit is simple, you need to carry out a few steps to build it and then put the whole thing together, including building the handle and attaching the circuit to it.

Assembling your metal detector circuit

The circuit in this project sends a detector signal and processes the signal that comes back to light up the LED. Here are the steps involved:

1. **Place the TDA0161 (IC1), 2N3904 (Q1) and four terminal blocks on the breadboard, as shown in Figure 2-4.**

   To help, we show the transistor pinout (Q1) in Figure 2-5.

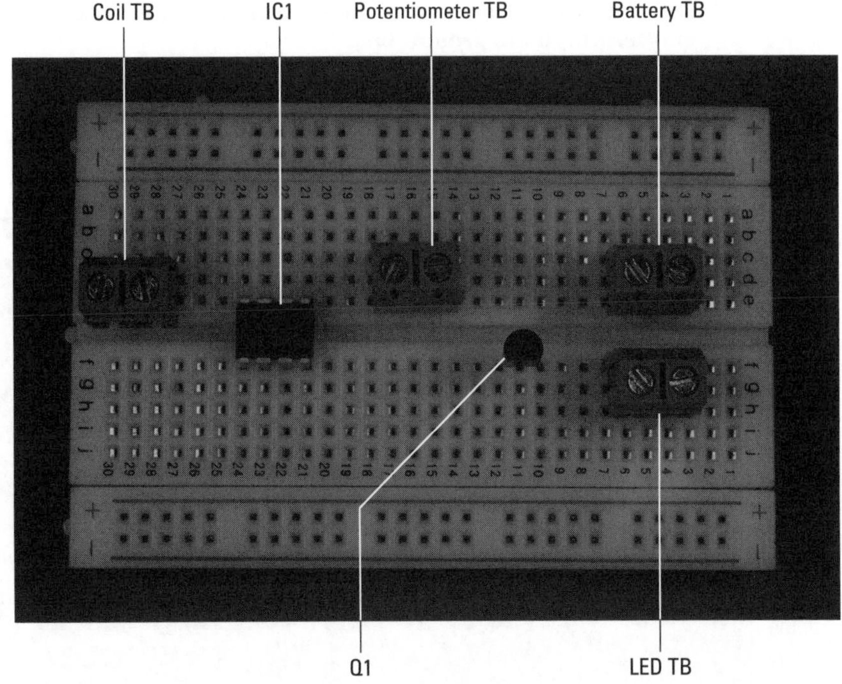

Figure 2-4: Install the IC (pin 1 goes top right), transistor and terminal blocks (TB) on the breadboard.

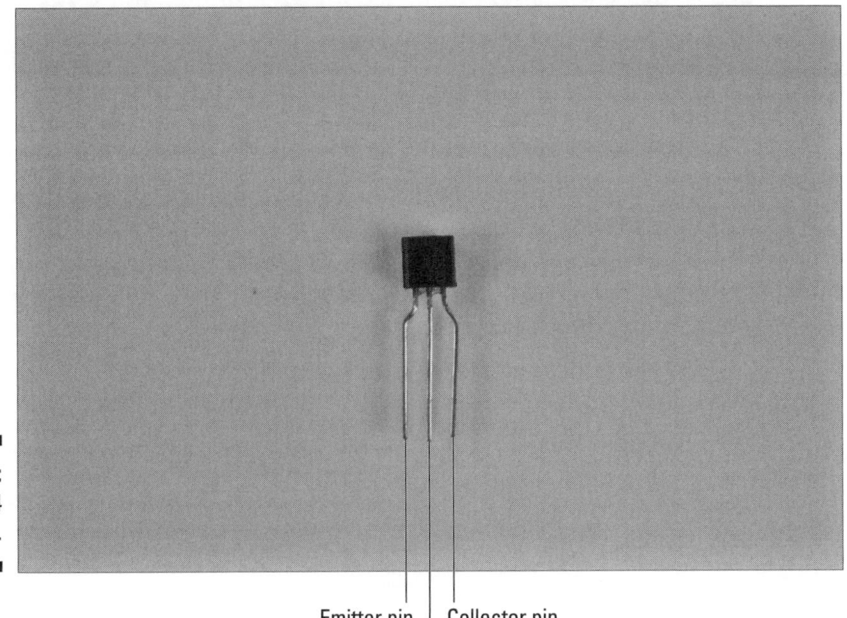

Figure 2-5: The 2N3904 pinout.

2. **Insert wires to connect the battery terminal block and the transistor emitter pin to the ground bus.**

   Then insert a wire between the two ground buses to connect them, as shown in Figure 2-6. Two shorter wires connect components to the ground bus; the long wire on the right connects the two ground buses.

Figure 2-6: Connect the components to the ground bus and then connect the two ground buses.

3. **Insert wires to connect IC1 and the battery terminal block to the +V bus.**

   Then insert a wire between the two +V buses to connect them, as shown in Figure 2-7.

4. **Insert wires to connect the IC and discrete components, terminal block for the coil (L1), terminal block for the potentiometer (R2) and terminal block for the LED.**

   We show these connections in Figure 2-8.

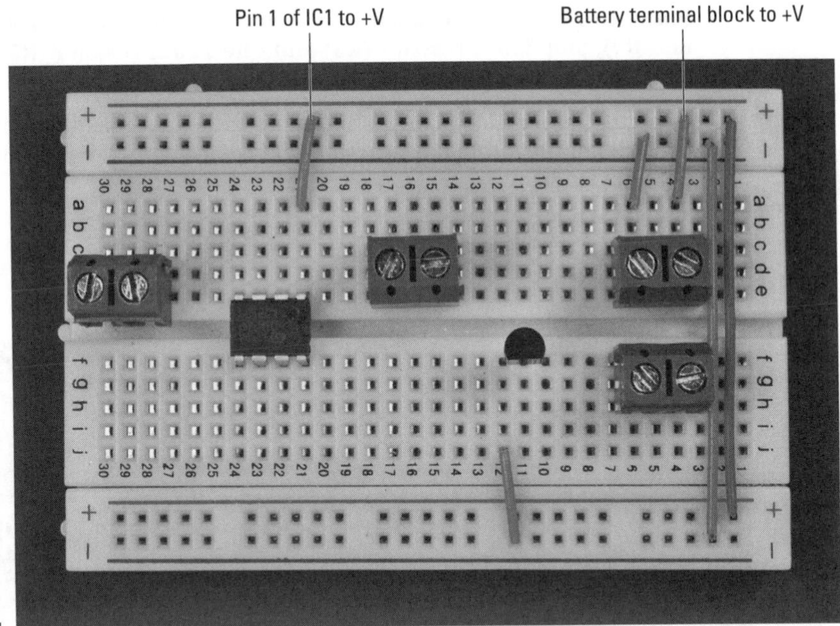

Figure 2-7: Connect the components to the +V bus.

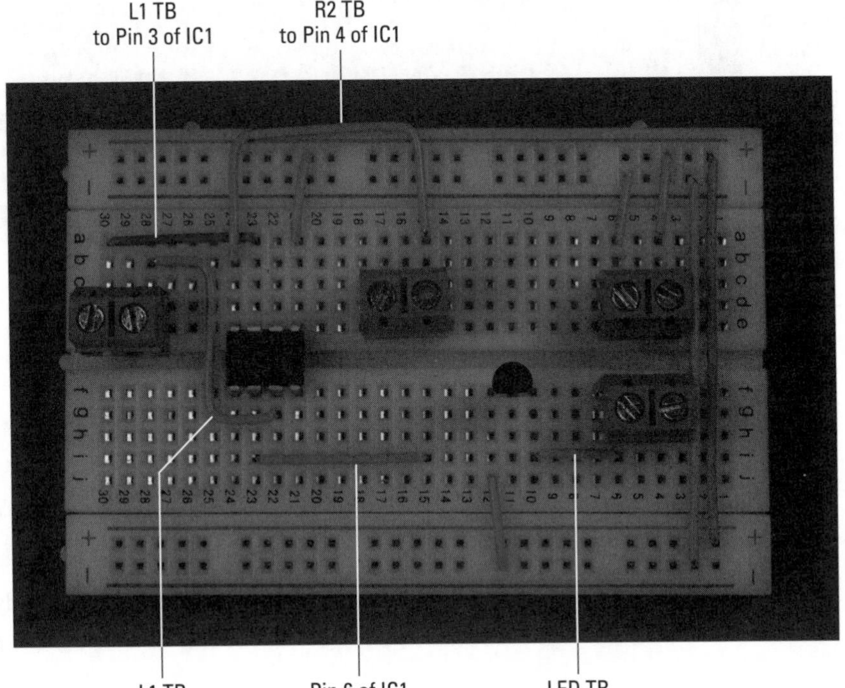

Figure 2-8: Hook up the ICs, terminal blocks (TB) and discrete components.

5. Insert two 0.0047 μF capacitors (C1 and C2), two 1 kΩ resistors (R1 and R4), one 330 Ω resistor (R3) and one 120 Ω resistor (R5) on the breadboard, as shown in Figure 2-9.

Clipping the wire leads on components can cause small bits of metal to fly through the air. Make sure that you wear your safety goggles when clipping leads!

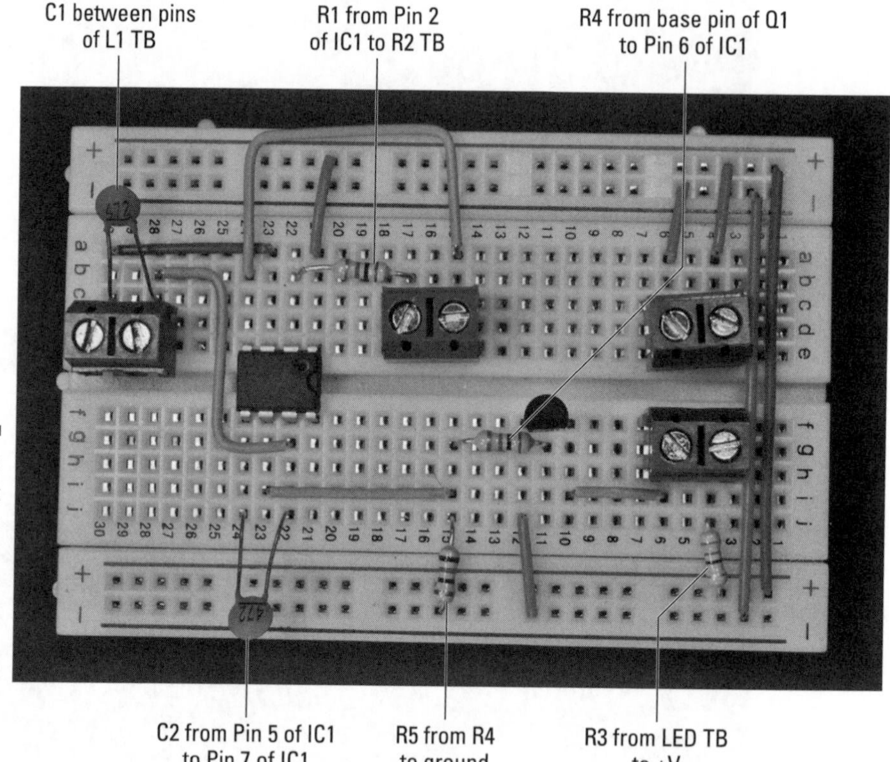

Figure 2-9: Insert resistors and capacitors on the breadboard.

Building the box to house the circuit

The box containing the circuit needs holes so that you can attach the LED, the potentiometer dial (for adjusting resistance on the IC) and the on/off switch, as well as various wire connections.

Follow these steps to get the metal detector circuit enclosure ready:

1. **Drill holes in the box where you plan to mount the LED, potentiometer, audio jacks, on/off switch and clamp.**

   TIP

   We put the on/off switch and the potentiometer on one side of the box, the LED on one end and the audio jacks on the bottom, as Figure 2-10 shows. But the placement is up to you.

   We always advise that you use safety goggles when drilling.

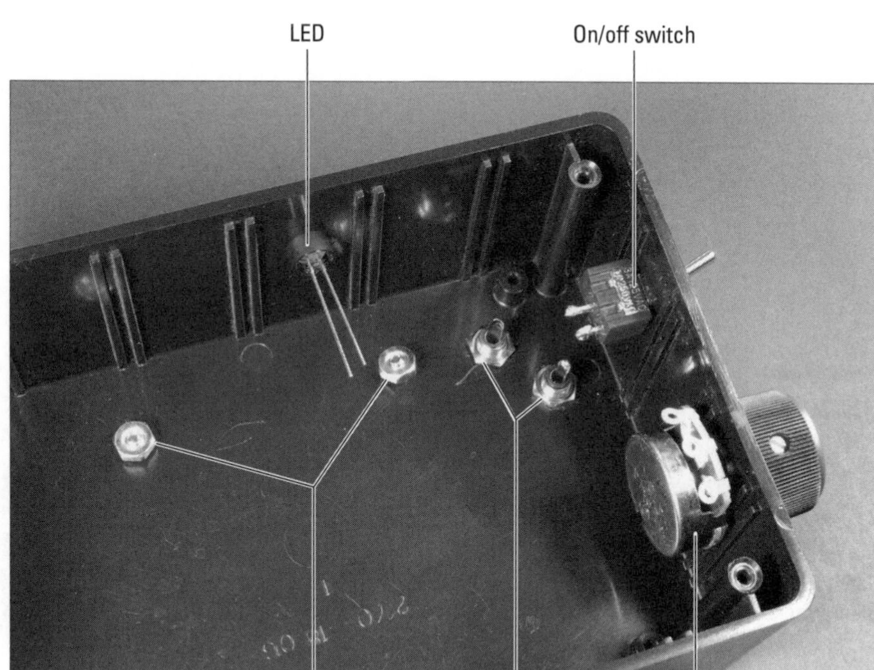

Figure 2-10: Box with on/off switch, phono jacks, potentiometer and LED in place.

2. **Slip the threads of the phono jacks through the drilled holes and secure with the nuts provided with the jacks.**
3. **Slip the shaft of the on/off switch through the drilled hole and secure with the nut provided.**
4. **Slip the shaft of the potentiometer through the drilled hole and secure with the nut provided.**
5. **Place the knob on the potentiometer shaft and secure with the set screw.**

6. **Feed the top half of the LED socket through the drilled hole from outside the box and insert the LED into the top half of the socket from inside the box.**
7. **Position the bottom half of the socket over the leads and snap onto the top half of the socket to secure the LED.**
8. **Solder the black wire from the battery pack to one lug of the on/off switch and solder a 20 cm black wire to the remaining lug of the on/off switch, as shown in Figure 2-11.**

Figure 2-11: Wires soldered to on/off switch, potentiometer, phone jacks and LED.

9. **Solder a 20 cm wire to the centre potentiometer lug and another 20 cm wire to the left potentiometer lug, as shown in Figure 2-11.**
10. **Solder a red 20 cm wire to the long lead of the LED and a black 20 cm wire to the short lead of the LED, as shown in Figure 2-11.**
11. **Slip a 2.5 cm piece of heat-shrink tubing over each solder joint and use a hair dryer to secure them in place.**
12. **Solder a 20 cm wire to the lug on each of the phono jacks, as shown in Figure 2-11.**

See Book I, Chapter 7 for advice about safe soldering if you're not very experienced in this art.

Putting it all together

After you've built your box and circuit, you're ready to introduce them to each other. Follow these steps:

1. **Attach Velcro to the breadboard and the box and secure the breadboard in the box.**
2. **Attach Velcro to the battery pack and the box and secure the battery pack in the box.**
3. **Insert the wires from the LED, potentiometer, on/off switch and battery pack to the terminal blocks on the breadboard, as shown in Figure 2-12.**
4. **Secure the wires with wire clips where needed.**

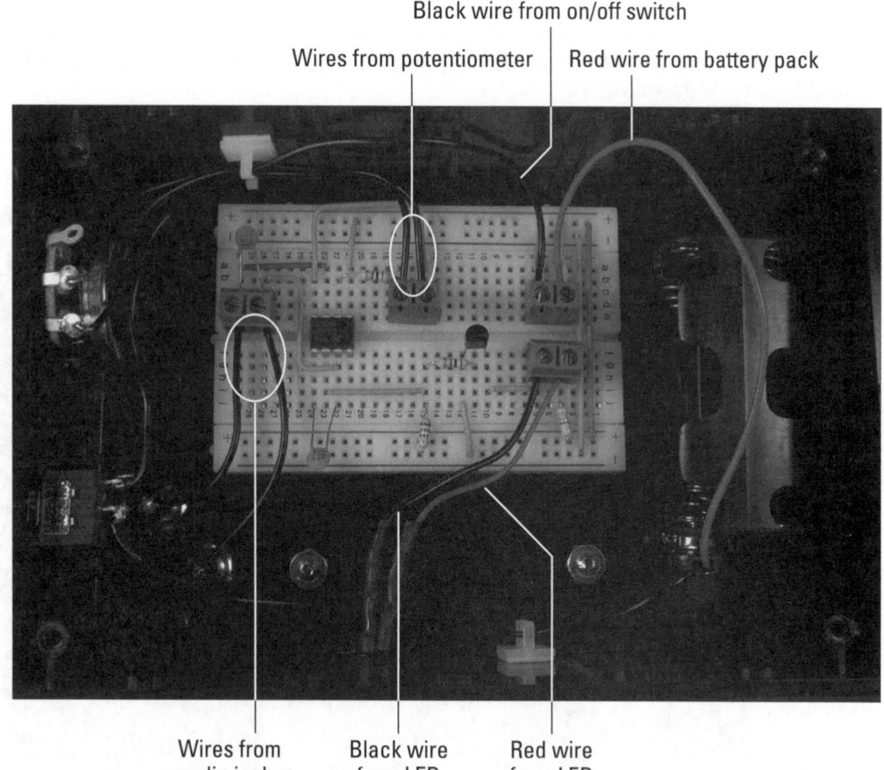

Figure 2-12: Connect the LED, potentiometer, on/off switch, coil and battery pack to the breadboard.

As you insert the various wires in Step 3, cut each of them to a sufficient length to reach the assigned terminal block and strip the insulation from the end of the wires.

Handling the handle

To wander around easily while pointing the metal detector at suspected deposits of metal, you need a handle:

1. **Glue a 20 cm length of 2.5 cm diameter PVC pipe into one end of a 45° angle PVC pipe fitting, facing up (see Figure 2-13).**
2. **Glue the other 45° PVC pipe fitting onto the other end of the PVC pipe, facing down (see Figure 2-13).**
3. **Glue a 7.5 cm length of 2.5 cm diameter PVC pipe into the open end of one of the 45° pipe fittings to form the coil end of the metal detector.**
4. **Glue a 15 cm length of 2.5 cm diameter PVC pipe into the open end of the other 45° fitting, which becomes the handle end of the metal detector.**
5. **Glue the 2.5 cm PVC cap on the open end of the 15 cm PVC pipe.**

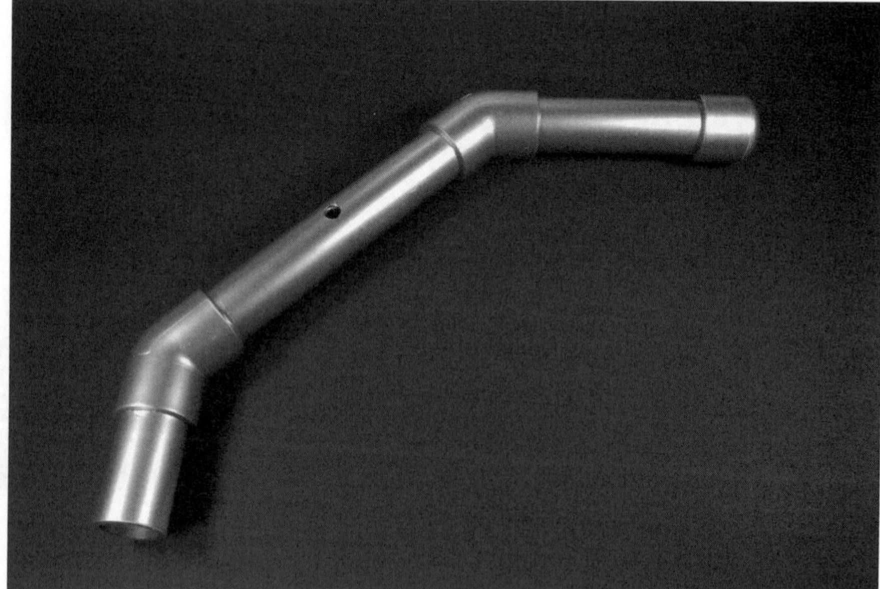

Figure 2-13: PVC pipe and fittings made into a handle for the metal detector.

6. Drill a 1 cm hole (for feeding the wires from the inductor to the box containing your circuit) in the middle of the long section of PVC pipe on the side that's going to be on your left when you're holding the metal detector (see Figure 2-13).
7. Solder 30 cm wires to each of the two inductor leads, and slip a 2.5 cm segment of heat-shrink tubing over each solder joint, using a hair dryer to secure the tubing in place (see Figure 2-14).

Figure 2-14: Wires attached to inductor.

8. Twist together the free ends of the wires from the inductor and feed them through the PVC pipe from the coil end until the end of the wire strand reaches the 1 cm hole.
9. Form a hook shape with a piece of 0.7–0.8 mm diameter wire and pull the wires through the 1 cm hole.
10. Insert the inductor in the end of the 2.5 cm PVC pipe, as shown in Figure 2-15, and use some glue to secure the inductor in the pipe.
11. Cut the wires to allow 7.5 cm to extend from the 1 cm hole in the pipe and attach each wire to a right-angle phono plug, as shown in Figure 2-16, using a plug that requires soldering to the wire or one that uses a screw to secure the wire, as we do.

Figure 2-15: Inductor in the end of the PVC pipe.

Figure 2-16: Phono plugs attached to wires from inductors.

12. **Press the clamp onto the 2.5 cm PVC pipe and attach the box to the clamp with the 8-32 screws and nuts (Figure 2-17 shows the box attached to the handle).**

13. **Plug the right-angle phono plugs into the phono jacks, as shown in Figure 2-17.**

Figure 2-17: Box attached to the handle and phono plugs in place.

If you're left-handed, consider placing the hole in Step 3 on the side that's going to be on your right when you're holding the detector. Doing so allows you to hold the handle in your left hand and operate the switches with your right hand.

We show the finished metal detector in Figure 2-18.

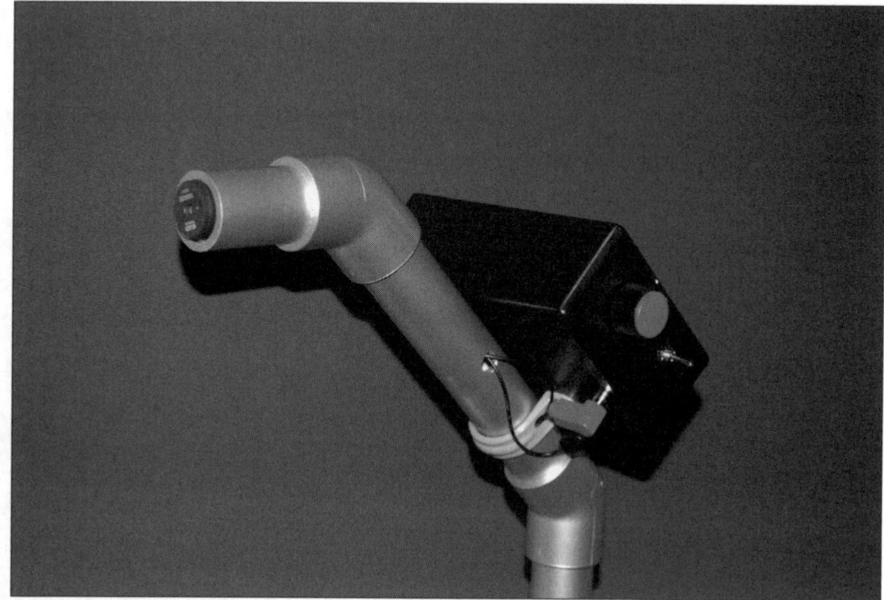

Figure 2-18: The finished metal detector.

Trying Out Your Detector

You probably have a fortune in ancient coins waiting to be discovered in your garden, and so when you've built your detector don't waste any time – get outside and see what you can find! Here's how to make your gadget work:

1. **Insert the batteries.**

   Secure the lid on the box with the screws provided and flip the on/off switch to 'On'.

2. **Hold the coil away from any metal, turn the potentiometer knob so that the LED is on and then turn the knob slightly in the other direction until the LED turns off.**

   Doing so calibrates the IC so that small changes in the oscillating signal that runs through the coil trigger it.

3. **Experiment with your detector by holding it near to different items containing metal.**

   You can usually detect coins and keys in trouser pockets as well as various types of tools and nails at a distance of about 1.3 cm. You should be able to detect larger metal objects (such as that space shuttle in your garden shed) at a distance of about 2.5 cm. That's not very powerful we admit, but it's a metal detector and you made it!

If you don't get the right results, here are some remedies to try:

- Check that all the batteries are fresh, tightly inserted in the battery pack and all facing the right direction.
- Check that no wires or components have come loose.
- Compare your breadboard against the photos to make sure that all the wires and components are connected correctly.

Metal detectors are incredibly addictive (well, maybe not, but they're fun to mess around with). Here are some other things to try with your detector:

- Adapt your detector so that it activates a buzzer, by simply replacing the LED in the circuit with a buzzer.
- Make a more powerful detector that can find coins several centimetres under the sand on the beach. Check online for other metal detector circuits with more oomph (for example, `www.thunting.com` specialises in metal detectors you can use for hunting treasure).

Chapter 3

Making Light Dance to the Music

In This Chapter

- Taking a peek at the project
- Creating the circuit
- Lighting up to music

No doubt you've seen the various lights that change with the mood of the music at a disco, maybe in a school hall or at a wedding. One of the lights reacts to the sound from the record, flashing with the beat, subsiding in the quieter passages and so on.

In this chapter, we combine light and sound electronics to build a light that works in a similar way, but which you can use with any kind of music (you don't have to drag your 1970s' flares from the back of the wardrobe, unless you really want to). We show you how to set up two rows of lights that illuminate to different frequencies of sound. When you put on a fast, loud or heavily accented piece of music – say, swing or reggae – the lights dance all around. Every piece of music has its own, unique effect.

We arrange the lights like notes on a music staff, but you can put them in any arrangement you like. By working through this fun project, you get to know more about frequency filters, operational amplifiers and how music can light up the dance floor.

Illuminating the Big Picture: Project Overview

In this chapter, you create a display of light-emitting diodes (LEDs) that light up in response to high- or low-frequency sounds. You can see the LED musical notes arrangement in the finished display in Figure 3-1.

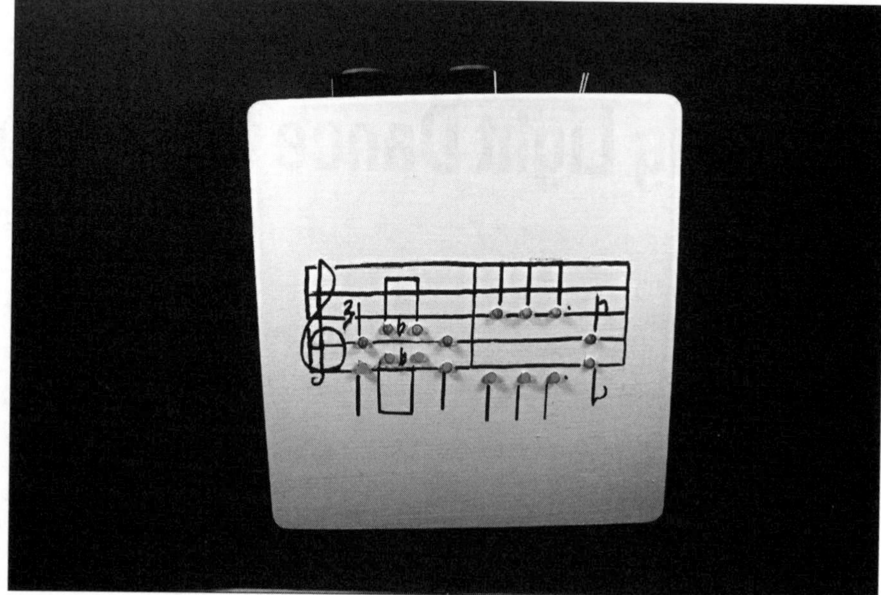

Figure 3-1: The final product: Dance to the Music!

Here's an overview of what you're going to be up to in the Dance to the Music project:

1. **Put together an electronic circuit to turn on the LEDs in response to sounds – half the LEDs flash to high-frequency sounds and the other half flash to low-frequency sounds.**
2. **Create a template for the musical notes, which you place on a wooden box and into which you drill holes for the LEDs.**
3. **Wire two groups of LEDs and resistors.**
4. **Apply the juice (in other words, pop in the batteries and flip the switch) and then play some music; the circuit sends current to each group of LEDs in response to the sound.**
5. **Turn up the volume, check out what each of your CDs does to the lights and enjoy the show!**

Scoping out the schematic

Music and light don't just happen: you have to start with a plan – or in this case, a schematic. You're going to put together one large breadboard and two LED arrays for this project.

Take a look at the schematic for the breadboard in Figure 3-2.

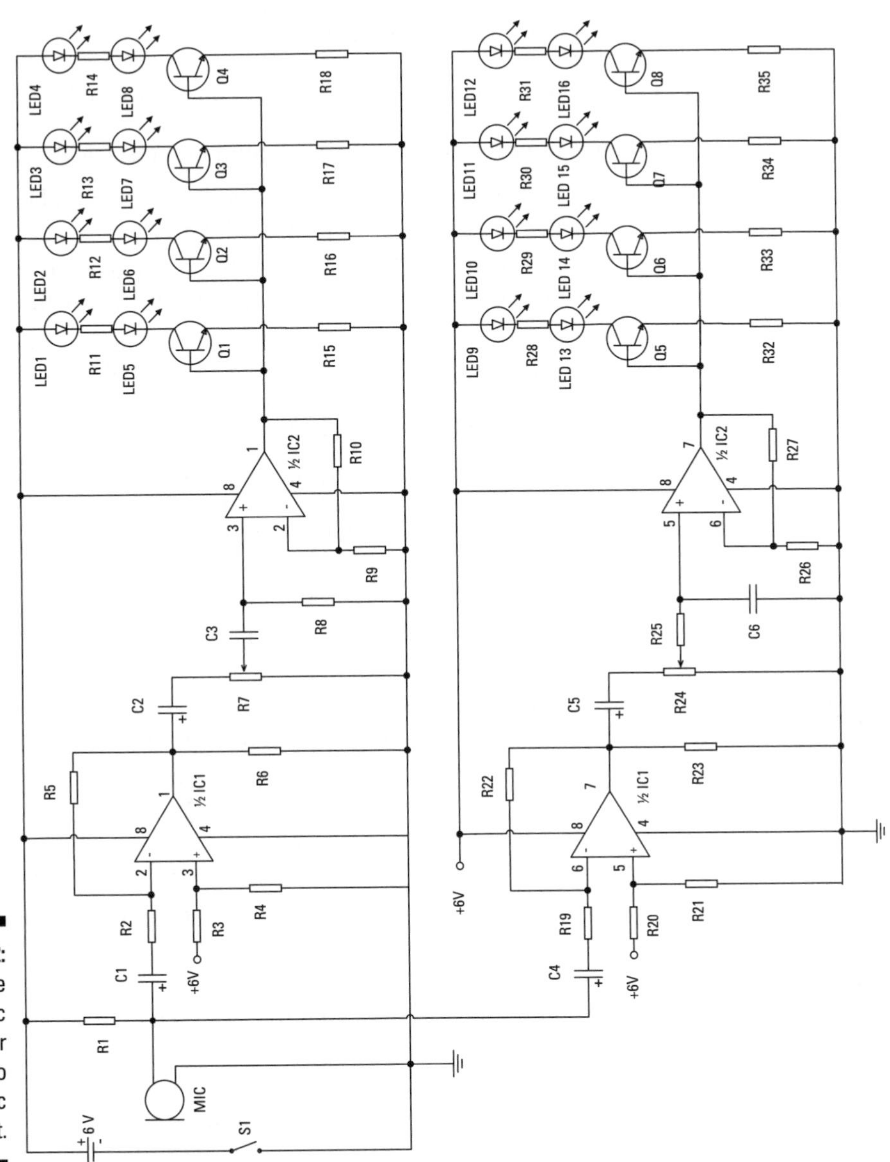

Figure 3-2: The schematic for your Dance to the Music project.

To make your musical score light up in response to the music, you need to create a circuit that uses a microphone, as well as two operational amplifier integrated circuits (ICs; check out Book III, Chapter 1 for details) in combination with resistors, capacitors and transistors (which we describe in Book II, Chapters 2, 3 and 6, respectively). Working together, they control which group of LEDs lights up to a particular frequency of music.

Here is a tour of the schematic elements that you use to control your project:

- **An electret microphone** transforms sound into electrical signals.
- **R1** connects the microphone to positive voltage and supplies about 4.5 volts (V) required by the microphone to function.
- **C1** and **C4** are capacitors that block the direct current (DC) voltage on the input signal and allow the alternating current (AC) signal to pass.

  The circuit splits between capacitors C1 and C4: the signal processed by the upper half of the circuit powers the LEDs blinking in response to high-frequency sound; the signal flowing through the lower half of the circuit powers the LEDs blinking in response to low-frequency sound.
- **IC1** is an operational amplifier (op amp) that amplifies the signal from the microphone. The IC contains two op amps; one half of IC1 is used in the upper circuit and one half in the lower circuit.
- **R2** and **R5** in the upper circuit and **R19** and **R22** in the lower circuit set the gain for each side of IC1. R5 is 50 times R2, and so a signal processed by the op amp is amplified approximately 50 times. R5 is 50 times R2, and R22 is 50 times R19.

  Gain is the amplitude of the voltage out divided by the amplitude of the voltage in: in other words, how much more juice goes out than comes in.
- **R3, R4** and **R6** in the upper circuit and **R20, R21** and **R23** in the lower circuit provide a DC bias to the op amp that allows the full AC signal to be amplified. If these resistors weren't present, the portion of the AC signal coming into the op amp with voltage less than 0 V would be lost.

  Bias involves applying voltage that's above ground to a portion of the circuit to amplify the positive and negative sides of a signal. Without DC bias, you lose part of the signal.
- **C2** and **C5** remove any DC bias from the signal coming out of the op amps.
- **R7** and **R24** are potentiometers that allow you to adjust the sensitivity of the circuit in relation to the music's volume.
- **C3** and **R8** function as a high pass filter and **R25** and **C6** function as a low pass filter. These filters are what make higher-frequency sounds light up LEDs 1–8 and lower-frequency sounds light up LEDs 9–16. For details, see the nearby sidebar 'Just passing through: Filters'.

Just passing through: Filters

The portion of the circuit made up of C3 and R8 is an *RC high pass filter.* With this filter, signals above a certain frequency – determined by the value of C3 and R8 – pass through more easily than signals below that frequency. The strength of signals below this key frequency is therefore reduced. This type of filter is made up of the capacitor in line with the signal path and the resistor between the output of the signal and ground.

Correspondingly, the portion of the circuit made up of R25 and C6 is an *RC low pass filter.* Here, signals below a certain frequency – determined by the value of R25 and C6 – pass through more easily than signals above that frequency. The strength of signals above this key frequency are reduced. This type of filter has the resistor in line with the signal path and the capacitor between the output of the resistor and ground.

For both types of filters, increasing the value of either (or both) the resistor or capacitor lowers the value of the key frequency. Lowering the value of either (or both) the resistor or capacitor increases the value of the key frequency.

- **IC2** is an op amp that amplifies the signal that passes through the filter. The IC contains two op amps; one half of IC2 is used in the upper circuit and one half of IC2 in the lower circuit.
- **R9** and **R10** in the upper circuit and **R26** and **R27** in the lower circuit set the gain of the op amp. R10 is 200 times R9, and so a signal processed by the op amp is amplified approximately 200 times. R10 is 200 times R9, and R27 is 200 times R26.
- **Q1, Q2, Q3** and **Q4** in the upper circuit and **Q5, Q6, Q7** and **Q8** in the lower circuit are 2N3904 transistors whose bases are connected to the output of the op amps in IC2. When the output of the op amp reaches about 0.7 V, the transistors turn on and current flows through the LEDs.

  The circuit doesn't mean a thing if you don't set up the lights for it to control, which is where the two groups of LEDs and resistors come in. They include:

  - **LED1–LED8** to light the display for the high-frequency circuit and **LED9–LED18** to light the display for the low-frequency circuit.
  - **R11–R18** in the upper circuit and **R28–R35** in the lower circuit are resistors, which in series with the LEDs limit the current running through the LEDs to approximately 10 milliamps.

Following some wiring tips

Behind the musical score we drew on the box that holds the circuit, there are a lot of resistors and LED leads soldered together. To make sure that leads that get bent and touch don't cause a short circuit, you need to protect them. Instead of electrical tape, we suggest that you use liquid electrical tape to coat the exposed leads.

Wires run between the circuit board resting on the bottom of the box and the LEDs that you insert in the top of the box. Be sure to cut your wires long enough so that when you open the box (which moves the LEDs in the top farther away from the circuit in the bottom of the box) you don't rip out the wires. Also be careful that you leave room to tuck those long wires inside the box when closed so that they don't poke out the sides or get caught in the hinges. Stranded wires work better than solid wires because they're more flexible.

Perusing the parts

You can purchase all the electronic parts that you need to build the circuit and assemble the box containing the LEDs at your nearest electronics shop or through an online vendor.

The circuit that transforms music into your dancing light show involves the following parts, several of which are shown in Figure 3-3:

- One 2.2 kΩ (kohm) 5% 1/4 W carbon or metal film resistor (R1)
- Eight 220 Ω 5% 1/4 W carbon or metal film resistors (R11–R14, R28–R31)
- Eight 100 Ω 5% 1/4 W carbon or metal film resistors (R15–R18, R32–R35)
- Two 10 kΩ panel-mounting D-shaped shaft, rotary-type potentiometers (R7, R24)
- Four 47 kΩ 1% 1/4 W carbon or metal film resistors (R3, R4, R20, R21)
- Two 100 kΩ 1% 1/4 W carbon or metal film resistors (R5, R22)
- Two 2 kΩ 1% 1/4 W carbon or metal film resistors (R2, R19)
- Three 5 kΩ 1% 1/4 W carbon or metal film resistors (R6, R8, R23)
- Two 1 kΩ 1% 1/4 W carbon or metal film resistors (R9, R26)
- Two 220 kΩ 1% 1/4 W carbon or metal film resistors (R10, R27)
- One 10 kΩ resistor (R25)
- One 0.001 microfarad (μF) ceramic capacitor (C3)
- Two 0.1 μF ceramic capacitors (C1, C4, C6)
- Two 10 μF electrolytic capacitors (C2, C5)

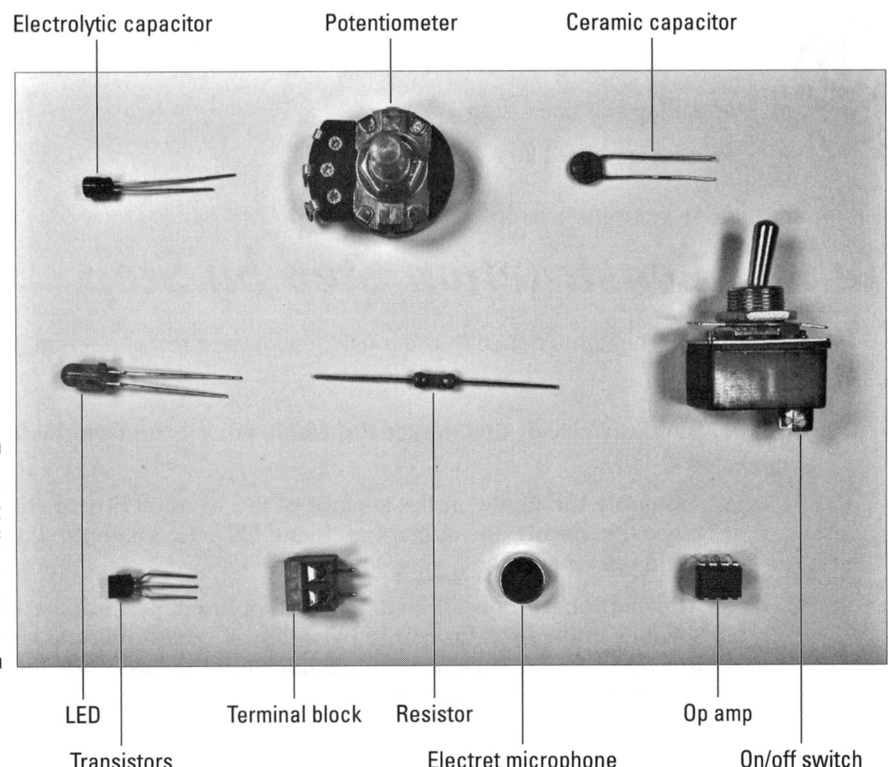

Figure 3-3: Important pieces of the Dance to the Music project.

- Eight green size 5mm LEDs (LED1–LED8)
- Eight red or orange size 5mm LEDs (LED9–LED16)
- Two LM358 op amps (IC1 and IC2)
- One electret microphone

We use the Horn part #EM9745 electret microphone (available from Digikey at `www.digikey.co.uk`) for two reasons: high sensitivity and a reasonable size for easy handling. You can use other electret microphones, though; plenty are available to choose from online.

- Two 830-contact breadboards
- One 4-AA battery pack with snap connector
- Eleven 2-pin terminal blocks (5.08mm pitch)
- Two knobs (for the potentiometer)
- Eight 2N3904 transistors (Q1–Q8)
- One panel mount SPST toggle switch

- A wooden box

  You may find one at a local craft supply shop that's just the right size.
- A metre of black 20 AWG wire
- A metre of red 20 AWG wire

Taking Construction Step by Step

To make the lights dance to the music, you have a few general tasks to perform:

1. **Build the circuit that makes the whole circuit run (see the following section).**
2. **Assemble the lights on the surface of the wooden box in which you place the circuit (the subject of 'Using LEDs to create the lights' later in this chapter).**
3. **Attach knobs, an on/off switch, a microphone, potentiometers and a speaker to the box. Insert the breadboard containing the circuit in the box and connect up everything (check out the later 'Adding the rest of the gubbins' section for details).**

Building the circuit

Time to go one-on-one with your breadboard. Grab your schematic (from the earlier Figure 3-2), tape it to the wall and get going.

Here are the steps involved in building the circuit:

1. **Place the two LM385 ICs (IC1 and IC2) and 11 terminal blocks on the breadboard, as shown in Figure 3-4.**

   As you can see, each terminal block has connections for two wires. Here's what the wires from each block are connected to:

   - Wires from one of the terminal blocks on the right side of Figure 3-4 go to the battery pack.
   - Wires from the other terminal block on the right go to the microphone.

• Wires from the terminal blocks in the centre go to the potentiometers.

• Wires from the terminal blocks on the left side of the breadboard go to the LEDs.

2. **Place the eight 2N3904 (Q1–Q8) transistors on the breadboard, as shown in Figure 3-4.**

   Insert each transistor lead into a separate breadboard row with the collector lead to the left side (see Figure 3-4), the base lead in the centre and the emitter lead to the right. The pin designations of the 2N3904 transistors are shown in Figure 3-5.

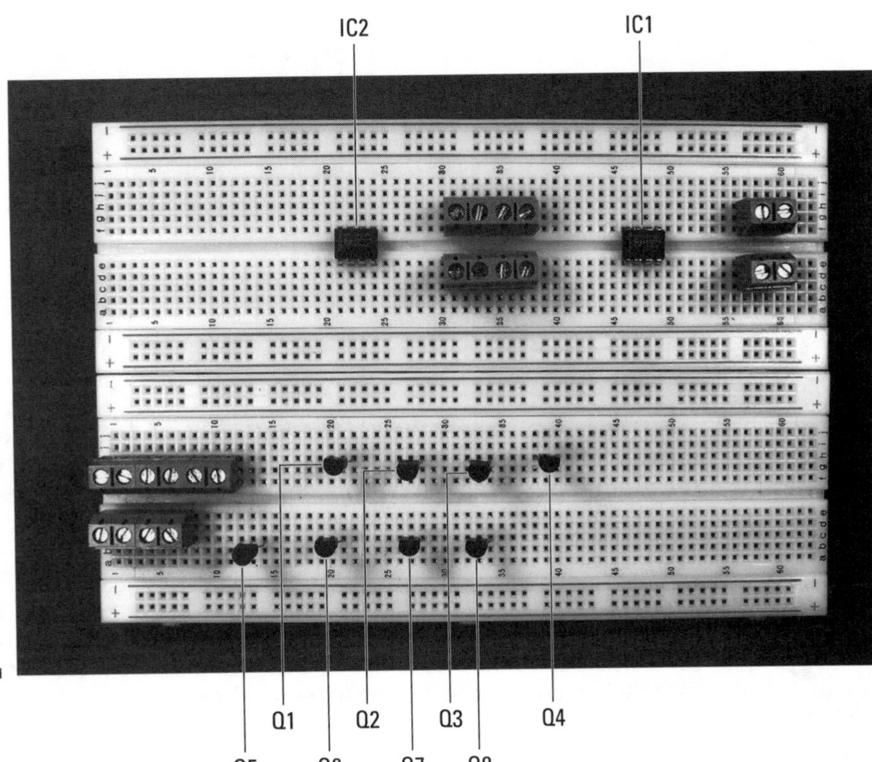

Figure 3-4: Place the LM385 ICs, 2N3904 transistors and terminal blocks.

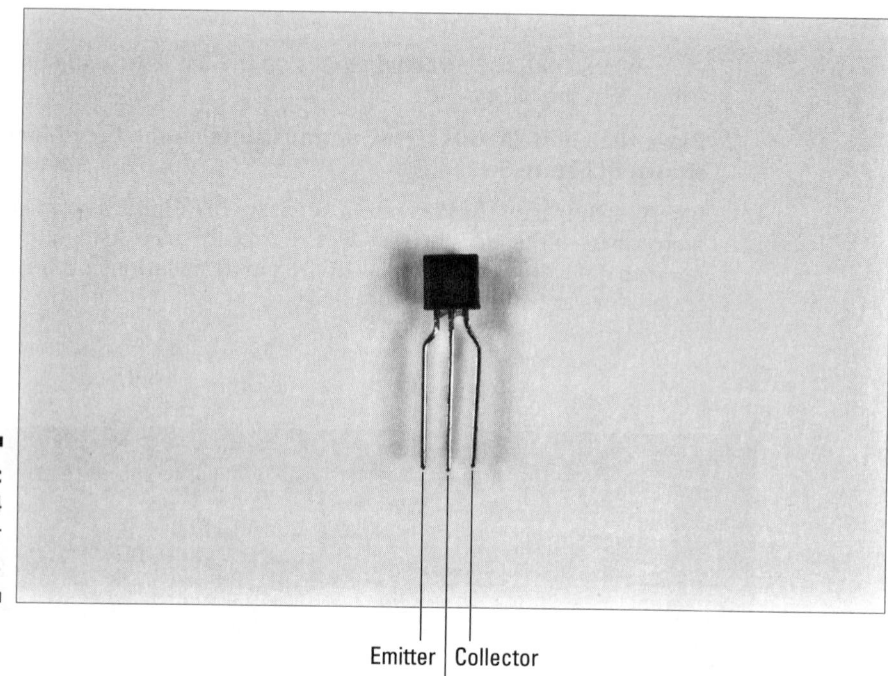

Figure 3-5: The 2N3904 transistor pinout.

3. **Insert wires to connect the ICs, the battery pack terminal block, the microphone terminal block and the potentiometer terminal blocks to the ground bus. Then insert wires between the ground buses to connect them to each other, as shown in Figure 3-6.**

   The ground buses are designated by a negative (–) sign on this breadboard.

4. **Insert wires to connect the ICs, the battery terminal block and one of the LED terminal blocks to the +V bus. Then insert wires between the +V buses to connect them, as shown in Figure 3-7.**

   The +V buses are designated by a + sign on this breadboard.

Figure 3-6: Shorter wires connect components to the ground bus; the two long wires on the right connect the ground buses.

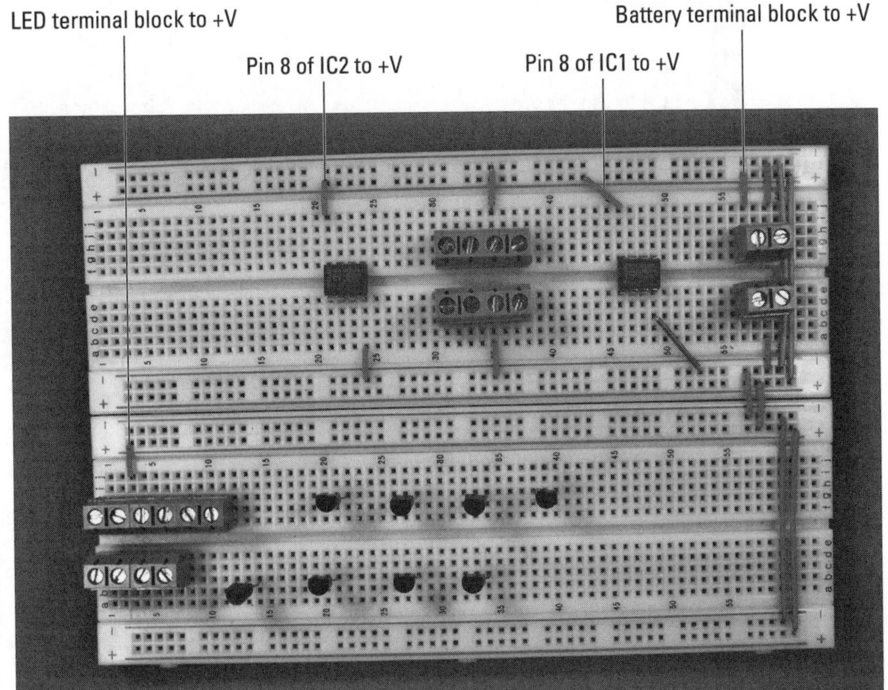

Figure 3-7: Connect components to the +V bus.

5. **Insert wires to connect the ICs, terminal blocks and discrete components.**

   Figure 3-8 shows you the details.

6. **Insert wires to connect IC2 to the transistors and the base pins of the transistors to each other.**

   We show you how in Figure 3-9.

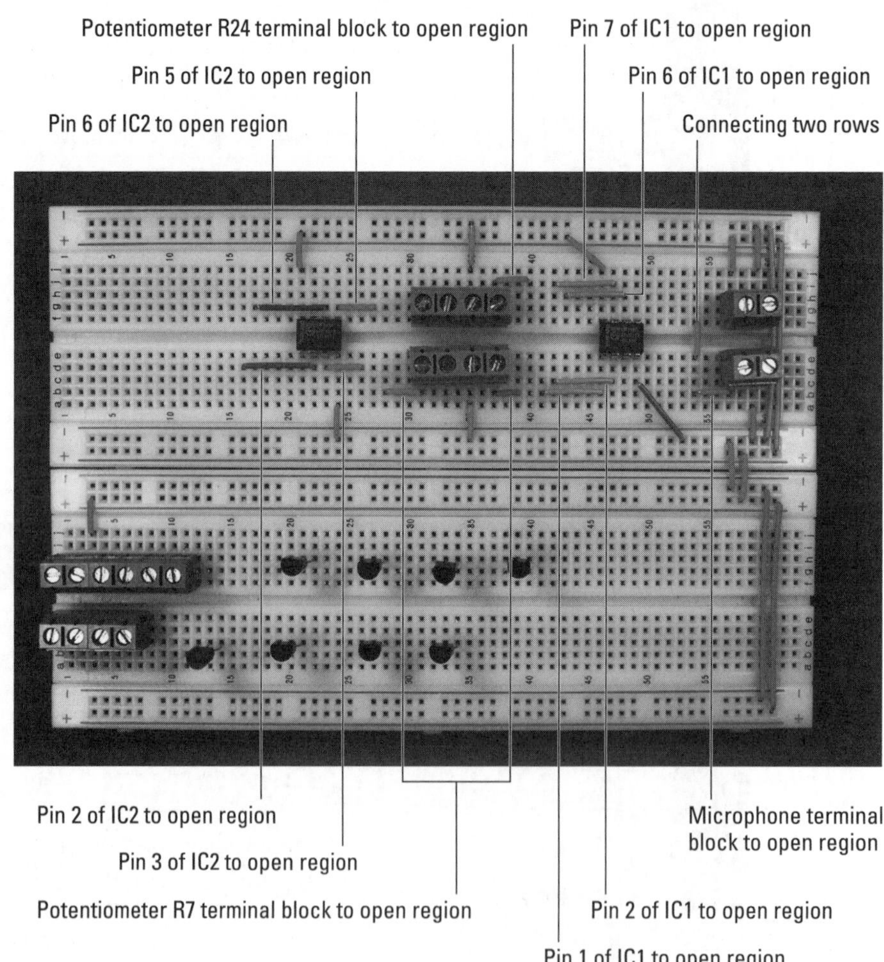

Figure 3-8: Hook up the IC, terminal blocks and discrete components.

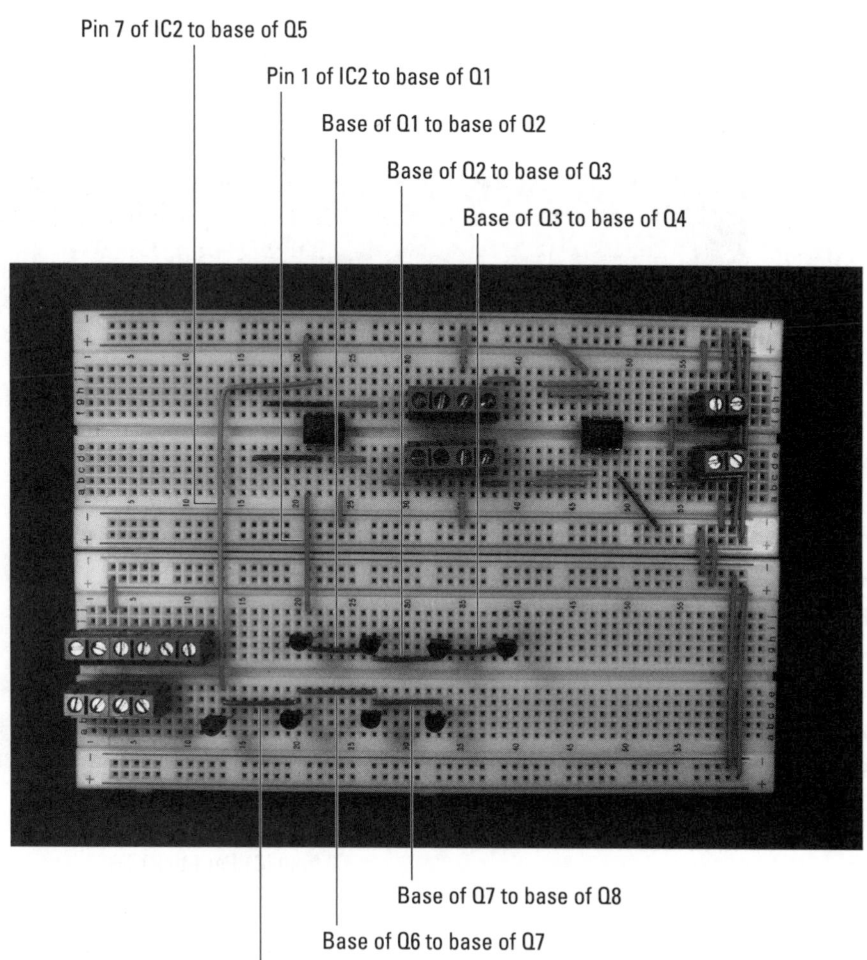

Figure 3-9: Connect IC2 to transistors.

7. **Insert three 0.1 μF capacitors (C1, C4 and C6), two 10 μF capacitors (C2 and C5) and one 0.001 μF capacitor (C3) on the breadboard, as shown in Figure 3-10.**

   Use the schematic in the earlier Figure 3-2 and the photo in Figure 3-10 to place each component correctly. For example, the schematic shows that the + side of C2 is connected to Pin 1 of IC1 and that the other side of C2 is connected to potentiometer R7. Therefore, you insert the long lead of C2 in the same row as the wire connected to Pin 1 of IC1 and the short lead in the same row as the wire connected to the terminal block for potentiometer R7.

   Be sure to insert the electrolytic capacitors the correct way around for safety reasons (see Book VIII, Chapter 4). The caps have the negative pin marked on them.

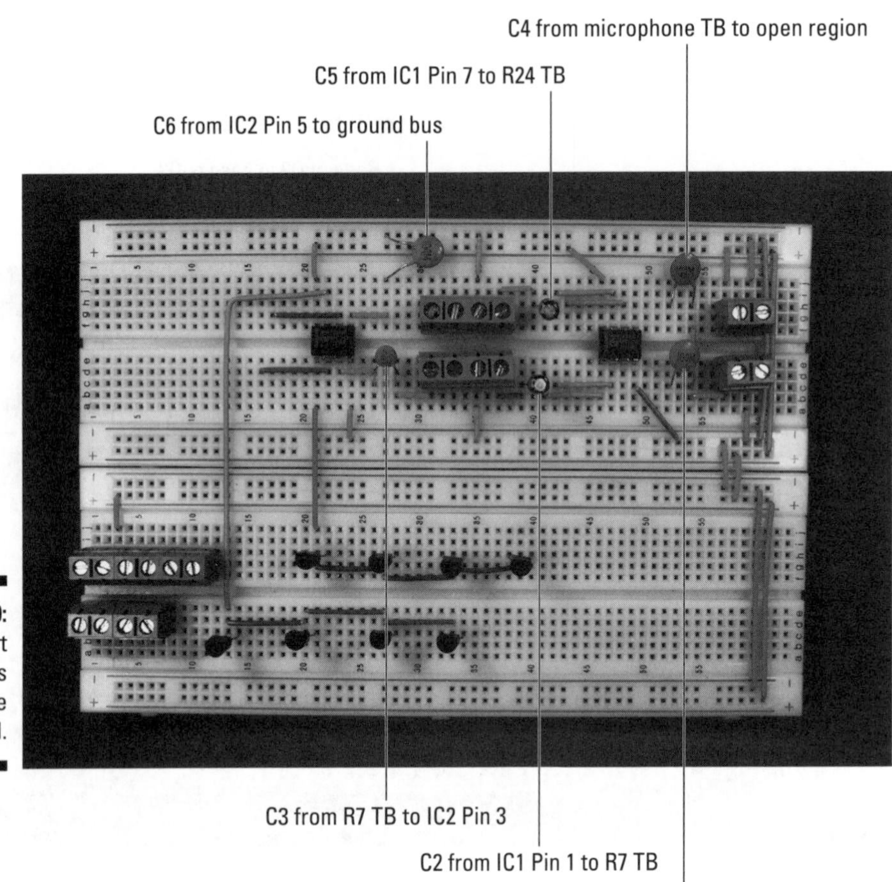

Figure 3-10: Insert capacitors into the breadboard.

8. **Insert eight 100 Ω resistors (R15–R18, R32–R35) from the emitter pins of each transistor to the ground bus.**

   Let Figure 3-11 be your guide.

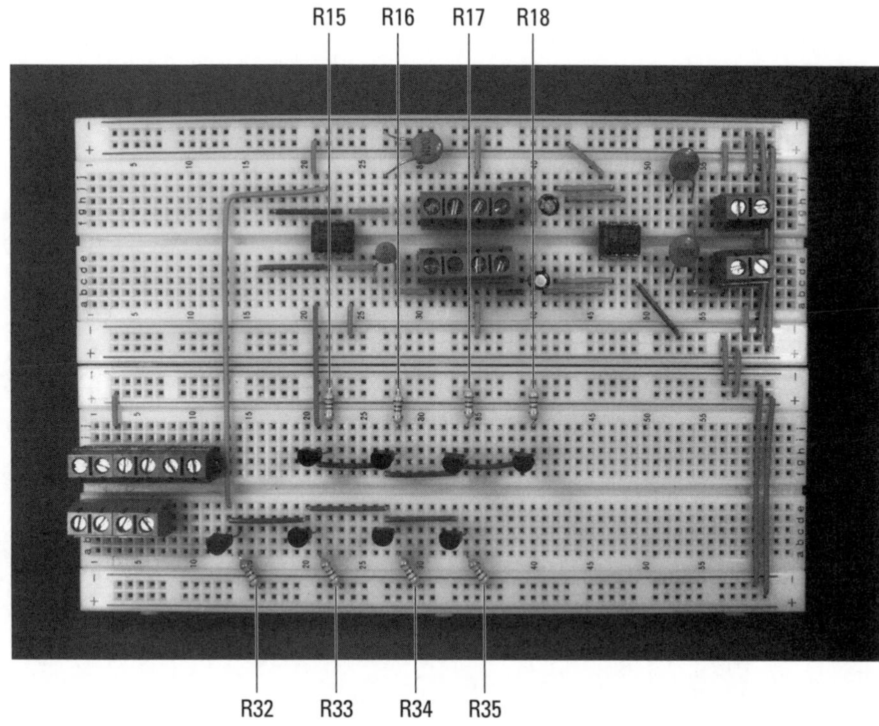

Figure 3-11: Insert resistors onto the breadboard.

9. **Insert more resistors as follows (see Figure 3-12):**
   - One 2.2 kΩ resistor (R1)
   - One 2 kΩ resistor (R2)
   - Two 47 kΩ resistors (R3 and R4)
   - One 100 kΩ resistor (R5)
   - Two 5 kΩ resistors (R6 and R8)
   - One 1 kΩ resistor (R9)
   - One 220 kΩ resistor (R10) on the breadboard

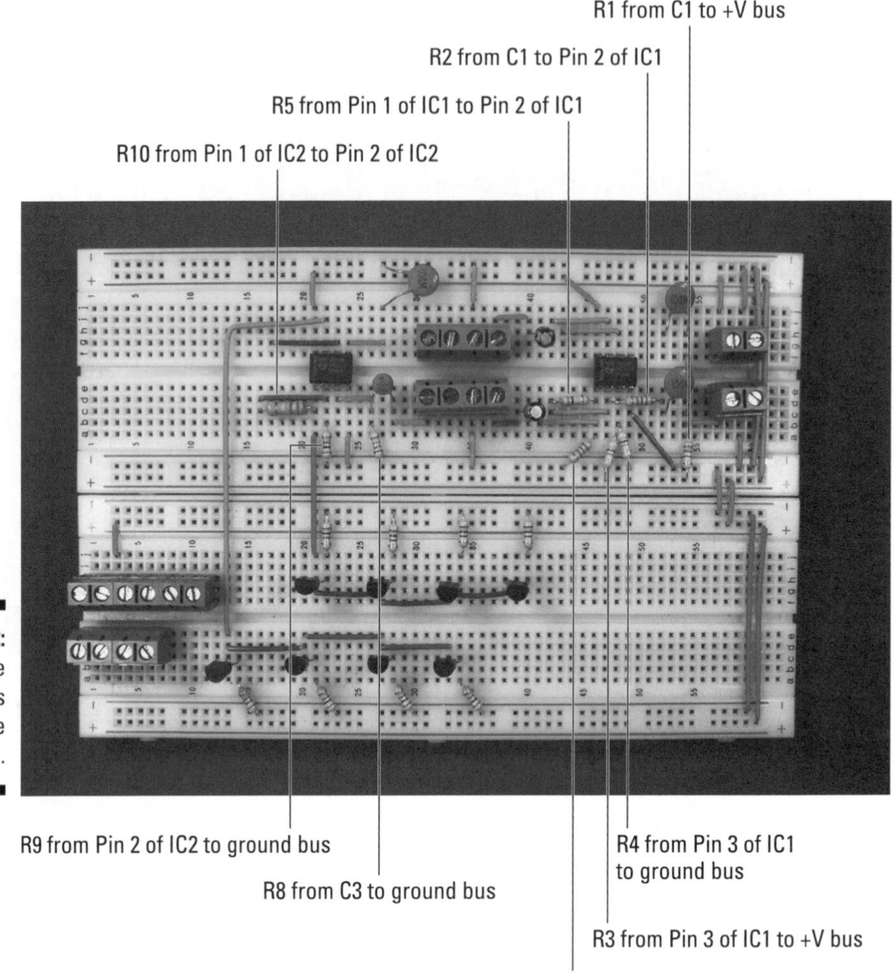

Figure 3-12: Insert more resistors onto the breadboard.

10. Insert the other resistors (as shown in Figure 3-13):

- One 2 kΩ resistor (R19)
- Two 47 kΩ resistors (R20 and R21)
- One 100 kΩ resistor (R22)
- One 5 kΩ resistor (R23)
- One 1 kΩ resistor (R26)

- One 10 kΩ resistor (R25)
- One 220 kΩ resistor (R27) on the breadboard

11. **Insert wires to connect the collector pins of the transistors to the terminal blocks.**

    Follow the photo in Figure 3-14.

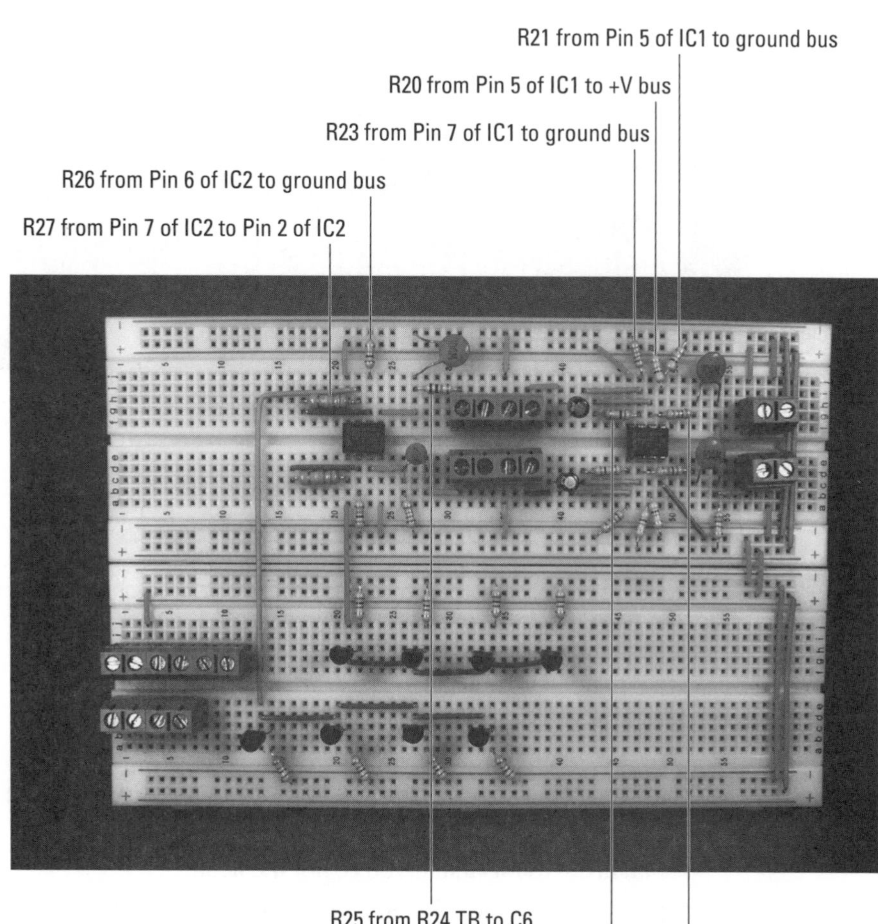

Figure 3-13: Insert the remaining resistors onto the breadboard.

Figure 3-14: Connect the collector pins of the transistor to terminal blocks.

Using LEDs to create the lights

The brains of the circuit that you create in the preceding section exist to make the light display work. You can now assemble those lights to create your Dance to the Music display:

1. **Select a series of musical notes.**

   Use eight high notes and eight low notes.

2. **Draw a musical staff on the top of the wooden box in pencil and then draw a dot for the spot where each LED is to go.**

   See Figure 3-15 for help if you're not musical!

3. **Drill test holes in a piece of scrap wood to determine the size of drill bit that you need to use to give a close press-in fit for the LEDs.**

   We used a 5 mm drill bit.

4. **Drill holes for the LEDs.**

   Use the locations you mark in Step 3.

5. **Draw the musical notes and staff on the wooden box.**

   Use a permanent marker or paint brush.

6. **Insert LEDs in the drilled holes.**

   Figure 3-15 shows the box at this stage.

7. **Attach resistors between the pairs of LEDs, as shown in Figure 3-16.**

   Connect the 220 Ω R11–R14 and R28–31 resistors to the short lead on one LED of each pair (four pairs of each colour) and to the long lead on the other LED of the pair.

8. **Solder the resistors to the leads and clip the leads just above the solder joint.**

   Clip only the leads to which you've soldered resistors. Figure 3-16 shows how the LEDs and resistors look at this point. Figure 3-17 shows a close-up to help you see the soldering more clearly.

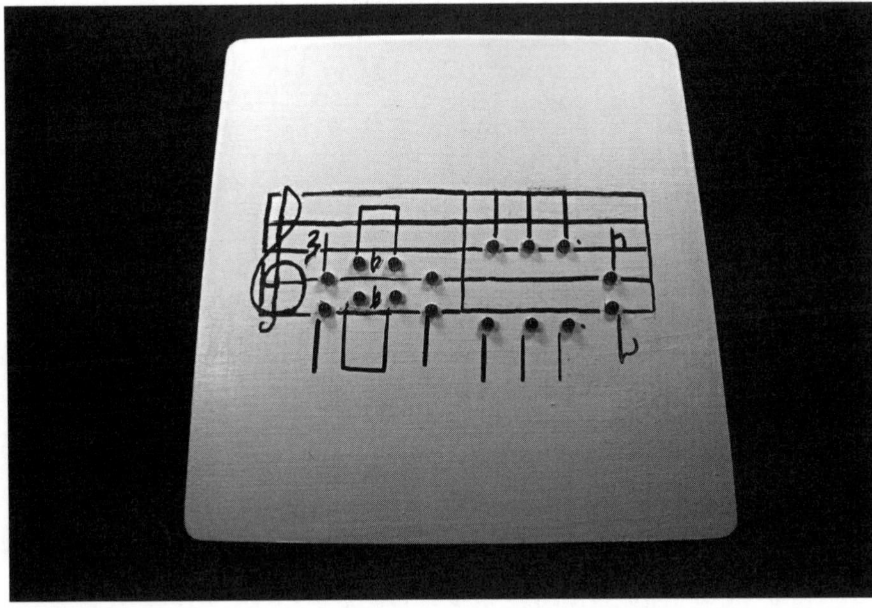

Figure 3-15: A bit of a favourite song created with a marker and LEDs.

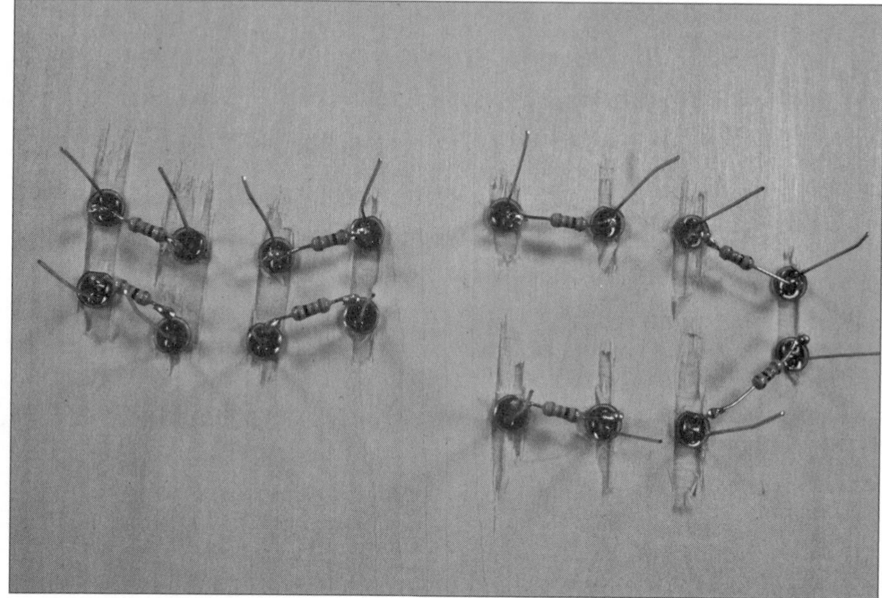

Figure 3-16: Resistors soldered and leads clipped.

Figure 3-17: A close-up of the soldering job.

Be sure to heed all the safety precautions about soldering that we provide in Book 1, Chapter 7. For example, don't leave your soldering iron on if you have to step away to accept a pizza delivery. And just in case a bit of solder has an air pocket that may cause it to pop, wear your safety goggles whenever you solder or clip leads and wires.

9. **Connect and solder short lengths of 20 AWG red wire between the remaining long LED leads.**

   Then attach a 30 cm piece of 20 AWG red wire to the remaining long lead of the first LED pair, as shown in Figure 3-18.

   These wires are the +V bus for the LED arrays; you attach the 30 cm wire to a terminal block to supply voltage to the bus.

10. **Clip the LED leads just above the solder joint.**

    Clip only the leads to which you've soldered the red wires; you need the remaining leads in the next step.

11. **Connect and solder a 30 cm length of 20 AWG black wire to the remaining short LED lead on each pair of LEDs, as shown in Figure 3-19.**

    You connect each of these eight wires to a terminal block. Figure 3-20 shows these connections in close-up.

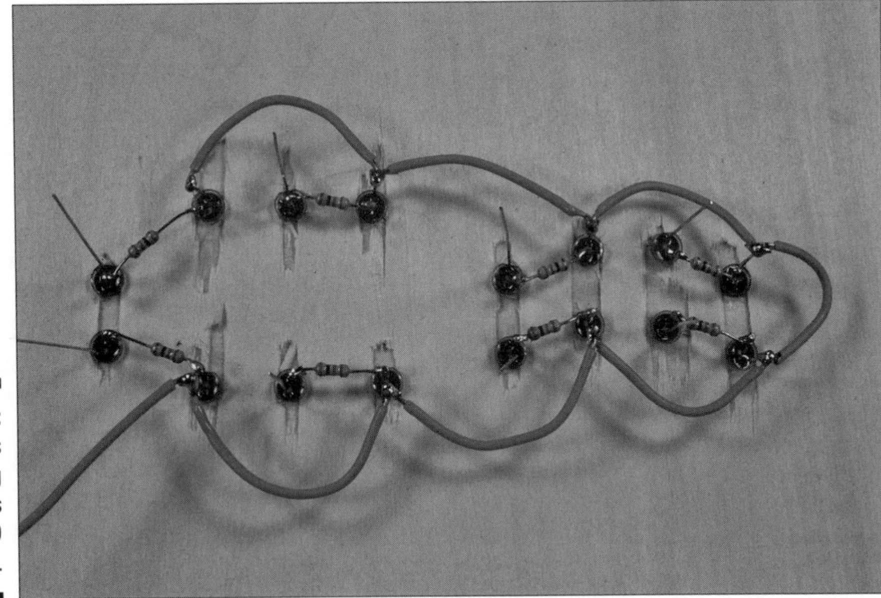

Figure 3-18: Red wires forming the +V bus for the LED arrays.

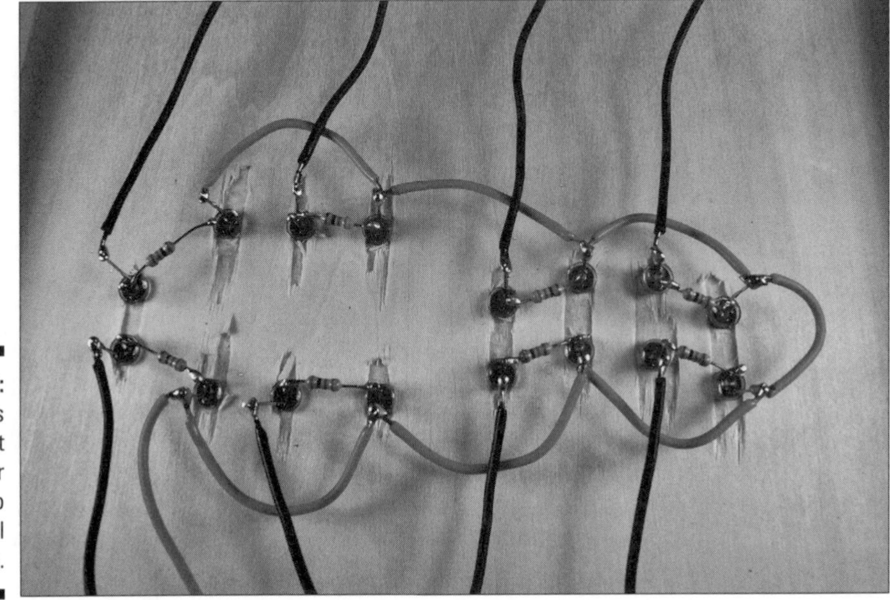

Figure 3-19: Black wires to connect each pair of LEDs to terminal blocks.

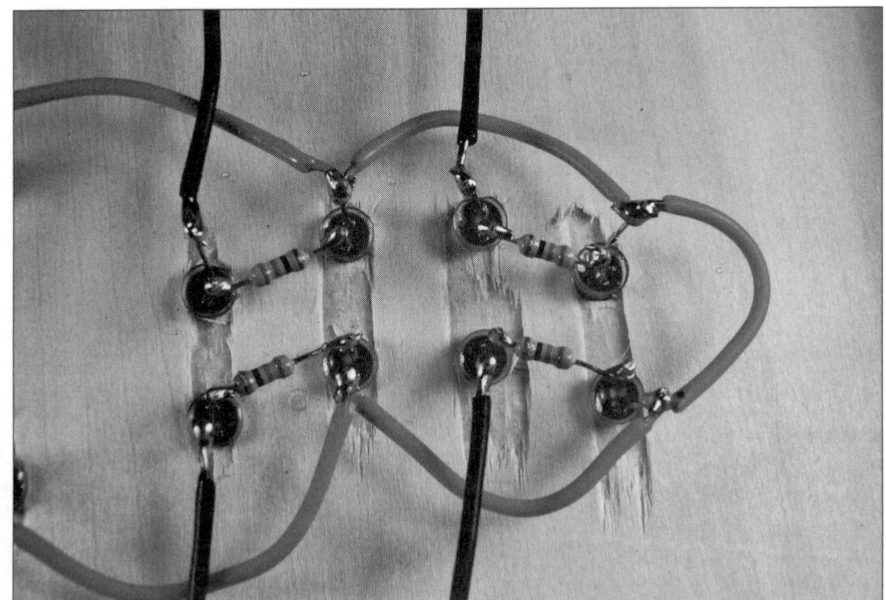

Figure 3-20: A close-up of all the connections.

12. **Clip the LED leads just above the solder joint.**

Make sure that the LED leads and solder joints don't touch each other. Coat them with liquid electrical tape to help prevent any shorts that can occur if you bend or push the wires together.

Adding the rest of the gubbins

After you've taken care of the circuit and LEDs (in the two preceding sections), you still have a lot of other bits and pieces sitting on your workbench, such as the microphone, potentiometers, a switch and so on. You need to assemble these parts before your project can make the music dance:

1. **Solder a 30 cm black wire to the microphone ground contact and a 30 cm red wire to the microphone +V contact.**

   Figure 3-21 identifies the microphone contacts.

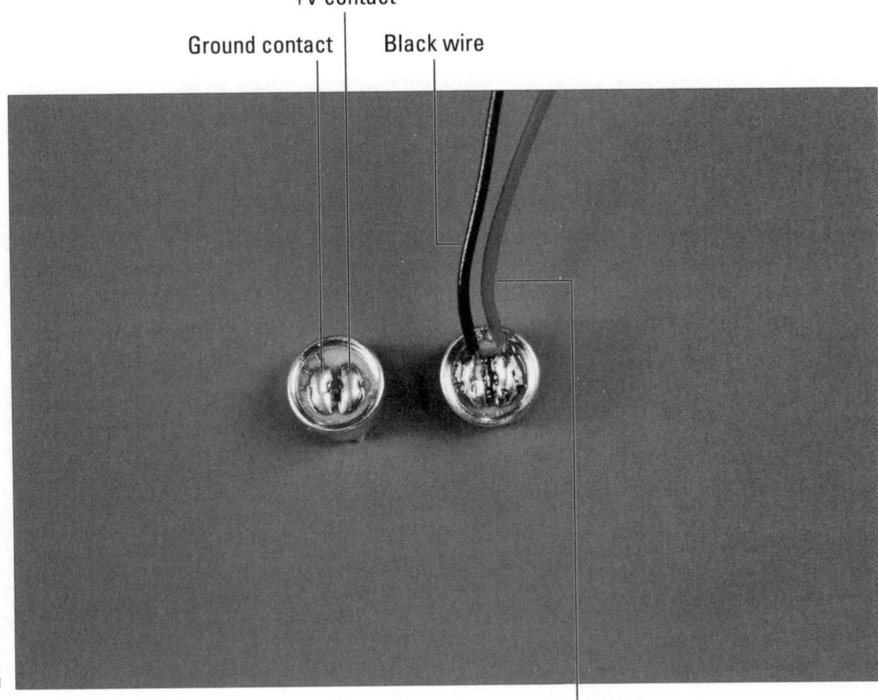

Figure 3-21: Solder wires to the microphone.

2. **Drill holes in the box where you plan to insert the microphone and on/off switch.**

   We put the potentiometers and on/off switch on one side of the box and the microphone on another side, but the placement is up to you. Choose a drill bit size for the microphone hole so that the microphone neatly passes through it. Figure 3-22 shows you where we place these components.

   TIP

   Make sure that you use safety goggles when drilling and clamp the box to your worktable!

3. **Slip the shaft of the on/off switch through the drilled hole.**

   Secure with the nut provided.

4. **Slip the shaft of the potentiometers through the drilled holes.**

   Secure with the nuts provided.

5. **Slip a knob over the shaft of each potentiometer.**

   Tighten with the set screw provided.

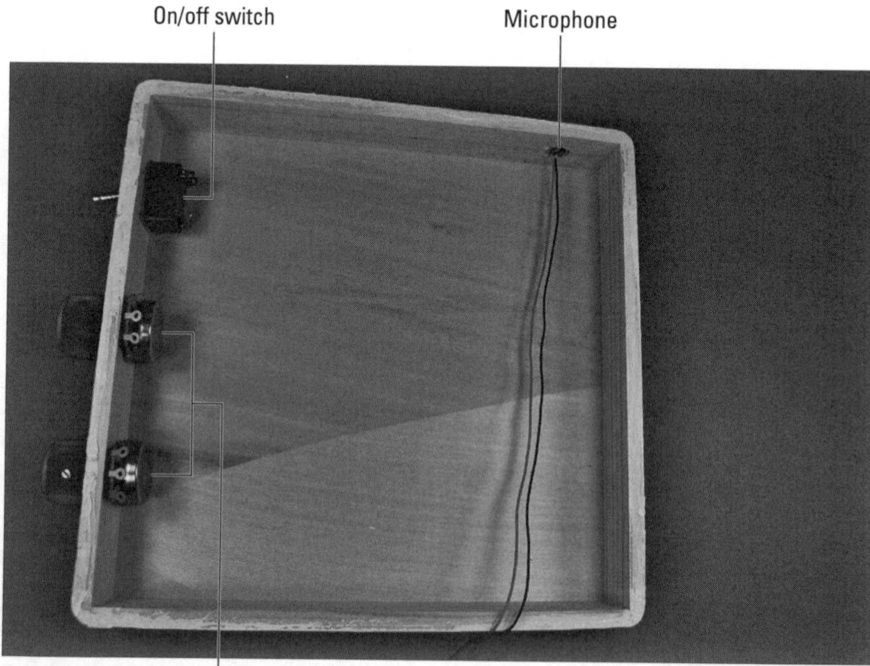

Figure 3-22: Box with on/off switch, potentiometers and microphone in place.

The tread on potentiometers is about 6 mm long, and so if the wall of your wooden box is 6 mm thick you can't use the nut to secure the potentiometer. Instead, check to make sure that the potentiometer shaft extends far enough beyond the box to allow the knob set screw to tighten on the shaft. If the shaft extends far enough, glue the face of the potentiometer to the box, making sure that you don't get any glue on the rotating shaft of the potentiometer. If the shaft doesn't protrude quite enough, use a small chisel to remove some wood on the inside of the box to let the potentiometer shaft extend a little farther before you glue.

6. **Slip the microphone into its drilled hole with a press fit.**

   Figure 3-22 shows the on/off switch, potentiometers and microphone in place in the box.

7. **Solder 30 cm black wires to each of the three potentiometer lugs.**

   Figure 3-23 shows you how.

8. **Solder the black wire from the battery pack to one lug of the on/off switch and solder a 30 cm black wire to the remaining lug of the on/off switch.**

   Figure 3-23 shows the switch after the wires are soldered.

Black wire from battery pack to on/off switch

Black wire to on/off switch

Wires to potentiometers

Figure 3-23: Wires soldered to the on/off switch and the potentiometers.

9. **Attach Velcro to the breadboard and the box.**

   Secure the breadboard in the box.

10. **Attach Velcro to the battery pack and the box.**

    Secure the battery pack in the box.

11. **Insert the wires from the LEDs, battery pack and on/off switch to the terminal blocks on the breadboard, as shown in Figure 3-24.**

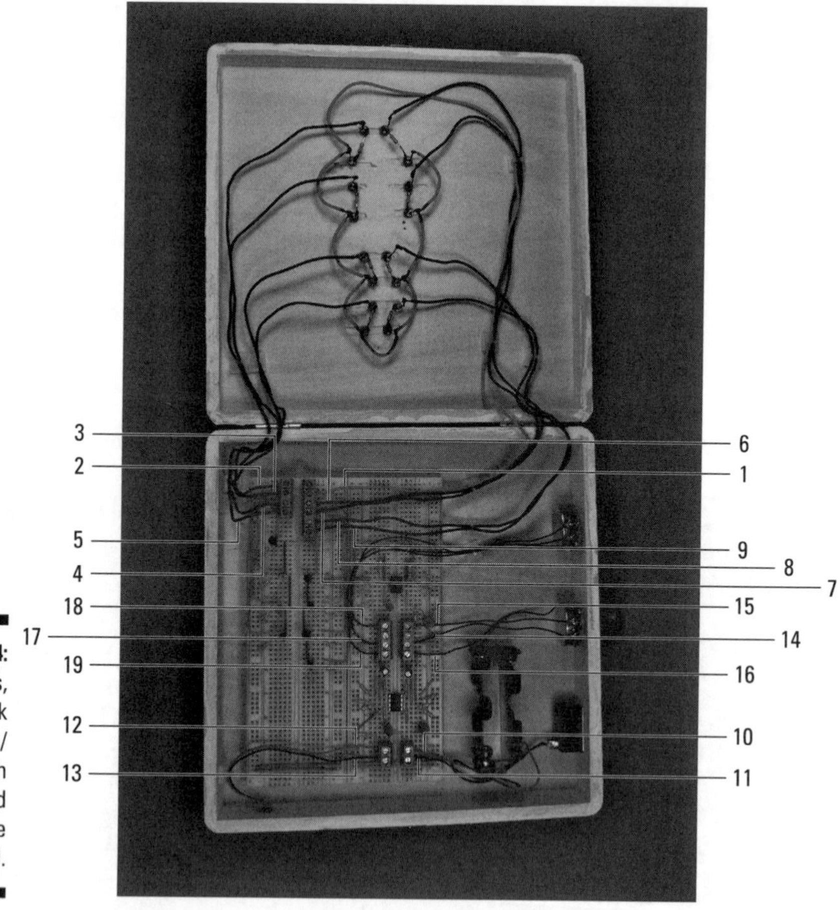

Figure 3-24: LEDs, battery pack and on/off switch connected to the breadboard.

Use the following as a key to the numbered callouts in Figure 3-24:

1: Red wire from LED +V bus

2–5: Wires from pair of red LEDs

6–9: Wires from pair of green LEDs

10: Red wire from battery pack

11: Black wire from on/off switch

12: Red wire from microphone

13: Black wire from microphone

14 and 17: Wires from right potentiometer lug

15 and 18: Wires from centre potentiometer lug

16 and 19: Wires from left potentiometer lug

12. **Cut each wire as you insert them to the length needed to reach the assigned terminal block, and strip the insulation from the end of the wire.**

    Secure the wires with wire clips.

Trying Out Your Dance to the Music Display

Now you can get the project up and running and see whether you think it's as much fun as we do.

You can use any sort of music you like, but we find that music with lots of instruments – such as a swing band with lots of brass – works best. Also, music with an upbeat tempo moves along and gets the lights switching on and off faster, which makes the effect better. Ella Fitzgerald singing 'Take the A Train' works well. (Ask your parents; they may know it!)

Here are the simple steps to get this project going:

1. **Pop the batteries into the battery pack.**
2. **Flip the on/off switch to on.**
3. **Put on some music.**

That's it! Watch the lights go on and off in response to the high and low frequencies in the music. You adjust the sensitivity of the LEDs by turning the potentiometers.

Here are the obvious things to check out if you're having a problem:

- Check that all the batteries are fresh and tight in the battery pack and that they all face the right direction.
- If one or two LEDs aren't working, replace them.
- If two LEDs in series with each other aren't functioning, you may have reversed the long and short leads of the LEDs; if so, just replace that pair of LEDs.
- You're playing Brahms's 'Lullaby' or something similar, which doesn't light up a single LED. Switch to Motörhead's 'Ace of Spades' or Sly and the Family Stone's 'Dance to the Music'.

By now, you're probably jumping around the living room, playing every CD you have to see what they do to the lights and wanting more, more, more. Here are a few different ways to take this project further:

- You can change from a musical staff and notes to any kind of shape you want to define with your LEDs. You can have two stars or a sun and moon, for example.
- You can use a band pass filter to add more layers of frequency. For example, between your high pass filter and low pass filter (see the earlier sidebar 'Just passing through: Filters'), you add two more band pass filters to hit intermediate frequencies and have four sets of LEDs going off in response to music.
- You can miniaturise your circuit to pin it on your shirt or take it with you to parties. You can get a smaller circuit by using smaller LEDs (for example, replacing the 5mm LEDs with 3mm LEDs). Or you can use a different method of building the circuit called a *dead bug circuit.* Imagine an IC turned on its back with its little prongs sticking up in the air, and you get an idea of what we mean. This method doesn't involve a breadboard but makes a connection directly to the LED.

Chapter 4

Hacking a Toy to Make a Talking Puppet

In This Chapter

- Preparing to build your project
- Bread-boarding the circuit
- Having fun with your puppet

Did you have a talking doll as a child – perhaps a Barbie who asked 'Would you like to go shopping?' or an action figure who declared 'Hold your fire until I give the order'? All you had to do was pull a string in the doll's back to hear half a dozen phrases in rotation. The trouble was that those recordings repeated over and over again, driving mad any adult within earshot. If only you could've made them say something different once in a while.

Well, today you can – with a little electronics. Engineers with a sense of humour even run toy-hacking workshops where you chop and change bits of toys to make them do new and surprising things. Action Man takes on a whole new character when with Barbie's voice box he declares 'I just love being a fashion model'. And how great to hear Barbie bark out the command 'Mortar attack, dig in'!

In this chapter, you discover how to work with a sound synthesiser chip, an amplifier and a speaker to produce almost any sound you can imagine, while also picking up a bit about how amplification works.

You can place a sound chip in just about any object and use switches to activate sounds. For this project, we choose a hollow hand puppet of Merlin, because it lets you insert the project breadboard and switches easily (and because we're fans of the television series). But you can use anything you like for your talking doll.

Talking about the Big Picture: Project Overview

In this chapter's project, you create a talking hand puppet, doll or whatever else you manage to squeeze the circuit into. You can program the synthesiser chip inside the puppet with any messages you like, from 'Cancel Doctor Who' to 'Bring back Doctor Who'!

The words or sounds are activated when you shake one of the puppet's hands or press its nose. Alas, adding the electronics to the puppet makes it unsuitable for use on your hand, but you can't have everything.

To program the sounds, you connect the board to a computer using an RS232 cable, which means that you need an old computer with an RS232 port and running the Windows XP operating system.

Using a cheap, old computer means that you can dedicate one to your electronics projects like this. You could add parallel ports to your more up-to-date computer using an expansion card or USB adaptor (which are available from stores such as Maplin or to order online - just search for 'serial port card' or 'USB serial port'). To install a card you'd have to follow the manufacturer's instructions in opening up the back of your desktop computer and inserting it into a PCI or PCI-express slot. The adaptor just plugs into a USB port but you may have to install a piece of driver software for it to work. All in all, using a cheap, old computer is much less hassle!

You can see our finished talking puppet (okay, you have to imagine the talking part) in Figure 4-1.

Here are the necessary tasks for creating your own talking toy:

1. **Put together an electronic circuit containing a sound synthesiser chip, an amplifier and a speaker.**
2. **Connect the circuit to your computer to program the sound synthesiser with sentences, music or whatever strange sounds you'd like to play around with (a burping Barbie, anyone? Sorry, this project brings out the child in us!).**
3. **Place the box containing the circuit inside the puppet and connect switches in the puppet to the circuit.**

After these steps, when someone presses the puppet's hand or nose, it plays whatever you program in the sound synthesiser for that particular switch.

Although the hand puppet we work with in this chapter may seem like a toy, it's not intended for young children. The wires and small electronic components can be swallowed and you certainly don't want youngsters playing with batteries.

Figure 4-1: Meet Magic Merlin.

Scoping out the schematic

You have to put together only one breadboard for this project. We present the schematic for the board in Figure 4-2.

Here's a tour of the schematic elements:

- **IC1** is a SpeakJet sound synthesiser model. You can connect this integrated circuit to your computer and program it to generate electrical signals by using software supplied by the manufacturer. These signals correspond to words that form sentences, sounds that create music or various cool sound effects. The software allows you to program eight sounds, each controlled by one of Pins 1–4 and 6–9 of the IC. The SpeakJet is available in the UK from Active Robots (www.active-robots.com).
- **C1** is a capacitor (something we introduce in Book II, Chapter 3) that filters noise from the +4.5V supply to IC2 (see the relevant bullet point later in this list).

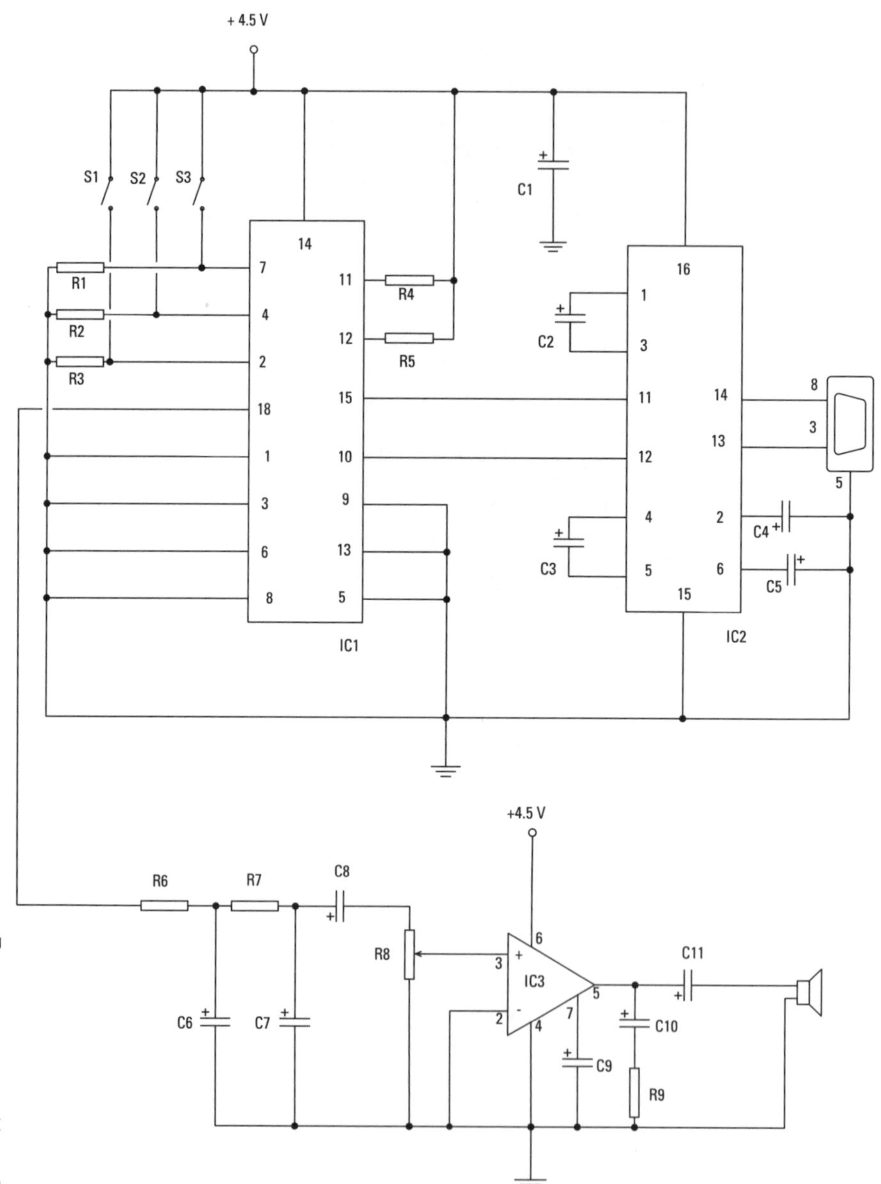

Figure 4-2: The schematic of the talking puppet circuit.

- **Switches S1, S2 and S3** control the voltage on Pins 2, 4 and 7 on IC1. The switches are normally *open,* which means that each pin is normally connected to ground. When you push one of the switches, the voltage on the corresponding pin rises to +4.5 volts (V). When you release the switch, the voltage on the pin returns to ground. Because we program the SpeakJet to trigger when the voltage on a pin changes from high (+4.5 V) to low (ground), the sound which that pin controls is triggered when you press and release the corresponding switch. We use three switches to control three sounds, because our puppet has three handy spots for placing switches.
- **Resistors R1, R2** and **R3** limit the current running to ground when you press switches S1, S2 or S3 (see Book II, Chapter 2 for an introduction to resistors).
- **IC2** is a MAX232 driver/receiver chip that converts the signals from your computer. These signals aren't generated with the correct voltage for this circuit, and so signals have to be converted so that the SpeakJet chip can use them. This chip also converts signals from the SpeakJet into signals that your computer can use.
- **C2, C3, C4** and **C5** are capacitors that fill a function that the Max232 designers call *charge pump capacitors.* These parts are required to make the IC2 function properly.
- **R6** and **R7, C6** and **C7** are two resistors and two capacitors, respectively, that form a *low pass* filter, which eliminates high-frequency noise prior to the signal reaching the amplifier.
- **C8** is a capacitor that removes any direct current (DC) offset from the output of IC1.
- **R8** is a potentiometer that controls the sound volume.
- **IC3** is an LM386N-1 audio amplifier that takes the electrical signal generated by the SpeakJet when you push and release one of the switches in the puppet and then amplifies the electrical signal to create enough power to drive the speaker.
- **C9** is a capacitor that improves the stability of the LM386 amplifier to prevent problems such as oscillation.
- **C10** acts as a current bank for the output. This capacitor drains when sudden surges of current occur and refills with electrons when the demand for current is low.
- **C11** is a capacitor that removes any DC offset from the output of the LM386 amplifier.

Noting some construction issues

The tactile switches that we use in this project have very tiny leads that are meant to be surface-mounted in an assembly line. Assuming that you don't have an assembly line handy, you have to use needle-nose pliers to crimp wire around the tiny leads to hold the wires in place while you solder them.

The DB9 connector that you use has a small metal tube or cup to which you can solder each wire connection. The easiest way to solder a wire to one of these tubes is to melt some solder into the tube, reheat the tube and then insert the bare wire end into the melted solder.

When inserting electrolytic or tantalum capacitors into the breadboard, pay attention to the polarity. Inserting the capacitors the wrong way can damage the capacitors and possibly other components in your circuit. The longer lead of the capacitor is the + side. The schematic in the earlier Figure 4-2 shows you which direction to insert the capacitor. For example, the + lead of capacitor C2 goes towards Pin 1 of IC2 and the other lead goes toward Pin 3 of IC2.

In order to feed wires from the tactile switches to a location where they can be connected to the electronics box via the shortest path, we had to cut a few holes in the fabric of the puppet. Don't worry: he didn't feel a thing.

Perusing the parts

Your biggest shopping decision for this project is what to place the project in. The hand puppet idea is nice because it's hollow and has a personality, as opposed to using just a plain wooden box. In theory, however, you can put the insides of the project into any object. For example, you can cut open a stuffed toy, take out the stuffing and put the project inside.

If you take our route and choose a hand puppet, do make sure to pick one that you like with enough room to contain the electronics enclosure and with openings that allow you to place switches in the hands, face or other areas. After you choose your housing unit, shop for the following parts for the project itself (several of which we show in Figures 4-3 and 4-4):

- Two 33 kΩ 5% rated, 0.5W, carbon or metal film resistors (R6 and R7)
- Five 1 kΩ 5% rated, 0.5W, carbon or metal film resistors (R1, R2, R3, R4, R5)
- One 10 kΩ rotary, knob-operated, 1/2 W or 1 W rated potentiometer (R8)
- One 10 Ω 5% rated, 0.5W, carbon or metal film resistor (R9)

- Two 0.01 microfarad (μF) ceramic capacitors (C6 and C7)
- Five 1 μF tantalum or electrolytic capacitors (C1, C2, C3, C4, C5)
- One 0.047 μF ceramic capacitor (C10)
- Two 10 μF electrolytic capacitors (C8 and C9)
- One 100 μF electrolytic capacitor (C11)
- One battery pack for three AAA batteries

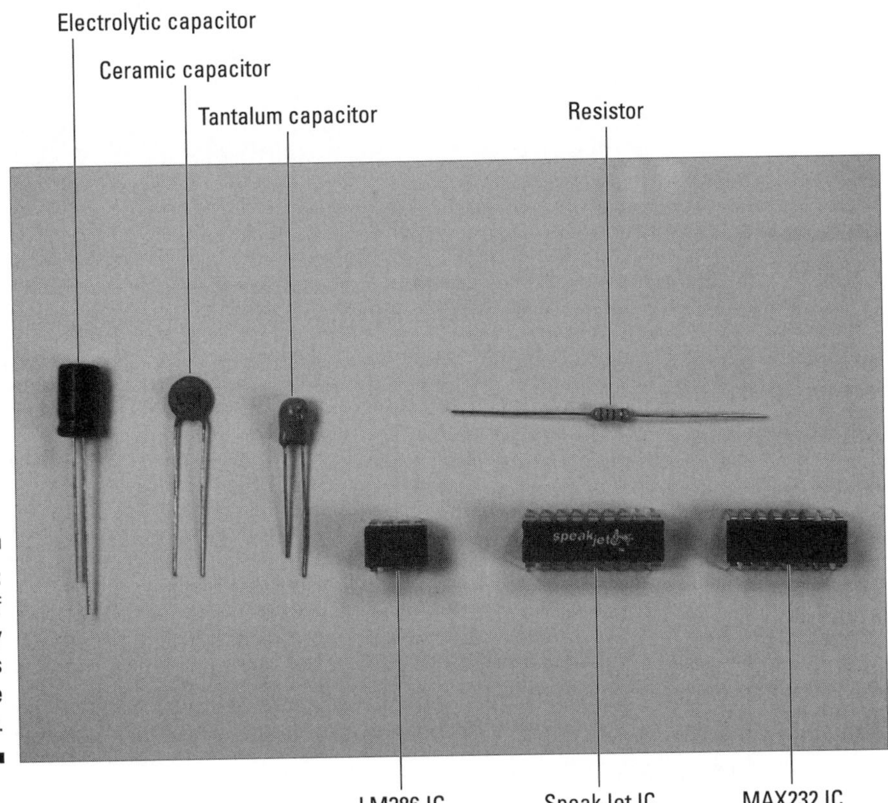

Figure 4-3: Many of the key components for the project.

We use three AAA batteries in this project to supply 4.5 V (a four-battery pack would make the supply voltage about 6 V, above the maximum supply voltage allowed for the SpeakJet IC).

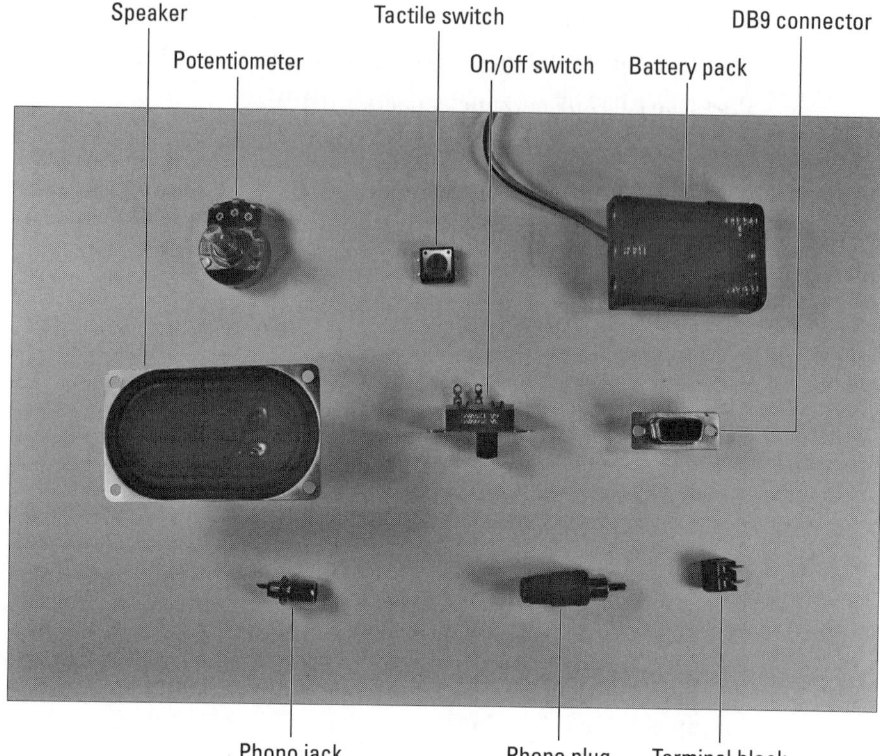

Figure 4-4: Key components, part 2!

- One SpeakJet sound synthesiser IC1

  The UK distributor for SpeakJet is Active Robots (`www.active-robots.com`).

- One LM386N-1 amplifier IC3

  Of the many versions of the LM386 amplifier, we choose this one because it works with the supply voltage of 4.5 V used by this circuit.

- One MAX232 driver/receiver IC2
- One single-pole, single-throw (SPST) slide switch, used as the on/off switch
- One 830-contact breadboard
- Eight 2-pin terminal blocks (5.08mm pitch)
- One knob (for potentiometer)
- One 8 Ω, 1 W speaker

- Three tactile switches (S1, S2, S3)

  Many of these tactical switches are very small – as tiny as 6 mm square. We use 12 mm x 12 mm switches (part #TS6424T2602AC) made by Mountain Switch and available from Mouser (`uk.mouser.com`).
- An enclosure – large enough to take the breadboard with the speaker (see later Figure 4-15 to see how they are arranged inside)
- Six phono plugs
- Six phono jacks
- One DB9 female connector, solder bucket type
- One DB9 serial port cable
- Four 1.3 cm 6-32 flathead screws
- Four 6-32 (approximately 4.7625 mm) nuts
- An assortment of different lengths of prestripped, short 0.7 mm diameter AWG wire
- An old computer with a RS232 serial port to attach the DB9 serial cable in order to program the sounds

  Look for one running Microsoft Windows XP (we provide more information on using old computers such as this sort in Book VI, Chapter 4).

Taking Construction Step by Step

Building your talking puppet requires you to complete the following tasks:

1. **Create the circuit (see the following section).**
2. **Drill various holes in a box and place the circuit in that box (which is the subject of 'Making the box puppet-friendly' later in this chapter).**
3. **Program the sound synthesiser (check out the later section 'Programming fun sounds').**
4. **Insert the box and switches into the puppet (flip to the later 'Hooking up your puppet' section).**

Creating your puppet's circuit

Gather together your breadboard and other parts to build the circuit that helps your puppet talk. Here are the steps involved:

1. **Place the SpeakJet IC (IC1), the MAX232 IC (IC2), the LM386 IC (IC3) and eight terminal blocks on a breadboard, as shown in Figure 4-5.**

   As you can see, you connect two wires to each terminal block. The wires from these eight terminal blocks go to the battery pack, the DB9 connector, the on/off switch, the three tactile switches in the puppet, the speaker and the potentiometer.

2. **Insert wires to connect the ICs and the terminal blocks to the ground bus and then insert a wire between the two ground buses to connect them to each other, as shown in Figure 4-6.**

   Fourteen shorter wires connect components to ground bus; the long wire on the left connects the two ground buses.

3. **Insert wires to connect the ICs and the terminal blocks for the battery pack and the three tactile switches (S1, S2, S3) to the +V bus.**

   Figure 4-7 shows the connections.

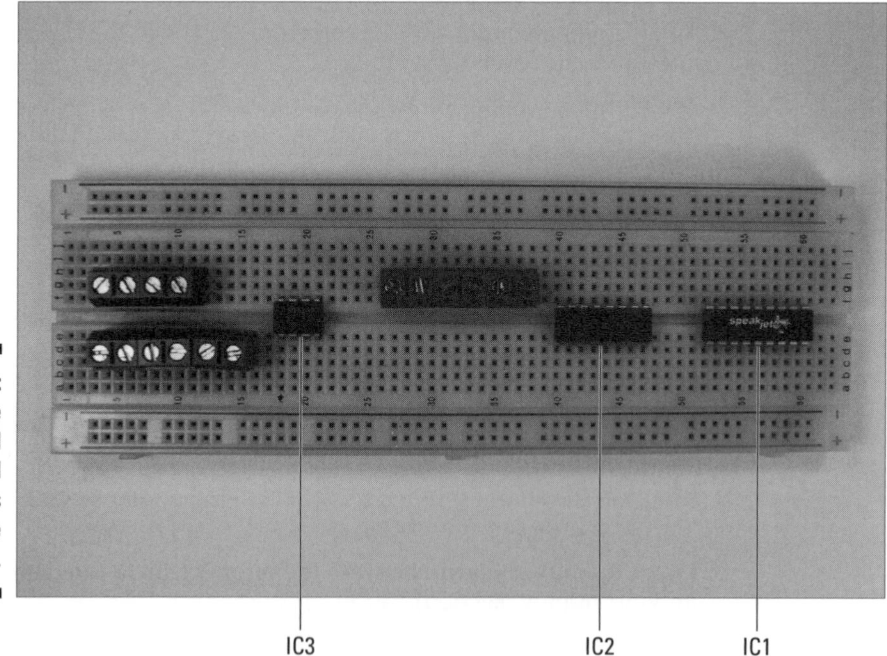

Figure 4-5: Place the ICs and terminal blocks on the breadboard.

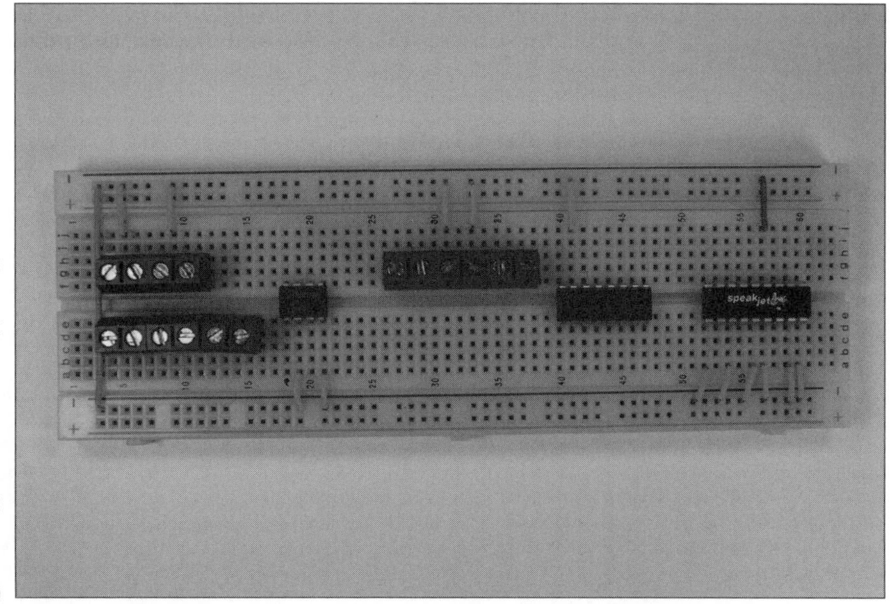

Figure 4-6: Connect the ICs and terminal blocks to ground and then connect the ground buses.

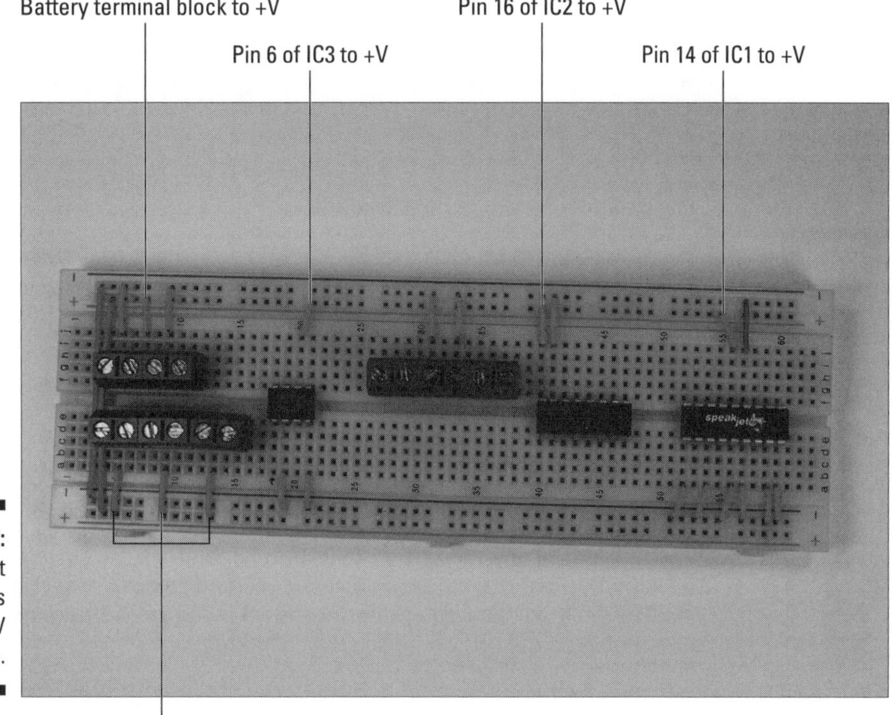

Figure 4-7: Connect components to the +V bus.

4. **Insert wires to connect the ICs; terminal blocks for the potentiometer (R8) and tactile switches (S1, S2, S3); and discrete components.**

   Follow Figure 4-8 as a guide.

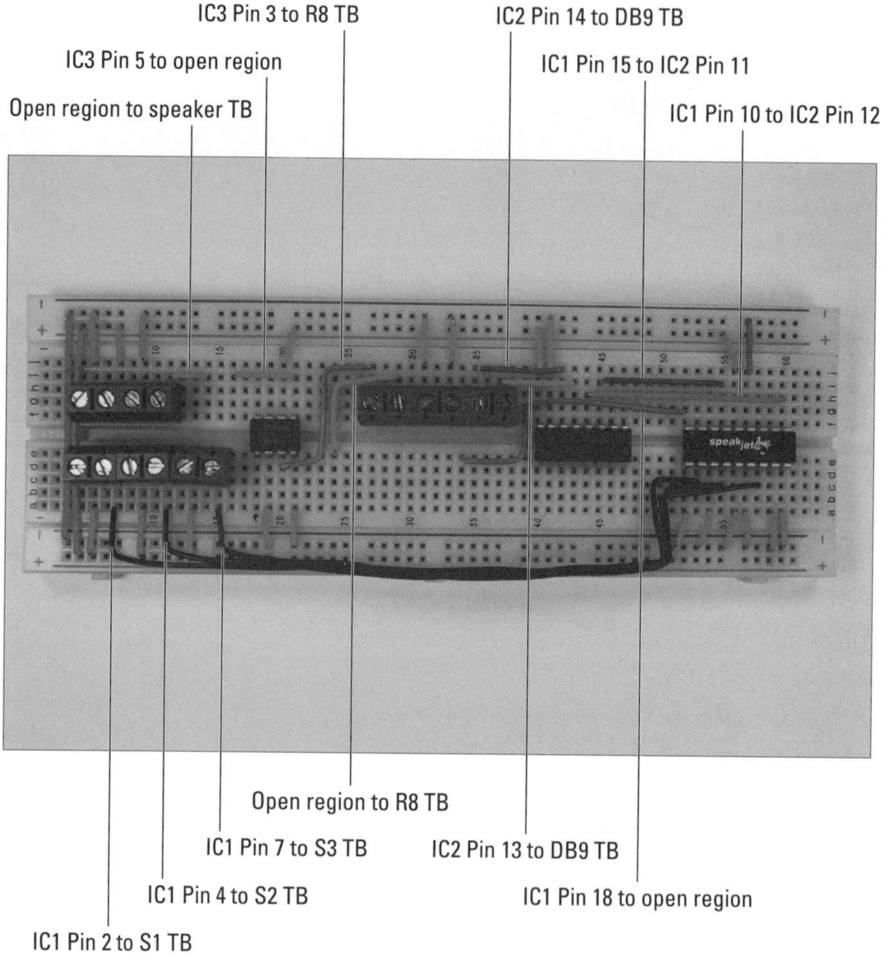

IC3 Pin 3 to R8 TB

IC2 Pin 14 to DB9 TB

IC3 Pin 5 to open region

IC1 Pin 15 to IC2 Pin 11

Open region to speaker TB

IC1 Pin 10 to IC2 Pin 12

Open region to R8 TB

IC1 Pin 7 to S3 TB

IC2 Pin 13 to DB9 TB

IC1 Pin 4 to S2 TB

IC1 Pin 18 to open region

IC1 Pin 2 to S1 TB

Figure 4-8: Hook up the ICs, terminal blocks (TB) and discrete components.

5. **Insert discrete components on the breadboard, as shown in Figure 4-9.**

   When inserting electrolytic or tantalum capacitors, be sure to check the schematic in the earlier Figure 4-2 to see where to insert the longer, positive (+) lead. The components shown in Figure 4-9 and indicated by the numbered callout comprise the following:

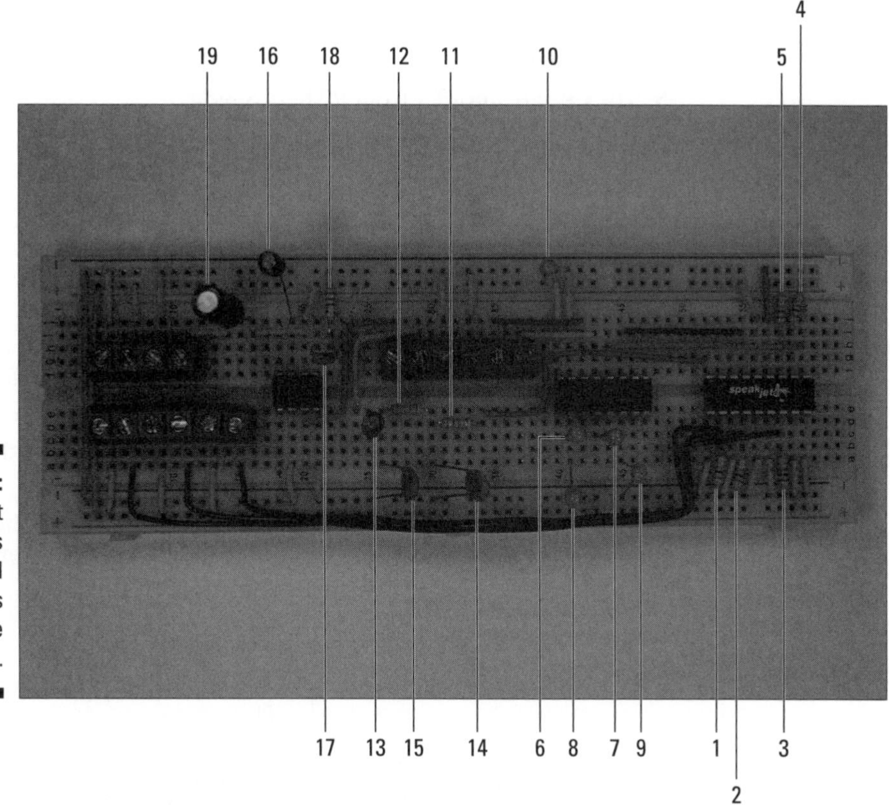

Figure 4-9: Insert resistors and capacitors on the breadboard.

1: 1 kΩ resistor R1 from IC1 Pin 7 to ground

2: 1 kΩ resistor R2 from IC1 Pin 4 to ground

3: 1 kΩ resistor R3 from IC1 Pin 2 to ground

4: 1 kΩ resistor R4 from IC1 Pin 11 to +V

5: 1 kΩ resistor R5 from IC1 Pin 12 to +V

6: 1 μF capacitor C2 from IC2 Pin 1 to Pin 3

7: 1 μF capacitor C3 from IC2 Pin 4 to Pin 5

8: 1 μF capacitor C4 from IC2 Pin 2 to ground

9: 1 μF capacitor C5 from IC2 Pin 6 to ground

10: 1 μF capacitor C1 from +V to ground

11: 33 kΩ resistor R6 from wire to IC1 Pin 18 to an open region

12: 33 kΩ resistor R7 from R6 to an open region

13: 10 μF capacitor C8 from R7 to wire to R8 terminal block

14: 0.01 μF capacitor C6 from R6 to ground

15: 0.01 μF capacitor C7 from R7 to ground

16: 10 μF capacitor C9 from IC3 Pin 7 to ground

17: 0.047 μF capacitor C10 from IC3 Pin 5 to open region

18: 10 Ω resistor R9 from C10 to ground

19: 100 μF capacitor C11 from wire to IC3 Pin 5 to wire to speaker terminal block

Clipping the wire leads on components can cause small bits of metal to fly through the air. Make sure that you wear your safety goggles when clipping leads!

6. **Solder 30 cm wires to the solder cups for Pins 3, 5 and 8 of the DB9 connector.**

The location of these pins is shown in Figure 4-10. Although the manufacturer prints the pin numbers next to the solder cups, you may need a magnifying glass to read the pin numbers on the connector.

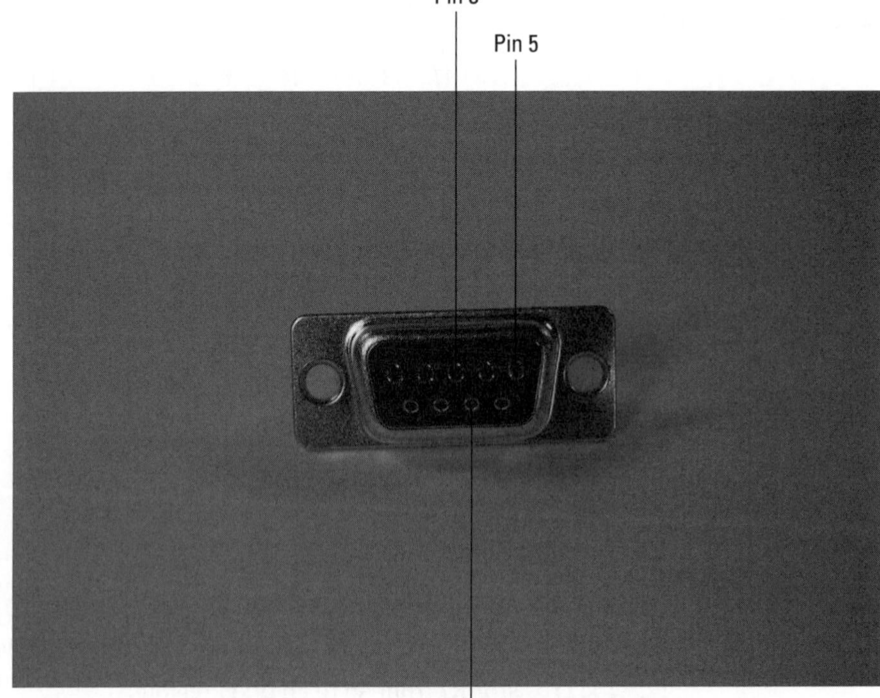

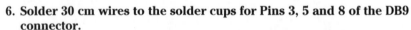

Figure 4-10: Solder cups on back of the DB9 connector.

Be sure to heed all the safety precautions about soldering that we describe in Book I, Chapter 7. For example, don't leave your soldering iron on and unattended. And just in case a bit of solder has an air pocket and pops, wear your safety goggles!

7. **Slip a 2.5 cm length of heat-shrink tubing along each wire until it covers the soldered joints.**

   Heat the tubing with a hair dryer to secure it and protect the joints.

Making the box puppet-friendly

In this section, we describe preparing the box so that you can insert the guts of the project into it and string wires to the puppet. This job involves some drilling, cutting and assembling.

Follow these steps to get the project box wired up:

1. **Drill holes in the box for the phono jacks, potentiometer, speaker, DB9 connecter and on/off switch in appropriate locations (along the lines in Figure 4-11).**

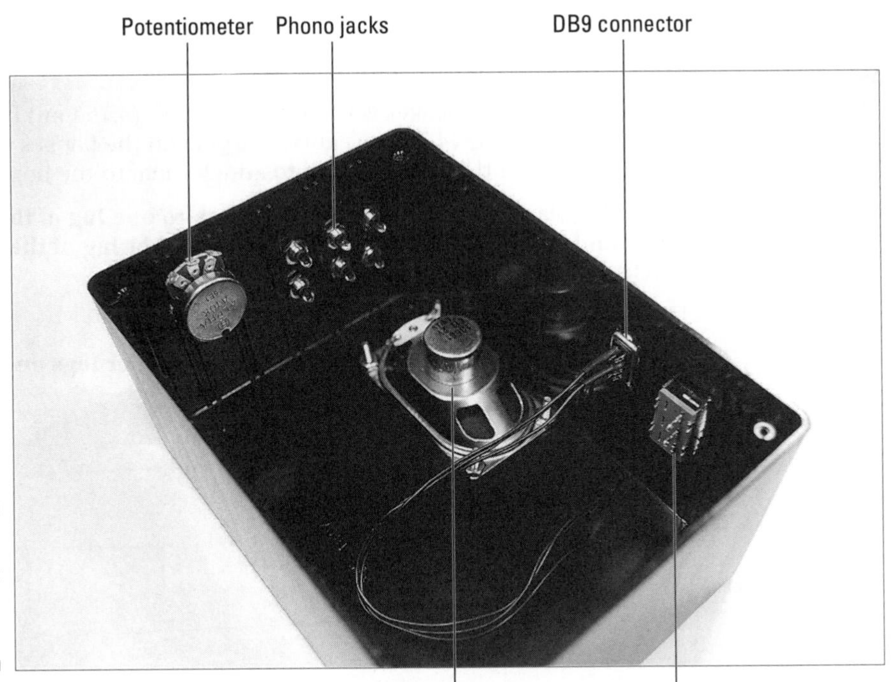

Figure 4-11: Mount components on the box.

We use a 1.3 cm drill bit for the phono jacks and an 8 mm bit for the potentiometer. To secure the speaker, we drill four holes using a 2.4 mm bit. You may need to use different bit sizes, depending on the parts you use.

Wear your safety goggles whenever you drill holes or cut wires. Also, the drill bit can bind as it goes through the plastic, causing the box to turn with the drill if it's not properly secured. Don't tempt fate: use a vice or other method to secure the box while you're drilling.

2. **Cut openings for the DB9 connector, on/off switch and the speaker.**

   We drill a pilot hole and then use a coping saw to cut the openings. The openings don't have to be the exact shape of the part. For the DB9 connector and the switch, make the openings big enough for the body of the part to fit through. The openings can even be a little oversized as long as enough plastic is left for you to secure the part's flange. For the speaker, cut an opening about 1 cm inside the outline of the speaker. All you need is an opening big enough to let the sound travel.

3. **Insert the phono jacks, potentiometer, speaker, DB9 connector and on/off switch.**

   Follow Figure 4-11 as a guide.

4. **Secure the phono jacks and potentiometer with the nuts provided with each part.**

   The outside of the box is shown in Figure 4-12.

   You can secure the speaker with four 1.3 cm 6-32 (4.76 mm) flathead screws and four 6-32 (4.76 mm) nuts. Use glue on the flanges of the DB9 connector and on the on/off switch to attach them to the box.

5. **Solder the black wire from the battery pack to one lug of the on/off switch and solder a 20 cm black wire to the other lug of the on/off switch.**

   Use Figure 4-13 to guide you.

6. **Solder a 15 cm black wire to each of the two solder lugs on the speaker.**

   Again, Figure 4-13 shows you what you need to see.

Figure 4-12: A view from the outside.

Figure 4-13: Wires soldered to the on/off switch and speaker.

7. **Solder a 20 cm wire to each of the three potentiometer lugs.**

   See Figure 4-14.

8. **Connect 20 cm wires to each of the audio jacks and solder them.**

   Figure 4-14 illustrates this aspect as well.

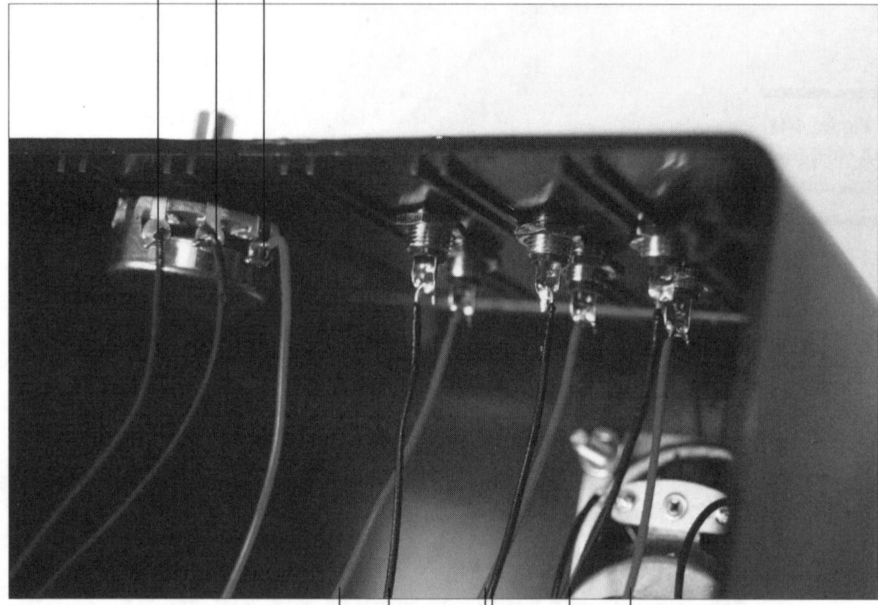

Figure 4-14: Solder wires to connect external components to the breadboard.

9. **Attach the breadboard and the battery pack to the box.**

   Using Velcro is the best way.

10. **Cut the wires from the phono jacks, potentiometer, on/off switch, battery pack and DB9 connector.**

    Do so to a length that allows you to arrange the wires neatly within the enclosure.

11. **Strip insulation from the ends of cut the wires, insert them in the terminal blocks and secure with wire clips, as shown in Figure 4-15.**

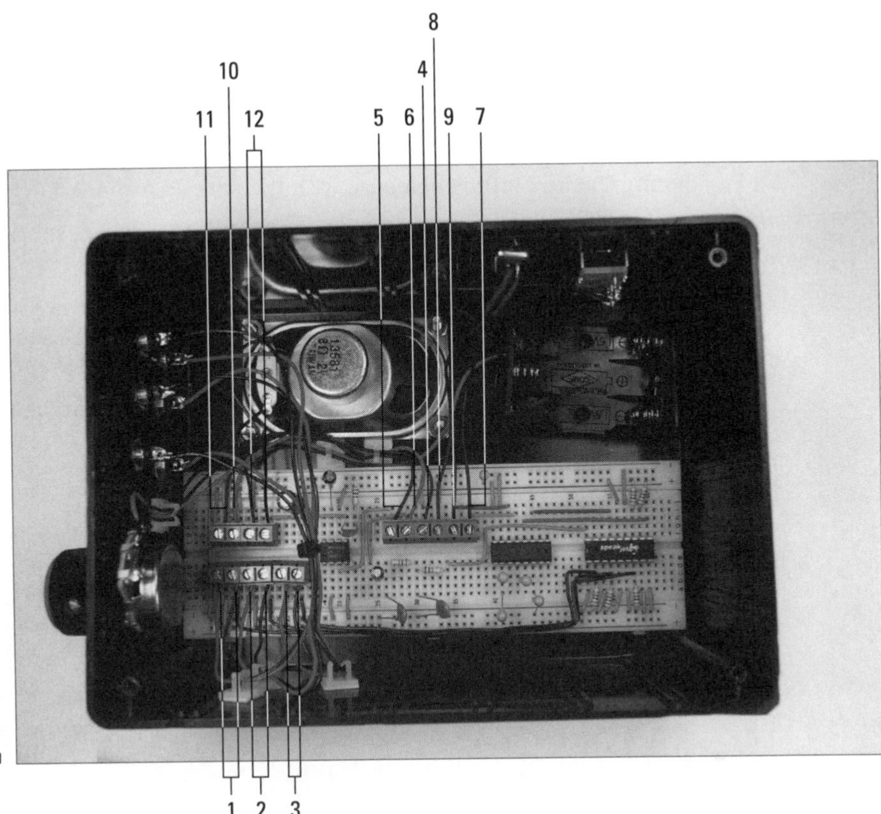

Figure 4-15: Insert wires into terminal blocks.

The components shown in Figure 4-15 and indicated by the numbered callout include:

1: Wires from left phono jack

2: Wires from centre phono jack

3: Wires from right phono jack

4: Wire from left potentiometer lug

5: Wire from centre potentiometer lug

6: Wire from right potentiometer lug

7: Wire from Pin 3 of DB9 connector

8: Wire from Pin 5 of DB9 connector

9: Wire from Pin 8 of DB9 connector

10: Red wire from battery pack

11: Black wire from on/off switch

12: Speaker wires

12. **Place the on/off switch in the off position and insert batteries in the battery pack.**

    Secure the lid on the enclosure with the screws provided.

13. **Get out your multimeter (check out Book I, Chapter 8, for how to use this tool).**

    Search for two contact pins on the tactile switches that are normally *open* (infinite resistance between the contact pins) and *closed* (nearly zero resistance between the two pins) when the switch is pressed.

14. **Solder a 30 cm wire to each of the two pins on each tactile switch.**

    The result needs to look like Figure 4-16.

Figure 4-16: Solder wires to tactile switches.

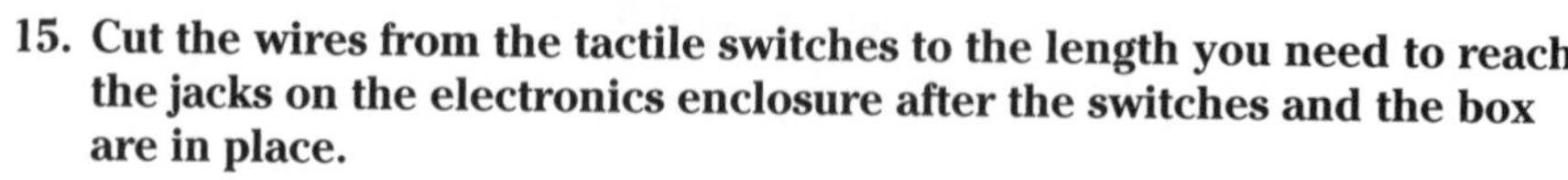

15. **Cut the wires from the tactile switches to the length you need to reach the jacks on the electronics enclosure after the switches and the box are in place.**

Leave about an extra 10 cm of length to allow for some shifting as you stuff (sorry, place gently) the box in the puppet.

16. **Unscrew the top from each phono plug, slip the top over the wire and solder each wire from the tactile switches to the centre lug of the plug.**

    Figure 4-17 shows you what the result looks like.

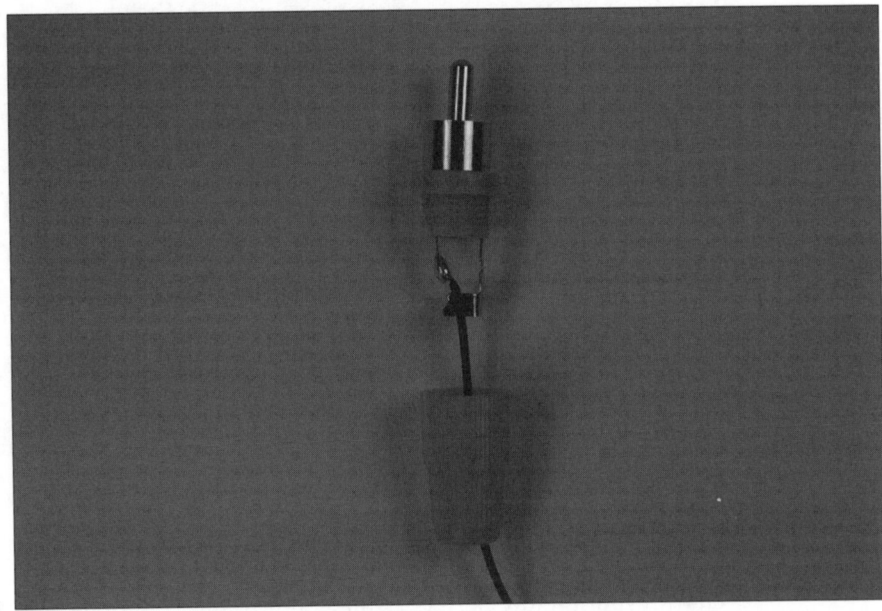

Figure 4-17: Solder wires to phono plugs.

17. **Screw the covers of the phono plugs back in place and feed each tactile switch to the proper location.**

    We place one switch in each hand and a switch under the puppet's nose.

18. **Feed the phono jacks behind the puppet, along the lines that we show in Figure 4-18.**

You may want to secure the switches in place in some fashion. We use a strip of adhesive-backed Velcro (somewhat wider than the switch) for the switch behind the nose. We stick the Velcro to the back of the switch, put the switch in place and press the adhesive back to the fabric, right under the puppet's nose. Another idea is to use a few stitches of thread to stop the switch from moving.

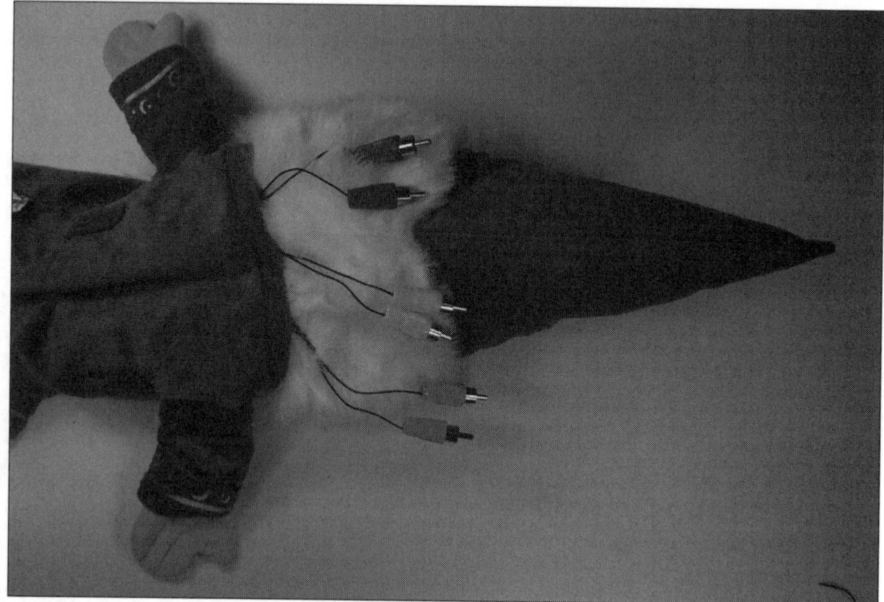

Figure 4-18: Place tactile switches in the puppet.

Programming fun sounds

Even with your picture-perfect puppet and circuit, your doll can't speak forth unless you program sounds into the sound synthesiser chip. For this task, you have to study a bit on the software provided by the manufacturer. SpeakJet doesn't provide a lot of documentation for using the software, but we feel that the chip is the best bet for the puppet because it offers so many cool options for creating sounds.

The steps that we provide here can be only an introduction to programming the SpeakJet. The SpeakJet user guide is available from the Downloads section in its product page on the Active Robots website (`www.active-robots.com/speakjet-sound-synthesizer.html`). If you need more information, try the SpeakJet discussion forum at `http://groups.yahoo.com/group/speakjet`. You can get the PhraseALator software that you use to program the chip and its manual from the Downloads area of the `www.speakjet.com` website.

By using this chip and software to program SpeakJet, you can get some pretty neat functionality. For example, at your disposal is a built-in set of 72 speech elements, 43 sound effects, 3 octaves of musical notes and 12 touch tones. By mixing and matching these and controlling the pitch, rate, bend and volume settings, you can produce just about any sound, phrase or musical tones you want.

After you install the software on your computer, you can simply assign preloaded sounds to the circuit of this project, which we cover in the next set of steps. Or, you can get fancy and start programming sounds or even words of your choosing (see the later sidebar 'Making custom sounds'). If you want to get all fancy, you need to play around for some time to suss out the software (which is beyond the scope of this book). With a little trial and error, you can get the hang of it.

We keep things simple and use the sounds that the manufacturer preloads. Here are the steps involved:

1. **Open the PhraseALator software and click the Event Configuration button.**

   Figure 4-19 shows the SpeakJet Event Configuration that we load onto the chip. The items in the 'Phrase# to Play' column in this figure indicate phrases that the manufacturer preloads.

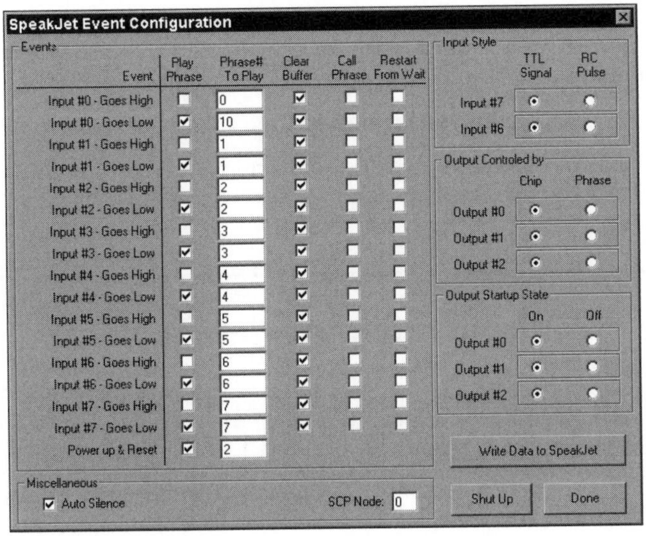

Figure 4-19: The Event Configuration screen of the Phrase-ALator software program.

2. **Select the check boxes in the 'Play Phrase' column only for input events listed as Goes Low.**

   Doing so means that the SpeakJet activates when you first press and then release the switch. See the following minitable for help in choosing which check boxes you need to select. (The simple route is to just select all the Goes Low inputs, as shown in Figure 4-19.)

The event number in the SpeakJet software Event Configuration window doesn't always match the pin number in the SpeakJet IC – that would be too easy. Here's a list of event numbers and corresponding pin numbers. We wire switches to Pins 2, 4 and 7, and so we use Events 6, 4 and 2.

Event	*Pin*
0	9
1	8
2	7
3	6
4	4
5	3
6	2
7	1

3. **Connect the box to your computer by using a serial port cable to program the SpeakJet.**

   Check over your circuit to make sure that no wires are loose or touch another wire – which can cause a short – before hooking them up to your computer.

4. **Connect the box to your computer.**

   Click the Write Data to SpeakJet button in the Event Configuration screen of the SpeakJet program (as shown in Figure 4-19).

Making custom sounds

We know that you're the inquiring kind who wants to program custom phrases as soon as possible, and so here's an overview of the process to get you started:

1. **Open the EEPROM Editor screen in the PhraseALator software.**
2. **Select the check box for the Phrase# that you want to use for a custom sound.**
3. **Select the sounds that you want to include in a phrase in the Phrase Editor screen that opens.** Be sure to check out all the words available from the library.
4. **Click Done in the bottom right of the screen after you make a custom phrase.** Doing so takes you back to the EEPROM Editor screen, where you can see that the selected sounds have been added to the phrase.
5. **Close the EEPROM Editor screen and open the Event Configuration screen.**
6. **Enter the phrase number that you created in the Phrase# to Play box for one of the input #s.** See the earlier minitable in this section to see how input numbers correspond to pin numbers.
7. **Connect the serial cable and then click the Write Data to SpeakJet button to program the chip.**

5. **Finish programming the chip.**

   Disconnect the box with the electronics from the computer.

Hooking up your puppet

No doubt you're dying to hear your puppet have its say, and so follow these steps to bring your newly loquacious friend to life:

1. **Plug the phono plugs from one of the tactile switches into the box and press the switch, adjusting the potentiometer to get the sound level right.**
2. **Unplug the phono plugs and insert the electronics box in the puppet, taking care not to move the tactile switches out of position.**
3. **Insert each phono plug in a phono jack and tuck the wires out of sight, as shown in Figure 4-20.**

Figure 4-20: The electronics concealed in the puppet.

Playing with Your Puppet

To operate your puppet, press a hand or nose and listen to the words of wisdom flow. If something doesn't work, here are two obvious problems to address:

- Check that all the batteries are fresh, tightly inserted in the battery pack and all facing the right direction.
- Check to see that no wires or components have come loose.

Adding sound to objects has endless potential, as manufacturers of stuffed toys and dolls have discovered. If you like this project, you can vary it or take it further, as follows:

- Use a different puppet, stuffed animal or plastic toy.
- Create a music box that plays when you press a switch.
- Add more switches for people to push. You can use up to eight of the event pins for sound.
- Add light to sound. For example, you may want to put a light on your doll's head that illuminates every time it speaks. See Book II, Chapter 5, for ideas about working with LEDs.

If you want to program the SpeakJet only once, you don't need to keep the MAX232 with your project. Instead, you can move the SpeakJet to a smaller breadboard, which allows you to fit the board into a smaller puppet or toy.

Index

• Symbols and Numerics •

• A •

• B •

• C •

• E •

• G •

• H •

• I •

• J •

• L •

• M •

• O •

• P •

• S •

• T •

• W •

• Z •